Stochastic Hydrology and its Use in Water Resources Systems Simulation and Optimization

NATO ASI Series

Advanced Science Institutes Series

A Series presenting the results of activities sponsored by the NATO Science Committee, which aims at the dissemination of advanced scientific and technological knowledge, with a view to strengthening links between scientific communities.

The Series is published by an international board of publishers in conjunction with the NATO Scientific Affairs Division

A Life Sciences	Plenum Publishing Corporation
B Physics	London and New York
C Mathematical	Kluwer Academic Publishers
and Physical Sciences	Dordrecht, Boston and London
D Behavioural and Social Sciences	
E Applied Sciences	
F Computer and Systems Sciences	Springer-Verlag
G Ecological Sciences	Berlin, Heidelberg, New York, London,
H Cell Biology	Paris and Tokyo
I Global Environmental Change	

NATO-PCO-DATA BASE

The electronic index to the NATO ASI Series provides full bibliographical references (with keywords and/or abstracts) to more than 30000 contributions from international scientists published in all sections of the NATO ASI Series.
Access to the NATO-PCO-DATA BASE is possible in two ways:

– via online FILE 128 (NATO-PCO-DATA BASE) hosted by ESRIN,
Via Galileo Galilei, I-00044 Frascati, Italy.

– via CD-ROM "NATO-PCO-DATA BASE" with user-friendly retrieval software in English, French and German (© WTV GmbH and DATAWARE Technologies Inc. 1989).

The CD-ROM can be ordered through any member of the Board of Publishers or through NATO-PCO, Overijse, Belgium.

Series E: Applied Sciences - Vol. 237

Stochastic Hydrology and its Use in Water Resources Systems Simulation and Optimization

edited by

J. B. Marco

Universidad Politécnica de Valencia,
Valencia, Spain

R. Harboe

Asian Institute of Technology,
Bangkok, Thailand

and

J. D. Salas

Colorado State University,
Fort Collins, Colorado, U.S.A.

Springer Science+Business Media, B.V.

Proceedings of the NATO Advanced Study Institute on
Stochastic Hydrology and its Use in Water Resources Systems Simulation and
Optimization
Peñíscola, Spain
18–29 September 1989

Library of Congress Cataloging-in-Publication Data

Stochastic hydrology and its use in water resources systems simulation
and optimization / edited by J.B. Marco, R. Harboe, J.D. Salas.
      p.   cm. -- (NATO ASI series. Series E, Applied sciences ; vol.
237)
    "Lectures and ... papers presented at the joint NATO-Advanced
Study Institute meeting on 'Stochastic hydrology and its use in
water resources systems simulation and optimization' which was held
at Peñiscola, Spain, in September 1989"--Pref.
    Includes index.
    ISBN 978-0-7923-2288-7   ISBN 978-94-011-1697-8 (eBook)
    DOI 10.1007/978-94-011-1697-8
    1. Hydrology--Statistical methods--Congresses. 2. Stochastic
processes--Congresses.   I. Marco, J. B. (Juan B.) II. Harboe, R.
III. Salas, J. D. (Jose D.) IV. Series: NATO ASI series. Series E,
Applied sciences ; no. 237.
GB565.2.S7S83 1993
551.48'01'5192--dc20   93-19300

ISBN 978-0-7923-2288-7

Printed on acid-free paper

TABLE OF CONTENTS

PART III
INVITED PAPERS ON WATER RESOURCES SYSTEMS OPTIMIZATION AND SIMULATION

PART IV
CONTRIBUTED PAPERS ON WATER RESOURCES SYSTEMS OPTIMIZATION AND SIMULATION

PREFACE

This is a book of proceedings containing the invited lecture papers, some selected contributing papers and case studies presented at the joint NATO Advanced Study Institute (ASI) on "Stochastic Hydrology and its Use in Water Resources Systems Simulation and Optimization" which was held at Peñiscola, Spain. A total of 79 participants from seventeen countries attended the Institute.

The main objective of this NATO ASI was to discuss the state of the art on stochastic methods and techniques for hydrological data analysis, data generation and forecasting which are used in the simulation and optimization of water resources systems. Since hydrological processes, such as precipitation and streamflow, are generally stochastic, some degree of randomness and uncertainty are involved in the design of hydraulic structures and in the planning and management of water resources systems.

A great deal of work has been done on the subject of stochastic hydrology during the past three decades. Likewise, considerable work exists on simulation and optimization of water resources systems. These proceedings discuss in detail the advances made on these subjects.

The contents of this book are organized into four sections:

- Lectures on Stochastic Hydrology
- Papers on Stochastic Hydrology
- Lectures on Water Resources Systems Analysis
- Papers on Water Resources Systems Analysis

The section on lectures consists of the lecture papers delivered by a selected group of invited speakers plus the comments from two invited panelists. The section on papers contains those from a group of contributing speakers along with case study papers.

The NATO ASI was organized with the support of the Scientific Affairs Division of NATO, the Department of Hydraulic Engineering and Environment of the Polytechnical University of Valencia, Spain, the Water Resources Department of Ruhr University, Germany, and the Hydraulic Science and Engineering Program, Department of Civil Engineering of Colorado State University, USA. The organizers would like to express their sincere gratitude to Dr L.V. Da Cunha for his continuous support in the preparation of the meeting and in the publishing of this book of proceedings.

The meeting was also sponsored by the Direccion General de Obras Publicas of the Ministry of Public Works of Spain. Thanks are due to its former General Director, Dr. Jose Luis Rubio, and to the President of the Confederacion Hidrografica del Jucar, Dr. Jose Carles, for their support and encouragement for carrying out the meeting. Likewise, our gratitude goes to the Diputacion Provincial de Castellon for providing some of the meeting facilities at Peñiscola.

The assistance from our younger colleagues and students Dr. Rafael Garcia-Bartual, Patricia Barreda and Carlos Lazaro and from the secretary Mrs. Pilar Almor, was extremely important in making the meeting a successful event.

Lastly, our special thanks go to all the invited speakers who donated their valuable time for writing, lecturing and participating in discussions during the NATO ASI. Likewise, our thanks to all the contributing speakers, to the members of the panel of discussions, and to all the participants in general, who formally and informally contributed to the success of the Institute.

Juan B. Marco

Valencia, Spain
March 1993

PART I

INVITED PAPERS ON STOCHASTIC HYDROLOGY

GENERAL INTRODUCTION TO APPLICATION OF STOCHASTIC HYDROLOGY IN WATER RESOURCES

VUJICA YEVJEVICH
Professor Emeritus of Civil Engineering
Colorado State University
Fort Collins, Colorado 80523
USA

ABSTRACT. Addition of stochastic hydrology to classical mathematical hydrologic statistics in the last three decades has meant new tools for more accurate solutions of various complex planning, design and operation problems in water resources development. Stochastic hydrology is conceived in this writing as studies and corresponding mathematical modeling of spatial-temporal hydrologic stochastic processes. Rush to develop models of these processes by researchers has often neglected the in-depth investigations of reliability of basic data and particularly of the structure of hydrologic inputs and outputs of water resources systems, namely their general composition of tendency, intermittency, periodicity and stochasticity. This neglect concerned not only details of these four characteristics but also their interactions. Several reasons of practical and general significance may justify efforts for mathematical modeling of hydrologic stochastic processes, the main one being the feasibility to simulate potential future samples of systems inputs and outputs, as well as their relationships, all for the purpose of solving more accurately complex water resources problems. Optimization of water resources decisions is benefited significantly by various forms of advanced hydrologic information, one of these forms being the hydrologic spatial-temporal stochastic models.

1. Historical Development of Statistical Hydrology

1.1. TWO TYPES OF HISTORICAL HYDROLOGIC DEVELOPMENTS

A careful review of contributions to hydrology in the past century reveals two basic types of development. First, there is a clear emergence of utilitarian hydrology (also called applied, engineering or practical hydrology), and the recent emergence of theoretical hydrology (also called basic, scientific, analytical-mathematical hydrology), as the two opposite poles of the entire hydrologic spectrum of activities and approaches. The other type of development is the division of hydrologic investigations into physical or deterministic hydrology and statistical or stochastic hydrology.

3

J.B. Marco et al. (eds.),
Stochastic Hydrology and its Use in Water Resources Systems Simulation and Optimization, 3–23.
© 1993 *Kluwer Academic Publishers.*

Principles and methods of utilitarian hydrology will be held in focus in
this introduction by considering stochastic hydrology in the light of its
application to simulation and optimization in water resources planning and
operation. Principles and methods of advanced application of mathematical
statistics, probability theory and stochastic processes in hydrology will
define the range of discussions in this introduction, since the present
concept of stochastic hydrology was an outgrowth and generalization of the
classical statistical hydrology.

1.2. EARLIEST STATISTICAL TREATMENT OF HYDROLOGIC DATA

First professionals of hydrology, who basically conceived, established,
treated and processed observations of hydrology-related variables, have
introduced into hydrologic practice the existing intuitive concept that all
basic hydrologic variables are random in nature (more or less governed by
laws of chance). The spatial-temporal variation of these variables were
consequently conceived as stochastic processes, though this latter term
came into hydrologic terminology much later. This intuitive concept of
randomness has been basically induced by hydrologic extremes, apart from
the evident chaos in the sequence and magnitude of many hydrologic
phenomena. Chance characteristics in the amount and timing of floods and
droughts were experienced by people in the earliest stages of development
of the human race, but particularly emphasized in modern times as well as
in the initial stages of industrial revolution. Early applications of
statistical hydrology did not require physical insights on how hydrologic
processes were generated, as long as one had correctly and sufficiently
observed their variations in space and time. Once data were available as
samples, statistical approaches to their analyses did not require in-depth
studies of their physical or stochastic generating mechanisms. Recently,
when the advanced stochastic mathematical modeling of hydrologic processes
came into being, physical and stochastic generating mechanisms and
structural analysis of processes became the lightning rods for improvement
and correct modeling of hydrologic stochastic processes.

Initial techniques of statistical analysis of hydrologic data consisted
mainly of four types: (1) sampling from the observed samples, such as
singling out the maximum or minimum values of water volumes or discharges
within the year, conceived as the new samples (for further analysis of
frequencies of these extreme values); (2) computation of frequency
distribution curves or graphs of well defined random variables, with or
without fitting the probability distribution functions, and their eventual
extrapolations beyond the range of observed values (an example is Fig. 1);
(3) development of linear or nonlinear correlative associations, with
measurements of the degree of these associations, and estimation of
corresponding regression functions between two or more hydrologic random
variables, for the purposes of extending or transferring information (an
example is Fig. 2); and (4) conducting statistical inferences on hypotheses
related to estimated parameters, probability distributions, correlative
associations and regression functions, with the use in final analysis of
confidence and/or tolerance limits (an example is Fig. 3). These areas of
application of statistics have often been considered by practitioners as
the classical hydrologic statistics. It has served well the practice for
over a century of dynamic water resources development in industrial

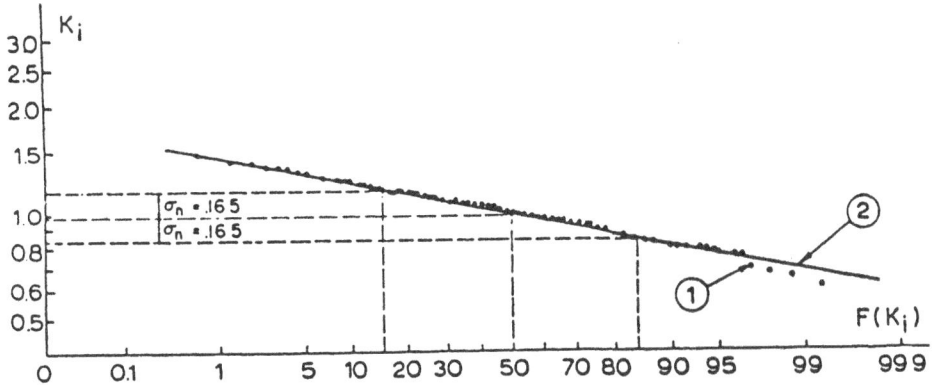

Figure 1. Fitting of straight line in logarithmic-probability scales to frequency distribution of annual flows of the Rhine River at Basle, Switzerland, for the period 1808-1957 (150 years), with flows in modular coefficients (K_i = flow divided by average flow).

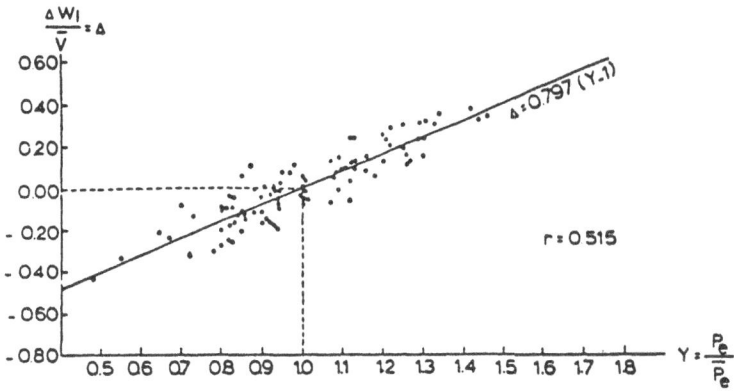

Figure 2. Simple linear regression of relative change of water carryover from year to year in river basin as $\Delta = \Delta W/V$, versus the relative annual effective precipitation on river basin, $Y=P_c/\bar{P}_c$, for the St. Lawrence River at Ogdensburg, New York, USA.

countries and elsewhere, and as long as the related water systems have been relatively simple. This practice led several hydrologists to innovations in statistical applications, with such techniques as the first generation (statistical simulation) of new hydrologic samples, based on observed data, or the first application of transformed scales of graphs of frequency or probability distributions, so that frequency distributions were plotted close to straight lines and the probability distribution functions as straight lines (Hazen, 1914; Sudler, 1927).

6

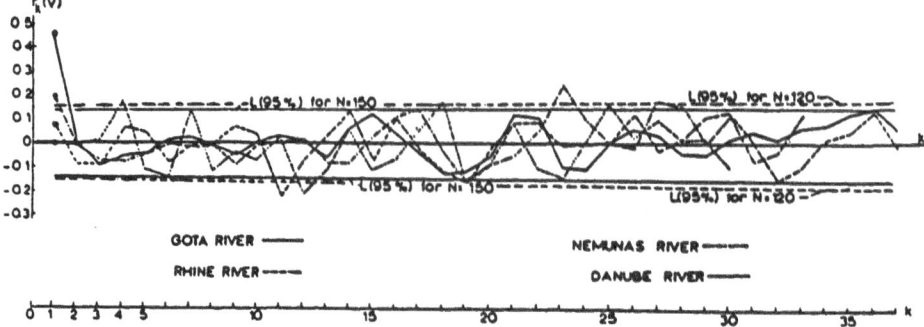

Figure 3. Testing correlograms for time dependence of annual flows of four rivers with long records: Göta River, Sweden (150 years); Nemunas River, Lithuania, USSR (132 years); Rhine River, Switzerland (150 years); and Danube River, Iron Gate, Rumania (120 years), with two confidence (tolerance) limits at the 95 per cent level, for two sample sizes: 150 (max) and 120 (min).

1.3. SPACE-TIME CHARACTERISTICS OF HYDROLOGIC PROCESSES

Systems approach to solving problems in water resources planning and operation by using the input-response-output method helped very much the identification of needed hydrologic information in most advanced forms for solution of these problems. Inputs are either natural or man-affected natural hydrologic processes of the inferred spatial-temporal flow characteristics. Responses of natural, artificial or mixed natural-artificial water resources systems represent some of the classical hydrologic topics which help to compute systems outputs for given systems inputs or vice versa. Outputs are governed by water demands and availability of flowing or stored water to meet these demands. Observed spatial-temporal hydrologic processes, as inputs or outputs of systems, provide the necessary data for extraction and transfer of information for current utilitarian objectives of decision making in water resources.

The best example of systems inputs, responses and outputs are water regulating reservoirs or other storage capacities, and lakes and aquifers not controlled at their outlets. Inputs are dominated by natural inflows or recharges, and by a combination of natural flow and upstream regulated outflow from storage capacities. Responses are governed by hydrologic conditions of outlets of non-regulating lakes and aquifers, or by the operational patterns of water demand releases. Outputs are dependent on demand as well as on the surplus or deficit of water in reservoirs and other storage capacities.

Inputs and outputs as the spatial-temporal, periodic-stochastic processes are sampled by observations at the sets of points in space (along lines, across areas, over volumes of the environment) and along time, either as continuous recordings or as observations at fixed discrete time points or as cumulative values over the fixed intervals. These samplings have led to concepts and techniques of investigation of time series at given spatial points, and distributions of main characteristics (usually of parameters of time series) over these spatial points. This practical hydrologic approach to spatial-temporal stochastic processes differed somewhat from the classical statistical multivariate approaches, used in three versions: (1) multivariate in the sense of studying joint properties of several variables (in case of hydrology such variables as precipitation, evaporation, infiltration, runoff, etc.); (2) multivariate in the sense of investigating jointly process characteristics at a set of spatial points; and (3) multivariate in the sense of considering variable values at a set of time points as multivariates, say within the cycle of the day or the year, or over the points of a sample of time series.

Apart from temporal and spatial properties of inputs and outputs, dependence of outputs on inputs or vice versa represents a characteristic that is most often neglected in practice, if not also in theory. However, that dependence is often not negligible. Similarly, practice has treated outputs differently from inputs. For inputs the past observations are used to derive general properties of hydrologic properties, as the most reliable approach to determine what would be the most likely characteristics of future inputs. Unfortunately, this sound and only feasible approach in deriving reliable information on inputs was not equally well applied to outputs. Though as many years of data may be available on the outputs observed (recorded outflows, energy produced, water supplied or consumed, etc.) as on inputs, rarely past outputs were structurally investigated and mathematically modeled, and still less often these models or results were used to predict the potential future outputs or to simulate (generate) the potential future output samples. These two aspects, dependence between inputs and outputs, and the lack of treatment of output information similarly as input information, may represent significant deviations in practice from the most rational and correct treatment of spatial-temporal properties of inputs and outputs.

1.4. HYDROLOGIC TEMPORAL PROCESSES

Considering inputs and outputs as temporal processes at fixed spatial points, their investigation consists generally of assessment of data reliability (accuracy), analysis of structural properties and composition, and mathematical modeling. Stated in a nutshell, most attention by researchers has been given to modeling in the recent three decades, much less to structural composition and properties, and least of all to developing powerful techniques for analysis of various types of errors, inconsistency and nonhomogeneity in data for the purposes of error corrections and the final assessment of data reliability. This reliability is then a basic measure of accuracy in the results of structural analysis and mathematical modeling. Likely, the main reason for this neglect is the fact that complexities increase from modeling to in-depth structural analysis, and from this analysis to assessment of accuracy and specificities of distortions in data, which may not repeat themselves in the future, or not in the same form.

Analysis of random, systematic (or inconsistency) and sampling errors, as well as nonhomogeneity in data (conceived as changes in nature by humans

8

and natural disruptions) require both the complex theoretical (mainly statistical) works as well as the eventual additional practical field works, with very tedious, detailed analyses of all sources of erros and/or changes, in order to identify them, eventually mathematically model and remove from data. The fact is that these errors and changes rarely repeated themselves in the same form or magnitude. Also at issue is whether some changes can be taken into account only by considering a couple of major time series parameters as having trends or jumps along the series, or if this approach leaves significant additional stochastic residuals within the series even when trends and jumps in these parameters are removed. This is an area of stochastic hydrology and practical hydrologic statistics which was relatively neglected both by practitioners and by researchers.

Structure of hydrologic time series of water inputs and water outputs consists mainly of one or more of these four basic structural properties and components. They are: (1) tendency, usually assumed to be only in major parameters, composed of trends and jumps or slippages; (2) intermittency in the process, mainly consisting in hydrology of intermittent sequences of zero and non-zero values; (3) periodicity and almost periodicity, mainly of cycles of the day and the year, but also of some others, usually conceived as periodicity or almost periodicity in major series parameters; and (4) stochasticity, as random variation of process around its averages, as the component without which the process would be considered deterministic.

On one extreme of examples of the structure of hydrologic time series are processes which have only the stochastic component, assuming there is no tendency in parameters. Best illustrations of such simple stochastic processes are inputs of homogeneous annual precipitation or annual river flow (an example is Fig. 4), provided no trends or jumps in the main

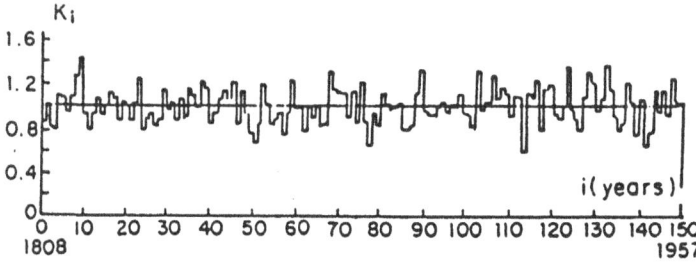

Figure 4. Time series of annual flows of the Rhine River at Balse, Switzerland, in modular coefficients (K_i as flow divided by average flow).

parameters of mean and standard deviation are found, no zero values occur, no periodicity in these parameters is found and no causes for their occurrence have been identified. On the other extreme of examples of the structure of hydrologic time series are cases in which all the above four structural properties are exhibited. An example of this structure and composition are daily precipitation series in which the systematic errors in data may have a trend in the mean, zero/nonzero values represent intermittency, main parameters are periodic (seasonality in precipitation) and there is a significant stochasticity, usually inherent in daily values

(an example is Fig. 5).   Similarly to inputs, outputs range in structure
from trend-stochastic time series to complex trend-intermittent-periodic-
stochastic processes.   An example of the former are annual water supply
time series to a city (Fig. 6), and an example of the latter is complex
water delivery time series to an irrigation project (Fig. 7).

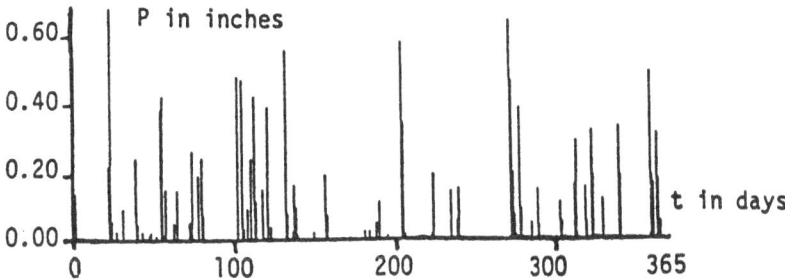

Figure 5. Daily precipitation series for the year 1944 at Durango,
Colorado, USA, illustrating the basic structure of intermittency,
periodicity and stochasticity.

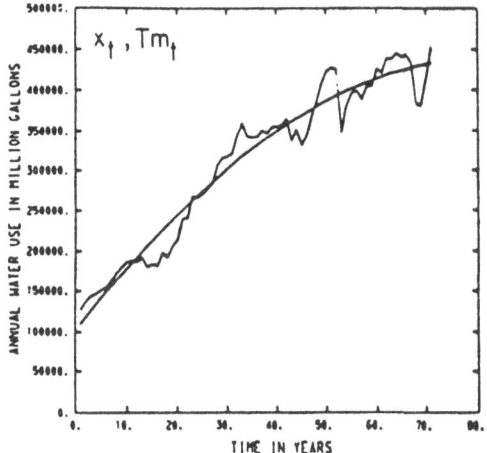

Figure 6. Annual water deliveries in million gallons and the fitted
quadratic trend for New York City, USA (1989-1968).

   Structural analysis means that a time series may be decomposed in
components, or its properties such as intermittency specially treated. For
the former, the clear understanding of components and the relationship
among them must result from the analysis.   For the latter, there is a need
for proper treatment of special characteristics such as the specificities
of each particular intermittency.
   The approach to the structural analysis of hydrologic time series, as
well as the interpretation of inferred properties with their explanation
by physical and stochastic processes, have not been often treated by

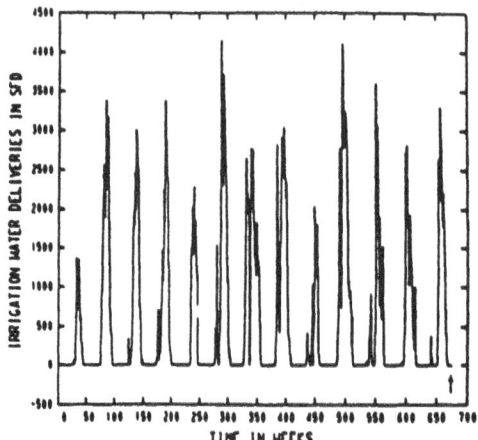

Figure 7. Weekly irrigation water deliveries from Carter Lake, Colorado, USA (1957-1969), illustrating intermittency, periodicity and stochasticity.

systematic and in-depth investigations. Many questions do not yet have rigorous answers. For instance, do trends and jumps occur only in parameters, or is there a large amount of stochasticity in tendency so that it is not removed by removing trends in parameters, and trend residuals increase stochastic variation of a time series? Similarly, are only basic time series parameters periodic, or how significant may periodicity be in parameters of the higher order statistical moments? What makes autocorrelation and autoregression parameters of stochastic dependence models periodic, and how does this periodicity change with the increase of the lag of these parameters? Several similar questions may be pertinent showing that structural analysis and study of interactions of structural components and properties have been somewhat neglected in hydrologic research in comparison with attention paid to mathematical modeling of time series. Here may exist a contradiction, namely structural analysis significantly improves drawing of hypotheses which are needed in modeling, while the rush to model often neglects the phase of structural analysis which makes models more realistic and closer to the true physical properties of time processes.

A fundamental principle of time series analysis and modeling in hydrology is based on the combination of deterministic functions among various forms of the random variable whose time series is investigated and the singled-out independent stochastic component. Deterministic functions may be completely missing where the original time series is an independent stochastic process. Various forms of original random variable are the detrended series in case of tendency, or resulting random variable(s) in the case of modeling intermittency, or the series with removed periodicity in parameters, etc. Therefore, six types of functions may be used in modeling complex hydrologic time series: (1) trend functions of parameters; (2) many necessary functions, deterministic or stochastic, used in modeling intermittency; (3) periodic functions of parameters of detrended series; (4) complex functions of modeling stochastic dependence of detrended and deseasonalized time series (trends and periodicity removed), with the resultant independent stochastic component of given order of stationarity; (5) the probability distribution function (PDF) of resulting independent stochastic component; and (6) the equation which represents the relationship between the previous five types of functions. In principle, only the probability distribution function of independent random component or noise is a stochastic component, all other functions are the deterministic relationships representing transformations of this component.

The opposite process to step-by-step modeling is the use of models to generate (simulate) new samples or a long sample of the process. One starts with the random noise and its PDF by first generating its sample(s). Then, by using various deterministic functions one adds to these samples the stochastic dependence, periodicity and trends, as well as the necessary treatment to simulate intermittency. In practice, one tries to reduce the independent stochastic component to the white noise, or to the normal, independent and standardized random variable. The reason for this effort is the simplicity in treating white noise (statistical techniques are well established and the generation of their random numbers is quite standardized as software on most computers).

Stated in a nutshell, hydrologic stochastic models are usually composed of a set of mathematical equations, including PDF of white or other types of noise. Two examples, Equations 1 and 2, are presented here for the final relationship of other equations of the model, one for natural water inputs, however without trend and intermittency, and the other for water resources system outputs with trend in major parameters but without intermittency. The first example may be related to the general daily or monthly periodic-stochastic runoff input models of perennial rivers without trends in parameters. Stochastic dependence is assumed of the ARMA type model (autoregressive-moving average) of the second order stationarity of resulting independent noise (Salas et al., 1980), with the model given as

$$x_{v,t} = m_v + s_v \left\{ \sum_{j=1}^{p} \phi_{j,v} y_{v,t-j} - \sum_{j=0}^{q} \theta_{j,v} z_{v,t-j} + z_{v,t} \right\} \tag{1}$$

where $z_{v,t}$ = independent stochastic component (noise); $y_{v,t} = (x_{v,t} - m_v)/s_v$ = the ARMA dependent stochastic component; $x_{v,t}$ = original periodic-stochastic process; $\phi_{j,v}$ and $\theta_{j,v}$ = periodic autoregressive and moving average coefficients, respectively; p and q = orders of AR and MA terms of the ARMA model, m_v and s_v = periodic means and periodic standard deviations at positions over the cycle of the original process, with $v = 1, 2, \ldots, w$, and w = number of discrete time intervals in the main cycle of periodicity. When parameters are not periodic, symbol v is removed, and parameters become constants over the entire series.

Similarly, general daily, weekly or monthly trend-periodic-stochastic model relationship of outputs (water use or delivery) time series may be expressed for autoregressive models of stochastic dependence (Salas and Yevjevich, 1972) as

Dependence structure

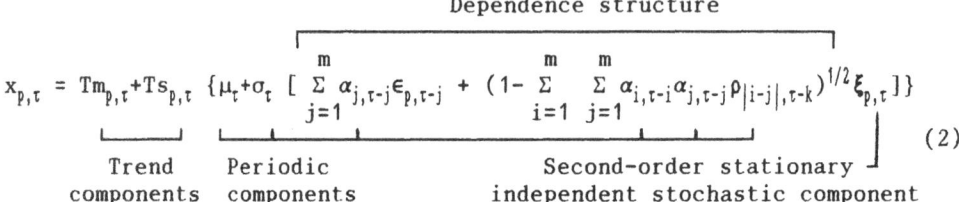

$$x_{p,\tau} = Tm_{p,\tau} + Ts_{p,\tau} \{\mu_\tau + \sigma_\tau \ [\ \sum_{j=1}^{m} \alpha_{j,\tau-j} \epsilon_{p,\tau-j} + (1 - \sum_{i=1}^{m} \sum_{j=1}^{m} \alpha_{i,\tau-i} \alpha_{j,\tau-j} \rho_{|i-j|,\tau-k})^{1/2} \xi_{p,\tau}]\} \quad (2)$$

Trend components	Periodic components	Second-order stationary independent stochastic component

in which k = i if i<j, and k = j if i>j, and where $Tm_{p,\tau}$ and $Ts_{p,\tau}$ are trend fitted functions for the mean m and standard deviation s; ρ = sequence of years (ρ=1,2, ...,n); τ = intervals as fractions of the year as τ = 1,2, ...,w; w = number of intervals within the year; m_τ and s_τ = periodic functions (fitted Fourier series) to the mean and standard deviation; $\epsilon_{p,\tau-j}$ = dependent stochastic component with removed trend and periodicity in the mean and standard deviation (standardized dependent stochastic component), j = 1,2, ...,m, with m = order of simple autoregressive dependence model; α_j = autoregressive coefficients; ρ_j = corresponding autocorrelation coefficients; and $\xi_{p,\tau}$ = independent standardized stochastic component (noise), assumed to be also made second-order stationary process. The τ-subindex in $\xi_{p,\tau}$ tells that the third or even higher order parameters may still contain trend and periodicity, with these trends and periodicities usually assumed negligible and therefore neglected, or are assumed non-negligible and further modeled, provided the sample size and data accuracy permit this modeling.

Models of Equations 1 and 2 may be simple or complex, depending on the type of series, length of time intervals t or τ used, type of stochastic dependence, significance of periodicity and number of periodic parameters, periodicity and trend in autocorrelation and autoregression coefficients, parameters of higher order moments, etc. Equation 1 has three other expressions as part of the set of model equations, namely periodic functions of the mean and standard deviation and PDF of the independent stochastic component. Equation 2 has five additional expressions, those of trend and periodicity in the mean and standard deviation and PDF of the independent stochastic component. With the ten equations of models of Equations 1 and 2 input and output of a water resources system (say a reservoir) may be mathematically described or modeled. If input and output are mutually dependent, a correlative dependence function between their independent stochastic components, as the 11th equation, will complete the modeling. Then these 11 equations permit simulation (generation) of input and output in the form of a set of new samples or one long sample of each.

1.5. HYDROLOGIC SPATIAL PROCESSES

A basic assumption in the investigation of spatial distributions of periodic-stochastic processes of hydrologic variables, or distributions over a set of points in space, is that a point has the same form of mathematical model expressions as all other points. Only the values of model parameters vary from point to point. Under this assumption, the problem becomes that of finding how parameters of these models change along a line, across an area or over a volume. In most hydrologic processes the mean and standard deviation are the two major parameters that nearly always vary spatially. The other parameters, such as autocorrelation, regression, skewness and Fourier coefficients as well as parameters of equations of periodicity and trend may also vary significantly over the space. Estimates of parameters at the set of points in question represent the basic information on spatial distributions of these parameters, with distributions obtained either by drawing isolines of parameters, or by fitting spatial distribution functions in one, two or three dimensions, whichever may be the case in practice. Figure 8 gives an example of interpolated isolines of distribution of the estimated parameter values at a large set of spatial points.

Each independent stochastic component of Equations 1 and 2 is most often spatially dependent variable. For two points in space, at a given distance and distance straight line orientation toward North (azimuth), a correlative association may be found, with the simple correlation coefficient measuring the strength of association usually as functions of the distance and the distance line azimuth. These functions decrease as the distance increases. The effect of orientation is more complex than the effect of distance; it usually depends on prevailing winds and movement of moist air masses. These functions slowly converge to zero correlation coefficient, and then randomly (sampling variation) oscillate around zero, or even become negative values for larger and larger distances. The rate of convergence depends on the time interval used. The smaller the interval, the faster the convergence. The function of correlation coefficient on distance and orientation may be fitted to its estimated values, as Fig. 9 and Equation 3 demonstrate. Each point would have its own equation, so that its

$$r = \exp(-0.00418d) \tag{3}$$

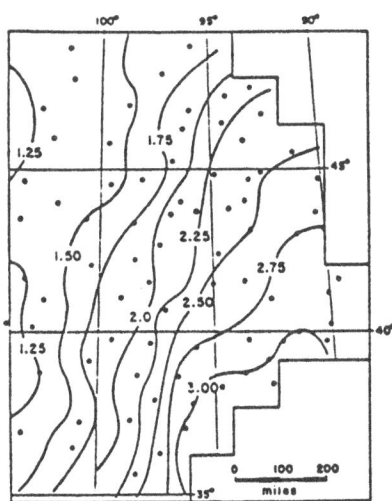

Figure 8. Isolines of the 30-year general monthly mean of precipitation over the Great Plains of the United States of America.

14

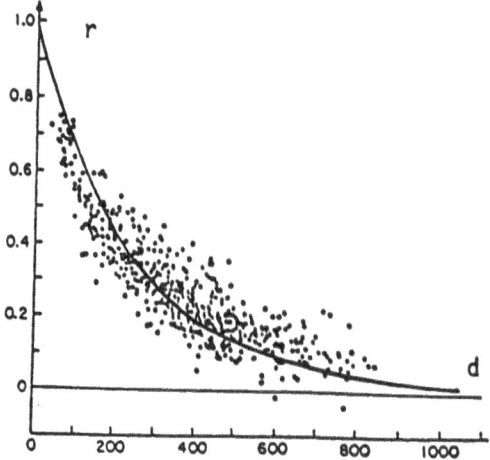

Figure 9. Relationship of the lag-zero cross correlation coefficient, r, of the stochastic components of monthly precipitation of pairs of stations for the Great Plains of the United States of America, and the interstation distance, d, with the fitted function.

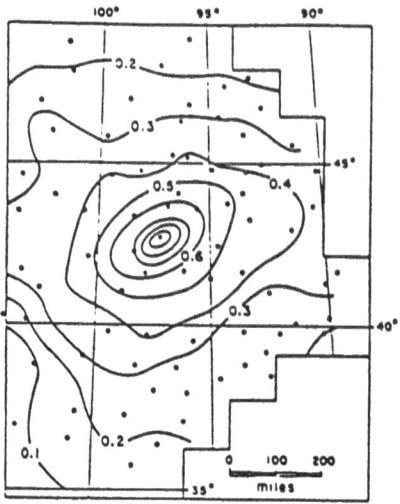

Figure 10. Isolines of the correlation coefficient between stochastic component of monthly precipitation series of the Great Plains, USA, for a fixed station versus the surrounding stations, showing some asymetry of isolines also due to the dependence on the azimuth of the stations connecting straight lines.

parameters vary from point to point, and thus can be regionalized as spatial distributions, or only one equation can be used for all points with regionally constant parameters. Figure 10 shows isolines of correlation coefficient between one station in the Great Plains and all other surrounding stations for independent stochastic components of monthly precipitation. Isolines are elongated, showing that the azimuth of straight lines between correlated stations also influences the spatial dependence.

Modeling spatial properties of hydrologic periodic-stochastic processes becomes more and more important in practice as needs arise for generation of samples at a set of points, such as inputs and outputs for a set of reservoirs mutually dependent in their operation, by preserving spatial distributions of model parameters and spatial correlative association of independent stochastic components. For solving problems in planning and operation of 15 dependent reservoirs in a region by simulation, when the application of this method is justified, simulation of inputs and outputs requires all four types of modeling: (1) of time series at each point; (2) of mutual dependence between independent stochastic components of corresponding inputs and outputs; (3) of spatial distributions of model parameters, estimated under 1 and 2; and (4) of spatial dependence of independent stochastic components of each input and output.

1.6. INFORMATION ON WATER QUANTITY AND WATER QUALITY VARIABLES

Utilitarian hydrology has two basic tasks in enabling reliable water resources planning and operation: (1) extraction of information on variables and their spatial-temporal, periodic-stochastic processes from a bank (pool) of observed data; and (2) transfer of information from variables, points and ranges of available data to variables, points and ranges of unavailable data, which need information. For these purposes there is a need for information transfer functions or relationships between related variables, related points of a given variable, and for the same variable the extrapolation of information on its PDF from the range of observed variable values to ranges of unobserved values (most often for extremes). Extraction and transfer of information are equally important for both water quantity and water quality variables. For transfers, relationships between water quality and water quantity variables are often a crucial factor. This case is very delicate in practice, since relationships between these variables are either very weak or very complex.

Old approaches to extraction and transfer of information did not have techniques available to measure quantity of information for extraction or transfer, nor techniques to measure how much information is really extracted or transferred by a given method of extraction or transfer. By using the concept of entropy and entropy coefficients (analogous to simple correlation coefficients), and similar measures based on entropy, it is feasible at present to compute and compare both the extractable and transferrable information with the extracted and transferred information, respectively (Harmancioglu et al., 1986). If a statistical technique shows that a large portion of available information is not either extracted or transferred, a better technique should be used or searched for.

2. Reasons for Mathematical Modeling of Hydrologic Stochastic Processes

2.1. CONDENSATION OF INFORMATION

The bulk of hydrologic observations are most often stored in data banks of computers at present. One may claim that this storage is relatively inexpensive, sufficient, with easy access and with proper backups basically safe. In the general present trends to condensate information, another approach to obtaining a further economy in storing information contained in data would be through mathematical modeling of processes which these data represent. This would mean a replacement of hundreds of tables of data stored on disks and/or in memories of computers, or hundreds of pages of raw data of classical files, by a set of mathematical expressions and the corresponding sets of estimated parameters of these equations. Say, one page of equations and parameters, stored in files or in data banks, instead of 80 pages of raw data of daily runoff series, for 80 years of observations, in ordinary files or in the corresponding storage on computer disks. Then, when one needed a sample or many samples of the process modeled, simulation or data generation by computers would provide the necessary information.

This type of condensation and storage of data in the form of models and estimated parameters, rather than keeping the original data in master data files as is done at present with data bank storage for final backup, but basically for verification purposes when needed, was not introduced into practice for four basic reasons: (1) modeling did not go as far in accuracy and in general feeling or reliability as to convince practitioners that simulated samples have the same chance to occur in the future with the same or close reliability as the observed time series; (2) mathematical models of series of small time intervals, such as hourly or daily time series, are often so complex, and therefore controversial, that no consensus was feasible among researchers and practitioners for models to be conceived and considered as containing the same and full information on all the series aspects as the observed historic sample (for example, the extreme observed values as compared with the extreme values of generated samples); (3) observed series often contain trends or jumps that are difficult to reliably estimate and remove, making model parameters biased or subject to errors, if not also the models themselves; and (4) an attitude, deeply ingrained with practitioners, namely that nothing could replace the observed series data in planning and design of water resources structures and systems.

2.2. HYDROLOGIC INFORMATION FOR UTILITARIAN OBJECTIVES

One practical form of presenting information for various uses in practice is in the form of models and estimates of their parameters. In the computer age, inputs into computers may be faster by using models than by using the bulk of data. As an example, it may be easier to enter into a computer a PDF of a random variable, for extrapolation purposes, than points of its frequency curve or of the raw data.

Hydrologic information, condensed and presented for subsequent uses in planning, design and operation of water resources systems and structures, is usually presented in the form of tables, curves and equations. In the computer age, switches between these three forms may be relatively easy, so models may be subsequently transformed into tables or families of curves or simple curves.

2.3. UNDERSTANDING STRUCTURE OF HYDROLOGIC STOCHASTIC PROCESSES

In general, modeling should be preceded by structural analysis of temporal and spatial stochastic processes, such as the analysis of errors, trend types and composition, intermittency, periodicity, type of stochastic dependence, stochasticity, patterns of spatial dependence and spatial parameter distributions, etc. The reason is that the structure dictates the relationship among its components as well as the properties of each component. By simulating new samples from these models, and by comparing them with the observed sample which was used in modeling, checks may be carried out as to whether models are structurally identical to the original sample or not.

Modeling may help investigations, especially of water outputs, of how important various components of the process are say in water supply of cities, irrigation or hydropower production. One may ask, for practical design of future water demand, how much of the total variation of past output series is percentage-wise explained by the trend in the mean and standard deviation, by the periodicity in these two parameters or in other parameters, and how much by the resulting stochasticity after trends and periodicities in these parameters are removed. For example, when the stochastic component comes out to be of a relatively small importance, say the total variance of a process is explained less than 10 per cent by the stochastic component, as in the case of the water supply of large cities with significant summer urban irrigation in arid regions, the future prediction of water demand may be much less affected by stochasticity in demand than by variations in trends and periodicities of major parameters. In such cases, classical methods of projecting future demand may safely be used instead of simulation of future demand output time series. When the case with stochasticity is opposite, namely it explains a large portion of the total variance of output, simulation may have some advantage in comparison with the classical method of demand projection.

2.4. INFERENCES ON HYDROLOGIC HYPOTHESES

For accepting or rejecting hypotheses on extracted or transferred hydrologic information (information on estimated particular values and parameters of random variables, estimated curves or models, etc.), it is necessary to compute probability distributions of these estimates or of testing parameters. Most hydrologic spatial-temporal processes of inputs and outputs are complex when their composition is governed by the trend-intermittency-periodicity-stochasticity structure, so that rarely are the PDF of these estimates or of testing parameters known or easy to compute. In these cases, experimental (Monte Carlo, stochastic simulation, samples generation) methods may be used to produce the most accurate but approximate frequency distributions of these estimates or testing parameters. By generating m samples of size N of a complex process, by using its estimated mathematical model and estimated model parameters, there will be m estimates of tested or testing parameters. Thus a sufficiently large number m of results of generated samples will yield the frequency distribution of each tested or testing parameter. Then this frequency distribution may be considered very close to the unknown or difficult-to-compute probability distribution function.

As an example, if one computes the 50-year return (recurrence) period flood from a flood frequency curve or fitted PDF at a river cross section, he or she may also be interested to find the confidence limits at a given probability level either for that 50-year flood or the confidence curves for the entire flood frequency curve. By generating m=100 samples of river flows, 100 flood frequency curves can be produced for selected sample sizes, and from them 100 50-year return period floods. From frequency distribution thus obtained for 50-year floods, the 5% and 95% flood values may be determined as the 90% confidence limits. Similarly, confidence limits may be determined for the other return period (or percentile) floods, so that confidence curves could be obtained for the entire original flood frequency distribution.

In the computer age, with good software programs available, any advanced PC would easily help to obtain the fast and inexpensive solutions of inference problems by using samples generation approach. Definitely, reliability of these inferences could not be better than information contained in the originally observed sample, accuracy of data, structural spatial-temporal analysis of time series, mathematical models and estimates of model parameters.

2.5. SIMULATION OF SAMPLES FOR PLANNING, DESIGN AND OPERATION

The basic reason for mathematical modeling of hydrologic spatial-temporal processes is the potential of models to increase accuracy in water resources solutions and decisions, by generating (simulating) samples of inputs and outputs of a system. When past information on inputs and outputs is compared with observed samples of inputs and outputs of the future, it always shows substantial differences among samples. It is only necessary to split a historic sample in two or more subsamples to demonstrate not only the differences between the estimated subsample parameters, but also the differences in clusterings of wet and dry seasons and years. Therefore, the unknown future is best investigated by a simulated set of equally likely future samples of inputs, outputs and the results of their applications.

The main condition for using m pairs of simulated inputs and outputs in the solution of a water resources or environmental problem is that that approach yields either the most accurate result or is the only approach leading to a reasonable result. This approach then needs three basic characteristics: (1) proof that it is needed as the best solution of a problem; (2) proper selection of simulation sample size at a point or a set of points; and (3) the accuracy of final results of application of simulated samples is prescribed by selecting sample errors through selection of the number m of samples to be simulated. It is then implicitly assumed that time structure and dependence, spatial distributions of parameters and spatial dependence, and dependence between independent stochastic components of inputs and outputs, are well preserved in both models and by the generated samples, namely within the prescribed sampling errors. This trinity: spatial structure of inputs and outputs, temporal structure of inputs and outputs, and dependence of inputs and outputs, represents then basic properties to be preserved in models and in simulated samples of hydrologic stochastic processes.

3. Simulation and Optimization in Water Resources

3.1. HYDROLOGIC STOCHASTIC SIMULATION

In principle, professionals have used physical, electrical or numerical simulations in water resources disciplines to mimic how prototypes, or in this case water resources engineering structures, would perform in planning, design and operation phases of a structure. Numerical simulations are either physical deterministic or physical stochastic (or pure statistical). These latter simulations are briefly discussed here.

Each model of hydrologic stochastic processes usually contains an independent or several independent stochastic components. If the process is more complex than a simple independent stationary process, also one or more deterministic functions are necessary to take into account transformations of independent stochastic component(s) to describe stochastic dependence, periodicity, intermittency and eventual trend in parameters. Therefore, the stochastic numerical simulation of samples of these processes is also composed of two activities: (1) generation of random independent numbers following a known PDF, such as uniform, normal, lognormal, gamma, and any other, even an empirical frequency distribution, with the total number equal to mN, where m = number of generated (simulated) samples, and N = sample size; and (2) transformations by the deterministic function of these independent random numbers into stochastically dependent random numbers, as well as further transformations of these latter dependent random numbers by deterministic functions to take into account periodicity, intermittency and trend in the series. This approach applies equally to inputs and outputs, so that their temporal, spatial and mutual dependence structures are preserved.

If sampling errors in estimated parameters of models are considered or inferred to be relatively high, i.e. original data are of small sample sizes or of a small number of spatial points, which underlie modeling and parameter estimation, the simulation techniques may produce sampling frequency distributions of major estimated parameters and draw the confidence limits around the estimated parameters. As estimates of various parameters of the same sample often are mutually dependent, the selection of joint confidence limits for these dependent estimates needs a very careful treatment. When sets of confidence limits to these mutually dependent estimates of parameters are selected (say a set of p confidence limits), repeated simulation of m samples of size N will produce the new p sets of samples, or for a total of pmN generated sample values of confidence limit input and output series. Optimizations would then produce results for these p sets of confidence limit samples, similarly as to the results of the set of m samples of the model basic estimates of parameters. These p sets of results may then be conceived as the confidence limits to the basic optimization results. These results furthermore enable the analysis of sensitivity of final optimization results to eventual errors in estimated model parameters.

In simple cases, models may be unnecessary, if there is at least an estimate of frequency distribution of independent random variables. Generation of uniform random numbers and the use of a conversion table or relationship curve between the numbers of uniform distribution and the corresponding numbers (numbers of the same probability of nonexceedence) of inferred frequency distribution leads to simple numerical generation of samples without mathematical modeling. As soon as the time series of hydrologic processes become more complex, also with the spatial characteristics and input-output dependence involved, mathematical modeling becomes unavoidable.

Then the generated samples become the input information for the solution of water resources problems carried out by optimization, conventional or simple techniques already available. In this use of generated or simulated samples a question never seems to be far from surfacing, namely whether one should trust the simulated samples as the current water resources practice is apt to trust the observed historical samples. Regardless of potential contrary statements, it is quite likely that practitioners will not trust equally well the generated samples as they trust the observed samples. This attitude may be the result of two quite different factors: (1) practitioners may exhibit some skepticism about the goodness of a model, especially because of the proliferation of models and often disagreements among researchers and modellers on a model's validity; and (2) the psychological factor, namely that the "manipulation of historic data" in forms of modeling, estimation and simulation could not be as good as the original data. This attitude leads planners and operators of water resources systems to assume that the future will likely be identical to the past, namely to the already available historic sample. This also leads to only one solution to a problem by using a pair of observed input and constructed output or demand, instead of a set of m solutions, and their PDF estimated through their frequency distribution of these m solutions. The simple example is the needed storage capacity for the over-the-year flow regulation, where the historic sample and estimated demand often give only one storage value, while the generation of m pairs of samples of inputs and outputs would produce PDF or frequency distribution of needed storage capacities for given over-the-year regulation horizon.

In some cases, one may select a couple of representative samples out of m generated pairs of input and output samples. Say, such samples as those having approximately the same mean and variance as the historic sample, or the mean and variance as averages of m generated samples. Also, the selection may include the samples which have the mean and variance close to the two confidence limits of the mean and variance of the already selected samples, say for the confidence levels of 5% and 95% of joint sampling distribution of the mean and variance. When the autocorrelation of net input series (input minus output series) is important in the solution of a problem, such problems as the selection of needed storage capacity, apart for the mean and variance, the selection of generated samples may also include a suitable measure of autocorrelation. Then, apart from an average solution, one may produce with the confidence limits to these major parameters also the approximate confidence limits to that mean solution, instead of producing m solutions for m pairs of generated samples.

3.2. OPTIMAL WATER RESOURCES SOLUTIONS

In a simplified way, optimal stochastic solutions of water resources problems may be divided into two major groups: (1) solutions based on the PDF, the probability distribution function, or the frequency distribution curve of the well defined controlling random variable to a problem, by which solutions in the form of a deterministic, unique value of the variable(s) which serves as the main dimension or quantity of a structure or machine capacity, divides PDF into two parts, satisfactory or unsatisfactory performance, with their two probabilities being complementary to unity; (2) solutions to problems which use the entire spatial-temporal, observed or simulated samples of inputs and outputs, and by applying the most appropriate optimization techniques lead to the optimal decisions in planning, design and operation of water resources structures and systems. The former is the classical risk-bound approach to water resources decisions. The latter is the contemporaneous technology, introduced by newest developments in various types of optimization (such as linear, dynamic, non-linear, hierarchical decision and other techniques). Both groups require stochastic or statistical hydrologic information, however either by using simple modeling of PDF, or by applying the complex spatial-temporal input-output modeling, respectively.

In using the optimization techniques in water resources a distinction should be made between planning and design of structures and systems, on one side, and operation of these structures and systems, on the other side. Optimization in the planning phase often starts from the assumed characteristics and dimensions of a structure or system. By using simulated inputs and outputs, as well as the corresponding optimization techniques, the results obtained are calculated as production, benefit, cost, risk, and various dimensions and magnitudes which were not assumed apriori. By varying dimensions or characteristics of a structure or a system, these results also vary, giving them the information for the final decisions on dimensions of a structure or on all the characteristics of a system. In essence, explicit solutions are used, namely dimensions or characteristics are assumed and varied with each assumed set producing results. Implicit solutions, namely the direct computation of optimal decisions, are feasible only in simple cases, such as using a PDF to obtain the optimal solution for a decision.

The situation is or may be different with the operation of water resources structures and systems, namely the authorization to construct a structure or develop a system usually commits the distribution of benefits, costs and risks, either by law or by contracts or agreements. These a priori decisions then become the constraints in operation, leaving relatively little room for optimization of operational decisions. An example of needed optimization in operation may be hydropower for which the prime or firm power is usually fully committed in advance, leaving only the available continuous secondary part-year power, or the highly intermittent tertiary power to be eventually optimized in temporal blocks of continuous power supply.

3.3. GENERAL APPLICATION OF STOCHASTIC HYDROLOGY IN WATER RESOURCES

The methods of utilitarian hydrology usually supply the extracted or transferred information on hydrologic variables relevant to each particular water resources problem or decision. This is the clasical statistical information of hydrology, which has in reality been responsible for shaping the utilitarian approach to hydrology. The introduction of stochastic modeling of spatial-temporal hydrologic processes came into practice by the three parallel developments: (1) modern water resources developments have increased the complexity of their structures and systems by extending the original single purpose-single structure-single water source approach to solutions of water problems into a versatile structure and system of multipurpose-multistructure-multisource approach and techniques of complex systems characteristics; (2) development in the last 30 years or so of advanced modeling of complex hydrologic, input-output, spatial-temporal stochastic processes, with the potential of simulating new samples or finding solutions to various intricate problems, not only because of advances in statistics and stochastic processes, in general, but also in numerical computational techniques and needs for these models even in their very complex forms; and (3) advancements in computer technology which still continue at a relatively fast pace.

The basic approach to the application of stochastic hydrology to water resources systems then becomes oriented to finding the proper methods and conditions for this application to various water resources problems, especially by using the advanced techniques of simulation or generation of samples based on the estimated stochastic models, and required by the optimization methods. It seems that the practice has already produced a sufficient number of cases of the use and misuse of stochastic hydrology in water resources, so that sorting of conditions for the proper and improper uses of these techniques deserves a corresponding analysis.

4. Conclusions

The following conclusions to this application of stochastic hydrology in water resources are:

(1) The recent marriage between the classical hydrologic statistics and the contemporaneous stochastic hydrology have made the utilitarian hydrology much more beneficial as well as important to water resources and environmental protection than previously, basically because of the potential supply of more accurate information on hydrologic variables and processes.

(2) While the attention to the modeling of hydrologic, spatial-temporal stochastic processes has been significant in the last three decades, the same cannot be claimed for the basic analysis of the quality of hydrologic data, such as the analysis of random, systematic and sampling errors and non-homogeneity in data, and for the in-depth structural decomposition and analysis of these processes with the physical justification of basic findings.

(3) While the analysis and modeling of temporal and spatial properties of input and output series may have been satisfactory from the practical point of view, the same cannot be said for the investigation of the mutual

dependence (correlation) between the input or supply and the output or demand (delivery) time series, and less so for the use of the historic output or supply time series of water resources systems to model these series mathematically with the use of these models to eventually simulate the future demand or output series, similarly as is currently done for input series.

(4) While the past extraction and transfer of hydrologic information for purposes of water resources decision making did not ask and checked how much hydrologic information was available and has been extracted or transferred by a given technique, the present techniques based on the concept and coefficients of entropy enable both the assessment of how much information is available in each case for extraction or transfer as well as how much a given technique has done it.

(5) Modeling mathematically the complex spatial-temporal hydrologic processes enables the simulation of potential future samples of inputs and outputs of water resources systems with their use in more accurate solutions of some complex problems.

(6) As in any new technical method, the simulation or generation of new samples has been appropriately used in many ways, but has also been misused with both needing a critical and careful assessment.

5. References

Harmancioglu, N.B., V. Yevjevich and J.T.B. Obeysekera (1986) 'Measures of information transfer between variables', in H.W. Shen, V. Yevjevich, J.T.B. Obeysekera and D.G. Decoursey (eds), Multivariate Analysis of Hydrologic Processes, Colorado State University (H.W. Shen), Fort Collins, Colorado, pp. 481-499.

Hazen, Allen (1914) 'Storage to be provided in impounding reservoirs for municipal water supply', Trans. Amer. Soc. of Civil Engineers, Vol. 77, Paper 1308, p. 1539.

Salas, J.D., J.W. Delleur, V. Yevjevich and W.L. Lane (1980) 'Applied modeling of hydrologic time series', Water Resources Publications, Littleton, Colorado.

Salas, J.D. and V. Yevjevich (1972) 'Stochastic structure of water use time series', Colorado State University, Hydrology Paper Series No. 52, Fort Collins, Colorado.

Sudler, C.E. (1972) 'Storage required for regulation of streamflow', Amer. Soc. of Civil Engineers, Vol. 91, p. 622.

PHILOSOPHY OF MODEL BUILDING

K.W. HIPEL
*Departments of Systems Design Engineering and Statistics
and Actuarial Science, University of Waterloo
Waterloo, Ontario
Canada N2L 3G1*

ABSTRACT.   General principles in model building are described and
methodologies for fitting stochastic models to time series are explained.
The basic idea in modelling a time series is to identify a simple model,
which has as few model parameters as possible in order to provide a good
statistical fit to the data.   In practice, this can be carried out by
following the identification, estimation and diagnostic check stages of
model construction.   An automatic selection criterion, such as the Akaike
information criterion, can be used to enhance this flexible approach to
model building.

1.   Introduction

To better understand and control his environment, mankind uses models.   In
order to be sufficiently accurate and realistic, a model must be able to
capture mathematically the key characteristics of a system being studied.
At the same time, a model must be designed in a straightforward manner so
that it can be easily understood, manipulated and interpreted.   Because of
the great complexity of water resources systems, models are extensively
developed and applied in water resources.   Of particular importance is the
water resources field of stochastic hydrology, in which stochastic or time
series models are used for purposes such as simulating and forecasting
hydrological systems.
    The major objectives of this paper are to outline general principles and
philosophies in model building and to explain how these basic concepts are
embedded into a comprehensive procedure for applying time series models to
hydrological observations.   In Section 2, basic modelling philosophies and
principles are explained.   To carry out a systematic data analysis study,
the exploratory and confirmatory data analysis procedures of Section 3 can
be   followed.   At   the   confirmatory   data   analysis   phase,   a   formal
mathematical model can be fitted to a time series by adhering to the
identification,   estimation   and   diagnostic   check   stages   of   model
construction described in Section 4.   Additionally, an automatic selection
criterion such as the Akaike information criterion (Akaike, 1974) can
enhance the model building stages.   Knowing the general classifications of

J.B. Marco et al. (eds.),
Stochastic Hydrology and its Use in Water Resources Systems Simulation and Optimization, 25–45.
© 1993 *Kluwer Academic Publishers.*

families of models according to the key statistical properties which they
can best describe, can assist a practitioner in deciding upon which group
of models he or she should consider for modelling the particular data sets
being examined. Table 1 in Section 5 lists classifications for time series
models used in stochastic hydrology. The application in Section 6 shows
how model building can be easily carried out in practice. Finally, in
Section 7 some appropriate conclusions are drawn about the art and science
of model building.

2. Modelling Philosophies

2.1. OVERVIEW

Hydrologists are aware of certain types of problems which arise when
modelling natural time series and these issues are outlined in Section
2.2. Because the practitioner is usually confronted with selecting the most
suitable model from a large set of possible models for fitting to a given
time series, the general topic of model discrimination is addressed in
Section 2.3. When choosing the most appropriate model, the fundamental
modelling principles of Section 2.4 can be satisfied by following the three
stages of model building described in Section 4. Model construction takes
place at the confirmatory data analysis stage explained in Section 3.

2.2. HYDROLOGICAL UNCERTAINTIES

Engineers are concerned with the role that uncertainty plays in the design,
analysis, operation and control of water resource systems. When a
stochastic or time series model which is fitted to a hydrological time
series is to be employed in various water resources applications, three
types of uncertainties have been delineated (Kisiel and Duckstein, 1972;
Wood and Rodriguez-Iturbe, 1975; Vicens et al., 1975; Wood, 1978). Firstly,
there is natural uncertainty which is the uncertainty inherent in the
natural phenomenon itself. By fitting a suitable time series model to the
time series measured from the phenomenon under consideration, it is hoped
that this natural uncertainty will be reflected in the mathematical
structure of the model. Because the parameters of the model must be
estimated statistically from a finite amount of historical data, the second
kind of uncertainty is labelled parameter uncertainty. Finally, due to the
fact that a particular model of the phenomenon may not be the "true" or
best model, this creates a third category of uncertainty which is model
uncertainty. Since the latter two types of uncertainty are dependent upon
the available data, these have been jointly referred to as information
uncertainties (Vicens et al., 1975).

Traditionally, the field of stochastic hydrology has been mainly
concerned with the problem of natural uncertainty. A host of stochastic
models have been developed to model natural time series and many of these
models are discussed throughout the hydrological and statistical
literature. For instance, in addition to ARMA (autoregressive-moving
average) (Box and Jenkins, 1976) models, fractional Gaussian noise models
(Mandelbrot and Wallis, 1969) and approximations to fractional Gaussian
noise models have been suggested for modelling annual geophysical data

sequences. Parameter uncertainty can be measured by the standard errors for the parameter estimates and a procedure for incorporating parameter uncertainty into simulation studies is presented by McLeod and Hipel (1978). As reported by hydrological researchers (Vicens et al., 1975; Wood, 1978), little work has been done regarding the issue of model uncertainty. Consequently, within this paper methods are described for alleviating the problem of model uncertainty.

2.3. MODEL DISCRIMINATION

Model uncertainty arises because the practitioner must select the most appropriate model from the total array of models which are available for fitting to a given time series. Hence, discrimination procedures are required for choosing the most suitable model. The basic idea behind model selection is to choose a model from the set of models under consideration such that the selected model describes the data best according to some criterion. Ljung (1978) presents a unified description of model discrimination methods and other comprehensive articles can be found in the available literature (see for example Caines (1976, 1978) and Kashyap and Rao (1976)). Criteria for choosing the most suitable model include the capability of a model to satisfy graphical identification methods (see Sections 4.1 and 6.1), the requirement that the model residuals pass sensitive diagnostic checks (refer to Section 4.1), the ability of a model to forecast accurately, and the capability of a model to preserve important historical statistics (Hipel and McLeod, 1978). A particularly flexible approach to model discrimination is the Akaike information criterion (AIC) (Akaike, 1974) which is described in Section 4.2.

2.4. MODELLING PRINCIPLES

An attractive feature of the AIC is that it automatically accounts for certain fundamental modelling principles. One precept of stochastic model building is to keep the model as simple as possible. This can be effected by developing a model which incorporates a minimum number of parameters in order to adequately describe the data. Box and Jenkins (1976) recommend adhering to the principle of model parsimony (i.e. keeping the number of model parameters to a minimum) and this rule has also been of concern to hydrologists (see, for example, Jackson (1975); Tao and Delleur (1976); Hipel et al. (1977); McLeod et al. (1977) and Salas et al. (1980)). Besides designing a parsimonious model, a second modelling tenet is to develop a model that imparts a "good statistical fit" to the data. To achieve a good statistical fit, efficient estimates must be obtained for the model parameters and the fitted model must pass rigorous diagnostic checks to ensure that the underlying modelling assumptions are satisfied.

The principle of model parsimony has historical roots that go back far into the past. Aristotle, for example, postulated that nature operates in the shortest possible way. A 14th century English Franciscan monk by the name of William of Ockham developed a principle now known as Ockham's Razor. One version of his principle states that when faced with competing explanations, choose the most simple one. Another equivalent statement for Ockham's Razor is entities are not to be multiplied without necessity.

Bertrand Russell (1946), the famous 20th century British mathematician, found Ockham's Razor to be very informative in logical analysis. This is because there is only one explanation or description of something which is minimum, whereas there can be an infinity of explanations which bring in other entities. Russell went on to claim that adherence to the minimum necessary explanation or description ensures that the examination of hypotheses and evidence for and against them will remain coherent. Checkland (1981) provides a good explanation of Ockham's Principle in his book on the theory and practice of systems engineering.

3. Data Analysis

3.1. SCIENTIFIC INVESTIGATION

As noted by Box (1974), two main tasks in a given scientific investigation are:
   1. the "design problem" where one must decide upon the appropriate data to obtain at each stage of an investigation, and
   2. the "analysis problem" where models are employed for determining what the data entitles the investigator to believe at each stage of the investigation.
   In order to execute comprehensive analyses of the data, it is absolutely essential to determine properly the appropriate data to obtain at the design phase. No amount of skill and experience in data analysis can extract information which is not contained in the data to begin with. Accordingly, suitable data collection schemes are needed for carrying out a time series analysis investigation. Within the statistical and engineering literature, extensive research has been published about designing optimal data collection schemes across a network of stations. For example, Moss (1979) wrote an introductory paper for a sequence of twenty-four papers published in Volume 16, Number 6, 1979, of Water Resources Research. Lettenmaier et al. (1978) suggested designs for data collection schemes when one intends to employ the intervention model (Box and Tiao, 1975) to ascertain the effects of an intervention upon the mean level of a time series. Because both parametric univariate and multivariate time series models must be used with observations separated by equal time intervals, proper sampling is of utmost importance in time series analysis. If available measurements are not evenly spaced, appropriate data filling techniques can be utilized to estimate a series of evenly spaced data from the given information. The particular technique to employ for data filling depends upon the type and amount of missing data (see, for example, Coons (1957), Dempster et al. (1977), Brubacher and Wilson (1976), Baracos et al. (1981) and McLeod et al. (1983)). An advantage of employing nonparametric tests for detecting trends in multiple time series is that they can usually be used with unequally spaced observations (see, for instance, Hirsch and Gilroy (1985), Van Belle and Hughes (1984) and Hipel (1988)).
   When dealing with time series studies in water resources, often the data were measured over a long period of time and the people analyzing the collected data did not take part in designing the data collection procedure in the first place. Nevertheless, wherever possible, practitioners are

advised to actively take part in the design of the scheme for collecting the data which they will analyze.

3.2. EXPLORATORY AND CONFIRMATORY DATA ANALYSES

When analyzing a given set of data, there are usually two main steps (Tukey, 1977). The first step is referred to as "exploratory data analysis" and the purpose of this phase of data analysis is to discover the important statistical properties of the data by carrying out simple graphical and numerical studies. For example, in order to detect trends in water quality time series, McLeod et al. (1983) employed techniques such as a graph of each series against time, the box and whisker graphs of Tukey (1977), Tukey smoothing (Tukey, 1977) and the autocorrelation function (ACF).

The second major step to data analysis is called "confirmatory data analysis" and the purpose of this stage is to statistically confirm in a rigorous fashion the presence or absence of certain properties in the data. For instance, in their trend assessment study of water quality time series, McLeod et al. (1983) used exploratory data analysis techniques to detect the presence of trends in some of the time series due to known external interventions. Following this, intervention analysis (Box and Tiao, 1975) was utilized as a confirmatory data analysis tool to determine if there was a significant change in the mean levels of the series suspected of containing trends.

The exploratory and confirmatory stages of data analysis can be compared to the process which takes place after a crime is committed (Tukey, 1977). At the "exploratory stage" of investigating a crime, a sleuth uses forensic tools and his common sense to discover evidence about the crime. If the detective does not understand how to execute an investigation, he may fail to look in the proper places for the criminal's fingerprints. On the other hand, if the investigator has no fingerprint powder he will not detect fingerprints on most objects. In an analogous fashion, the statistical analyst requires both the tools of the trade and common sense.

In the criminal justice system, the suspected criminal is taken to court after the collection of evidence by the investigative bodies. Following the evaluation of the available evidence, the jury and judge must ascertain if the criminal is guilty based upon the current evidence. Likewise, in a statistical study, the purpose of the second main step of "confirmatory data analysis" is to quantitatively verify if suspected statistical characteristics such as different trends are actually present in the data. When enough evidence is available, the results of a "confirmatory data analysis" can be quite useful to the decision makers. For instance, when intervention analysis is employed in an environmental impact assessment study for properly confirming the presence of trends in water quality time series, the results can be used in court for forcing the polluters to adopt appropriate pollution abatement procedures.

Some exploratory techniques, such as a graph of the series against time and box and whisker graphs, do not require that the time series be evenly spaced over time. On the other hand, other exploratory data analysis techniques like Tukey smoothing and the ACF must be used with data points that are equally spaced over time. An assumption underlying virtually all of the time series or stochastic models which can be used in practical applications at the confirmatory data analysis stage is that the data sets to which they are fitted consist of observations separated by equal time intervals. Although it would be desirable to possess stochastic models which can readily handle time series consisting of any kind of unevenly spaced observations, currently no such practical models exist and, indeed, it may turn out to be mathematically intractable to develop these types of stochastic models. In practice, if the measurements are not evenly spaced, appropriate techniques, such as those referred to in Section 3.1, must be utilized to produce a series of equally spaced data that is estimated from the given information. When, for instance, there are only a few missing observations, intervention analysis can be used to simultaneously estimate the missing values and also model trends due to known interventions (Baracos et al., 1981). If there are large gaps in the data, one can use the seasonal adjustment approach of McLeod et al. (1983).

4. Model Construction

4.1. STAGES IN MODEL BUILDING

In many applications one may wish to ascertain the most appropriate stochastic model to fit a given data set at the confirmatory data analysis stage. No matter what type of stochastic model is to be fitted to a given data set, it is recommended to follow the identification, estimation and diagnostic stages of model construction (Box and Jenkins, 1976). At the identification stage, the more appropriate models to fit to the data can be tentatively selected by examining various types of graphs. Some of the identification information may already be available from studies completed at the exploratory data analysis step discussed in the previous section. Because there may be many different families of stochastic models which can be fitted to the time series under consideration, one must select the one or more families of models which are the most suitable to consider. The family selections can be based upon a sound physical understanding of the problem, output from the identification stage, and exploratory data analyses. Although sometimes it is possible to choose the best model from one or more families based solely upon identification results, in practice it is often not obvious which model is most appropriate and hence two or three models must be tentatively entertained. At the estimation stage, maximum likelihood estimates can be obtained for the model parameters and subsequently the fitted model can be diagnosed to ensure that the key modelling assumptions are satisfied. When considering linear stochastic models such as ARMA models, one should check that the estimated model residuals are not correlated, possess constant variance (i.e. homoscedasticity) and perhaps are approximately normally distributed. If the residuals are not white noise, the model should be redesigned by repeating the three phases of model construction. In practice, it has been found that a suitable Box-Cox transformation (Box and Cox, 1964) can rectify anomalies such as heteroscedasticity and non-normality. The specific tools utilized at the three stages of model construction are dependent upon the particular family of models being entertained. The logic underlying the traditional approach to model construction is displayed as a flowchart in Figure 1.

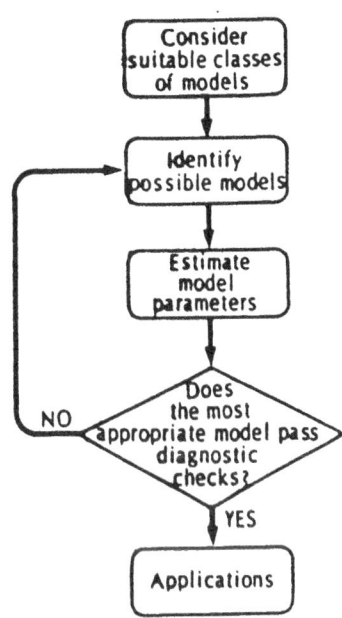

Figure 1. Model construction

4.2. AUTOMATIC SELECTION CRITERIA

A number of automatic selection criteria (ASC's) have been proposed for measuring the degree to which the objectives of good statistical fit and model parsimony are attained (see Hipel (1981) for a comprehensive list of ASC's). The most popular ASC's for use in model discrimination are the Akaike information criterion (AIC) (Akaike, 1974) and Schwarz's (1978) approximation of the Bayes information criterion (BIC). The AIC is defined as

$$AIC = -2lnML+2k \tag{1}$$

and Schwarz's approximation of the BIC is given by

$$BIC = -2lnML+klnn \tag{2}$$

where ML denotes maximum likelihood, k is the number of free parameters, and n is the number of entries in the time series. An attractive feature of the ASC's, such as the AIC and BIC, is that the ASC's automatically account for the two key modelling principles discussed in Section 2.4. The first term on the right hand side of both Equations 1 and 2 allows for the criterion of good statistical fit, while the second entry accounts for the doctrine of parameter parsimony. Because of the general form of Equations

1 and 2, when there are several competing models to choose from, the model that possesses the minimum value of the AIC and BIC, respectively, should be selected. The procedure of finding the model having the minimum value of the AIC is referred to as MAICE.

An ASC can be conveniently employed to enhance the three stages of model construction (Hipel, 1981). Figure 2 depicts how an ASC can be incorporated into the general model building process when considering suitable classes of models for fitting to the given data. Sometimes it is known in advance whether or not a data transformation, such as a specific kind of Box-Cox transformation, is required. For instance, when fitting seasonal models to average monthly riverflow data, usually the natural logarithmic transformation is the best type of Box-Cox transformation to invoke (Tao and Delleur, 1976). If it is not certain whether a data transformation is needed, the necessity for a transformation may be detected later by diagnostic testing of the model residuals. As shown in Figure 2, following data transformation considerations there are basically two approaches for employing an ASC in model construction. One method is to perform an exhaustive ASC study by calculating the ASC's for all possible models which are deemed worthwhile for fitting to a given data set. This requires specifying upper limits to the number of parameters which can be included in the various model components of a given type of stochastic model. The values of the ASC are then calculated for all possible combinations of the model parameters and the model with the minimum ASC is chosen. Although the selected model may adequately satisfy the important mathematical assumptions underlying the model, as indicated in Figure 2, it is always advisable to check for whiteness, homoscedasticity and perhaps normality of the residuals (Box and Jenkins, 1976; Salas et al., 1980; Hipel et al., 1977). When the residuals are not white, other models should be considered by specifying a more flexible range for the model parameters. If the residuals do not possess constant variance and perhaps are not normally distributed, then a suitable data transformation may alleviate the situation. The reader should bear in mind that the three diagnostic checks suggested in Figure 2 are appropriate for use with linear stochastic models such as univariate and multivariate ARMA models. When different assumptions are made regarding the innovations in a family of models, appropriate diagnostic checks can be used in place of those given in Figure 2. Lewis (1985) and McKenzie (1985) present models for which the innovations are not assumed to follow a normal distribution.

If the diagnostic check stage is skipped and information from the identification and estimation stages is ignored when employing the exhaustive enumeration procedure with an ASC, it is possible that the best model may be missed. For example, McLeod et al. (1977) and Hipel (1981) show that the most suitable type of ARMA model to fit the annual sunspot series is an autoregressive (AR) model of order 9 with the third to eighth AR parameters omitted from the model and the data transformed by a square root transformation. If diagnostic testing had not been carried out and the standard errors of the parameter estimates had been ignored, the most suitable model would not have been discovered.

An exhaustive ASC study may prove to be rather expensive due to the amount of computations. Consequently, as illustrated in Figure 2, an alternative approach is to estimate only the parameters and hence the ASC for a subset of models. For example, information from the identification

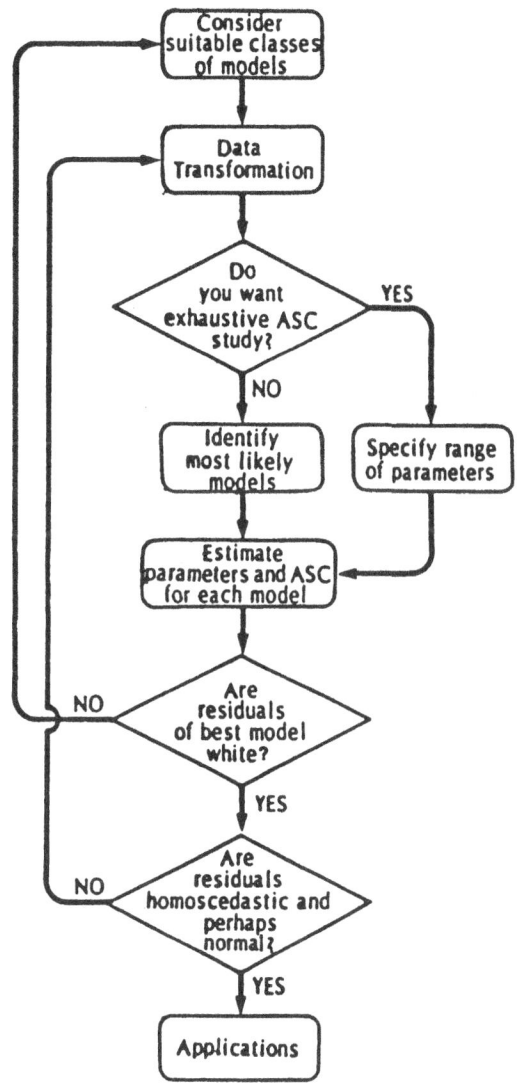

Figure 2. An ASC in model construction.

stage may indicate three tentative models to fit to the time series. The ASC is then calculated only for these three models and the model with the minimum ASC value is selected. If there are any problems with the residuals, appropriate action may be taken as shown in Figure 2. Otherwise, the chosen model can be employed in practical applications such as forecasting and simulation.

In the past, researchers and practitioners alike in water resources have tended to use models which are too complex and hence contain too many parameters. As an example, people have often assumed that one should use the most general form of the multivariate ARMA model in hydrological applications. This, in turn, has meant that there are too many parameters in the multivariate model and it is very difficult to obtain efficient estimates for the model parameters. Due to the aforesaid, Salas et al. (1985) and Camacho et al. (1985) have suggested employing the parsimonious CARMA (contemporaneous ARMA) model, for which Camacho et al. (1987) have developed an efficient estimation procedure. In general, it is usually advantageous to first consider a group of fairly simple models for describing data sets and then to construct a more complex model only when the simpler models do not adequately describe the data. Information obtained by building the simpler models can usually assist in designing the more complex models.

5. Categorizing Models

5.1. GENERAL PROBLEM

Due to the diversity and complexity of the problems which arise within the field of water resources, river basin management can be a challenging and arduous task. Not only must the complex physical processes of water resources problems be properly understood, but perplexing socio-economic factors must be considered as well if the water resources of a given area are to be effectively managed and utilized. Because so many different kinds of problems must be addressed, a wide range of models must be used to assist decision makers in water resources management. For example, conflict analysis (Fraser and Hipel, 1984; Fang et al., 1993) can be employed to model the decision making which takes place within the political arena. Stochastic and deterministic conceptual models can be employed for modelling and understanding the hydrological characteristics of water resources systems.

Models can be classified according to a wide variety of criteria. One approach is to categorize models according to the field in which they were originally developed. Thus, for example, one has systems engineering, economic, operational research, decision analysis, economic and hydrological sets of models. However, often models can be applied across disciplines or fields. For example, ARMA models have been applied in areas such as economics, hydrology and transportation engineering. Conflict analysis models have been employed in fields ranging from water resources engineering to international affairs. Because many systems models are independent of the field of applications, a better approach to model classification is to categorize models according to the main characteristics that they are designed to describe. In the next section, criteria are put forward for classifying time series models. These classification criteria are especially useful for deciding upon which types or kinds of time series models to employ for solving problems in stochastic hydrology.

5.2. CLASSIFYING TIME SERIES MODELS

5.2.1. *Introduction.* From Figures 1 and 2 it can be seen that the first step in model construction is to select suitable classes or families of models from which the most appropriate model to fit to a given set of data can be eventually chosen by following the identification, estimation and diagnostic check stages of model development. For example, when modelling annual hydrological time series, one may wish to consider the ARMA family of models (Box and Jenkins, 1976), the classes of nonGaussian models suggested by Lewis (1985) and fractional differencing models (Hosking, 1984). Certainly, if one is not aware that certain classes of models exist, one may not fit the most appropriate model to a given data set.

By definition, any model is an approximation to reality. Nonetheless, if one is cognizant of the important statistical characteristics of a data set, one can model the data as realistically as possible by considering one or more families of models which are capable of modelling these major characteristics. For instance, suppose that one wishes to model an average monthly riverflow time series. In order to model adequately the seasonally varying correlation structure and to preserve the stationary statistical properties within each season, one would have to consider the families of PAR (Salas et al., 1980) and PARMA (Vecchia, 1985; Jimenez et al., 1989) models. Note that another class of models, such as the family of seasonal autoregressive integrated moving average (ARIMA) models, would be incapable of modelling the aforementioned statistical characteristics. By fitting models that best describe the main statistical properties of the data being modelled, more accurate forecasts and simulated sequences can be obtained (Noakes et al., 1985; Thompstone et al., 1985).

In practice, one must be aware of the main statistical characteristics of the data which are to be modelled. This knowledge may be due to a sound physical understanding of the problem and to information obtained from exploratory data analyses. Given the important statistical characteristics to be modelled, one can determine the families of models which may be capable of modelling them. Consequently, it is convenient to classify each family of models according to the key statistical properties or criteria that it is designed to model.

5.2.2. *Criteria for Classifying Time Series Models.* In Table 1 a wide variety of criteria are presented by Hipel (1985) for classifying families of time series models. Usually, a given family is designed for modelling more than one statistical characteristic. However, no class of models exists that can simultaneously handle all of the criteria listed in Table 1. If one attempted to design a family of models that could describe too many criteria, the models would become too cumbersome due to an abundance of model parameters and it would be difficult, or perhaps impossible, to devise a scheme for economically obtaining efficient estimates of the model parameters. As a result, the practitioner must be cognizant of the key properties of both his data and the families of models that are available for fitting to the data. Detailed discussions of the criteria in Table 1 are given by Hipel (1985).

TABLE 1. Criteria for classifying time series models

physically based
discrete time (evenly and unevenly spaced data)
continuous time
continuous observations
discrete observations
Gaussian
nonGaussian
linear
nonlinear
seasonal
nonseasonal
stationary
nonstationary
short memory
long memory
univariate
multivariate
periodic
disaggregation
aggregation
parametric
nonparametric
time domain
frequency domain
Bayesian
state-space formulation (Kalman filter)

6. Model Building Application

6.1. INTRODUCTION

As noted in Section 4.1, the specific tools employed in model construction are dependent upon the family of models that one is employing for fitting to a data set. Consider, for instance, the case of fitting the most appropriate ARMA model (Box and Jenkins, 1976) to a nonseasonal or else deseasonalized hydrological time series. At the identification stage in Figures 1 and 2, a variety of graphs can be employed for deciding upon the order, p, of the AR parameters and the order, q, of the MA operators needed in the ARMA (p,q) model to describe the series. In particular, four identification graphs that can be employed for ARMA model design are the autocorrelation function (ACF), partial autocorrelation function (PACF), inverse autocorrelation function (IACF) (Cleveland, 1972) and inverse partial autocorrelation function (IPACF) (Hipel et al., 1977). The values of each of the foregoing functions are plotted against positive lag k along with 95% confidence intervals. The properties of the identification methods with respect to ARMA models are summarized in Table 2. If, for example, the values of the sample PACF and IACF are only significantly different from zero at lags 1 and 2, the values of the sample ACF and IPACF attenuate, then one would fit an ARMA(2,0) model to the series. Other identification methods developed for use with ARMA models include the R and S arrays of Gray et al. (1978) introduced into hydrology by Salas and Obeysekera (1982), the corner method (Beguin et al., 1980) and the extended sample autocorrelation function (Tsay and Tiao, 1984). The identification techniques in Table 2 have been extended for use with seasonal data (see, for instance, Hipel et al. (1977), and Hipel and McLeod (1993).

TABLE 2. Properties of four identification methods

Identification Method	Types of ARMA models		
	ARMA (p,o)	ARMA (o,q)	ARMA (p,q)
ACF	Attenuates	Truncates after lag q	Attenuates
PACF	Truncates after lag p	Attenuates	Attenuates
IACF	Truncates after lag p	Attenuates	Attenuates
IPACF	Attenuates	Truncates after lag q	Attenuates

At the estimation stage, the method of maximum likelihood can be employed to obtain statistically efficient parameter estimates of the ARMA model suggested at the identification stage. As pointed out in Section 4.1, flexible statistical tests are available for use at the diagnostic check stage to ensure that the ARMA model residuals are white, possess constant variance (i.e. are homoscedastic) and are approximately normally distributed.

The intervention model constitutes an extension of the basic ARMA family of models. Because this model is extremely useful in stochastic hydrology and environmental impact assessment, an intervention model is fitted to a hydrological time series in the next subsection by following the model building approach of Figure 2.

6.2. INTERVENTION ANALYSIS APPLICATION

Intervention analysis (Box and Tiao, 1975) is a stochastic modelling technique to rigorously analyze whether or not natural or man-induced interventions cause significant changes in the mean level of a time series. Although it is a mathematical tool which possesses many useful applications, intervention analysis is especially suitable for utilization in the geophysical and environmental sciences. For example, as explained in Chapter 19 of Hipel and McLeod (1993), the method has been employed to

38

determine the statistical effects of reservoir construction and operation upon the mean levels of the downstream flows for both nonseasonal and seasonal data, to find out how a devastating forest fire can alter the seasonal flow characteristics of a river, to ascertain the effectiveness of pollution abatement procedures for improving air and water quality, to determine the effects of changing the type of snow gauges used for measuring the amount of snowfall in the Canadian Arctic, to fill in missing data points, and to design data collection procedures.

Intervention models are closely related to transfer function-noise models and a detailed description of the mathematical structure of univariate and multivariate intervention models can be found in the published literature. When determining an intervention model for describing a given data set, the model building technique in Figure 2 can be employed. To demonstrate the efficacy of the MAICE procedure, the method is used to select an intervention model which describes the effects of the Aswan Dam on the average annual flows of the Nile River. The average annual flows of the Nile River at Aswan, Egypt, which are available from 1870 to 1945, are shown in Figure 3 in cubic metres per second (m^3/s).

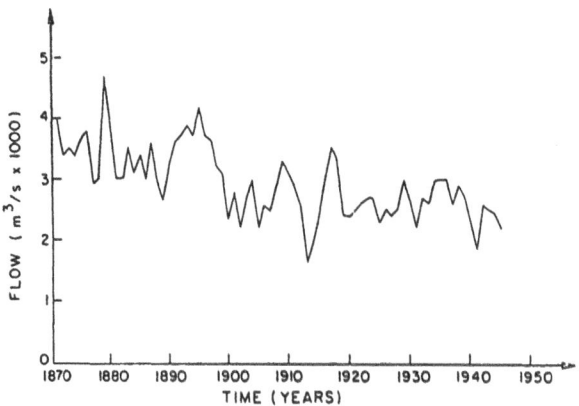

Figure 3. Average annual flows of the Nile River at Aswan.

During 1903 the Aswan Dam was completed and the reservoir was filled for the first time. As can be seen in Figure 3, there appears to be a drop in the mean level of the flows from 1903 onwards. The dam intervention which caused the drop in the average annual flows could be an accumulative effect of various physical factors (Hipel et al., 1975). The reservoir size caused an increase in evaporation and greater percolation into the underlying soil. In addition, some water was taken from the reservoir to be used for irrigation and domestic water supply. Finally, there were perhaps some flow measurement errors which were thought not to exceed 5% of the flow.

An intervention model for the Nile River was originally developed by Hipel et al. (1975). However, it is shown here and also by Hipel (1981) how the MAICE procedure simplifies the selection of the best model which is more plausible than the model suggested by Hipel et al. (1975).

Let the time series for the Nile flows be represented by z_t, where t=1870,1871,...,1945. A Box-Cox power transformation (Box and Cox, 1964) for the series is written as

$$z_t^{(\lambda)} = \begin{cases} \dfrac{(z_t^{\lambda}-1)}{\lambda} & \lambda \neq 0 \\ \ln z_t & \lambda = 0 \end{cases} \qquad (3)$$

where λ is the Box-Cox power parameter. The Nile intervention model for the $z_t^{(\lambda)}$ series can be written in a general format as

$$z_t^{(\lambda)} - \mu = v(B)\xi_t + N_t \qquad (4)$$

where μ is the theoretical mean of the entire $z_t^{(\lambda)}$ series, B is the backward shift operator defined by $B^j\xi_t = \xi_{t-j}$ where j is a non-negative integer, $v(B)$ is the transfer function for the intervention series, ξ_t is the step intervention series which possesses a value of zero before 1902 and a value of unity from 1903 onwards, and N_t is the noise term which can be modelled by an ARMA (p,q) model. By varying the choice of the Box-Cox parameter $\lambda, v(B)$, and N_t, different intervention models can be considered for modelling the Nile data. In Table 3 a range of representative intervention models from Equation 3 are listed. For each model, an X entry indicates the type of component contained in the model. Often riverflow data are transformed by natural logarithms and hence models with $\lambda=0$ and $\lambda=1$ are considered. One would probably suspect that a transfer function with the parameter ω_0 would be appropriate to reflect a step intervention. However, it is possible that there could be some initial transient effects which require a transfer function of the form $\omega_0/(1-\delta_1 B)$, where ω_0 and δ_1 are the transfer function parameters. For example, it may take two or three years for the ground water levels to reach a steady-state condition after the reservoir is filled. Identification procedures from Table 2 revealed that the noise term N_t is either ARMA(0,1) or ARMA(1,0). The white noise ARMA(0,0) model is also included in Table 3 for comparison purposes.

To obtain maximum likelihood estimates (MLE's) for the intervention model parameters, the method of McLeod (1977) was employed. The number of parameters k in the AIC formula in Equation 1 is due to the ARMA model parameters for N_t, and the additional parameters contained in the transfer function. From Table 3 it can be seen that the minimum value of the AIC occurs for model number 1. The MLE's for the parameters and the standard errors of Model 1 are listed in Table 4. The difference equation for this intervention model is written as

$$z_t - 3357.164 = -724.868\xi_t + (1+0.406B)a_t \qquad (5)$$

A plot of the residual ACF (McLeod, 1978; Hipel et al., 1977) in Figure 4 reveals that the estimated values fall within the 5% significance interval. Hence, the most appropriate intervention model, according to the MAICE procedure, possesses residuals that are white noise.

A comparison of the AIC values in Table 3 demonstrates that the models which assume an ARMA(0,0) term for N_t (i.e. Models 3,6,9 and 12) are much less desirable than the other models. Whenever an ARMA(0,1) noise term is used instead of an ARMA(1,0) component, it causes an improvement in the AIC value. The AIC entries in Table 3 also confirm that it is not necessary to take natural logarithms of the data. In addition, a comparison of the AIC values between Models 1 and 4 reveals that the type of transfer function causes a difference between the AIC values of less than unity. Although a transfer function of the form ω_0 is more preferred, both from a physical understanding of the problem and also the MAICE procedure, the fact of the matter is that $\omega_0/(1-\delta_1 B)$ in Model 4 is not radically different from ω_0 in Model 1. When the parameter estimates are substituted into the aforesaid two transfer functions, the steady-state gains for both models are quite close. Finally, when the MAICE procedure is not invoked, an inferior model may be chosen. Hipel et al. (1975) suggested that Model 8 be selected to model the Nile River while the results from Table 3 can be used to show that the plausibility of Model 8 versus Model 1 is much lower.

TABLE 3.   Intervention models for the Nile River

Model Number	Box-Cox Parameter λ		Transfer Function $v(B)$		Noise Term N_t			AIC
	1.0	0.0	ω_0	$\dfrac{\omega_0}{1-\delta_1 B}$	ARMA (0,1)	ARMA (1,0)	ARMA (0,0)	
(1)	X		X		X			906.70
(2)	X		X			X		907.71
(3)	X		X				X	941.61
(4)	X			X	X			907.47
(5)	X			X		X		907.91
(6)	X			X			X	943.37
(7)		X	X		X			908.07
(8)		X	X			X		908.75
(9)		X	X				X	941.62
(10)		X		X	X			908.55
(11)		X		X		X		908.57
(12)		X		X			X	943.22

TABLE 4.   Parameter estimates for the best Nile intervention model

Parameter	MLE (Standard Error)
ω_0	-724.868 (84.788)
θ_1	-0.406 (0.106)
μ	3357.164 (18.727)
σ_a^2	1.596×10^5

The model in Equation 5 can be used for applications such as forecasting and simulation. However, the intervention model can also be employed to describe statistically the change in the mean level of the Nile River due to the Aswan Dam. By subtracting the expected value of z_t in Equation 5 before the intervention from the expected value of z_t after 1902, the drop in the mean level is 724.87 m³/s. The average value of the flows prior to 1903 is 3370.10 m³/s, and consequently the percentage decrease in the mean level is 21.51 per cent. Because the MLE for ω_0 possesses a limiting normal distribution, confidence limits can be calculated. The 95% confidence limits can be determined by adding to and subtracting from $\hat{\omega}_0$, 1.96 times its standard error of 84.79. These limits show that the change in the average flows is probably not greater than 891.06 m³/s and not less than 558.68 m³/s. In terms of percentage change relative to the average flows prior to the intervention, the 95 % confidence interval is from 16.58 to 26.44 %, while the best estimate of the percentage drop in the average is 21.51 per cent.

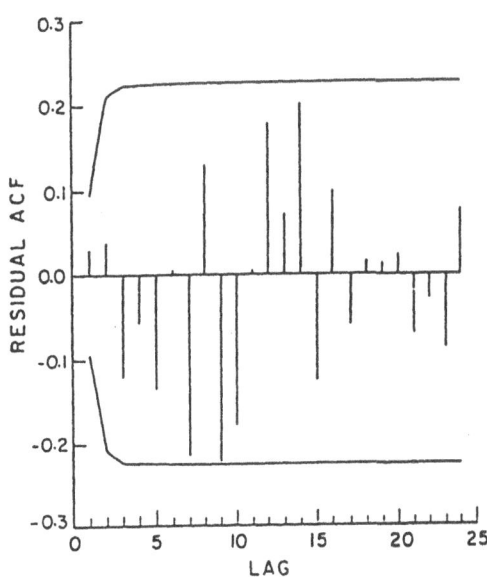

Figure 4. Residual ACF for the best intervention model for the Nile River.

7.   Conclusions

Figures 1 and 2 present a flexible and comprehensive model construction procedure for fitting time series models to hydrological data sets, as well as other kinds of time series. An attractive feature of the approach is that it is based upon sound model building principles. For example, the principles of model parsimony, as advocated in Ockham's Razor, and achieving a good statistical fit of the stochastic model to the data (see Section 2.4), are embedded into the overall model building procedure. In addition, these two principles are mathematically encoded into the ASC formulae in Equations 1 and 2.

In a practical application, a practitioner can employ exploratory data analysis techniques, usually in the form of informative graphs, in order to reach at least a qualitative understanding of the basic statistical characteristics of the data being studied. Subsequent to exploratory data analyses, the practitioner can select the best time series model to fit the data set at the confirmatory data analysis stage by following the three stages of model construction outlined in Figures 1 and 2. The application of Section 6.2 illustrates how the most appropriate intervention model can be selected using the MAICE procedure of Section 4.2.

As pointed out by Hipel (1985), selecting a suitable stochastic model to fit to a time series can be considered as both an art and a science. The art of model selection comes into play when the modeller uses his knowledge about the physical and statistical aspects of the problem to ascertain which families should be considered in the first place. Table 1 and related discussions should assist the decision maker in the art of family discrimination. Additionally, choosing the most appropriate tools to use in exploratory and confirmatory data analyses, as well as model construction, requires an art which is based upon common sense, good judgement and experience. By employing the most comprehensive, yet simple statistical methods, the efficacy of decisions made in the art of model choice and building can be rigorously checked from a scientific viewpoint. In practical applications, the foregoing approach usually involves interactive feedback between the art and science of time series modelling.

8. References

Akaike, H. (1974) 'A new look at the statistical model identification', IEEE Transactions on Automatic Control AC-19, pp 716-723.

Baracos, P.C., Hipel, K.W. and McLeod, A.I. (1981) 'Modeling hydrologic time series from the Arctic', Water Resources Bulletin 17, 3, pp 414-422.

Beguin, J.M., Gourieroux, C. and Monfort, A. (1980) 'Identification of a mixed autoregressive-moving average process: the corner method', in O.D. Anderson (Ed.), Time Series, North Holland, Amsterdam, pp 423-436.

Box, G.E.P. (1974) 'Statistics and the environment', Journal of the Washington Academy of Science 64, 2, pp 52-59.

Box, G.E.P. and Cox, D.R. (1964) 'An analysis of transformations', Journal of the Royal Statistical Society, Series B 26, pp 211-252.

Box, G.E.P. and Jenkins, G.M. (1976) Time Series Analysis: Forecasting and Control, Second Edition, Holden-Day, San Francisco.

Box, G.E.P. and Tiao, G.C. (1975) 'Intervention analysis with applications to economic and environmental problems', Journal of the American Statistical Association 70, 349, pp 70-79.

Brubacher, S.R. and Wilson, G.T. (1976) 'Interpolating time series with applications to the estimation of holiday effects on electricity demand', Journal of the Royal Statistical Society, Series C (Applied Statistics) 25, 2, pp 107-116.

Caines, P.E. (1976) 'Prediction error identification methods for stationary stochastic processes', IEEE Transactions on Automatic Control AC-21, 4, pp 500-506.

Caines, P.E. (1978) 'Stationary linear and nonlinear system identification and predictor set completeness', IEEE Transactions on Automatic Control AC-23, 4, pp 583-594.

Camacho, F., McLeod, A.I. and Hipel, K.W. (1985) 'Contemporaneous autoregressive moving average (CARMA) modeling in hydrology', Water Resources Bulletin 21, 4, pp 709-720.

Camacho, F., McLeod, A.I. and Hipel, K.W. (1987) 'Contemporaneous bivariate time series', Biometrika 74, 1, pp 103-113.

Checkland, P. (1981) Systems Thinking, Systems Practice, John Wiley, Chichester, United Kingdom.

Cleveland, W.S. (1972) 'The inverse autocorrelations of a time series and their applications', Technometrics 14, 2, pp 277-298.

Coons, I. (1957) 'The analysis of covariance as a missing plot technique', Biometrics 13, pp 387-405.

Dempster, A.P., Laird, N.M. and Rubin, D.B. (1977) 'Maximum likelihood from incomplete data via the EM algorithm', Journal of the Royal Statistical Society, Series B 39, 1, pp 1-38.

Fang, L., Hipel, K.W. and Kilgour, D.M. (1993) 'Interactive decision making: the graph model for conflict resolution', Wiley, New York.

Fraser, N.M. and Hipel, K.W. (1984) Conflict Analysis: Models and Resolutions, North-Holland, New York.

Gray, H.L., Kelley, G.D. and McIntire, D.D. (1978) 'A new approach to ARMA modelling', Communications in Statistics B7, 1, pp 1-77.

Hipel, K.W. (1981) 'Geophysical model discrimination using the Akaike information criterion', IEEE Transactions on Automatic Control AC-26, 2, pp 358-378.

Hipel, K.W. (1985) 'Time series analysis in perspective', Water Resources Bulletin 21, 4, pp 609-624.

Hipel, K.W. (Ed) (1988) 'Nonparametric approaches to environmental impact assessment', monograph published by the American Water Resources Association, Bethesda, Maryland.

Hipel, K.W., Lennox, W.C., Unny, T.E. and McLeod, A.I. (1975) 'Intervention analysis in water resources', Water Resources Research, 11, 6, pp 855-861.

Hipel, K.W. and McLeod, A.I. (1978) 'Preservation of the rescaled adjusted range, 2, Simulation studies using Box-Jenkins models', Water Resources Research, 14, 3, pp 509-516.

Hipel, K.W. and McLeod, A.I. (1993) Time Series Modelling for Water Resources and Environmental Systems, Elsevier, Amsterdam.

Hipel, K.W., McLeod, A.I. and Lennox, W.C. (1977) 'Advances in Box-Jenkins modeling, 1, model construction', Water Resources Research 13, 3, pp 567-575.

Hirsch, R.M. and Gilroy, E.J. (1985) 'Detectability of step trends in the rate of atmospheric deposition of sulphate', Water Resources Bulletin 21, 5, pp 773-784.

Hosking, J.R.M. (1984) 'Modelling persistence in hydrological time series using fractional differencing', Water Resources Research 20, 12, pp 1898-1908.

Jackson, B.B. (1975) 'Markov mixture models for drought length', Water Resources Research 11, 1, pp 64-74.

Jimenez, C., McLeod, A.I. and Hipel, K.W. (1989) 'Kalman filter estimation for periodic autoregressive-moving average models', Stochastic Hydrology and Hydraulics 3, 3, pp 229-242.

Kashyap, R.L. and Rao, A.R. (1976) Dynamic Stochastic Models from Empirical Data, Academic Press, New York.

Kisiel, C. and Duckstein, L. (1972) 'Model choice and validation', General Report, Proceedings of the International Symposium on Uncertainties in Hydrologic and Water Resource Systems, Tucson, Arizona, pp 1282-1308.

Lettenmaier, D.P., Hipel, K.W. and McLeod, A.I. (1978) 'Assessment of environmental impacts, part two: data collection', Environmental Management 2, 6, pp 537-554.

Lewis, P.A.W. (1985) 'Some simple models for continuous variate time series', Water Resources Bulletin 21, 4, pp 635-644.

Ljung, L. (1978) 'Convergence analysis of parametric identification methods', IEEE Transactions on Automatic Control AC-23, 5, pp 770-783.

Mandelbrot, B.B. and Wallis, J.R. (1969) 'Computer experiments with fractional Gaussian noises, Parts 1 to 3', Water Resources Research 5, 1, pp 228-267.

McKenzie, E. (1985) 'Some simple models for discrete variate time series', Water Resources Bulletin 21,4, pp 645,650.

McLeod, A.I. (1977) 'Improved Box-Jenkins estimators', Biometrika 64, 3, pp 531-534.

McLeod, A.I. (1978) 'On the distribution of residual autocorrelations in Box-Jenkins models', Journal of the Royal Statistical Society Series B, 40,3, pp 296-302.

McLeod, A.I. and Hipel, K.W. (1978) 'Simulation procedures for Box-Jenkins models', Water Resources Research 14, 5, pp 969-975.

McLeod, A.I., Hipel, K.W. and Camacho, F. (1983) 'Trend assessment of water quality time series', Water Resources Bulletin 19, 4, pp 537-547.

McLeod, A.I., Hipel, K.W. and Lennox, W.C. (1977) 'Advances in Box-Jenkins modelling 2, applications', Water Resources Research 13, 3, pp 577-586.

Moss, M.E. (1979) 'Some basic considerations in the design of hydrologic data networks', Water Resources Research 15, 6, pp 1673-1676.

Noakes, D.J., McLeod, A.I. and Hipel, K.W. (1985) 'Forecasting monthly riverflow time series', International Journal of Forecasting 1, pp 179-190.

Russell, B. (1946) History of Western Philosophy, Allen and Unwin, London.

Salas, J.D., Delleur, J.W., Yevjevich, V. and Lane, W.L. (1980) Applied Modelling of Hydrologic Series, Water Resources Publications, Littleton, Colorado.

Salas, J.D. and Obeysekera, J.T.B. (1982) 'ARMA model identification of hydrologic time series', Water Resources Research, 18, 4, pp 1011-1021.

Salas, J.D., Tabios III, G.Q. and Bartolini, P. (1985) 'Approaches to multivariate modeling of water resources time series', Water Resources Bulletin 21, 4, PP 683-708.

Schwarz, G. (1978) 'Estimating the dimension of a model', Annals of Statistics 6, 2, pp 461-464.

Tao, P.C. and Delleur, J.W. (1976) 'Seasonal and nonseasonal ARMA models', Journal of the Hydraulics Division', American Society of Civil Engineers 102, HY10, pp 1541-1559.

Thompstone, R.M., Hipel, K.W. and McLeod, A.I. (1985) 'Forecasting quarter-monthly riverflows', Water Resources Bulletin 21, 5, pp 731-741.

Tsay, R.S. and Tiao, G.C. (1984) 'Consistent estimates of autoregressive parameters and extended sample autocorrelation function for stationary and nonstationary ARMA models', Journal of the American Statistical Association 79, pp 84-96.

Tukey, J. (1977) Exploratory Data Analysis, Addison-Wesley, Reading, Massachusetts.

Van Belle, G. and Hughes, J.P. (1984) 'Nonparametric tests for trend in water quality', Water Resources Research 20, 1, pp 127-136.

Vecchia, A.V. (1985) 'Periodic autoregressive-moving average (PARMA) modelling with applications to water resources', Water Resources Bulletin 21, 5, pp 721-730.

Vicens, G.J., Rodriguez-Iturbe, I. and Schaake Jr., J.C. (1975) 'A Bayesian framework for the use of regional information in hydrology', Water Resources Research 11, 3, pp 405-414.

Wood, E.F. (1978) 'Analyzing hydrologic uncertainty and its impact upon decision making in water resources', Advances in Water Resources 1, 5, pp 299-305.

Wood, E.F. and Rodriguez-Iturbe, I. (1975) 'A Bayesian approach to analyzing uncertainty among flood frequency models', Water Resources Research 11, 6, pp 839-843.

MODELS FOR DATA GENERATION IN HYDROLOGY: UNIVARIATE TECHNIQUES

JOSE D. SALAS
Hydrologic Science & Engineering Program
Department of Civil Engineering
Colorado State University
Fort Collins, Colorado 80523
USA

BONIFACIO FERNANDEZ
Departamento de Hidraulica
Universidad Catolica de
Chile
Santiago de Chile
Chile

ABSTRACT. A number of stochastic models for representing seasonal hydrological time series are reviewed herein. Included is a description of the main statistical properties generally present in most seasonal series and the procedures for dealing with trends, periodicity, non-normality and correlation structure. Modelling approaches are viewed under the categories of direct modelling, disaggregation and aggregation approaches. In addition, models of intermittent and non-intermittent series are considered.

1. Introduction

At present, stochastic modelling of seasonal hydrological time series is an accepted approach for data generation and forecasting of hydrological events. Although almost from the beginning of the century the importance of statistical approaches for analyzing hydrological data was evident, it was not until the mid 1950's that formal development of stochastic modelling started with the introduction and application of autoregressive models for monthly rainfall (Hannan, 1955). Since then, extensive research efforts have been made towards improving the early concepts and models, providing physical (conceptual) justification of some models, introducing alternative models, and developing and applying improved estimation procedures as well as fitness tests for such models.

About thirty five years has elapsed since Hannan first suggested the lag-one autoregressive model for the modelling and simulation of monthly rainfall. Since then significant steps in this area are represented by the works of Thomas and Fiering (1962) for modelling annual and seasonal streamflows, Yevjevich (1963) and Roesner and Yevjevich (1966) for modelling annual runoff and monthly precipitation and streamflow, respectively; Matalas (1967) for multivariate lag-one autoregressive modelling and Yevjevich (1972a) for extending the lag-one autoregressive models to multilag with seasonal parameters. About the same time, the book of Box and Jenkins (1970) stirred interest in the application of a new class of models, namely the autoregressive-moving average (ARMA) models

47

J.B. Marco et al. (eds.),
Stochastic Hydrology and its Use in Water Resources Systems Simulation and Optimization, 47–73.
© 1993 *Kluwer Academic Publishers.*

which have gained wide application in stochastic hydrology (Carlson et al., 1970; O'Connell, 1971; McKerchar and Delleur, 1973; Hipel et al., 1977; Lettenmaier and Burges, 1977; Curry and Bras, 1978; Salas et al., 1980; Cooper and Wood, 1980; Rao et al., 1982; and Stedinger and Taylor, 1982). A landmark in the modelling of seasonal hydrological time series was the development of the disaggregation model (Valencia and Schaake, 1973) which allows the generation of seasonal hydrological samples by preserving statistics at both annual and seasonal time scales. Improvements and applications of disaggregation models have also been made (see for instance, Mejia and Rousselle, 1974; Todini, 1980; Lane, 1982; Santos and Salas, 1983; Valencia et al., 1983; and Stedinger and Vogel, 1984).

The thirty five years of research and application of stochastic approaches to hydrological time series modelling has led to extensive literature which has been reviewed by other investigators in the field. Excellent analysis and reviews have been conducted by Rodriguez-Iturbe et al., (1972), Klemes (1974), Jackson (1975), Matalas (1975), Lawrance and Kottegoda (1977) and DeCoursey et al. (1982). In addition, the books of Salas et al. (1980), Kottegoda (1980), Loucks et al. (1981) and Bras and Rodriguez-Iturbe (1985) include advances in the modelling of hydrological time series.

2. Time Series of Hydrological Processes

Although basic hydrological processes evolve on a continuous time scale, most analysis and modelling of such processes are made by defining them on a discrete time scale. In most cases, a discrete time process is derived by aggregating the continuous process within a given time interval while in other cases it is derived by sampling the continuous process at discrete points in time. For instance, a weekly streamflow series can be derived by aggregating the continuous flow hydrograph on a weekly basis while a daily streamflow series can be the result of simply sampling the flows of a stream once daily. In any case, most time series of hydrological processes are defined at hourly, daily, weekly, monthly, bimonthly, quarterly and annual time intervals. The term "seasonal" usually refers to time intervals of the order of months, however, in this report it will be used rather loosely, i.e. simply meaning that the time interval is a fraction of the year.

The plot of a hydrological series versus time gives a good indication of its main statistical characteristics. For instance, Figure 1 shows the daily, monthly and annual streamflow time series of the Boise River near Twin Springs, Idaho. The daily series shows a typical pattern in which during some days of the year the flows are low and in other days the flows are high. During the days of low flows the variability is small, while during high flows the variability is large. This characteristic behaviour of daily flows indicates that the daily mean and daily standard deviation vary periodically throughout the years. This is further substantiated if one observes the plot of ten years of monthly flows. Comparing the daily and monthly series, one can see that well defined recession curves are present in the daily series while this characteristic is less apparent in the monthly series. Likewise, the rising limb of the daily hydrograph is more "rough" (more variable) than that of the monthly hydrograph. In

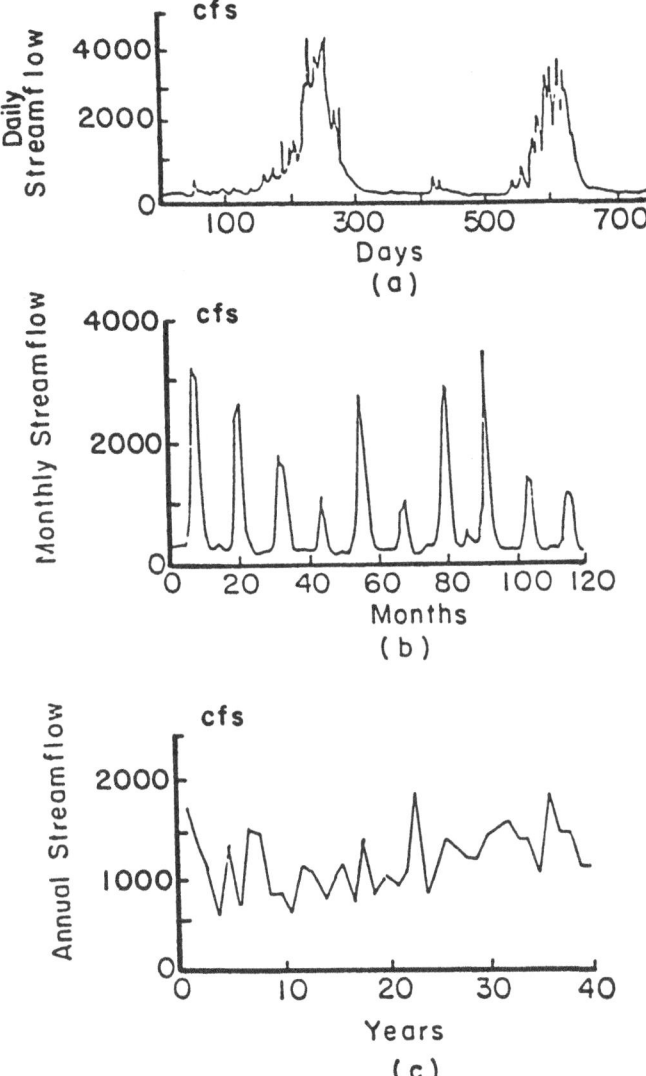

Figure 1. (a) daily, (b) monthly, and (c) annual streamflow time series for the Boise River near Twin Springs, Idaho.

general, such "roughness" diminishes as the time series interval increases.

Differences in the overall statistical characteristics of the time series of the Boise River can be more clearly observed from Table 1 in which results are shown for daily, weekly, monthly and annual series. The overall standard deviation, skewness coefficient and lag-one serial correlation coefficient decrease as the time interval increases. The last two rows of Table 1 measure the "directionality" of the hydrographs (Weiss, 1977; Fernandez and Salas, 1986). For instance, if $P(X_t > X_{t-1})$ is smaller than $P(X_t < X_{t-1})$, it means that the rising limb of the hydrograph is steeper than the recession limb. Table 1 indicates that this is the case for the daily and weekly series of the Boise River, but as the time interval reaches a month or longer, the two probabilities are practically equivalent. Therefore, one can say that there is some evidence that the daily and weekly streamflow series of the Boise River are directional, while the monthly and annual series are not. The series shown in Figure 1(c) indicates that the periodic pattern depicted by the daily and monthly series does not appear any more in the annual series. The standard deviation is now smaller than the mean, the skewness is close to zero and the serial correlation coefficient is also close to zero which may indicate that the series is approximately normal and independent. Thus, as the time interval is larger, the underlying time series becomes simpler to analyze and to model; conversely, as the sample time step is smaller, the amount of information contained in the sample is larger, but the characteristics of the series become more complex and the corresponding statistical modelling more difficult.

TABLE 1. Basic statistical characteristics of streamflow series of the Boise River near Twin Springs, Idaho, for the period 1921–1960

Statistical Characteristic	Time Series			
	Daily	Weekly	Monthly	Annual
Mean (cfs)	1172.7	1172.7	1172.7	1172.7
Standard Deviation (cfs)	1458.6	1423.1	1338.9	335.2
Skewness Coefficient	2.335	2.216	1.949	0.228
Lag-one Serial Correlation Coefficient	0.986	0.913	0.650	0.222
$P(X_t > X_{t-1})$	0.33	0.42	0.50	0.49
$P(X_t < X_{t-1})$	0.52	0.56	0.49	0.51

Although the foregoing Boise River example gives a good indication of the typical streamflow time series patterns for various time intervals, quite different looking hydrographs may occur for rivers of different environments as shown in Figure 2 for (a) the weekly flow series of the Greenbrier

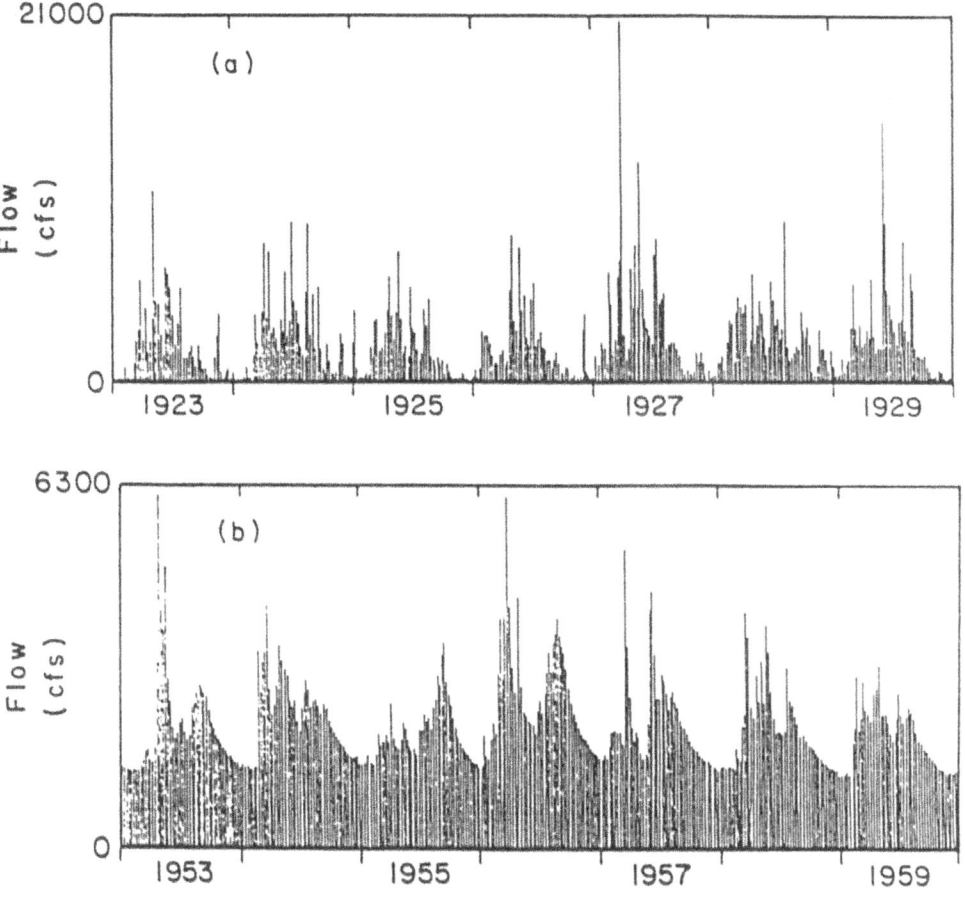

Figure 2. Weekly streamflow series of (a) the Greenbrier River near Alderson, W. Virginia, and (b) the McKenzie River at McKenzie Bridge, Oregon.

River near Alderson, W. Virginia, and (b) the McKenzie River at McKenzie Bridge, Oregon. While the McKenzie River flows show well defined and "smooth" recession limbs with a substantial base flow component, on the other hand the Greenbrier River flows show quite a "rough" recession, in fact the recession limb is almost as variable as the rising limb of the hydrograph series. In addition, in the latter series the base flow component is minimal. The simple observation of both weekly time series indicates that the McKenzie River weekly flow series is more "directional" than the Greenbrier River series. One can compute the skewness of the first differences as a measure of "directionality" (Fernandez and Salas, 1986). A nondirectional series (also called a reversible series) has zero skewness of the first differences. The coefficients of skewness of the first differences of the seasonally standardized weekly series of the McKenzie and Greenbrier Rivers are 0.985 and 0.313, respectively, which confirms the visual observation that the McKenzie River series is more directional than the Greenbrier River series. In the foregoing examples the variables defined either at annual, monthly, weekly or daily time intervals always had values greater than zero. However, there are some cases in which hydrological variables are intermittent, with zero and nonzero values. The common example is the case of short-term precipitation series of hourly, daily and weekly values. Depending on the climatic environment, monthly, seasonal and even longer time interval precipitation series may also be intermittent. For instance, Figure 3 shows the monthly precipitation series at the Hachita gauging station in New Mexico in which during one or more months per year the precipitation is zero. Likewise, ephemeral streams in semiarid and arid regions are intermittent with time series having zero and nonzero flows. Often an intermittent time series arises as a result of separating the original series by using a selected threshold. For instance, when flood flow series are analyzed by using the "partial series" approach, the flood sequence is derived by selecting those flows which are greater than a given base flow. In either case, an intermittent time series is made up of a mixed random variable, namely a discrete variable for the zero values and a continuous variable for the nonzero values.

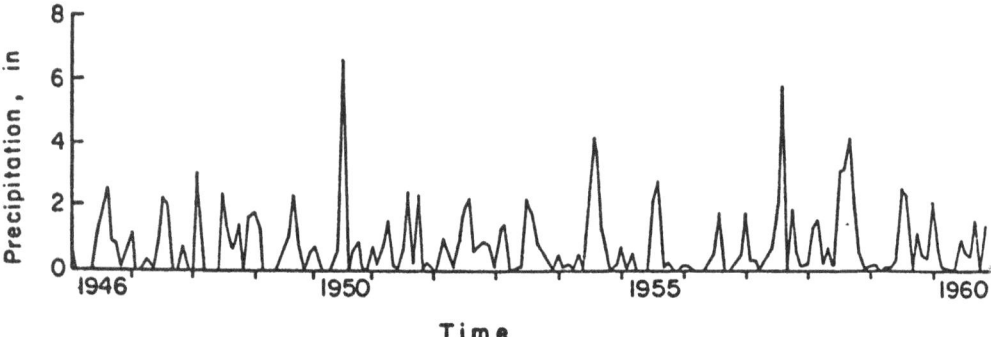

Figure 3. Monthly precipitation for station 29.3775 at Hachita, New Mexico (Roesner and Yevjevich, 1966).

For either intermittent or nonintermittent hydrological time series, some overall statistical characteristics can be defined, in addition to the mean, standard deviation, skewness coefficient and serial correlation coefficients. In particular, such characteristics can be related to storage and drought-related statistics (Salas et al., 1980). For instance, the mean accumulated deficit for a given time series of inflows and flow demand level is an important statistic related to the needed storage capacity of a potential reservoir. Likewise, for a selected threshold flow level, certain low-flow characteristics of a time series can be derived, such as the 5-day or 10-day low flow frequencies, the longest drought or the average drought length for the available sample size. The importance of such storage and drought-related characteristics in modelling hydrological time series in general and seasonal time series in particular has been reported by Mandelbrot and Wallis (1968), Wallis and Matalas (1972), O'Connell (1971), Yevjevich (1972b), Rodriguez-Iturbe et al. (1972), Kottegoda (1974), Hipel and McLeod (1977) and Hoshi and Burges (1978).

3. Periodic Properties of Seasonal Hydrological Time Series

Hydrological series defined at time intervals smaller than a year have distinct period patterns which are essentially due to the annual revolution of the earth around the sun which produces the so-called annual cycle in most hydrological processes. In general, seasonal hydrological time series exhibit periodic properties in mean, variance, covariance and skewness. The periodicity in the mean and in the standard deviation may be easily observed in the plot of the underlying hydrological time series. However, the periodicity in higher order moments is not so easy to observe and usually requires further mathematical analysis. Let us consider a seasonal hydrological time series represented by $X_{v,\tau}$ where $v = 1,2,\ldots,N$ denotes the year, $\tau=1,2,\ldots,\omega$ denotes the season and ω and N denote the number of seasons in a year and the number of years of record available, respectively. Seasonal statistics can be derived from $X_{v,\tau}$ such as the mean $\bar{X}_\tau$, standard deviation S_τ, skewness coefficient g_τ and lag-k correlation coefficient $r_{k,\tau}$. Plots of these statistics versus time interval τ will indicate whether these statistics appear to be periodic or whether they may be considered constant throughout the year. Figure 4 shows the plot of $\bar{X}_\tau$, S_τ, g_τ and $r_{1,\tau}$ for the weekly flows of the Greenbrier River. In this example the first three statistics appear to be clearly periodic while the periodicity in $r_{1,\tau}$ is less apparent.

In addition to observing the plots of seasonal statistical characteristics, the techniques of autocorrelation analysis and spectral analysis can be used for identifying cycles or periodic components of seasonal hydrological time series. Considering the (two-dimensional) periodic series $X_{v,\tau}$ simply by X_t, the lag-k autocorrelation coefficient r_k can be determined. The graph of r_k versus k is called the correlogram. Some examples of correlograms of monthly precipitation at Hachita, New Mexico and monthly streamflows (logarithms) for Middle Fork American River near Auburn, California, are shown in Figure 5 (a) and (c). Likewise, spectral analysis has been used to detect cycles in hydrological data (Roesner and Yevjevich, 1966; Quimpo, 1967). Briefly, the Fourier transform of the autocorrelation function provides a quantitative information, in the frequency domain, of the presence or absence of periodicity. Specifically, if the plot of the spectral density function against the frequency exhibits a slow decay of intervening peaks and troughs, it may be an indication that periodic components are present in the underlying time series. Figures 5(b) and (d) show spectral density functions for the same data of precipitation and streamflow referred to above.

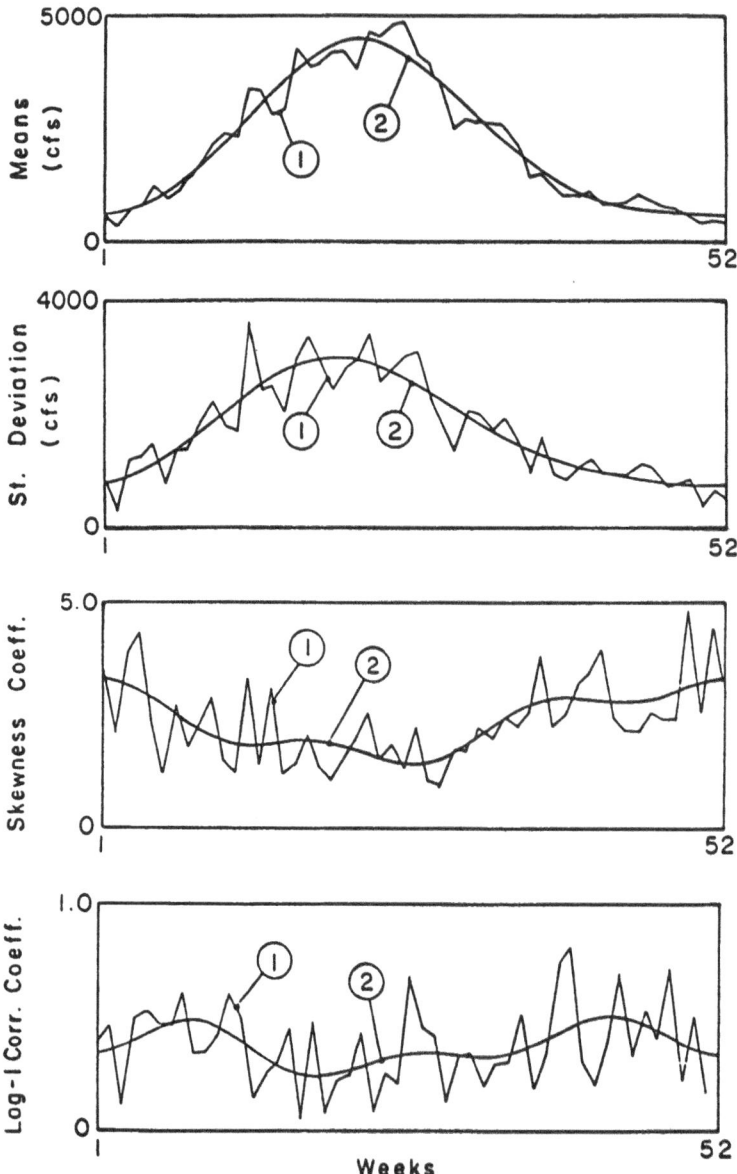

Figure 4. Seasonal mean, standard deviation, skewness coefficient and lag-one correlation coefficient of the weekly flows of the Greenbrier River for (1) the sample and (2) the fitted function based on Fourier series analysis.

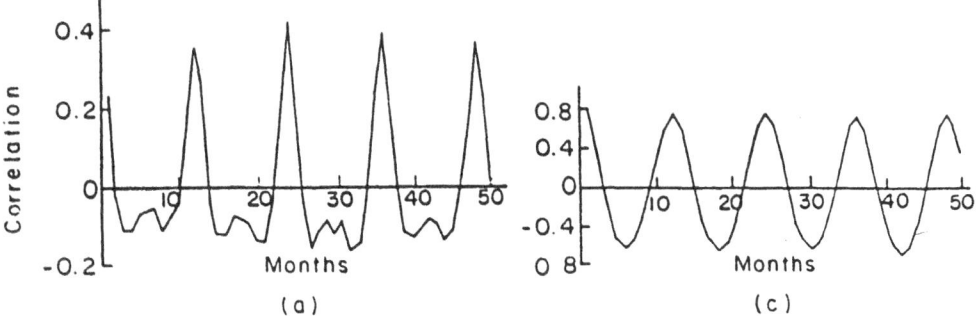

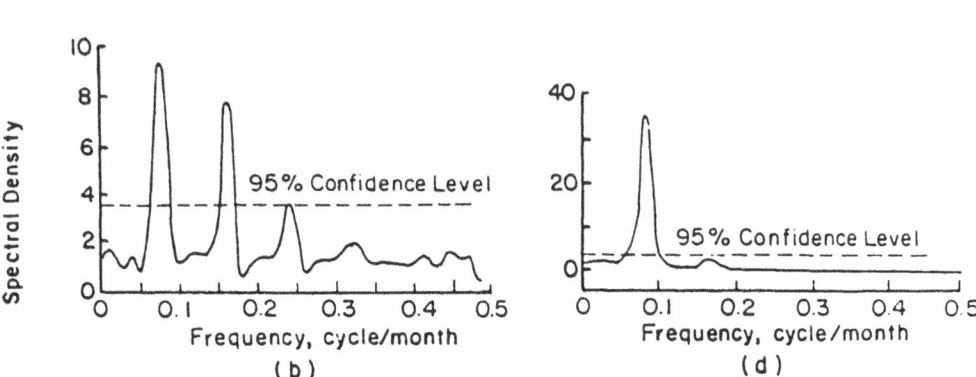

Figure 5. Correlogram and variance spectrum for the series of monthly precipitation of the Hachita station, New Mexico, (a) and (b), respectively and for the logarithms of monthly streamflow of the Middle Fork of the American River near Auburn, California, (c) and (d), respectively (after Roesner and Yevjevich, 1966).

4. Modelling of Seasonal Hydrological Series

There are a number of approaches which have been suggested in the literature for modelling hydrological time series defined at time intervals smaller than a year. Figure 6 summarizes the various steps usually followed, which essentially involve alternative ways of dealing with trends, periodicities and non-normality and how one can go about formulating the model, estimating the parameters and testing the fitness of such models. By direct modelling approach (Fig. 6) what is meant is that a model is written directly in terms of the seasonal variable of interest. For instance, if one wishes to simulate monthly flows by using

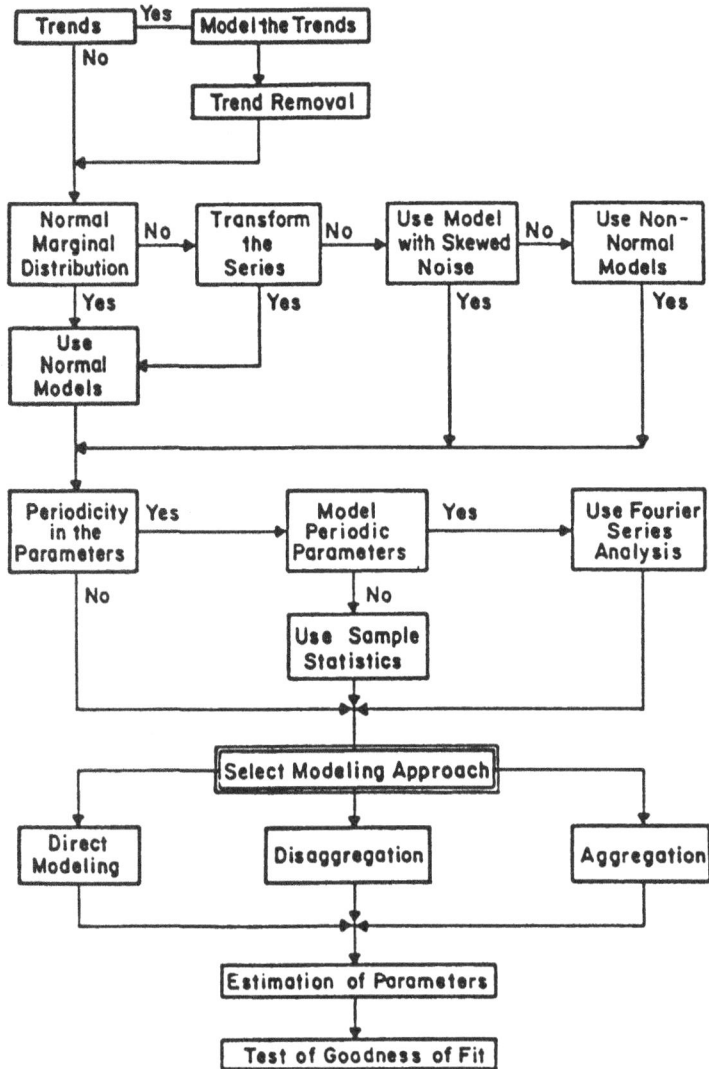

Figure 6. Schematic diagram for modelling seasonal hydrological time series.

an autoregressive model, the variable representing the monthly flows is included directly in the model. The disaggregation approach means that in order to generate say monthly flows, the annual flows are generated first, then they are disaggregated into monthly values by some appropriate scheme. On the other hand, the aggregation approach means that in order to generate say monthly flows, the corresponding weekly flows are first modelled, which subsequently leads to the model and parameters defined at the monthly time scale. The review included herein follows much the same steps as suggested in Figure 6 except that some conceptual arguments are included in the first subsection aimed at justifying the class of models included in the text.

4.1. CONCEPTUAL BASIS OF SEASONAL FLOWS MODELS

A number of conceptual (physically-based) simulation models for representing the hydrological cycle of watersheds have been suggested in the literature since the early 1960's. The same can be said about stochastic models. Extensive literature already exists on these two modelling approaches. However, less attention has been given to linking or bringing together both conceptual and stochastic modelling schemes. Some work on this subject can be found in Yevjevich (1964), Fiering (1967), Quimpo (1971), Moss and Bryson (1974), Spolia and Chandre (1974), O'Connor (1976), Pegram (1977), Selvalingan (1977), Eagleson (1978), Sorooshian and Dracup (1980), Salas et al., (1981), Kuczera (1982), Koch (1982), Sorooshian et al. (1983) and Alley (1984. The conceptual model considered herein is an extension of the nonseasonal simple lumped watershed model used by Salas et al. (1981) and closely follows the model configuration suggested by Salas and Obeysekera (1992).

Consider the conceptual representation of a natural watershed as in Figure 7, in which $X_{v,\tau}$ = the precipitation in the season τ of year v; $G_{v,\tau}$ = the groundwater storage at the end of the season τ in the year v; and $Z_{v,\tau}$ = the streamflow in the season τ of year v. It is assumed that there is no significant surface storage in the watershed and the entire precipitation amount contributes to evaporation, infiltration and surface runoff. Then, the amount $a_\tau X_{v,\tau}$ infiltrates, percolates, and reaches the groundwater storage and the amount $b_\tau X_{v,\tau}$ evaporates from the soil and plants. Thus $(1 - a_\tau - b_\tau) X_{v,\tau} = d_\tau X_{v,\tau}$ represents the surface runoff reaching the stream. The contribution of the aquifer to the stream is represented by $c_\tau G_{v,\tau-1}$ which is equivalent to assuming that the storage behaves like a linear reservoir. In general, the parameters a_τ, b_τ and c_τ are assumed to vary seasonally. In addition, the following conditions are required: $0 \leq a_\tau$, b_τ, $c_\tau \leq 1$ and $0 \leq a_\tau + b_\tau \leq 1$.

The seasonal streamflow $Z_{v,\tau}$ is made up of the surface runoff $d_\tau X_{v,\tau}$ and the groundwater contribution $c_\tau G_{v,\tau-1}$. Thus,

$$Z_{v,\tau} = d_\tau X_{v,\tau} + c_\tau G_{v,\tau-1} \tag{1}$$

The mass balance equation for the groundwater storage is written as:

$$G_{v,\tau} = G_{v,\tau-1} + a_\tau X_{v,\tau} - c_\tau G_{v,\tau-1} \tag{2}$$

Combining Equations 1 and 2 gives:

58

$$Z_{v,\tau} - \frac{c_\tau}{c_{\tau-1}} (1-c_{\tau-1}) \, Z_{v,\tau-1} = d_\tau X_{v,\tau} - [\frac{d_{\tau-1} \, c_\tau}{c_{\tau-1}}(1-c_{\tau-1})-c_\tau a_{\tau-1}]X_{v,\tau-1} \tag{3}$$

which is a periodic autoregressive moving average (PARMA) process if $X_{v,\tau}$ is a white noise process. It follows that under certain conditions Model 3 can be simplified to the well known Thomas-Fiering model, while under other assumptions (for instance, that the precipitation process is ARMA type) the streamflow model becomes a general PARMA process.

Note that in the above text no assumption was made concerning the marginal distribution of $X_{v,\tau}$. If $X_{v,\tau}$ is normal, then Model 3 and its simplifications or extensions fall into the class of PARMA models which requires normal marginal distribution. However, this is seldom the case in seasonal hydrological time series. Modelling periodic dependent hydrological time series with nonGaussian marginal distribution is a relevant problem in stochastic hydrology. Furthermore, in the previous formulation, if the groundwater contribution $G_{v,\tau}$ and/or the precipitation input $X_{v,\tau}$ are intermittent series, then the streamflow $Z_{v,\tau}$ can become an intermittent series as well. Modelling such a type of intermittent process requiring periodic dependence and arbitrary marginal distributions is another important subject of in stochastic hydrology.

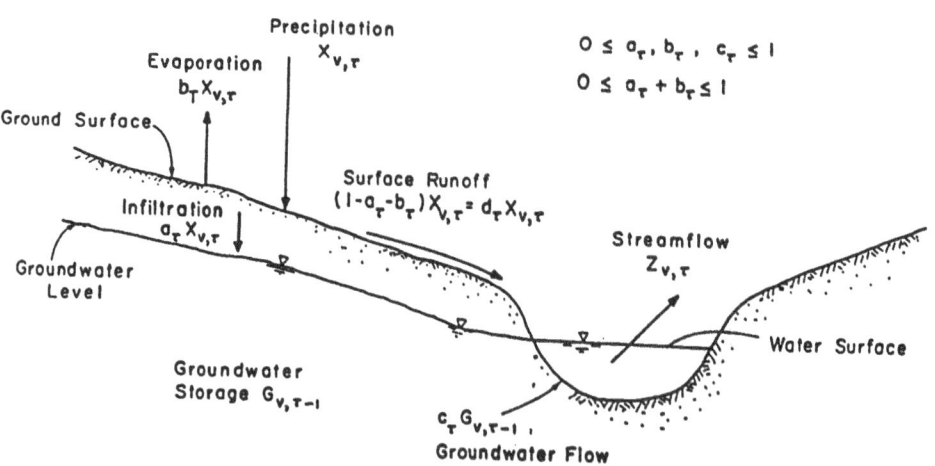

Figure 7. Schematic representation of precipitation-storage-streamflow processes for a natural watershed (after Salas and Obeysekera, 1992).

4.2. DEALING WITH TRENDS, PERIODICITY AND NON-NORMALITY

Hydrological time series can generally be represented by four components, namely: (1) long-term trends or other deterministic changes exhibited beyond the annual cycle, (2) seasonal or periodic changes of days, weeks or months within the annual cycle, (3) almost periodic changes, such as tidal and solar effects, and (4) stochastic or random variations. The first three components are usually considered to be deterministic in the sense that their future values can be known exactly, in advance. On the other hand, the latter component is stochastic or random, in the sense that future outcomes can be known only on a probability basis. Long-term changes which can be detected in hydrological time series are: linear or nonlinear trends, and gradual or abrupt jumps (slippages). In general, hydrological time series can exhibit trends and jumps in various statistical properties such as mean, variance, covariance and skewness. Appropriate techniques for removing and modelling series with trends and jumps can be found in Salas and Yevjevich (1972), Lettenmaier (1977), Hirsh and Slack (1984) and Meidment and Parzen (1984).

Much work in modelling seasonal hydrological time series has been based on the assumption that the underlying time series can be made stationary after some appropriate "transformation". For instance, a given series $X_{\nu,\tau}$ can be made stationary in the mean and variance by seasonal standardization, namely:

$$Z_{\nu,\tau} = (X_{\nu,\tau} - \mu_\tau) / \sigma_\tau \tag{4}$$

where μ_τ and σ_τ are the mean and standard deviation for season τ and $\tau=1,\ldots,\omega$ with ω the number of seasons in the year. Then, $Z_{\nu,\tau}$ is the "deseasonalized" series which is assumed to be stationary in the mean and in the variance. This has been a popular procedure in modelling seasonal hydrological series (Quimpo, 1967; Yevjevich, 1972a; Young and Jetmar, 1976; Tao and Delleur, 1974; Croley, 1975). However, the "deseasonalized" series $Z_{\nu,\tau}$ may still have some periodicity left in the correlation structure and higher order moments such as the skewness which cannot be removed by Equation 4.

The mean μ_τ and standard deviation σ_τ in Equation 4 can be determined directly by their corresponding sample estimates. However, when the time interval of the series is small, the sampling variability of a given seasonal statistic is large. This effect is even more important in the case of the seasonal skewness and seasonal correlation. Thus, in order to smooth out such sampling variability and to decrease the number of parameters involved, Fourier series analysis may be used for representing a seasonal statistic (Yevjevich, 1972a). Figure 4 gives an example of fitting the weekly mean, standard deviation, skewness coefficient and lag-one correlation coefficient of the Greenbrier River flows by using Fourier series. However, the use of Fourier series for representing seasonal statistics such as the mean and standard deviation, may distort other statistics such as the seasonal correlation (Rodriguez-Iturbe et al., 1971; Srikanthan and McMahon, 1981).

Annual hydrological time series can often be assumed to be approximately normal. However, series defined at time intervals smaller than a year, generally depart from normal. In fact, the skewness appears to increase as the time interval decreases. Although several probability density functions can be fitted to skewed data, the three-parameter log-normal and the gamma distributions appear to be the most widely used distributions in stochastic hydrology. Lettenmaier and Burges (1977) pointed out that, given the short length of the typical historical streamflow records, it is difficult to distinguish between these two distributions when they are fitted to historic data. The problem of skewed hydrological series has been approached by a number of methods. A widely used method has been to transform the streamflow series by using logarithmic or power transformations (Box and Cox, 1964) in order to obtain a series closer to normal. Another approach has been to find the statistical properties of the noise in such a way as to reproduce the skew of the series as suggested by Fiering (1967) for stationary and for seasonal first order autoregressive models. In addition, some alternative models which can accommodate skewed marginal distribution in their model structure have been advanced (Weiss, 1977; Gaver and Lewis, 1981; Fernandez and Salas, 1986; Sim, 1987).

In order to reproduce the skewness of hydrological series, several transformation procedures have been proposed. The most popular and simple procedure is based upon the assumption that the underlying marginal distribution can be well represented by a log-normal probability function. If the sequence $(X_{1,\tau}, X_{2,\tau}, \ldots, X_{\nu,\tau}, \ldots, X_{\omega,\tau})$ is assumed to be log-normal distributed, the relationship

$$Y_{\nu,\tau} = \log (X_{\nu,\tau} - a_\tau) ; \qquad \tau = 1, 2, \ldots, \omega \qquad (5)$$

transforms the original time series into normal. Then, the new series $Y_{\nu,\tau}$ is modelled by using a normal model. The problem of such transformation is that the properties reproduced by the model are those of the transformed series and not necessarily the properties of the original series. However, in order to correct the bias introduced by the transformation, Matalas (1967) presented some relationships between the moments of the original and the log-transformed variables for a stationary process. Burges (1970) and Hirsch (1979) extended such a procedure and provided similar relationships for the case of seasonal series. Although the use of the referred relationships allows the first three moments of the marginal distribution of the original series to be preserved, there is a difference in the autocorrelation structure of the X and Y processes. Thus, Matalas (1967) proposed a correction for the lag-one autocorrelation coefficient in the case of the stationary autoregressive process of order one and Hirsh (1979) and Fernandez and Salas (1986) gave a similar correction for the case of a periodic lag-one autoregressive model.

An alternative approach to transformations in modelling skewed seasonal hydrological time series is to use the lag-one autoregressive model with a skewed noise. In this regard, Thomas and Fiering (1962) proposed the Wilson-Hilferty (W-H) transformation to reproduce skewed noises in an AR(1) process. The use of the Wilson-Hilferty transformation implies an approximation to a Chi-square variable and is theoretically limited to small values of the skewness coefficient ($\gamma_\tau < 2$). Kirby (1972) proposed a modification of the W-H transformation which is useful for values of γ_τ up to 9.0. Similar W-H transformations were used by Beard (1965), Payned et al. (1969) and McGinnis and Sammons (1970). Likewise, Yevjevich (1972b) proposed using a skewed noise in multilag stationary and periodic autoregressive models based on fitting the empirical distribution of the noise with a given probability model.

4.3. DIRECT MODELLING APPROACHES

The most popular periodic model used in hydrology is that introduced by Hannan (1955) and Thomas and Fiering (1962) which considers an AR(1) model structure. It assumes that the variables are normal, or have been made normal by some appropriate transformation. The model can be written in the form:

$$Z_{v,\tau} = \phi_\tau \, Z_{v,\tau-1} + \epsilon_{v,\tau} \tag{6}$$

where ϕ_τ is the autoregressive coefficient of season τ and ϵ is the noise with mean zero and variance $\sigma_\tau^2(\epsilon)$. This model is denoted as PAR(1) since it is periodic autoregressive of order 1. Salas and Yevjevich (1972) and Croley (1972) used PAR(1), PAR(2) and PAR(3) models with periodic coefficients for modelling monthly streamflow and water use series. Similar studies with PAR(p) models have been reported by Croley and Rao (1976) and Delleur et al. (1976). Estimation of parameters of such models can be done by the method of moments by using the periodic counterpart of the Yule-Walker equation (Salas et al., 1980). For instance, for the PAR(1) model the parameters can be estimated by

$$\hat{\phi}_\tau = r_{1,\tau} \tag{7}$$

and

$$\hat{\sigma}_\tau^2(\epsilon) = 1 - r^2_{1,\tau} \tag{8}$$

in which $r_{1,\tau}$ is the (lag-one) correlation coefficients between the Z's for seasons τ and $\tau-1$.

An extension of periodic autoregressive models to incorporate moving average terms was proposed by Tao and Delleur (1976). The model, denoted PARMA(p,q), is defined as:

$$Z_{v,\tau} = \sum_{j=1}^{p} \phi_{j,\tau} \, Z_{v,\tau-j} - \sum_{i=1}^{q} \theta_{i,\tau} \, \epsilon_{v,\tau-i} + \epsilon_{v,\tau} \tag{9}$$

where the ϕ's are the periodic autoregressive coefficients, the θ's are the periodic moving average coefficients and ϵ is the noise. Since the marginal distribution of Z is assumed to be standard normal, the distribution of ϵ is also normal with mean zero and variance $\sigma_\tau^2(\epsilon)$. The method of moments can be applied to estimate the $(p+q+1)\omega$ parameters, which can be used as initial estimates for least squares or maximum likelihood estimation procedures. Expressions for estimating the periodic parameters of PARMA(2,0) and PARMA(3,0) models have been reported by Salas and Yevjevich (1972) and Croley (1972). A system of equations similar to the Yule-Walker equations for PARMA(p,q) models were proposed by Hirsch (1979) and Salas et al. (1982). The system of equations for the PARMA(1,1) model becomes:

$$\phi_{1,\tau} = r_{2,\tau} \,/\, r_{1,\tau-1} \tag{10}$$

$$\phi_{1,\tau} - \theta_{1,\tau} \, \sigma^2_{\tau-1}(\epsilon) = r_{1,\tau} \tag{11}$$

and

$$\theta_{1,\tau} = \phi_{1,\tau} + \frac{(1 - \phi_{1,\tau} r_{1,\tau})}{(\phi_{1,\tau} - r_{1,\tau})} - \frac{(\phi_{1,\tau+1} - r_{1,\tau+1})}{(\phi_{1,\tau} - r_{1,\tau})\theta_{1,\tau+1}} \tag{12}$$

for $\tau = 1, 2, \ldots, \omega$, which must be solved simultaneously to estimate the parameters. An algorithm for estimating maximum likelihood parameters for such PARMA models has been proposed by Vecchia (1983).

The foregoing Model 9 considers a constant order of the autoregressive and moving average parts for every season. However, in order to attain parsimony, it might be useful to vary the number of parameters from season to season. In this case p and q becomes p_τ and q_τ in Equation 9. McLeod and Hipel (1978a) considered a PAR(p_τ) model and used multiple linear regression for estimating the seasonal autoregressive parameters independently for every season.

One of the concerns in modelling periodic hydrological series is related to the definition of the season adopted. Usually (as in the aforementioned models), what is assumed is a certain number of equally spaced time intervals in a year and the season is defined by the length of such a time interval. In this procedure there is not a clear hydrological definition of the season. For instance, the unit month or week have more a social than hydrological meaning. A definition of seasons not necessarily of constant length might be attractive. This could be oriented to obtain more parsimonious models. An approach in this direction has been proposed by Thompstone (1983) for modelling weekly and monthly series. The basic idea behind this approach is to group seasons for which the parameters are not statistically different. This was done for the autoregressive process and the model was labelled Parsimonious Periodic Autoregressive (PPAR). As in the PAR(p_τ) model, the parameters are estimated by multiple linear regression of Z_τ regressed on $Z_{\tau-1}$, $Z_{\tau-2}$, $\ldots$ for each season and then grouped using a test for the comparison of the variance of the noise and the autoregressive coefficients.

The foregoing PAR and PARMA models assume that the underlying time series is normal. However, since seasonal hydrological time series are usually non-normal, some transformation is applied and the modelling is made in the transformed domain. Although the use of these linear-Gaussian models has been quite popular in hydrology, they present some shortcomings. When they are applied to transformed data, some corrections have to be incorporated in the estimation of parameters in order to reproduce the properties in the original (untransformed) domain. If the transformation needed to normalize the series is not simple or if the number of seasons is large, the required correction in parameter estimation becomes cumbersome. Additionally, if some characteristics beyond the covariance are important, the use of nonlinear and/or nonGaussian models might be required (Tong, 1983).

The modelling of periodic streamflow time series with skewed marginal distributions has been proposed by using time series models with gamma distribution and autoregressive dependence structure (Fernandez and Salas,

1986). This has been done by extending the stationary linear gamma model (Gaver and Lewis, 1981) and the stationary multiplicative gamma model (McKenzie, 1982) to accommodate periodicity in the parameters of the marginal distribution and in the autocorrelation. Let us assume that $X_{v,\tau}$ represents a hydrological series defined for season $\tau=1, \ldots, \omega$ and year $v=1, \ldots, N$ with ω and N the total number of seasons and number of years of data, respectively. Furthermore, let us consider $X_{v,\tau}$ to be a periodic series with a marginal three-parameter gamma distribution $G(\alpha_\tau, \beta_\tau, \lambda_\tau)$ with α, β and λ representing the scale, shape and location parameters, respectively. Then,

$$Z_{v,\tau} = X_{v,\tau} - \lambda_\tau , \quad \tau=1,\ldots,\omega; \tag{13}$$

has a two-parameter gamma distribution $G(\alpha_\tau, \beta_\tau)$. The model representing $Z_{v,\tau}$ can be written as

$$Z_{v,\tau} = \phi_\tau \, Z_{v,\tau-1} + Z_{v,\tau-1}^{\delta_\tau} \, W_{v,\tau} \tag{14}$$

where $Z_{v,0} = Z_{\mu-1,\omega}$, ϕ_τ is a periodic autoregressive coefficient, δ_τ is a periodic autoregressive exponent and $W_{v,\tau}$ is the noise process. Model 14 is denoted PGAR(1), periodic gamma autoregressive of order one. The parameters ϕ_τ and δ_τ are defined as a function of α_τ, β_τ and the lag-one autocorrelation coefficient $\rho_{1,\tau}$ as

$$\phi_\tau = 0 \text{ and } \delta_\tau = \rho_{1,\tau} \, [\frac{\beta_{\tau-1}}{\beta_\tau}]^{1/2} \qquad \text{for } \beta_\tau < \beta_{\tau-1} \tag{15}$$

and

$$\phi_\tau = \rho_{1,\tau} \, [\frac{\alpha_\tau}{\alpha_{\tau-1}}] \, [\frac{\beta_\tau}{\beta_{\tau-1}}]^{1/2} \text{ and } \delta_\tau = 0 \qquad \text{for } \beta_\tau \geq \beta_{\tau-1} \tag{16}$$

Likewise, the noise $W_{v,\tau}$ is either the noise of an additive gamma process or the noise of a multiplicative gamma process depending on the constraints $\beta_\tau < \beta_{\tau-1}$ or $\beta_\tau \geq \beta_{\tau-1}$, respectively (Fernandez and Salas, 1986). Model 14 was applied for modelling weekly streamflow series for several rivers in the United States. The results obtained indicated that such a PGAR(1) process compares favourably with respect to the normal-based models in reproducing the basic statistics usually considered for streamflow simulation.

Modelling of intermittent hydrological series in general and intermittent streamflow series in particular has attracted the attention of researchers in the field, although not to the same extent as modelling nonintermittent series. One of the major problems in using the current generation models to ephemeral streams in semiarid and arid regions is the modelling of zero flows. The most common approach to this problem is to use a binary generator to determine whether the flow is zero or nonzero and if the latter, to generate a nonzero flow. This approach, however, has problems related to the measure of serial dependence and the method of modelling it, as noted by Srikanthan and McMahon (1980). For ephemeral streams, the Markov model proposed by Thomas and Fiering (1962) for the generation of seasonal flows does not generate the necessary number of zero flows. The number of zero flows generated by this approach is found to depend to some extent on the coefficients of variation and skewness of the historical flows, yet there appears to be little correlation between the percentage generated and the percentage observed in the historical record. Besides, often the monthly means are found to be overestimated due to setting negative flows to zero.

Beard (1973) suggested a procedure for dealing with zero flows based on the recognition that conditions are more dry during some zero flow periods than during others. The degree of dryness is represented by artificial negative flows which can eliminate the discontinuity of the frequency curve caused by zero flows. Lee (1975) generated ephemeral flows by assuming: no correlation between periods of flow and no flow and a regression relationship between total streamflow volume and streamflow duration. He found that a Weibull distribution fitted the distribution of flow periods and time between flows, and the spread of the points about the regression line was taken into account using the observed deviation for the regression line and a Weibull-distributed random variable. The difficulty associated with this method is how to obtain the individual monthly flows after the total streamflow volume and streamflow duration have been computed. An excellent review and comparison of various procedures for modelling this type of flows have been made by Srikanthan and McMahon (1980).

Yakowitz (1972, 1973) proposed a model for daily river flows in an arid region where flows are intermittent and where steep ascension and recession of the hydrographs is observed. The model considers flow as a nonlinear function of previous flows plus a random component. The random component is a Markov process with a transition matrix being dependent on the month of the year. Yakowitz showed that the model adequately represented the daily streamflows of Rillito River; however, Lawrance and Kottegoda (1977) indicated its mathematical difficulty. Kelman (1977) developed a model of daily precipitation based on a censored AR(1) process. Let Z_t be an AR(1) process with mean μ and variance σ^2 and lag-1 serial correlation ρ. The (censored) intermittent process Y_t is defined as $Y_t = Z_t$ if $Z_t > 0$ and $Y_t = 0$ if $Z_t \leq 0$. The process Y_t was further modified by a power transformation $X_t = Y_t^{1/\alpha}$ to yield the precipitation process. Pseudo-maximum likelihood estimators of the parameters μ, σ, ρ and α were obtained. The model was successfully applied to eight daily precipitation data across the US. Kelman also used the concept of intermittent process for modelling daily streamflow.

Chang et al. (1984) used the discrete ARMA (DARMA) models developed by Jacobs and Lewis (1978a,b,c; 1982) for modelling daily precipitation series in Indiana. The precipitation data was transformed into a three-state variable and the DARMA model was applied to account for the dependence among the daily precipitation quantities and among the daily precipitation occurrence times. Seasonality was considered by breaking the year into four seasons and applying separate stationary models for each season. However, Chang et al. pointed out that: "The DARMA models in their present form have the disadvantage of being stationary. This property creates problems in the model's application to the annually periodic daily precipitation series". A procedure which can circumvent such limitations is outlined below.

Let us consider the model

$$Y_{v,\tau} = X_{v,\tau} \, Z_{v,\tau} \tag{17}$$

where v denotes the year; $\tau=1, \ldots, \omega$ denotes the season and ω is the number of seasons in the year. The process $X_{v,\tau}$ will be considered a periodic dependent Bernoulli process and $Z_{v,\tau}$ may be either an independent or dependent periodic process with any marginal distribution. Furthermore, it is assumed that $X_{v,\tau}$ and $Z_{v,\tau}$ are independent between them. Chebaane et al. (1992) developed Model 17 considering that $X_{v,\tau}$ is a periodic discrete autoregressive process (PDAR process) defined as

$$X_{v,\tau} = V_{v,\tau} \, X_{v,\tau-1} + (1 - V_{v,\tau}) \, U_{v,\tau} \tag{18}$$

where $X_{v,0} = X_{v-1,\omega}$ and $V_{v,\tau}$ and $U_{v,\tau}$ are each independent Bernoulli processes with periodic parameters. Furthermore, $Z_{v,\tau}$ of Model 17 is considered to be the PGAR(1) model defined by Equation 14. Model 17 has been applied to monthly intermittent flows of the Arroyo Trabuco Creek in California. It has been shown (Chebaane et al., 1992) that synthetic flows generated by the product PDAR-PGAR(1) model as defined above closely resemble their historical counterparts in terms of reproducing the percentage of zero flows in each month and the monthly mean, variance and lag-one autocorrelation.

4.4. DISAGGREGATION APPROACHES

The foregoing models and modelling techniques have been designed to preserve the statistical characteristics of the historical data available. For instance, models based on annual flows are directed to preserve annual statistical characteristics such as the mean, standard deviation, skewness and correlation structure. Likewise, models based on monthly flows aim to preserve monthly statistical characteristics such as the monthly mean, monthly standard deviation, monthly skewness and month-to-month correlation structure. However, the foregoing models may not preserve the statistical characteristics at both levels of aggregation. In other words, if say PAR models are used to generate seasonal flows, the historical seasonal statistics would be preserved. Now, if the generated seasonal flows are aggregated to obtain the corresponding annual flows, there is no assurance that the historical annual statistics would be preserved. In fact, often such annual statistics are not preserved.

The need to preserve long-term (annual) and short-term (seasonal) time series properties directed the development of disaggregation models. Although Harms and Campbell (1967) essentially proposed the first disaggregation approach, it was not until Valencia and Schaake (1973) proposed the so-called disaggregation model that it became a major technique for modelling hydrological time series. An alternative model is the so-called disaggregation by fragments proposed by Svanidze (1980). These disaggregation models enable the breakdown of a sequence of longer-time units (say, annual) into a sequence of shorter-time units (say, monthly or weekly). The sequence of annual flows can be generated using ARMA or other alternative models and disaggregated into seasonal flows. The generated series are supposed to preserve relevant statistical properties at both annual and seasonal levels. Disaggregation models are useful not only for disaggregating variables in time (say annual flows into monthly flows) but for disaggregation in space as well. After Valencia and Schaake (1973) proposed their model, further studies, modifications and applications of disaggregation models have been made in the past years (Mejia and Rousselle, 1976; Tao and Delleur, 1976; Curry and Bras, 1978; Lane, 1979; Srikanthan and McMahon, 1980; Loucks et al., 1981; Santos and Salas, 1983; and Stedinger and Vogel, 1984).

4.5. AGGREGATION APPROACHES

In disaggregation, the annual time series is first modelled and then disaggregated into seasonal values. On the other hand, in the aggregation approach (Vecchia et al., 1983; Obeysekera and Salas, 1986), the seasonal time series is modelled and then aggregated, which produces a specified compatible model for the annual series. With the symbols previously used, let $Y_{\nu,\tau}$ represent a hydrological process for season τ and year ν and X_ν the corresponding aggregated annual process, such that:

$$X_\nu = \sum_{\tau=1}^{\omega} Y_{\nu,\tau} \qquad (19)$$

Vecchia et al. (1983) illustrate the aggregation procedure by using a two-season PARMA(1,1) model like

$$Y_{\nu,1} = \phi_1 \, Y_{\nu-1,2} - \theta_1 \, \epsilon_{\nu-1,2} + \epsilon_{\nu,1} \qquad (20a)$$

$$Y_{\nu,2} = \phi_2 \, Y_{\nu,1} - \theta_1 \, \epsilon_{\nu,1} + \epsilon_{\nu,2} \qquad (20b)$$

After combining Equations 19 and 20 yields:

$$X_\nu - \phi_1 \, \phi_2 \, X_{\nu-1} = -\phi_1 \, \theta_2 \, \epsilon_{\nu-1,1} + (1 + \phi_2 - \theta_2)\epsilon_{\nu,1} +$$
$$+ (\phi_1 - \theta_1 - \theta_1\phi_2)\epsilon_{\nu-1,2} + \epsilon_{\nu,2} \qquad (21)$$

The right-hand side of (21) is recognized as a sum of two independent moving average processes each of order one, which is representable as a single moving average process also of order one. Hence, Equation 21 becomes:

$$X_\nu - \phi \, X_{\nu-1} = -\theta\eta_{\nu-1} + \eta_\nu \qquad (22)$$

where $\phi = \phi_1\phi_2$. The θ and the variance σ_η^2 of the noise η can be expressed as functions of the parameters of the seasonal models (20).

Proceeding in an analogous fashion, it can be shown (Vecchia et al., 1983) that for the case of an arbitrary number of seasons ω, each with a PARMA(1,1) model, the aggregated model at the annual level is a stationary ARMA(1,1) model. A general aggregation method and parameter estimation of the corresponding annual model has been suggested. The parameters of the annual model estimated using the relationship between the parameters of the annual and seasonal models were shown to be more efficient than those estimated directly from the annual series. A natural extension of the foregoing work for modelling and estimation of a given seasonal time series would be modelling the time series at a lower level first and estimating the parameters of the higher level model based on the relationship of the parameters between the two seasonal models.

5.   Summary and Conclusions

A number of stochastic models for modelling single-site seasonal hydrological time series have been reviewed herein.   Included is a description of the stochastic characteristics usually found in hydrological series and the approaches suggested for dealing with trends, periodicity, non-normality and correlation. Modelling approaches have been classified under the categories of direct modelling, disaggregation and aggregation. In addition, models of intermittent and nonintermittent series have been considered.

Seasonal hydrological series are more complex to analyze and to model due to the inherent periodic characteristics usually manifested in such statistics as the mean, variance, skewness and covariance.   While the periodic mean and variance can be removed from the original series by "seasonal" standardization, the periodic skewness can be accounted for either by transforming the original sample into normal or by assuming a model which can reproduce dependent non-normal variables. However, each approach has shortcomings that the modeller must be aware of. Likewise, the periodic covariance can be accounted for by the usual periodic autoregressive PAR(1) process or the periodic versions of ARMA processes in general, for instance the PARMA(1,1) process. While these PAR or PARMA processes require the underlying variable to be normally distributed, the PGAR(1) process can reproduce both periodic skewness and periodic lag-one covariance (in addition to periodic means and variances) without the need for transformations.   In addition, methods for modelling intermittent hydrological sequences have been developed within the framework of discrete ARMA processes. Finally, disaggregation approaches are also useful for the modelling and generation of seasonal series when reproduction of statistics are needed at more than one time level. Related to this is the so-called aggregation approach in which the idea is to model the series at a lower level first and then derive the model and properties of a higher level series. Further developments on these approaches will be beneficial for the modelling and simulation of water resources systems.

The emphasis in this paper has been on seasonal models without much discussion on model testing and verification. For these important topics the reader is referred to Salas et al. (1980) and Stedinger and Taylor (1982).

6.   Acknowledgements

The authors wish to acknowledge the NSF Grants CEE-8110782 on "Stochastic Modeling of Geophysical Time Series" and INT-8715102 on "US - Italy Cooperative Research on Prediction of Floods and Precipitation Events", for providing the support for the research leading to this paper.

7. References

Alley, W.M. (1984) 'On the treatment of evapotranspiration, soil moisture accounting and aquifer recharge in monthly water balance models', Water Resources Research 20(8), pp 1137-1149.

Bartolini, P. and Salas, J.D. (1985) 'Properties of multivariate periodic ARMA(1,1) models', presented at the International Symposium on Multivariate Analysis in Hydrologic Processes, Colorado State University, Fort Collins, Colorado, July 15-17.

Beard, L.R. (1965) 'Use of interrelated records to simulate streamflow', Journ. of Hydraulics Div., ASCE 91(HY5), pp 13-22.

Beard, L.R. (1973) 'Transfer of streamflow data within Texas', Texas Water Development Board, Texas, pp 1-24.

Box, G.E.P. and Cox, D.R. (1964) 'An analysis of transformations', Jour. Royal Stat. Soc., Series B, 26, pp 211-252.

Box, G.E.P. and Jenkins, G.M. (1970) 'Time series analysis forecasting and control', Holden-Day Press, San Francisco.

Bras, R.L. and Rodriguez-Iturbe, I. (1985) 'Random functions and hydrology', Addison-Wesley Publ. Co., Massachusetts.

Burges, S.J. (1970) 'Use of stochastic hydrology to determine storage requirements for reservoirs - a critical analysis', Report EEP-34, Stanford University, Stanford, California.

Carlson, R.F., MacCormick, A.J. and Watts, D.G. (1970) 'Application of linear models to four annual streamflow series', Water Resources Research, 6, pp 1070-1078.

Chang, T.J., Kavvas, M.L. and Delleur, J. (1984) 'Daily precipitation modeling by discrete autoregressive moving average processes', Water Resources Research, 20(5), pp 565-580.

Chebaane, M., Salas, J.D. and Boes, D.C. (1992) 'Modeling of monthly intermittent streamflow processes', Water Resources Papers, 105, 153 p., Colorado State University, Fort Collins, Colorado.

Cooper, D.M. and Wood, E.F. (1980) 'Model identification and parameter estimation for hydrologic input/output models', Proc. of Joint Automatic Control Conf., San Francisco, Paper No. FP6-B.

Croley, T.E., II (1972) 'Sequential stochastic optimization in water resources', Ph.D. Dissertation, Dept. of Civil Engineering, Colorado State University, Fort Collins, Colorado.

Croley, T.E. and Rao, K.N. (1976) 'A manual for hydrologic time series deseasonalization and serial dependence reduction', Rep. 199, Iowa Inst. of Hydr. Res., Univ. of Iowa, Iowa City.

Curry, K. and Bras, R.L. (1978) 'Theory and applications of the multivariate broken line, disaggregation and monthly autoregressive streamflow generators to the Nile River', TAP Report 78-5, MIT, Cambridge, Massachusetts.

DeCoursey, D.G., Schaake, J.C. and Seely, E.H. (1982) 'Stochastic models in hydrology, in hydrologic modeling of small watersheds', ASAE, Monograph No. 5, Edit. by C.T. Haan, H.P. Johnson and D.L. Brakensiek.

Delleur, V.W., Tao, P.C. and Kavvas, M.L. (1976) 'An evaluation of practicality and complexity of some rainfall and runoff time series models', Water Resources Research 12(5), pp 953-970.

Eagleson, P., Climate, Soil and Vegetation, 7. 'A derived distribution of annual water yield', Water Resources Research 14(5), pp 765-776.

Fernandez, B. and Salas, J.D. (1986) 'Periodic gamma autoregressive processes for operational hydrology', Water Resources Research 22(10), pp 1385-1396.

Fiering, M.B. (1967) 'Streamflow synthesis', Harvard University Press, Cambridge, Massachusetts, 139 pp.

Gaver, D.P. and Lewis, P.A.W. (1980) 'First-order autoregressive gamma sequences and point processes', Adv. Appl. Prob. 12, pp 727-745.

Gladishev, E.G. (1961) 'Periodically correlated random sequences', Soviet Math., Vol 2, pp 385-388.

Hannan, E.J. (1955) 'A test for singularities in Sydney rainfall', Austr. Jour. Phys. 8(2), pp 289-297.

Harms, A.A. and Campbell, T.H. (1967) 'An extension of the Thomas-Fiering model for the sequential generation of streamflow', Water Resources Research 3(3), pp 653-661.

Hipel, K.W., McLeod, A.I. and Lennox, W.C. (1977) 'Advances in Box-Jenkins modeling, 1. Modeling construction', Water Resources Research 13(3), pp 567-575.

Hirsch, R.M. (1979) 'Synthetic hydrology and water supply reliability', Water Resources Research 15(6), pp 1603-1615.

Hirsch, R.M. and Slack, S.R. (1984) 'A nonparametric trend test for seasonal data with serial dependence', Water Resources Research 20(6), pp 727-732.

Hoshi, K. and Burges, S.J. (1978) 'The impact of seasonal flow characteristics and demand patterns on required reservoir storage', J. Hydrol. 37, pp 241-260.

Jackson, B.B. (1975) 'The use of streamflow models in planning', Water Resources Research 11, pp 54-63.

Jacobs, P.A. and Lewis, P.A.W. (1978a) 'Discrete time series generated by mixtures, 1. Correlation and run properties', J.R. Stat. Soc. B, 40(1), pp 94-105.

Jacobs, P.A. and Lewis, P.A.W. (1978b) 'Discrete time series generated by mixtures, 2. Asymptotic properties', J.R. Stat. Soc. B, 40(2), pp 222-228.

Jacobs, P.A. and Lewis, P.A.W. (1978c) 'Discrete time series generated by mixtures, 3. Autoregressive process (DAR(P))', Tech. Rep. NPS 55-78-022, Nav. Postgrad. School, Monterey, California.

Jacobs, P.A. and Lewis, P.A.W. (1982) 'Stationary discrete autoregressive moving average time series generated by mixtures', Tech. Rep. NPS 55-82-003, Nav. Postgrad. School, Monterey, California.

Jones, R. and Brelford, W. (1967) 'Time series with periodic structure', Biometrika, 54(3 and 4), pp 7403-7408.

Kavvas, M.L. and Delleur, J.W. (1975) 'Removal of periodicities by differencing and monthly mean substraction', J. Hydrol. 26, pp 335-353.

Kelman, J. (1977) 'Stochastic modeling of hydrologic intermittent daily process', Hydrology Paper 89, Colorado State University, Fort Collins, Colorado.

Kirby, W. (1972) 'Computer-oriented Wilson-Hilferty transformation that preserves the first three moments and the lower bound of the Pearson type 3 distribution', Water Resources Research 8(5), pp 1251-1254.

Klemes, V. (1974) 'The Hurst phenomenon - a puzzle', Water Resources Research 10, pp 675-688.

Klemes, V. (1978) 'Physically based stochastic hydrologic analysis, in advances in hydroscience', V.T. Chow, Ed., Academic Press, New York, pp 285-352.

Koch, R. (1982) 'A physical-probabilistic approach to stochastic hydrology', Ph.D. Dissertation, Dept. of Civil Engineering, Colorado State University, Fort Collins, Colorado.

Kottegoda, N.T. (1974) 'Effects of skewness in three stochastic pentad river flow models on crossing properties of synthesized data', Water Resources Research 10(3), pp 446-456.

Kottegoda, N.T. (1980) 'Stochastic water resources technology', MacMillan, London, England.

Kuczera, G. (1982) 'On the relationship between the reliability of parameter estimates and hydrologic time series data used in calibration', Water Resources Research 18(1), pp 146-154.

Lane, W.L. (1979) 'Applied stochastic techniques (last computer package); user manual', Div. of Planning Tech. Services, Bureau of Reclamation, Denver, Colorado.

Lane, W.L. (1982) 'Corrected parameters estimates for disaggregation schemes, in statistical analysis of rainfall and runoff', Edit. by V. Singh, Water Resources Publications, Littleton, Colorado.

Lawrance, A.J. and Kottegoda, N.T. (1977) 'Stochastic modeling of riverflow time series', Jour. Statistical Society, Series A, 140(1), pp 1-47.

Ledolter, J. (1978) 'The analysis of multivariate time series applied to problems in hydrology', Jour. of Hydrology 36, pp 327-352.

Lee, S. (1975) 'Stochastic generation of synthetic streamflow sequences in ephemeral streams', IAHS Symp., Tokyo, pp 691-701.

Lettenmaier, D.P. and Burges, S.J. (1977) 'An operational approach to preserving skew in hydrologic models of long-term persistence', Water Resources Research 13(2), pp 281-290.

Loucks, D.D., Stedinger, J.R. and Haith, D.A. (1981) 'Water Resource systems planning and analysis', Prentice-Hall, New Jersey.

Maidment, D.R. and Parzen, E. (1984) 'Time patterns of water use in six Texas cities', ASCE Wat. Resources Plan. and Management 110(1), pp 90-106.

Mandelbrot, B.B. and Wallis, J.R. (1968) 'Noah, Joseph and operational hydrology', Water Resources Research 4(5), pp 909-918.

Matalas, N.C. (1967) 'Mathematical assessment of synthetic hydrology', Water Resources Research 3(4), pp 937-945.

Matalas, N.C. (1975) 'Developments in stochastic hydrology', Reviews of Geoph. and Space Physics, 13(3), pp 67-73.

McGinnis, D.F., Jr and Sammons, W.H. (1970) 'Discussion of daily streamflow simulation' by K. Payne, W.D. Neumann and K.D. Kerri, ASCE Jour. of the Hydraulics Div., 96(HY5), pp 1201-1206.

McKerchar, A.I.M. and Delleur, J.W. (1974) 'Application of seasonal parametric linear stochastic models to monthly flow data', Water Resources Research 10(2), pp 246-255.

McLeod, A.I., Hipel, K.W. and Lennox, W.C. (1977) 'Advances in Box-Jenkins modeling, 2. Applications', Water Resources Research 13(3), pp 577-586.

McLeod, A.I. and Hipel, K.W. (1978a) 'Simulation Procedures for Box-Jenkins models', Water Resources Research 14(5), pp 969-975.

McLeod, A.I. and Hipel, K.W. (1978b) 'Developments in monthly autoregressive modeling', Tech. Report 45-XM011178, Dep. of System Design Eng.,

Univ. of Waterloo, Ontario.

Mejia, J.M. and Rousselle, J. (1976) 'Disaggregation models in hydrology revisited', Water Resources Research 12(2), pp 185-186.

Moreau, D.H. and Pyatt, E.E. (1970) 'Weekly and monthly flows in synthetic hydrology', Water Resources Research 6(1), pp 53-61.

Moss, M.E. and Bryson, M.C. (1974) 'Autocorrelation structure of monthly streamflows', Water Resources Research 10(4), pp 737-744.

Obeysekera, J.T.B. and Salas, J.D. (1982) 'On the aggregation and disaggregation of streamflow time series', American Geophysical Union EOS 63(18), p 321.

O'Connell, P.E. (1971) 'A simple stochastic modeling of Hurst's law, in mathematical models in hydrology', Symposium, Warsaw (IAHS Pub. No. 100, 1974), Vol. 1, pp 169-187.

Pagano, M. (1978) 'On periodic and multiple autoregressions', Ann. Statistic, 6 pp 1310-1317.

Payne, K., Neumann, W.R. and Kerri, K.D. (1969) 'Daily streamflow simulation', ASCE Jour. of Hydrualics Div. 95(HY4), pp 1163-1179.

Quimpo, R.G. (1967) 'Stochastic modeling of daily river flow sequences', Hydrology Paper 18, Colorado State University, Fort Collins, Colorado.

Rao, A.R., Kashyap, R.L. and Mao, L.T. (1982) 'Optimal choice of type and order of river flow time series models', Water Resources Research 18(4), pp 1097-1109.

Rodriguez-Iturbe, I., Mejia, J. and Dawdy, D. (1972) 'Streamflow simulation, 1. A new look at Markovian models', Fractional Gaussian Noise and Crossing Theory, Water Resources Research 8(4), pp 921-930.

Rodriguez-Iturbe, I., Dawdy, D.R. and Garcia, L.E. 'Adequacy of Markovian models with cycle components for stochastic streamflow simulation', Water Resources Research 7(5), p 1127.

Roesner, L.A. and Yevjevich, V. (1966) 'Mathematical models for time series of monthly precipitation and monthly runoff', Hydrology Paper 15, Colorado State University, Fort Collins, Colorado.

Salas, J.D. and Yevjevich, V. (1972) 'Stochastic modeling of water use time series', Hydrology Paper No. 52, Colorado State University, Fort Collins, Colorado.

Salas, J.D. and Smith, R.A. (1980) 'Uncertainties in hydrologic time series analysis', ASCE Spring Meeting at Portland, Oregon, Preprint, pp 80-158.

Salas, J.D., Delleur, J.W., Jevjevich, V. and Lane, W.L. (1980) 'Applied modeling of hydrologic time series', Water Resources Publications, Littleton, Colorado.

Salas, J.D., Boes, D.C. and Smith, R.A. (1982) 'Estimation of ARMA models with seasonal parameters', Water Resources Research 18(4), pp 1006-1010.

Salas, J.D. and Obeysekera, J.T.B. (1992) 'Conceptual basis of seasonal streamflow time series models', Jour. Hydr. Eng., ASCE, 118(8), pp 1186-1194.

Salas, J.D., Obeysekera, J.T.B. and Smith, R. (1981) 'Identification of streamflow stochastic models', Jour. Hydr. Div., ASCE, 107(HY7), pp 853-866.

Santos, E. and Salas, J.D. (1983) 'A parsimonious step disaggregation model for operational hydrology', paper presented at the AGU Fall Meeting, San Francisco, California, EOS 64(45), p 706.

Sim, C.H. (1987) 'A mixed gamma ARMA(1,1) model for riverflow time series',

Water Resources Research 23(1), pp 32-36.

Sorooshian, S. and Dracup, J.A. 'Stochastic parameter estimation procedures for hydrologic rainfall-runoff models: correlated and heterostochastic error cases', Water Resources Research 16(2), pp 430-442.

Sorooshian, S., Gupta, V.K. and Fulton, J.L. 'Evaluation of maximum likelihood parameter estimation techniques for conceptual rainfall-runoff model: influence of calibration data variability and length on model credibility', Water Resources Research 19(1), pp 251-259.

Spolia, S.K. and Chander, S. (1974) 'Modeling of surface runoff systems by an ARMA model', Jour. of Hydrology, Vol. 22, pp 317-332.

Srikanthan, R. and McMahon, T.A. (1980) 'Stochastic generation of monthly flows for ephemeral streams', Jour. of Hydrology, Vol. 47, pp 19-40.

Srikanthan, R. and McMahon, T.A. 'Stochastic generation of monthly streamflows', Jour. Hydr. Div. of ASCE, 108(HY3), pp 419-441.

Stedinger, J.R. and Taylor, M.R. (1982) 'Synthetic streamflow generation, 1. Model verification and validation', Water Resources Research 18(4), pp 909-918.

Stedinger, J.R. and Vogel, R.M. (1984) 'Disaggregation procedures for generating serially correlated flow vectors', Water Resources Research 20(1), pp 47-56.

Svanidze, G.G. (1980) 'Mathematical modeling of hydrologic series', Water Resources Publications, Fort Collins, Colorado, 314 p.

Tao, P.C. and Delleur, J.W. (1976) 'Seasonal and nonseasonal ARMA models in hydrology', J. Hydr. Div., ASCE 102(HY10), pp 1591-1559.

Thomas, H.A., Jr. and Fiering, M.B. (1962) 'Mathematical synthesis of streamflow sequences for analysis of river basins by simulation, in: the design of water resources systems', Maas, A. et al., Harvard University Press, Cambridge, Massachusetts, pp 459-493.

Thompstone, R.M. (1983) 'Topics in hydrological time series modeling', Ph.D. Dissertation, Dept. of Systems Design, University of Waterloo, Waterloo, Canada.

Todnini, E. (1980) 'The preservation of skewness in linear disaggregation schemes', Journal of Hydrology 47, pp 199-214.

US Army Corps of Engineers (1973) 'HEC-4 monthly streamflow simulation', Sacramento, California.

Valencia, D. and Schaake, J.C. (1973) 'Disaggregation processes in stochastic hydrology', Water Resources Research 9(3), pp 580-585.

Vecchia, A.V. (1983) 'Aggregation and estimation for periodic autoregressive-moving average models', Ph.D. Dissertation, Departament of Statistics, Colorado State University, Fort Collins, Colorado.

Vecchia, A.V., Obeysekera, J.T., Salas, J.D. and Boes, D.C. (1983) 'Aggregation and estimation for low-order periodic ARMA models', Water Resources Research 19(5), pp 1297-1306.

Wallis, J.R. and Matalas, N.C. (1972) 'Sensitivity of reservoir design to the generating mechanism of inflows', Water Resources Research 8(3), pp 634-641.

Weiss, G. (1977) 'Shot noise models for the generation of synthetic streamflow data', Water Resources Research 13(1), pp 101-108.

Yakowitz, S.J. (1972) 'A statistical model for daily streamflow records with application to the Rillito River', in Proceedings of the Int. Symp. on Uncertainties in Hydrologic and Water Resources Systems, Vol. 1, Univ. of Arizona, Tucson, pp 273-283.

Yakowitz, S.J. (1973) 'A stochastic model for daily riverflow in an arid region', Water Resources Research 9, pp 1271-1285.

Yevjevich, V. (1963) 'Fluctuation of wet and dry years, part I, research data assembly and mathematical models', Hydrology Paper 1, Colorado State University, Fort Collins, Colorado.

Yevjevich, V. (1972a) 'Structural analysis of hydrologic time series', Hydrology Paper 56, Colorado State University, Fort Collins, Colorado.

Yevjevich, V. (1972b) 'Stochastic processes in hydrology', Water Resources Publications, Fort Collins, Colorado.

Young, G.K. and Jetmar, R.V. (1976) 'Modeling monthly hydrologic persistence', Water Resources Research 12(5), pp 829-835.

MODELS FOR DATA GENERATION IN HYDROLOGY: MULTIVARIATE TECHNIQUES

JOSE D. SALAS
Hydrologic Science and Engineering Program
Department of Civil Engineering
Colorado State University
Fort Collins, Colorado 80523

ABSTRACT.   Most approaches suggested for modelling and generation of multisite hydrological series fall within the general framework of multivariate ARMA models.   Formal modelling procedures suggest a three-stage iterative process, namely: model identification, parameter estimation and diagnostic checks.   In general, it is not an easy task, especially if high order vector ARMA models are used.   However, simpler models, such as the contemporaneous model, may be sufficient for most applications in water resources.   Modelling approaches based on these models are reviewed and examples of modelling bivariate and trivariate streamflow series are included.

1.   Introduction

The main purpose of this paper is to discuss stochastic techniques for modelling multivariate hydrological time series which are needed in the analysis of water resources systems. We will specifically deal with series defined at several points along a line, over an area or across space, or sets of mutually related series of various kinds defined at a point. These types of series are usually called multiple, multi-site, multi-point or in general multivariate time series (Salas et al., 1980). Some examples of multivariate time series in water resources are: series of annual or seasonal precipitation at several gauging stations in a region; streamflow series at various points along a river or at various rivers; different water quality variables at a particular river cross-section; series of precipitation, evaporation, inflows and outflows of a given reservoir system; and series of different kinds of water demands.

   Multivariate hydrological time series may be generally represented by four components: (1) long-term trends or other deterministic changes exhibited beyond the annual cycle, (2) seasonal or periodic changes of days, weeks or months within the annual cycle, (3) almost periodic changes, such as tidal and solar effects, and (4) stochastic or random variations. The first three components are considered to be deterministic while the latter component is stochastic. Long-term changes which may be detected in water   resources   time series are: linear   or   non-linear trends,   and

75

J.B. Marco et al. (eds.),
Stochastic Hydrology and its Use in Water Resources Systems Simulation and Optimization, 75–95.
© 1993 *Kluwer Academic Publishers.*

gradual or abrupt jumps (Yevjevich, 1972; Salas et al., 1980). Appropriate techniques for removing and modelling trends and jumps may be found in Yevjevich (1972), Salas and Yevjevich (1972), Hipel et al. (1975), Lettenmaier (1976), Hirsch et al. (1982), and Maidment and Parzen (1984). In addition, periodic mean, variance, covariance and skewness are inherent properties of hydrological series defined at time intervals smaller than a year. The stochastic characteristic of water resources time series is represented by the correlation structure in time and in space and by the underlying probability distribution of the noise.

2. Description and Review of Multivariate Models

Three basic types of dependence relationships may be found in hydrological time series: (1) contemporaneous, (2) unidirectional and (3) feedback relationships. When more than two series are analyzed, mixed relationships may also occur. Two or more series are contemporaneously related, only if their concurrent or contemporaneous values are dependent on each other. For example, streamflows at several stations in a region may have contemporaneous dependence, if their flows are not connected either by natural courses or by man-made intervention. The main source of dependence in this case is the regional precipitation, which is input to the corresponding watersheds. This is commonly exhibited by annual time series where only each year's values are significantly cross-correlated to each other. Likewise, annual or seasonal time series of precipitation at several stations in a region may be considered to be a set of contemporaneously correlated series.

A unidirectional relationship may be exhibited between streamflows sampled at two points along a river, i.e. the upstream station flows causes the downstream station flows, but not conversely. Other examples of unidirectional relationships are rainfall-runoff relations and streamflow-river chlorides relations. One-way causality may include three or more variables. One example is that of streamflow sampled at three gauging stations along a river. We would expect that flows of the uppermost station causes the flows at both downstream stations, while the flows of the station inbetween is affected only by the flows of the upstream station and causes the flows of the downstream station. Likewise, the flows of the lowermost station is affected by those of the upstream stations, but not conversely. In the foregoing examples each variable is assumed to be affected by its own past values. Feedback relationship exists if causality is present in both directions. Examples in this category may be the case of time series of precipitation and evapotranspiration for some large tropical watersheds or time series of precipitation, lake evaporation and lake storage for some large lakes. Furthermore, three or more time series may have mixed relationships in any combination of the foregoing types of dependence.

Various multivariate models have been proposed in the water resource literature during the past three decades in order to represent the dependence relationships found in water resources series in general and hydrological series in particular. In the light of our previous discussion models can be classified as contemporaneous ARMA models, transfer function ARMA models and vector ARMA models. Likewise, an extension of ARMA models may be defined in the form of the so-called ARMAX models. In addition, there are other models such as multiple regression, aggregation and disaggregation models, which are applicable to multiple series. Some concepts related to these models will be introduced and reviewed in the following sections.

2.1. MULTIVARIATE ARMA MODELS

Let us consider a set of n time series $x_t^{(i)}$, i=1, ..., n, each with zero mean. A multivariate ARMA (p,q) model for such series may be written as:

$$\underline{x}_t = \sum_{i=1}^{p} \Phi(i)\ \underline{x}_{t-i} + \sum_{i=1}^{q} \Theta(i)\ \underline{e}_{t-i} + \underline{e}_t \tag{1}$$

where $\Phi(i)$ is an n x n parameter matrix with elements $\phi^{jk}(i)$, $\Theta(i)$ is an n x n parameter matrix with elements $\theta^{jk}(i)$, $\underline{x}_t$ is an n x 1 matrix with elements $x_t^{(i)}$, i=1, ..., n and $\underline{e}_t$ is an n x 1 matrix with noise elements $e_t^{(i)}$, i=1, ...,n. The e's have zero means, are uncorrelated in time but are contemporaneously correlated amongst themselves or $E\{e_t^{(i)}e_t^{(j)}\} \neq 0$ and $E\{e_t^{(i)}e_{t'}^{(j)}\} = 0$ for $t \neq t'$.

A contemporaneous ARMA model results if the parameters Φ and Θ are both diagonal matrices or, $\phi^{jk}(i) = 0$ and $\theta^{jk}(i) = 0$ for $j \neq k$ (Salas et al., 1980; Camacho et al., 1983; Camacho et al, 1985a; Camacho et al., 1985b). A transfer-function model is derived from Equation 1, when the parameter matrices Φ and Θ are lower triangular or $\phi^{jk}(i) = 0$ and $\theta^{jk}(i) = 0$ for j<k (Tiao and Box, 1981). In general, Model 1 is called a vector model when the matrices Φ and Θ are defined as full matrices with non-zero elements (Dunsmuir and Hannan, 1976: Ledolter, 1978). In this case, the model implies feedback relationships. The vector (full) model may be used in general, and particular models may arise as a result of parameter estimation and diagnostic checks. Although this may be a logical concept, unfortunately, estimation and fitness tests for vector models, in most cases, are complex and cumbersome. Simpler models therefore, such as contemporaneous models, are quite attractive from the practical standpoint (Salas et al., 1985).

An extension of ARMA models has been presented by Cooper and Wood (1980) which is referred to as multiple input-output models. This class of models is basically an ARMA model with exogenous variables. They are called ARMAX (p,q,r) models which are written as

$$\underline{x}_t = \sum_{i=1}^{p} \Phi(i)\ \underline{x}_{t-i} + \sum_{i=0}^{q} \Theta(i)\ \underline{e}_{t-i} + \sum_{i=0}^{r} \Gamma(i)\ \underline{u}_{t-i} \tag{2}$$

where $\underline{x}_t$, Φ, Θ and $\underline{e}$ are as defined previously, $\underline{u}_t$ is a k-dimensional vector of exogenous variables and $\Gamma(i)$ is the corresponding n x k parameter matrix. In a series of papers Cooper and Wood (1980, 1982a, 1982b) presented identification, parameter estimation and testing of these models with applications to hydrology. The mathematical and statistical properties of multivariate ARMAX models have been studied by Hannan and Kavalieres (1984). The reader is referred to the original papers for details on ARMAX models. The ensuing text will concentrate on simpler multivariate ARMA models.

Contemporaneous ARMA models have been suggested because the general vector ARMA model is more difficult to estimate and test, especially for high order models in which the number of parameters may be large (Salas et al., 1980). This point is important in modelling water resources time series, since long historical data are not always available, so that the accuracy of the estimated parameters is questionable (Camacho et al., 1983). In contemporaneous models, the diagonalization of the parameter matrices allows "model decoupling" into component equations, so that the model parameters do not have to be estimated jointly, and univariate modelling procedures can be employed (Salas and Pegram, 1977). Once the parameters are estimated, the residuals of each component can be recovered. Such residuals are assumed to be contemporaneously correlated; the variance-covariance matrix of such residuals can be estimated and one can proceed to "whitening" them by some appropriate procedure, to arrive at uncorrelated white noise residuals, which may be used to check the adequacy of model fit.

Let us illustrate the foregoing concept by considering a bivariate contemporaneous ARMA (1,1) model as

$$
\begin{bmatrix} x^{(1)} \\ x^{(2)} \end{bmatrix}_t = \begin{bmatrix} \phi^{11} & 0 \\ 0 & \phi^{22} \end{bmatrix} \begin{bmatrix} x^{(1)} \\ x^{(2)} \end{bmatrix}_{t-1} + \begin{bmatrix} \theta^{11} & 0 \\ 0 & \theta^{22} \end{bmatrix} \begin{bmatrix} e^{(1)} \\ e^{(2)} \end{bmatrix}_{t-1} + \begin{bmatrix} e^{(1)} \\ e^{(2)} \end{bmatrix}_t
\tag{3}
$$

Each component may be expressed as:

$$
x_t^{(1)} = \phi^{11} x_{t-1}^{(1)} + \theta^{11} e_{t-1}^{(1)} + e_t^{(1)}
\tag{4}
$$

and

$$
x_t^{(2)} = \phi^{22} x_{t-1}^{(2)} + \theta^{22} e_{t-1}^{(2)} + e_t^{(2)}
\tag{5}
$$

Thus, the components are univariate ARMA (1,1) models where $e_t^{(1)}$ and $e_t^{(2)}$ are each serially uncorrelated but contemporaneously (spatially) correlated to each other. Once the parameters ϕ and θ are estimated (based on univariate estimation procedures), $\underline{e}_t$ can be derived from Equations 4 and 5. Finally, a simple bivariate model of the form:

$$
\begin{bmatrix} e^{(1)} \\ e^{(2)} \end{bmatrix}_t = \begin{bmatrix} b^{11} & 0 \\ b^{21} & b^{22} \end{bmatrix} \begin{bmatrix} \epsilon^{(1)} \\ \epsilon^{(2)} \end{bmatrix}_t
\tag{6}
$$

may be used for the e's where $\epsilon_t^{(1)}$ and $\epsilon_t^{(2)}$ are residuals with mean zero and variance one, which are uncorrelated in time and in space. The matrix of elements b^{ij}, $i,j=1,2$, can be estimated from the variance-covariance matrix of the e's (Salas et al., 1980).

The contemporaneous Model 3 can be rewritten in terms of the ϵ_t's. From Equations 3 and 6 we get:

$$
\begin{pmatrix} x^{(1)} \\ x^{(2)} \end{pmatrix}_t = \begin{pmatrix} \phi^{11} & 0 \\ 0 & \phi^{22} \end{pmatrix} \begin{pmatrix} x^{(1)} \\ x^{(2)} \end{pmatrix}_{t-1} + \begin{pmatrix} \theta^{11}b^{11} & 0 \\ \theta^{22}b^{21} & \theta^{22}b^{22} \end{pmatrix} \begin{pmatrix} \epsilon^{(1)} \\ \epsilon^{(2)} \end{pmatrix}_{t-1} + \begin{pmatrix} b^{11} & 0 \\ b^{21} & b^{22} \end{pmatrix} \begin{pmatrix} \epsilon^{(1)} \\ \epsilon^{(2)} \end{pmatrix}_t \tag{7}
$$

which is another form of ARMA model, in which the matrix of moving average parameters is lower triangular and the noises are uncorrelated in time and space (between them). Note that, written in this manner, the ARMA model cannot be uncoupled (thus, univariate procedures cannot be used). In addition, note that Model 7 now appears with one more parameter.

Contemporaneous modelling was also used for models other than ARMA models (Matalas and Wallis, 1971, 1976; Mejia et al., 1974; Curry and Bras, 1978). For instance, Matalas and Wallis (1971, 1976) considered the multivariate fractional Gaussian noise (FGN) model in which FGN models are fitted to the individual series and the residuals are modelled using a contemporaneous model as in Equation 6.

Particular cases of low order vector ARMA models have been used in the water resource literature. Matalas (1967) proposed the multisite AR(1) model given by

$$
\underline{x}_t = \Phi\,\underline{x}_{t-1} + B\,\underline{\epsilon}_t \tag{8}
$$

where Φ and B are assumed to be full matrices. This model preserves both lag-zero and lag-one cross-covariance matrices. If Model 8 is applied to seasonal series, each series must be standardized (seasonally) prior to applying the model. A simple contemporaneous AR(1) model will result from Equation 8 when Φ is a diagonal matrix. Matalas pointed out that this can be done if the lag-one cross-correlation coefficients are not considered important. Pegram and James (1972) extended this concept to multivariate AR(p) models. Likewise, O'Connell (1974) studied the multivariate ARMA(1,1) model for streamflow time series with long term persistence. The model is defined as:

$$
\underline{x}_t = \Phi\,\underline{x}_{t-1} + B\,\underline{\epsilon}_t + C\,\underline{\epsilon}_{t-1} \tag{9}
$$

where Φ is a full matrix and either B or C are triangular matrices. O'Connell also suggested diagonalizing the Φ matrix, and B and C are both assumed to be lower triangular matrices. The non-zero diagonal elements of Φ are estimated based on the lag-one autocorrelation and Hurst coefficients, derived from each series.

The foregoing models apply to stationary series or to any time series which can be made stationary. For instance, multivariate models (8) and (9) may be applied to seasonal hydrological series if the residual series after seasonal standardization does not have a periodic autocovariance structure. Otherwise, a non-stationary model is more appropriate. For example, a general vector periodic autoregressive PAR(p) model with seasonally varying parameters was given by Salas and Pegram (1977) as:

$$
\underline{x}_{\nu,\tau} = \sum_{i=1}^{p} \Phi_\tau(i)\,\underline{x}_{\nu,\tau-i} + B_\tau\,\underline{\epsilon}_{\nu,\tau} \tag{10}
$$

where $\Phi_\tau(i)$ and B_τ are periodic parameter matrices and the mean of $x_{\psi,\tau}$ is zero. This model is suitable for reproducing a periodic autocovariance.

2.2. OTHER MULTIVARIATE MODELS

Fiering (1964) introduced multivariate analysis for multisite streamflow synthesis using principal component analysis. Beard (1965) adopted a multivariate regression model for multisite monthly streamflow synthesis. Prior to fitting the model, three transformations are made to the original flows, which are assumed to be approximately log-Pearson Type III distributed. Firstly, a normalization by log-transformation is applied to the original flows $y_{\psi,\tau}$. Then the log-transformed series is seasonally standardized. Subsequently, the Wilson-Hilferty transformation is applied to obtain a normal variable $z_{\psi,\tau}$. Finally, a multivariate regression model is used as:

$$z_{\psi,\tau}^{(j)} = \sum_{\substack{i=1 \\ i \neq j}}^{n} \alpha_\tau(i) \, z_{\psi,\tau}^{(i)} + \alpha_\tau(j) \, z_{\psi,\tau-1}^{(j)} + \beta_\tau(j) \, u_{\psi,\tau} + \epsilon_{\psi,\tau}^{(j)} \sqrt{1 - R_\tau^2(j)}$$

(11)

where $\alpha_\tau(i)$ and $\beta_\tau(j)$ are regression coefficients for stations i and j, respectively, j is the index of the dependent station, n is the number of stations, $R_\tau(j)$ is the multiple correlation coefficient with respect to station j, $\epsilon_{\psi,\tau}^{(j)}$ is an independent standard normal deviate, and $u_{\psi,\tau}$ is equal to the sum of transformed flows of all stations for the six months preceding the antecedent month $\tau-1$, such that

$$u_{\psi,\tau} = \sum_{i=1}^{n} \sum_{k=1}^{6} z_{\psi,\tau-k-1}^{(i)}$$

(12)

The model adopted by Beard (1965) also appeared in the HEC-4 monthly streamflow simulation package, except that the term u was not included (US Corps of Engineers, 1973). The multiple regression model (11) without the term u was also used by Payne et al. (1969) for multisite daily streamflow synthesis. However, they suggested a correction of the Wilson-Hilferty transformation due to the boundedness property of gamma variables. Following this lead, McGinnis and Sammons (1970) gave the final form of the Wilson-Hilferty transformation.

Another model, closely linked with vector AR models, was presented by Moreau and Pyatt (1970). This model is for generating weekly flow patterns of several stations which are conditioned on monthly flows. The means, variances and covariances of the weekly and monthly flows are preserved. The model has the same form as the multivariate AR(1) model, except for the definition of the variable x_t in Equation 8. It has been noted by Salas et al. (1980) that Moreau and Pyatt's model may be considered as an "aggregation" model since, in the process of generating weekly flows, it preserves certain statistics at the monthly level. However, a major limitation of the model is the large number of parameters involved. A way to circumvent such limitation may be to set a model structure for series at both weekly and monthly levels, as suggested by Vecchia et al. (1983) and Obeysekera and Salas (1986).

The need to preserve annual and seasonal time series properties inspired the development of disaggregation models. Harms and Campbell (1967) proposed essentially the first disaggregation approach, however, Valencia and Schaake (1973) proposed the so-called disaggregation model which has become a major technique for modelling hydrological time series. These models break down a sequence of longer-time units into a sequence of shorter-time units. For instance, a sequence of annual flows may be generated using ARMA or other alternative models and disaggregated into monthly flows. The generated series are supposed to preserve relevant statistical properties at both annual and monthly levels. Disaggregation models are useful not only for disaggregating variables in time (say annual flows into monthly flows) but, for disaggregating in space as well (Salas et al. 1980).

3. Modelling Procedures

Multivariate time series modelling comprises three basic stages, namely: model identification, parameter estimation and diagnostic checking (Ledolter, 1978; Hipel and McLeod, 1980; Salas et al., 1980; Tiao and Box, 1981; Camacho et al., 1983; Camacho et al., 1985a; Camacho et al., 1985b; Salas et al., 1985). This strategy, originally formalized by Box and Jenkins (1976), is an iterative procedure of model building to ensure satisfactory model fitting and utilization. The first task is to plot the available time series data. In any modelling exercise time series plots may quickly show some characteristics such as trends, periodicities, temporal and spatial relationships and degree of randomness. Likewise, prior transformations must be made (if needed) in order to normalize the data. The model identification stage is divided into model structure specification and model order determination. The first step involves deciding the type of dependence relationship to be incorporated in the model and whether to use a model with constant or periodic parameters. The second step is concerned with identifying the appropriate values of p and q of the ARMA (p,q) model. After having identified some tentative models, the parameter estimation follows. The diagnostic checks are basically tests applied to the model residuals (innovations) to check · some assumptions about them (i.e. independence, normality). Implementing the model according to its intended utilization provides a more rigorous test of validating the model (Stedinger et al., 1985). If the model is found inadequate, the identification stage and parameter estimation stage are repeated. Each of these stages are now discussed in more detail.

3.1. MODEL IDENTIFICATION

3.1.1. *Model Structure.* In this step the structural form of the model or the dependence representation and parameterization of the model is sought. Three dependence mechanisms and relationships are: contemporaneous, triangular and feedback relationships. These dependence mechanisms are related to particular representation and parameterization of ARMA models. Contemporaneous models are appropriate for series exhibiting contemporaneous (lag-zero) relationship, transfer-function models are for series with directional relationship and vector models apply to series with feedback relationships. If feasible, the appropriate dependence mechanism must be selected based on the physical configuration and knowledge of the system being modelled (Finzi et al., 1975). A practical approach is to assume simpler models first. For instance, for a particular case, one may first assume that the model is contemporaneous ARMA (p,q). Then, determine p and q, estimate the parameters and check the fitness of the model. If the checks reject the assumed model, then either a transfer-function model or a full vector model must be considered (Salas et al., 1985).

The relationship between two time series may be studied by examining the cross-correlation structure of their univariate model residuals (Haugh and Box, 1977; Pierce and Haugh, 1977). It is assumed that an appropriate univariate model is fitted to the individual series to derive the residuals. Then, the cross-correlation function between the two residual series is examined. A contemporaneous relationship may exist if the cross-correlation function has the shape as in Figure 1a, while a cross-correlation function as in Figure 1b may suggest a transfer-function (unidirectional relationship). On the other hand, the cross-correlation of Figure 1c may suggest a vector (feedback) relationship. The limits in Figure 1 may be given by $\pm$ $2/\sqrt{N}$, where N is the number of data points (Tiao and Box, 1981). This approach is limited to a bivariate relationship and stationary models.

It is often the case that models with periodic parameters may be needed for seasonal water resources time series. Series which are seasonally standardized may have residuals with periodic correlation structure. Therefore, one must decide whether to use a model with constant or periodic parameters. In general, periodicities may be assessed by a combined graphical-statistical analysis or by Fourier analysis. Graphical-statistical analysis involves plotting a given statistical characteristic versus time, making a judgement about its pattern and testing whether such characteristics are statistically different from each other. Fourier's series analysis involves, assessing the significance of the variance explained by each harmonic (component cycle) or using the so-called periodogram analysis. Details of such analyses may be found in Salas et al. (1980).

3.1.2. *Model Order.* Several techniques for model order determination have been suggested in the literature. They include cross-correlation analysis, partial correlation analysis, likelihood ratio test and Akaike-type information criteria. These techniques can be used for contemporaneous, transfer-function or vector ARMA models.

For a stationary multivariate time series, it is known that the population lag-k cross-correlation matrix $M(k)$ of a moving average process $MA(q)$, is zero for all k>q. On the other hand, $M(k)$ of $AR(p)$ and $ARMA(p,q)$ models are infinite in extent and decay gradually to zero as k increases (Ledolter, 1978; Tiao and Box, 1981). Therefore, the $MA(q)$ model may be identified from the sample cross-correlation matrix. The order q is determined when the estimates of the matrices $M(k)$ are statistically equal to zero for k=q+1, q+2, ... Tiao and Box (1981) suggest the limits $\pm 2/\sqrt{N}$ to test the null hypothesis of no correlation and to display the results by assigning the + sign to values above the upper limit, the - sign to values below the lower limit and the . sign to values within the limits. Alternatively, the student's t-test (Benjamin and Cornell, 1970) may be

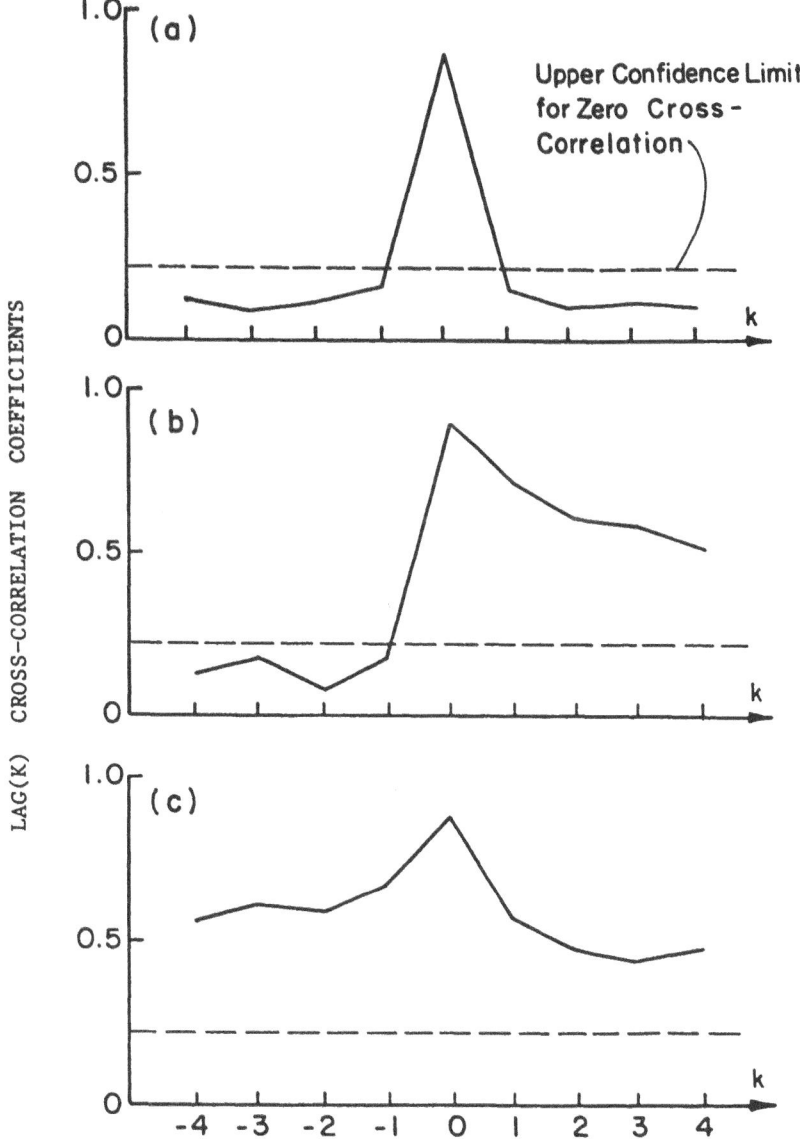

Figure 1. Cross-correlation function of model residuals for (a) contemporaneous, (b) transfer-function and (c) vector (feedback), relationships (after Salas et al., 1985).

used to test the significance of cross-correlations and the Anderson (1941) test for testing the significance of autocorrelations.

Likewise, partial correlation analysis may be useful for model order determination. Whereas the cross-correlation function of an AR(p) model is infitine in extent, its partial correlation function is zero after lag p. Jenkins and Alavi (1981) gave the approximate variance of the diagonal elements of the sample estimates of the partial correlation matrix P(k) as 1/N, under the assumption that k>p for an AR(p) model. Thus, under the null hypothesis that P(k) is zero, the test limits can be taken as $\pm 2/\sqrt{N}$. Likewise, the partial correlation function for both MA(q) and ARMA(p,q) models decays gradually towards zero. Thus, when both cross-correlation and partial correlation functions tend to be infinite in extent, then a mixed ARMA(p,q) model must be used as opposed to MA(q) or AR(p) models. Actually, determining the orders p and q of a multivariate ARMA(p,q) model is not an easy task. Ledolter (1978) and Newbold (1981) indicate the practical difficulties in choosing the order of ARMA(p,q) models. However, Tiao and Box (1981) developed a criteria based on the so-called R- and S-arrays of Gray et al. (1978) which, at least theoretically, could point out the order of an ARMA(p,q) model.

The likelihood ratio test is an approach that may be used to determine the order of AR(p) models. This involves testing the null hypothesis that the parameter matrix $\Phi(p)=0$ against the alternative $\Phi(p)\neq 0$. The likelihood ratio test statistic U is first computed as (Tiao and Box, 1981)

$$U = \det SS_e(p) / \det SS_e(p-1) \tag{13}$$

where $SS_e(p)$ is a matrix of product residuals computed from

$$SS_e(p) = \sum_{t=p+1}^{N} [\underline{x}_t - \Phi(1)\underline{x}_{t-1} - \ldots - \Phi(p)\underline{x}_{t-p}] [\underline{x}_t - \Phi(1)\underline{x}_{t-1} - \ldots - \Phi(p)\underline{x}_{t-p}]^T \tag{14}$$

where $\underline{x}_t$ is a vector of size n x 1 and N is the number of observations. Then, the statistic

$$\chi(p) = -(N - p - np - 1/2) \log_e U \tag{15}$$

is, under the null hypothesis, asymptotically chi-square distributed with n^2 degrees of freedom. In Equation 14 the parameter matrix Φ must be estimated beforehand. For this purpose, the Yule-Walker equation

$$\Phi(1)M(k-1) + \Phi(2)M(k-2) + \ldots + \Phi(p)M(k-p) = M(k); \text{ for } k = 1,\ldots, p \tag{16}$$

may be used to estimate the Φ's where M(k) is substituted by the lag-k sample cross-correlation matrix.

Tiao and Box (1981) and Camacho et al. (1983), also suggested analyzing the extent to which the fit is improved as the order of an AR(p) model is increased. This is done by examining the diagonal elements of the residual variance-covariance matrices corresponding to successive fitted AR models. For instance, if the diagonal elements of the variance-covariance matrix of an AR(p) model are smaller than those of AR(1), AR(2), ..., AR(p-1) or about the same as those of AR(p+1), AR(p+2), ..., then the AR(p) model is selected. The sample residual variance-covariance matrix, denoted by $\Sigma_{ee}(p)$, may be computed from

$$\Sigma_{ee}(p) = SS_e(p) \, / \, (N-p) \tag{17}$$

where $SS_e(p)$ is given by Equation 14.

The Akaike Information Criteria (AIC) proposed by Akaike (1974) is another technique for selecting the order of AR(p) models. For an n-variate AR(p) model fitted to N observations, the AIC takes the form (Priestley, 1981):

$$AIC(p) = (N-p) \, \log \, \det \, \Sigma_{ee}(p) + 2 \, n \, p^2 \tag{18}$$

where $\Sigma_{ee}(p)$ is given by Equation 17. The model with the minimum AIC is the one selected. Parzen (1977) also proposed an alternative AIC-type criterion. Likewise, a modification of the Akaike criterion, which is applicable to multivariate ARMA(p,q) models, was given by Hannan (1980). However, the terms involved are somewhat cumbersome to estimate for higher-order ARMA models.

Various statistical tools are available for identifying the order of multivariate ARMA models. However, one must be aware of the fact that their actual usefulness may be quite limited when applied to hydrological time series with short samples. A practical approach may be to assume low order models first, say ARMA(1,0), ARMA(2,0) or ARMA(1,1) models, and then proceed with the modelling approach. Then, the assumed order of the models may be increased as the need arises. In addition, the foregoing tools are limited to stationary series (models). Therefore, the suggestion of assuming successively low order models becomes more relevant when one is dealing with multisite seasonal hydrological series.

3.2. PARAMETER ESTIMATION

Moments and maximum likelihood are among the most popular methods of parameter estimation which have been suggested in the literature. These methods will be briefly discussed in this section. Firstly, in relation to the multisite AR(1) model of Equation 8, the parameters Φ and B may be estimated by the method of moments as:

$$\Phi = M(1) \, M^{-1}(0) \tag{19}$$

and

$$B \, B^T = M(0) - M(1) \, M^{-1}(0) \, M^T(1) \tag{20}$$

where $M(0)$ and $M(1)$ are the lag-zero and lag-one cross-covariance matrices respectively, and the superscripts T and -1 denote the transpose and the inverse of a matrix, respectively. Matalas suggested using principal component analysis to solve for B in Equation 20. However, a more direct solution was suggested by Young and Pisano (1968). Likewise, for the ARMA(1,1) model (9) the parameters are related to covariances by (O'Connell, 1974)

$$\Phi = M(2) \, M(1)^{-1} \tag{21}$$

$$C\ B^T = \Phi M(0) - M(1) \tag{22}$$

and

$$B\ B^T + C\ C^T = M(0) - \Phi\ M^T(1) + [\Phi\ M(0) - M(1)]\ \Phi^T \tag{23}$$

While matrix Φ can be estimated directly, however solving for B and C requires numerical techniques.

For estimation based on contemporaneous ARMA (p,q) models with constant or periodic parameters, each series is fitted by appropriate ARMA (p,q) models whereby univariate time series analysis and procedures are employed. Then, the univariate model residuals are recovered and fitted to a multivariate model as in Equation 6. Non-normalities in the data are handled by the usual normalizing transformations. Further advances on such modelling procedures have been suggested by Camacho et al. (1983, 1985a, 1985b); Stedinger et al. (1985); Bartolini and Salas (1985); Bartolini et al. (1988) and Haltiner and Salas (1988). Several methods of estimating the parameters of contemporaneous ARMA (1,1) models for annual streamflow series were compared by Stedinger et al. (1985). They found that auto-regressive and moving average parameters are best estimated by using the univariate maximum likelihood method. The variance-covariance matrix of the model residuals can then be estimated using the model parameters based on the method of moments.

Now consider the vector PAR(p) model with periodic parameters as in Equation 10. It may be shown that the periodic model parameter matrices $\Phi_\tau(i)$ i=1, ..., p are related to the correlation matrices $M_\tau(j)$ by (Salas and Pegram, 1977):

$$\begin{bmatrix} \Phi_\tau(1) \\ \Phi_\tau(2) \\ \cdot \\ \cdot \\ \cdot \\ \Phi_\tau(p) \end{bmatrix} = \begin{bmatrix} M_{\tau-1}(0) & M_{\tau-1}^T(1) & \cdots & M_{\tau-1}^T(p-1) \\ M_{\tau-1}(1) & M_{\tau-2}(0) & \cdots & M_{\tau-2}^T(p-2) \\ \cdot & \cdot & & \cdot \\ \cdot & \cdot & & \cdot \\ \cdot & \cdot & & \cdot \\ M_{\tau-1}(p-1) & M_{\tau-2}(p-2) & \cdots & M_{\tau-p}(0) \end{bmatrix}^{-1} \begin{bmatrix} M_\tau(1) \\ M_\tau(2) \\ \cdot \\ \cdot \\ \cdot \\ M_\tau(p) \end{bmatrix} \tag{24}$$

where $M_\tau(j) = E[X_{v,\tau}\ X^T_{v,\tau-j})$ and the matrix B_τ is given by

$$B_\tau\ B_\tau^T = M_\tau(0) - \sum_{i=1}^{p} \Phi_\tau(i)\ M_\tau^T(i) \tag{25}$$

Equations 24 and 25 may be used to estimate the model parameters by substituting the covariance matrices by their corresponding sample estimates. Note that for estimation of contemporaneous or transfer-function models, Equations 24 and 25 are valid with appropriate elements of the Φ matrices set equal to zero. In addition, for nonseasonal or constant parameter models, the subscripts τ are simply deleted.

The moment estimation of PARMA(p,q) models (with periodic parameters) is more complex than for PAR(p) models. For instance, let us consider the vector PARMA(1,1) model

$$\underline{x}_{v,\tau} = \Phi_\tau(1)\underline{x}_{v,\tau-1} - \Theta_\tau(1)\ \underline{\epsilon}_{v,\tau-1} + B_\tau\ \underline{\epsilon}_{v,\tau} \tag{26}$$

It may be shown that the following relations hold (Salas et al., 1985):

$$\Phi_\tau(1) = M_\tau(2) \; M_{\tau-1}^{-1}(1) \tag{27}$$

$$\Phi_\tau(1) \; B_{\tau-1}^{\mathsf{T}} = \Phi_\tau(1) \; M_{\tau-1}(0) - M_\tau(1) \tag{28}$$

and

$$B_\tau B_\tau^{\mathsf{T}} + \Theta_\tau(1)\Theta_\tau^{\mathsf{T}}(1) = M_\tau(0) - \Phi_\tau(1)M_\tau^{\mathsf{T}}(1) + [\Phi_\tau(1)M_{\tau-1}(0) - M_\tau(1)]\Phi_\tau^{\mathsf{T}}(1) \tag{29}$$

Equation 27 gives the estimate for $\Phi_\tau(1)$, while Equations 28 and 29 must be solved numerically to obtain $\Theta_\tau(1)$ and B_τ. Numerical solutions for stationary ARMA(1,1) models are given by O'Connell (1974) and Bras and Rodriguez-Iturbe (1985). Haltiner and Salas (1988) provides a method for moment estimation when the PARMA(1,1) model is contemporaneous.

Wilson (1973) studied the likelihood function of ARMA(p,q) models with constant parameters and presented an approximate maximum likelihood estimation procedure. To illustrate his approach, let us rewrite the ARMA(p,q) model with constant parameters as

$$\underline{e}_t = \underline{x}_t - \Phi(1)\underline{x}_{t-1} - \dots - \Phi(p)\underline{x}_{t-p} + \Theta(1)\underline{e}_{t-1} + \dots + \Theta(q)\underline{e}_{t-q} \tag{30}$$

where $\underline{e}_t$ is the white noise vector (nx1) with zero means and variance-covariance matrix G. Equation 30 can be used to obtain the values of $\underline{e}_t$ recursively. However, when the recursion is started, there are quantities on the right hand side which are unknown. In practice, the recursion can be started at t=p+1, setting the terms $\underline{e}_p$, ..., $\underline{e}_{p-q+1}$ equal to zero. If the series sample size is N, then the effective number of observations is N-p.

Under the assumption of joint normality of $(\underline{e}_{p+1}, \underline{e}_{p+2}, \dots, \underline{e}_N)$ the logarithm of its likelihood function is (Wilson, 1973)

$$LL(\Phi,\Theta,G) = -0.5n(N-p)\log2\pi - 0.5(N-p)\left\{\log|G| + \frac{1}{N-p}\sum_{t=p+1}^{N} \underline{e}_t^{\mathsf{T}}G^{-1}\underline{e}_t\right\} \tag{31}$$

where $|G|$ is the determinant of G and n is the number of time series. Since the first term of the right hand side of Equation 31 is a constant, maximizing $LL(\Phi,\Theta,G)$ is equivalent to minimizing the second term. This can be achieved by setting its derivatives to zero and solving the equations for Φ, Θ and G. Since the resulting equations are non-linear, the solution must be found numerically. An approximate solution can be obtained by minimizing the function

$$\left| \frac{1}{N-p}\sum_{t=p+1}^{N} \underline{e}_t(\Phi,\Theta) \; \underline{e}_t^{\mathsf{T}}(\Phi,\Theta) \right| \tag{32}$$

which essentially gives least squares estimators. Ledolter (1978) also described an approximate maximum likelihood procedure for the ARMA(p,q) model of Equation 30 by minimizing the (objective) function

$$F(\Phi,\Theta,G) = \log |G| + \frac{1}{N-p} \sum_{t=p+1}^{N} \underline{e}_t^T G^{-1} \underline{e}_t \qquad (33)$$

which is essentially the second term of Equation 31. Likewise, Lettenmaier (1980) used Equation (33) as the objective function for estimating the parameters of the ARMA(1,1) model, as well as a modified function, in an attempt to incorporate some measure of long term persistence. In addition, Cooper and Wood (1982b) proposed a recursive maximum likelihood scheme for estimating the parameters of ARMAX models. Interested readers are referred to the original papers for further information about such specific procedures.

The log-likelihood function given in Equations 31, 32 and 33 are only "conditional" likelihood functions; consequently, they provide only approximate maximum likelihood parameters (Tiao and Box, 1981). This is because of the manner of starting the recursion in Equation 30, where the model residuals $\underline{e}_p, \ldots, \underline{e}_{p-q+1}$ are taken to be zero. An optimal start is the method of back forecasting as in Box and Jenkins (1976) for univariate estimation. However, Wilson (1973) pointed out that this latter approach is difficult to apply to the multivariate case. For vector AR(p) models with periodic parameters, an approximate maximum likelihood estimation procedure was devised by Salas and Pegram (1977). They used the PAR(p) model as in Equation 10 except that the periodic means and standard deviations of the original data are also included. To reduce the number of parameters to be estimated, they suggested fitting trigonometric series to the model parameters. In this manner, the coefficients of the trigonometric series become the parameters to be estimated. Likewise, they suggested restricting the autoregressive parameters to be diagonal matrices (i.e. a contemporaneous relationship) to reduce further computational burden.

3.3. DIAGNOSTIC CHECKS

Diagnostic checks are necessary in order to see whether the model has been misspecified and to search for model improvements. Since, the residuals of ARMA models are assumed to be stationary white noise series (i.e. independent and normally distributed), checking the validity of this assumption is usually sufficient for diagnosing model adequacy. Generally, statistical tests for independence and normality of residuals are employed. Other diagnostic checks may include model implementation as well as testing the robustness of the model. For instance, the model may be implemented according to its intended utilization, such as data augmentation, generation or forecasting, and the model performance is examined. Robustness may be applied to see if the model preserves properties not explicitly parameterized in the model (e.g. Hurst coefficient, drought characteristic, etc.). The text below mainly elaborates on diagnostic checks applied to model residuals.

First, the plot of residual series against time gives a quick picture of any remaining non-stationarities, periodicities and correlations. The residuals of a multivariate model may be tested for normality considering each residual series individually. For this purpose, a number of tests are available in the literature, such as the skewness

test, the chi-square test and the Kolmogorov-Smirnov test. These tests
assume that the series tested are independent. Thus, they should be
applied in tandem with tests for independence of residuals. As suggested
by Tiao and Box (1981), cross-correlations falling within $\pm 2/\sqrt{N}$ (where
N is the number of observations) are considered to be zero. In addition,
the student's t-test for testing cross-correlations being zero and the
Anderson test or the Portmanteau lack of fitness test for testing
autocorrelations may be used (Box and Jenkins, 1976).

Another useful check is model overfitting, which involves, fitting a
higher order model than the one originally prescribed in the identification
stage. This procedure becomes advantageous if we know the direction in
which the model is likely to be violated (e.g. not enough autoregressive
or moving average terms in the model). The likelihood ratio test, the
Akaike information criterion (AIC), Parzen's criterion or Hannan's modified
AIC (mentioned previously in the section) may be used in selecting the best
among competing models (i.e. prescribed and overfitted models). The latter
three information-criteria tests are especially attractive since they
simultaneously address minimum variance and model parsimony properties.

4. Example of Multivariate Modelling

An example is presented in this section to illustrate some of the points
discussed earlier. Consider a stationary bivariate contemporaneous
ARMA(1,1) model written as

$$
\begin{bmatrix} x_t^{(1)} \\ x_t^{(2)} \end{bmatrix} = \begin{bmatrix} \phi^{11} & 0 \\ 0 & \phi^{22} \end{bmatrix} \begin{bmatrix} x_{t-1}^{(1)} \\ x_{t-1}^{(2)} \end{bmatrix} - \begin{bmatrix} \theta^{11} & 0 \\ 0 & \theta^{22} \end{bmatrix} \begin{bmatrix} e_{t-1}^{(1)} \\ e_{t-1}^{(2)} \end{bmatrix} + \begin{bmatrix} e_t^{(1)} \\ e_t^{(2)} \end{bmatrix} \tag{34}
$$

where the x's are assumed to have zero means and unit variances, and the
e's are contemporaneously correlated normal variables with zero means and
variance-covariance matrix G. The model residuals are represented by
Equation 6. Model 34 can be "uncoupled" and univariate estimation
procedures can be used. This is followed by estimation of the b's in
Equation 6.

Equation 34 is first uncoupled resulting into two univariate ARMA(1,1)
models. The corresponding moment estimators of ϕ^{ii} and θ^{ii}, i=1,2 are given
by (Box and Jenkins, 1976; Salas et al., 1980):

$$
\phi^{ii} = m^{ii}(2) \, / \, m^{ii}(1) \tag{35}
$$

and

$$
\theta^{ii} = (-\beta \pm \sqrt{\beta^2 - 4\alpha^2}) \, / \, 2\alpha \tag{36}
$$

where $\alpha = \phi^{ii} - m^{ii}(1)$, $\beta = 2\phi^{ii}m^{ii}(1) - (\phi^{ii})^2 - 1$ and $m^{ii}(k)$ is the estimated
lag-k autocorrelation of the series $x^{(i)}$. The solution of Equation 36
must be chosen such that $|\theta^{ii}| < 1$. The B parameter matrix is given by

$$
B \, B^T = G \tag{37}
$$

where G is the variance-covariance (symmetric) matrix of the residuals e of Equation 34. Assuming that B is a lower triangular matrix, its elements may be found by:

$$b^{11} = \sqrt{g^{11}} \qquad b^{12} = 0$$
$$b^{21} = g^{12}/b^{11} \qquad b^{22} = \sqrt{g^{22} - (b^{21})^2} \qquad (38)$$

By using the method of moments, the elements of G can be shown to be (Stedinger et al., 1985):

$$g^{ij} = m^{ij}(0) \ (1 - \phi^{ii} \ \phi^{jj}) \ / \ (1 - \phi^{ii} \ \theta^{jj} - \theta^{ii} \ \phi^{jj} + \theta^{ii} \ \theta^{jj}) \qquad (39)$$

where $m^{ij}(0)$ is the estimated lag-zero cross-correlation of $x_t^{(i)}$ and $x_t^{(j)}$.

Synthetic samples were generated based on the Model 34 and 6 with parameters as shown in Table 1. The b's were obtained from Equation 38 based on the assumed lag-zero cross-correlation matrix M(0) with elements $m^{11}(0) = m^{22}(0) = 1$, and $m^{12}(0) = m^{21}(0) = 0.4$, and values of g^{ij} determined from Equation 39. A total of 100 samples of size N=100 were generated. For each sample, the parameter set was estimated and, based on the 100 sample estimates, the mean, bias and mean square error of the estimated parameters were obtained. The results are shown in Table 1. They indicate that the parameters are reasonably well reproduced although some small negative biases are observed.

TABLE 1. Comparison of population and estimated parameters for the model used in Example 1

Parameter	Population	Estimates		
		Mean	Bias	MSE
ϕ_{11}	0.400	0.378	-0.022	0.177
ϕ_{22}	0.500	0.470	-0.030	0.079
θ_{11}	0.170	0.137	-0.033	0.216
θ_{22}	0.150	0.167	0.017	0.087
b_{11}	0.970	0.953	-0.017	0.001
b_{21}	0.378	0.375	-0.003	0.007
b_{22}	0.848	0.831	-0.017	0.003

(MSE = mean square error)

5. Remarks

A number of approaches has been suggested for modelling and generation of multiple hydrological time series. In the early stages, suggested approaches were based on principal component analysis, as well as multiple regression analysis. Subsequently, structural forms which may be classified within the general framework of ARMA models have become widely accepted.

Most approaches in the water resources literature suggest a three-stage iterative model building, namely: model identification, parameter estimation and diagnostic checks. Identifying "the model" consists of determining the structure of the model, whether it involves constant or periodic parameters and finding the order of the model. Considering the general class of multivariate ARMA models, simpler versions, such as the contemporaneous model, are quite attractive in modelling water resources time series.

If the time series at hand are seasonal, one needs to define which model parameters may be assumed to be constant and which must be seasonal. In this regard, it is not just a matter of subsituting the parameters of a stationary model by a periodic function or using different stationary models for different seasons. If the covariance characteristics are seasonal, then one must use a model with periodic parameters.

Several methods are available for determining the order of multivariate models. They include: cross-correlation analysis, partial-correlation analysis, likelihood ratio test and Akaike-type information criteria. Often though, low order models are sufficient to represent most characteristics of interest of hydrological time series. Then, a practical approach may be to assume low order models first and to change the order if the need arises.

A number of estimation approaches based on the moments and maximum likelihood methods have been reviewed herein. In general, they are somewhat cumbersome for the full vector ARMA model and for multivariate models with periodic parameters. Therefore, simpler contemporaneous ARMA models offers a viable alternative for modelling hydrological time series. In these models, the diagonalization of the parameter matrices allows "model decoupling" into component equations, such that univariate estimation procedures can be employed. This procedure allows fitting constant parameters models to some sites and periodic parameters models to others. Likewise, it allows fitting different order of models at different sites. Once the parameters are estimated, the residuals of each component are recovered and a simple lag-zero model will provide the spatial correlation structure for such residual series. Diagnostic checks are necessary in order to see whether the selected model is appropriate and whether it is better than other competing models. Since ARMA models assume that the residual series are normal and uncorrelated in time and in space, appropriate statistical tests must be applied in order to verify such assumptions. Likewise, statistical criteria must be used for comparison among competing models, so that the selected model is parsimonious in the number of parameters.

This paper has concentrated on reviewing and applying canonical forms of multivariate ARMA models. However, multivariate modelling of hydrological time series extends to other forms such as disaggregation approaches (Valencia and Schaake, 1972) or aggregation approaches (Vecchia et al., 1983, Bartolini et al., 1988). Both approaches aim at reproducing relevant statistics at several time intervals (for instance, annual and monthly levels). The former reproduces and generates variables at the higher level first, then disaggregates them into lower levels. On the other hand, the latter approach models and generates at lower levels first, and higher level models and variables are obtained by aggregation. The main concept in this latter approach is that the model and parameters of the lower level variable determine the corresponding model and parameters of the higher level variable. Therefore, model identification and parameter estimation must be done, considering the linkage of lower and higher levels.

6. Acknowledgments

The author wish to acknowledge the NSF Grants CEE-8110782 on "Stochastic Modeling of Geophysical Time Series" and INT-8715102 on "US-Italy Cooperative Research on Prediction of Floods and Precipitation Events" for providing the support for the research leading to this paper.

7. References

Akaike, H. (1974) 'A new look at the statistical model identification', IEEE Trans. Automatic Control AC-19, pp 716-722.

Anderson, R.L. (1941) 'Distribution of serial correlation coefficient', Ann. Math. Statist. 8, pp 1-13.

Bartolini, P. and Salas, J.D. (1985) 'Properties of multivariate periodic ARMA(1,1) models', Presented at the International Symposium on Multivariate Analysis in Hydrologic Processes, Colorado State University, Fort Collins, Colorado, July 15-17.

Bartolini, P., Salas, J.D. and Obeysekera, J. (1988) 'Multivariate periodic ARMA(1,1) processes', Water Resour. Res. 24(8), pp 1237-1246.

Beard, L.R. (1965) 'Use of interrelated records to simulate streamflow', Jour. of Hydraulics Div., ASCE 91(HY5), pp 13-22.

Benjamin, J.R. and Cornell, C.A. (1970) 'Probability, statistics and decision of civil engineers', McGraw-Hill, Inc., New York.

Box, G.E.P. and Jenkins, G.M. (1976) 'Time series analysis forecasting and control', Holden-Day Press, San Francisco.

Bras, R.L. and Rodriguez-Iturbe, I. (1985) 'Random functions and hydrology', Addison-Wesley Publ. Co., Massachusetts.

Camacho, F., McLeod, A.I. and Hipel, K.W. (1983) 'The use and abuse of multivariate time series in hydrology', Tech. Rep. 83-12, Dept. of Statist. and Actuarial Sciences, the University of Western Ontario, Canada.

Camacho, F., McLeod, A.I. and Hipel, K.W. (1985a) 'Contemporaneous autoregressive-moving average (CARMA) modeling in water resources' in Special AWRA Monograph on Time Series Analysis in Water Resources, K.W. Hipel (editor).

Camacho, F., McLeod, A.I. and Hipel, K.W. (1985b) 'Developments in multivariate ARMA modeling in hydrology', presented at the International Symposium on Multivariate Analysis of Hydrologic Processes, Colorado State University, Fort Collins, Colorado, July 15-17.

Cooper, D.M. and Wood, E.F. (1980) 'Model identification and parameter estimation for hydrologic input/output models', Proc. of Joint Automatic Control Conf., San Francisco, Paper No. FP6-B.

Cooper, D.M. and Wood, E.F. (1982a) 'Identification of multivariate time series and multivariate input-output models', Water Resources Research 18(4), pp 937-946.

Cooper, D.M. and Wood, E.F. (1982b) 'Parameter estimation of multiple input-output time series models: application to rainfall-runoff processes', Water Resources Research 18(5), pp 1352-1364.

Curry, K. and Bras, R.L. (1978) 'Theory and applications of the multivariate broken line, disaggregation, and monthly autoregressive streamflow generators to the Nile River', Tech. Rep. No. 78-5, Massachusetts Institute of Technology, Cambridge, Massachusetts.

Dunsmuir, W. and Hannan, E.J. (1976) 'Vector linear time series models', Adv. Appl. Prob. 8, pp 339-364.

Fiering, M.B. (1964) 'Multivariate technique for synthetic hydrology', Jour. of Hydraulics Div., ASCE 90(HY5), pp 844-850.

Finzi, G., Todini, E. and Wallis, J.R. (1975) 'Comment upon multivariate synthetic hydrology', Water Resources Research 11(6), pp 844-850.

Gray, H.L., Kelley, G.D. and McIntire, D.D. (9178) 'A new approach to ARMA modeling', Comm. Statist. B7, pp 1-77.

Haltiner, J.P. and Salas, J.D. (1988) 'Development and testing of a multivariate seasonal ARMA(1,1) model', Jour. of Hydrology, 104, pp 247-272.

Hannan, E.J. (1980) 'The estimation of the order of an ARMA process', Ann. Statist. 8(5), pp 1071-1081.

Hannan, E.J. and Kavalieris, L. (1984) 'Multivariate linear time series models', Adv. Appl. Prob. 16, pp 492,561.

Harms, A.A. and Campbell, T.H. (1967) 'An extension of the Thomas-Fiering model for the sequential generation of streamflow', Water Resources Research 3(3), pp 653-661.

Haugh, L.D. and Box, G.E.P. (1977) 'Identification of dynamic regression (distributed lag) models connecting two time series', Jour. of Amer. Statist. Assoc. 72(357), pp 121-130.

Hipel, K.W., Lennox, W.C., Unny, T.E. and McLeod, A.I. (1975) 'Intervention analysis in water resources', Water Resources Research, 11, pp 855-861.

Hipel, K.W. and McLeod, A.I. (1980) 'Perspectives in stochastic hydrology', in Time Series (Proceedings of the International Conference held at Nottingham University), O.D. Anderson (editor), North Holland, pp 73-102.

Hirsch, R.M., Slack, J.R. and Smith, R.A. (1982) 'Techniques of trend analysis for monthly water quality data', Water Resources Research, 18, pp 107-121.

Jenkins, G.M. and Alavi, A.S. (1981) 'Some aspects of modeling and fore-casting multivariate time series', Jour. of Time Series Analysis 2(1), pp 1-47.

Ledolter, J. (1978) 'The analysis of multivariate time series applied to problems in hydrology', Journal of Hydrology 36, pp 327-352.

Lettenmaier, D.P. (1976) 'Detection of trends in water quality data from records with dependent observations', Water Resources Research, 12, pp 1037-1046.

Lettenmaier, D.P. (1980) 'Parameter estimation for multivariate streamflow synthesis', Proc. of Joint Automatic Control Conf., San Francisco, Paper FA6-D.

Loucks, D.D., Stedinger, J.R. and Haith, D.A. (1981) 'Water resource systems planning and analysis', Prentice-Hall, New Jersey.

Maidment, D.R. and Parzen, E. (1984) 'Time patterns of water use in six Texan cities', ASCE Water Resources Plan and Management 110(1), pp 90-106.

Matalas, N.C. (1967) 'Mathematical assessment of synthetic hydrology',

Water Resources Research 3(4), pp 931-945.

Matalas, N.C. and Wallis, J.R. (1971) 'Statistical properties of multivariate fractional noise processes', Water Resources Research 7(6), pp 1460-1468.

Matalas, N.C. and Wallis, J.R. (1976) 'Generation of synthetic flow sequences', in System Approach to Water Management, A.K. Biswas (Ed.), McGraw-Hill, Inc., New York.

McGinnis, D.F. Jr. and Sammons, W.H. (1970) 'Discussion of Daily Streamflow Simulation', by K. Payne, W.D. Neumann and K.D. Kerri. ASCE Journal of the Hydraulics Division 96(HY5), pp 1201-1206.

Mejia, J.M., Dawdy, D.R. and Nordin, D.F. (1974) 'Streamflow simulation 3. The broken line process and operational hydrology', Water Resources Research, 10(2), pp 242-245.

Moreau, D.H. and Pyatt, E.E. (1970) 'Weekly and monthly flows in synthetic hydrology', Water Resources Research 6(1), pp 53-61.

Newbold, P. (1981) 'Some recent developments in time series analysis', Inter. Statist. Review 49, pp 53-66.

Obeysekera, J. and Salas, J.D. (1986) 'Modeling of aggregated hydrologic time series', Jour. Hydrology, 86, pp 197-219.

O'Connell, P.E. (1974) 'Stochastic modeling of long-term persistence in streamflow sequences', Ph.D. Dissertation, Civil Engineering Dept., Imperial College of Science and Technology, London.

Parzen, E. (1977) 'Multiple time series modeling: determining the roles of approximating autoregressive schemes', in: Multivariate Analysis, IV, P. Krishnaiah (Ed.), North-Holland, Amsterdam, pp 283-295.

Payne, K., Neumann, W.R. and Kerri, K.D. (1969) 'Daily streamflow simulation', ASCE Jour. of Hydraulics Div. 95 (HY4), pp 1163-1179.

Pegram, G.G.S. and James, W. (1972) 'Multilag multivariate autoregressive model for the generation of operational hydrology', Water Resources Research 8(4), pp 1074-1076.

Pierce, D.A. and Haugh, L.D. (1977) 'Causality in temporal systems: characterizations and a survey', Jour. of Econometrics 5, pp 265-294.

Priestley, M.B. (1981) 'Multivariate series, prediction and control', in Spectral Analysis and Time Series, Vol. 2, Academic Press, London.

Salas, J.D. and Yevjevich, V. (1972) 'Stochastic modeling of water use time series', Hydrology Papers N. 52, Colorado State University, Fort Collins, Colorado.

Salas, J.D. and Pegram, G.G.S. (1977) 'A seasonal multivariate multilag autoregressive model in hydrology', Proc. of the Third International Symposium on Theoretical and Applied Hydrology, Colorado State University, Fort Collins, Colorado.

Salas, J.D., Delleur, J.W., Yevjevich, V. and Lane, W.L. (1980) 'Applied modeling of hydrologic time series', Water Resources Publ., Littleton, Colorado.

Salas, J.D., Tabios, G. and Bartolini, P. (1985) 'Approaches to multivariate modeling of water resources time series', in Special AWRA Monograph on Time Series in Water Resources, K.W. Hipel (Ed.).

Stedinger, J.R., Lettenmaier, D.P. and Vogel, R.M. (1985) 'Multisite ARMA(1,1) and disaggregation models for annual streamflow generation', Water Resources Research 21(4), pp 497-509.

Stedinger, J.R. and Vogel, R.M. (1984) 'Disaggregation procedures for generating serially correlated flow vectors', Water Resources Research

20(1), pp 47-56.

Tao, P.C. and Delleur, J.W. (1976) 'Multistation, multiyear synthesis of hydrologic time series by disaggregation', Water Resources Research 12(6), pp 1303-1312.

Tiao, G.C. and Box, G.E.P. (1981) 'Modeling multiple time series with applications', Jour. of the Amer. Statist. Assoc. 76(376), pp 802-816.

Todini, E. (1980) 'The preservation of skewness in linear disaggregation schemes', Journal of Hydrology 47, pp 199-214.

US Army Corps of Engineers (1973) 'HEC-4 monthly streamflow simulation', Sacramento, California.

Valencia, R.D. and Schaake, J.C. Jr. (1973) 'Disaggregation processes in stochastic hydrology', Water Resources Research 9(3), pp 580-585.

Vecchia, A.V., Obeysekera, J.T.B., Salas, J.D. and Boes, D.C. (1983) 'Aggregation and estimation for low-order periodic ARMA models', Water Resources Research 19(5), pp 1297-1306.

Wilson, G.T. (1973) 'The estimation of parameters in multivariate time series models', Jour. Royal Statist. Soc., Series B 35(1), pp 76-85.

Yevjevich, V. (1972) 'Stochastic processes in hydrology', Water Resources Publ., Fort Collins, Colorado.

Young, G.K. and Pisano, W.C. (1968) 'Operational hydrology using residuals', ASCE Jour. of the Hydraulic Div. 94 (HY4), pp 909-923.

AGGREGATION AND DISAGGREGATION MODELLING

WILLIAM L. LANE
Bureau of Reclamation
P.O. Box 25007
Denver, Colorado 80225
USA

ABSTRACT. Stochastic models that simultaneously treat a time series at several time scales have become popular for both research and practical applications. This paper covers two multilevel hydrological time series methods. Aggregation begins with models for the smaller time increment and infers the appropriate models and parameters for the aggregated time series. These models show promise for improved time series model choices and more reliable parameter estimates. Disaggregation is stochastic generation that starts with a previously generated aggregate series and subdivides or disaggregates that series into a finer scale time series. Disaggregation modelling has proven to be a very practical approach, especially for multisite and multivariate analysis.

1. Introduction

Two special forms of stochastic models in the last two decades have become important for the study and modelling of hydrological times series. These two models, aggregation and disaggregation, differ from other existing and more traditional models by their unique treatment of the data on two or more levels or scales. This distinction brings about several new problems but also tremendous advantages to the practitioner applying stochastic techniques to water resources modelling.

Disaggregation models were developed first and their use has grown rapidly during the last 15 years. These approaches are known mainly for being practical. To appreciate these models, one first has to examine the reasons for the creation of these models. In the early development of stochastic time series modelling, a single time series was analyzed, studied and modelled. Typically, this would be a time series of annual data at a single site. Later, two series were modelled together. This typically would be annual data at two sites on two different rivers. First, annual data was studied, then later seasonal data at one site, and then multivariate multiseasonal data. In each of these evolutions, the complexity and the number of separate data series that were being simultaneously treated expanded. Likewise, the number of parameters quickly became a major problem. Additionally, the use of more and more

97

J.B. Marco et al. (eds.),
Stochastic Hydrology and its Use in Water Resources Systems Simulation and Optimization, 97–116.
© 1993 *Kluwer Academic Publishers.*

complex models, such as long term memory models, posed complexities that were not manageable with existing models. Disaggregation models offered an approach for reducing the complexity of the modelling while still allowing the use of long term memory models and the use of multivariate times series. They also allowed for the modelling of seasonal data. The goal of any disaggregation model is to generate data while preserving statistical properties of the data at more than one level. Disaggregation modelling is presently a practical approach to solving the myriad problems associated with multivariate multiseasonal generation of synthetic data.

Aggregation modelling is a more recent area of study. It was initiated around ten years ago with most work done in the last five years. Technically, it is really an area of study rather than a generating technique. Again as with disaggregation, the concern is with data at more than one level. In this case, the process is to study the effects of aggregating or combining data. The classic example is a study where the seasonal model is assumed as known. The form and parameters of the corresponding model for the annual data are mathematically derived. The annual data of course is completely defined by the seasonal data. Likewise, an annual time series model is completely defined by the seasonal time series model. While this type of study may not be generational modelling, it is still potentially of great importance. It is a process to determine theoretically the form of compatible models at several levels. In many respects aggregation is the reverse or mirror image of disaggregation. As such it may provide a more theoretical background for the selection, form and parameters estimation procedures for disaggregation modelling. In many ways, aggregation studies could be considered to be theoretic studies that are needed to help in producing future improvements to disaggregation modelling techniques. These studies may provide the underpinnings for more theoretically sound disaggregation modelling.

A reasonable approach is to recognize the value of both aggregation and disaggregation and to work both together in a form of synergism. The two areas have many mathematical similarities and share the same concerns and problems. Both lend themselves best to a matrix notation and rely on the same study of covariances and statistical preservation. Jointly, the two form what could best be termed multilevel modelling, which is a growing and exciting area for both research and practice.

This paper is organized as follows. First is a discussion of multilevel modelling in general. Then aggregation modelling is covered, followed by disaggregation modelling and finally conclusions.

2. Multilevel Modelling

In multilevel modelling, information or knowledge is transferred or preserved from one level to another. Properties are being preserved at more than one level, say both monthly and annually. The levels may also be referred to as scales or horizons. In both disaggregation and aggregation, the preservation of the statistical properties of both the annual and the monthly data are the goal. The levels may also be in a spatial sense. There may also be multiple layers such as the case of daily data that aggregates to form weekly data which aggregates to form monthly data. Usually only two layers are treated. The aggregated series is also sometimes called the upper, coarse scale or base series, while the disaggregated series is also called the lower, fine scale or subseries. Both may at times be called generated series depending upon what generating process is used. Also, they may both be called original series depending on whether the upper or lower layer series is analyzed first.

TABLE 1. Interchangeable Terminology

Lower Series	Upper Series	Levels
fine	coarse	scales
disaggregated	original	horizon
original	aggregated	layer
subseries	base series	step

2.1. ELEMENTARY MULTILEVEL MODELS

To examine multilevel modelling we will first start with a model without complications and slowly expand to the more complex applications. Multilevel modelling is really a special form of multivariate modelling. Virtually all modelling in this field is based upon the style and concepts put forth by Matalas (1967a) in what must be considered a classic paper in stochastic hydrology. Therefore, it is a logical place to begin a mathematical analysis of multilevel processes. The multivariate model is usually given in the form of a matrix equation to conserve space and to be clear and concise. Starting with an equation in the form:

$$Y = AX + BE \qquad (1)$$

where Y is an m by one matrix of annual values for the current time period for m sites, X is an m by one matrix of annual values for the previous time period, A and B are m by m matrices of model coefficients, and E is an m by one matrix of random components. The values in Y and X are standardized values obtained by subtracting the appropriate mean values from the original or actual values. The Matalas form is easily utilized to generate data for many situations. It is in a linear dependency model form, or more specifically a multivariate linear autoregressive model. If A is taken to be diagonal, the approach degenerates into one used by many where the serial correlation in each series is treated separately from the contemporary cross correlations (Matalas, 1967).

To use this same form of equation in the form of a multilevel analysis, consider the case of monthly data at a single site. With the same equations, Y and X become column matrices composed of monthly data at all sites. This is similar to an approach taken by Moreau and Pyatt (1970). Now let us examine the number of parameters. A and C have 12 by 12 dimensions for a total of 300 parameters (2 times 12 times 12 plus 12 for the mean values). This model will preserve all of the covariances and variances between the seasonal values. It can be shown that values are also automatically preserved for the annual series. It should be noted that in doing this exercise some common assumptions were made. These deal with linearity of dependency, distribution form of values and possible transformations, stationarity and the basic dependency structure of the time series (time series model). Often the data are reduced first to a zero mean and less often to a unit variance. The data are usually transformed to follow the normal distribution. If this problem is expanded to five sites with weekly data, the number of parameters increases to nearly 70,000. This is obviously too many parameters. One practical approach is reducing the number of parameters by using a disaggregation approach. A second, but currently less practical approach, is to use a monthly model such as an AR model on a monthly basis.

With any approach, some assumptions will have to be made with regard to a time series model that allows for a model structure with less parameters. Some of the observed correlations (dependencies) will then be ignored or, more correctly, indirectly and approximately preserved. The choice of model must be coordinated with the observed actual data. Theoretic approaches, such as aggregation studies, provide guidance to model changes and preservation of covariances.

2.2. PROBLEMS OF SCALE

There are some problems associated with selecting models for stochastically generating data at two different time scales. Preservation of the means, variances and contemporaneous correlations are not a significant problem. From a stochastic time series analysis standpoint, the problems come in the modelling or preservation of dependency serially between values at lagged time intervals and also in the preservation of the distribution shape. These are the aspects of hydrological time series that are altered with changes in the level or scale. This problem is analogous to the problems in rainfall runoff modelling or in frequency analysis when the scale is changed. In those fields, problems of time scale and data analysis at several levels are just beginning to receive attention.

Why are there differences in time series behaviour when looking at the same data at differing time units? First, there are obviously some differences in the physical causes that dominate the persistence found in hydrology data. Thus the cause of the serial correlation changes with the change in scale. At short time increments, weather persistence has a high level of effect. Season also has an effect. Basin size will be ignored in this discussion but also has a scale effect. In going to longer time periods, the individual storm and weather persistence is reduced, but the effects of seasonal tendencies and the carryover of soil moisture are of course still very strong. With annual data, the effect of seasons and harmonic behaviour is removed. Carryover on dependency due to the short term weather and minor upper level soil moisture is reduced to a great extent. Dependency is due more to overall soil moisture and groundwater influences.

As a result of differing physical causes, or more correctly differences in the relative magnitude of the different physical causes for the dependency, there is a different domination of the serial correlation process. At the very short time periods, a very strong AR (autoregressive) behaviour would be expected, perhaps with some MA (moving average) behaviour resulting in ARMA (autoregressive moving average) behaviour. As the time scale increases, the MA component would also increase in influence, perhaps peaking at the weekly time scale and then dropping off with increasing time scale. At all time scales, the AR behaviour would tend to account for the largest portion of serial correlation, With annual data, high frequency short duration influences are filtered out. At this scale, cycles are no longer expected.

2.3. TIME SERIES MODEL CHOICE

One very important point needs to be emphasized regarding model choice. From an information standpoint, a great deal of knowledge is gained by the choice of a model. Note that this information may be either good or bad. Assuming the choice is valid, the model greatly restricts the behaviour of the generated data and will improve the results. The choice confines and limits what a stochastic model will generate. Model choice includes the distribution forms chosen, the correlation model, and especially any parameters that are assumed to take on certain values (such as zero) or are defined in terms of other values. The model choice should be based on experience, tempered with physical reasoning and theoretical studies in statistics and time series modelling.

3.   Aggregation Modelling

3.1. INTRODUCTION TO AGGREGATION MODELLING

Aggregation modelling has been a relatively recent addition to the field of stochastic hydrology and its true impact is not yet known. It may actually provide new models for time series. It may provide improved parameter estimates or it may simply provide some theoretical support for modelling assumptions and explanation for time series behaviour that has been observed in nature.

Aggregation modelling is a study technique by which an assumption is made regarding the basic form of the original or lower level time series and the properties of the aggregated time series are calculated. For example, by studying the monthly time series models, one can infer or learn something regarding the behaviour of the annual time series. Past studies have tended to show promise in two very important areas. One, if the lower level time series is assumed to take a certain form, the form of the upper level series can be determined. The second, also relying on the same assumption, is that the parameters for an annual model possibly may be better estimated. The study of aggregation modelling will tend to provide important theoretical reasoning needed to understand and make use of the observations of time series at differing time levels. Both temporal and spatial aggregation are possible and the study of both are important to the study of transfer of knowledge.

Thus far, the aggregation studies have been limited mainly to the study of preservation and behaviour of the autocorrelations. Study has not yet been directed towards the effects of aggregation on the distribution of the values. Aggregation has been a study of how a simple model defined at the lower level propagates and changes as the data is aggregated. For example, daily data aggregates to weekly data or monthly data and those aggregate to give seasonal data and on to annual data. If the correct model is used, obviously all the aggregated models will also be correct and will fit the actual data. One important fact is that the lower level time series defines exactly the upper level time series. That is, the lower level contains all of the information. With that accepted as a basic fact, it follows that the correct lower level model defines all time series models for any and all aggregated time scales. On the other hand, even if the model is incorrect, the aggregation study may still be very valuable. Deficiencies experienced in the aggregated model are due to correlation structures that are not accounted for in the low level model. Study of those deficiencies may provide a key to finding the necessary model structure.

3.2. HISTORICAL BACKGROUND IN AGGREGATION MODELLING

The first work to provide leadership in the study of aggregation was Kavvas et al. (1977) They provided the basic direction that has been taken for practically all subsequent study of aggregation. They correctly noted that the stochastic structure of aggregated times series can be derived directly from the structure of the short time series. Their study concentrated on the derivation of the aggregated series lag covariance structure. Although they do not actually derive time series models for the upper level series based upon the original lower level series, they do provide the basic concept since the complete covariance structure does define the time series model. They also make the point "One can then construct the time series models corresponding to these larger intervals from the theoretically calculated covariances and spectral functions without ever constructing the data for larger intervals." They also make a good point that the aggregation process is in fact a many to one process whereby the aggregated time series is defined exactly by the original series. In disaggregation, the process is not a direct transformation. That is, there are many (an infinite number) sets of disaggregated values that will aggregate to give the same higher level value.

In calculating the aggregated series from the original series, there are several possible schemes. One simply is to add up the original values in each year. However, there are approaches that are more appealing to the mathematically inclined. For example, one can create a series which is the cumulative sum of all previous original values and then by taking differences at the appropriate lag and choosing resulting values at the proper interval arrive at exactly the same aggregated time series. Another approach (Kavvas et al., 1977), starting with the original monthly series, is to pass a moving sum through the series and then, by selecting every twelfth value, obtain the aggregated series. With only some fairly simple reasoning, this moving sum approach provides an interesting intuitive insight into what properties the aggregated series might have. From a statistical standpoint, the moving sum series is in fact the same as moving average, the two differing only by a scale constant. If a moving average is taken of a set of serially independent monthly data, the moving average or aggregated series would follow an MA (moving average) model. Likewise, if the moving average was taken for an autoregressive original series, the result would be expected to follow an ARMA model. Now if every twelfth value was selected, the result might still be ARMA but with only a very weak MA component since the only averaged portion that remains is the influence carried from one year to the next by the first month which is dependent only upon the last month of the previous year.

Obeysekera and Salas (1982) showed that starting with a seasonal AR model would result in an ARMA(1,1) model for the annual series. One must however be cautious and point out that the conclusion is based upon the assumption that all of the annual dependency is due only to a simple seasonal model.

Vecchia et al. (1983) continued on the same line but generalized further by considering periodic ARMA(1,1) models for the monthly data. Again it can be shown that the aggregated annual series is indeed ARMA(1,1). An interesting result was their finding that parameter estimation for the annual series model was improved by using the monthly data and the monthly time series models. Analysis was performed using only simulated data. From a practical standpoint, some caution must be exercised because of the model assumptions that are implicit in these approaches. It was noted that the aggregated ARMA(1,1) was often very close to an AR(1). This is an important result for two reasons. One is that it supports simple intuitive reasoning for model behaviour changes with scale. Second, although it is based on results with simulated data, it seems to agree with observations concerning actual annual data. Perhaps with future research, the MA portion could be directly tied just to the monthly modelling structure that spans the boundaries from one year to the next. Vecchia et al. also noted that the aggregation process, even with very highly serially correlated monthly models, does not necessarily produce strong serial dependence at the annual level. Again this is in keeping with actual observations.

Rao et al. (1985) continued along the same line. Again, the AR portion of the annual ARMA seems to dominate. They primarily looked at forecasting where it is shown that a completely correct monthly model will outperform the simple annual model that is based on the same monthly model. This is hardly surprising since the only source for dependency is largely ignored in the annual model. From a practical standpoint, it would be very surprising to find an annual time series model actually used for forecasting without at least some additional terms for preserving dependency with the last month or months of the previous year. This fairly long paper, while largely relying on previous work by others, helps to provide background. Rao et al. (1985) also cautioned that they felt that it would not be possible to find relationships between very different time scales because of the large number of values being averaged (or equivalently summed). This seems to imply some doubt that the true lower time scale model could indeed be correctly assumed that would account for all of the dependency present at the large time scales.

The cornerstone of aggregation modelling is best stated by Obeysekera and Salas (1986): "It is only natural to think that, if one could reasonably model a hydrological process defined in continuous time, then all aggregated processes derived from it could be (at least theoretically) defined." They continue in this paper to examine the stationarity and invertibility conditions that govern the aggregated time series given a seasonal series that is periodic AR(1) or periodic ARMA(1,1). Their paper makes what is perhaps the clearest derivation of the model for the aggregated time series. They also make use of actual data on the Niger River so this paper contains one of the first actual tests of the results of previous aggregation studies. The results were not too encouraging. The annual correlograms, when based upon the assumed seasonal models, differed greatly from the actual annual correlogram. The results were better when based upon a periodic ARMA(1,1) than when based upon periodic AR(1). Results quickly became much worse when the number of seasons in a year were increased. The annual data showed much higher correlation than could be predicted by the assumed seasonal models.

Further basic studies are contained in Bartolini et al. (1988) where the previous studies are extended to apply to multivariate ARMA(1,1) original series. The estimation of parameters for multivariate ARMA(1,1) models has always been prone to serious computational difficulties. Therefore these models are often simplified to CARMA(1,1) models that are computationally much easier but neglect all of the lagged cross dependencies between the different series. These models were focused on in this paper. In general, the ARMA(1,1) and CARMA(1,1) (contemporaneous autoregressive moving average) models aggregate to form ARMA(1,1) models. This paper relied on synthetic data, but subsequent studies are planned for utilizing actual data.

3.3. GENERAL AGGREGATION MODEL

As far as a general aggregation model is concerned, there is only one model and it is not actually a time series model. It can be written as

$$X = A Y \tag{2}$$

where X is the aggregated series at one time, Y is the corresponding subseries values for the same time in a column matrix and matrix A is a row matrix composed of ones. If the problem is a multivariate problem, with m sites and n seasons, then the dimension would be one by m for X, n by m for Y, and one by n for A. In aggregation modelling we are really studying the changes in the time series and associated times series models as it is transformed from the original, Y series to the X series. This model is useful however because it is indeed the basis for determining these changes. This model is also useful for verifying and showing how expected values, means, variances and covariances are transformed from one level to another. For notation, $Sxy(1)$ is defined to be the expected value of the matrix product of X with the transpose of Y from the previous time period before. Then, using the general aggregation model, it can be shown that $Sxy(1) = A \ Syy(1)$ where $Syy(1)$ is likewise defined. This type of derived relationship is the basis of aggregation studies. The matrix products are sometimes called the moment matrices. They are moments about zero. If the data have first been standardized to means of zero, they will contain covariances. It should be noted in passing that this exact relationship given by Equation 2 only holds for data that has not been transformed. In most actual applications, the time series models are fitted to data that have first been reduced to normal distribution.

All aggregation studies to date have assumed the form of the monthly or seasonal time series model and have examined what that assumption tells us about the aggregated series. This is a relatively straightforward approach although the mathematics can become a substantial chore. The process becomes more involved as the assumptions about the lower level series are relaxed and as the model is further complicated. For example if the lower level model is assumed to be a simple AR(1) model, the task is relatively simple. It would be much more complicated if the lower level model were a multivariate periodic ARMA(1,1) model. All of the dependency structure applied thus far is linear in form. As such it lends itself very well to data which already is normally distributed. However, that is not very realistic and will undoubtedly become a major stumbling block in aggregation studies, much as it has in other time series modelling and generating approaches.

3.4. PROPERTIES OF THE AGGREGATION MODEL

It is worthwhile to examine the preservation of statistics as time series are aggregated. Starting with the equation of the general aggregation model (Equation 2), it is easy to take expected values and show that the mean of the aggregated time series is automatically preserved as long as the means of all of the seasons are preserved:

$$E[X] = A\ E[Y] \tag{3}$$

where E[ ] is the expected value operator. If both sides of the equation are post multiplied by the transpose of itself, we obtain

$$E[XX'] = A\ E\ [YY']\ A' \tag{4}$$

or

$$Sxx = A\ Syy\ A' \tag{5}$$

using the moment matrix notation just given and where the prime is notation for a matrix transpose. Examination shows that the moment for the aggregated value at a site is equal to the sum of all of the corresponding lower level moments for the same site. Likewise it can be shown that lagged moments in the aggregated series are simply equal to the corresponding lagged moments for the lower level series. If the discussion is limited to a single site, all of the moments in Syy will need to be preserved by the lower level model in order to preserve Sxx in the upper level. However, most lower level time series models will not preserve all of the moments in Syy. Rather, the models will only directly preserve a small fraction of these moments. By "directly preserving", this requires that the time series model has the structure (parameters involving the relevant variables) for the corresponding dependency and that the moments are used in the estimation of the parameters. If the lower level time series model has indeed the right structure (contains all of the dependency structure in the time series), then all of the moments in Syy will be preserved. A few will be directly preserved and will have associated with the parameters, while all of the others will be indirectly preserved. While it is not often done, it is possible to calculate the indirectly preserved moments (see Lane, 1982b). If the indirectly preserved lower level moments are compared to the sample moments, while taking into account the expected sampling variation, a judgement may be made as to the adequacy of the lower level time series model for preserving upper level characteristics.

3.5. AGGREGATED TIME SERIES BEHAVIOUR

While aggregation is a relatively new area that is virtually all research without practical application to date, some trends and conclusions can be noted. First, ARMA models seem to be a result for the aggregated series regardless of the form of the assumed lower level time series model. In some regards this is disappointing since multivariate ARMA models present some very difficult parameter estimation problems that have not been solved despite more than a decade of attempts. Next, the AR portion of the ARMA model for the aggregated series seems to be very dominant over the MA portion. This is encouraging since, if the MA effect is small and can be neglected, an AR model is very much easier to apply. Third, results to date indicate that the lower level model may need to be much more complicated than AR or ARMA in order to adequately preserve upper level properties. In general, the annual correlations are anticipated to be much higher than would be obtained from the aggregation process for simple AR(1) or AR(2) lower level models, especially when several seasons are used such as at the monthly time scale. Fourth, as has long been known in the area of flood frequency and is supported by theoretic arguments, the distribution of aggregated time series tends in the direction of the normal probability distribution. Fifth, the model type and the model parameters can be expected to change in a smooth and gradual and somewhat predictable manner as the time scale or level changes.

3.6. EXAMPLE OF AGGREGATION MODELLING

For an example, assume that there are two seasons to the year and the lower level model is AR(1). To simplify the calculations and not be caught up with minor housekeeping problems, the seasonal data is standardized to zero mean and unit variance. Further, the seasonal lag one serial correlation is taken to be constant for both seasons with a value of r. If X is the annual series and Y is the seasonal series, then the seasonal covariance matrix is Syy and its elements are given by

$$Syy = \begin{matrix} 1 & r \\ r & 1 \end{matrix} \qquad (6)$$

Accordingly, the variance of the annual series is given as

$$Sxx = 2(1+r) \qquad (7)$$

which is simply the sum of all of the elements of Syy. The annual lagged covariance can be easily calculated in terms of the seasonal moments. One way is to create an expanded version of Syy as a seasonal covariance matrix that includes all of the covariances pertinent to the annual covariances of interest. Seasonal covariances are required for two full years lag for the annual lag one covariance and three full years for the annual lag two covariance. As an example, the seasonal covariance matrix for the annual lag one case, which we shall notate as Syy(0+1), is as follows

$$Syy(0+1) = \begin{matrix} 1 & r1 & r2 & r3 \\ r1 & 1 & r1 & r2 \\ r2 & r1 & 1 & r1 \\ r3 & r2 & r1 & 1 \end{matrix} \qquad (8)$$

where r2 indicates the lag two covariance (in this simple case using an AR(1) seasonal model, $r2 = r^{**}2$ and $r3 = r^{**}3$, where "**" indicates exponentiation). It can be noticed that the matrix Syy(0+1) can be segregated into four separate submatrices. The upper left and lower right two by two matrices are in fact Syy(0) while the remaining upper right and lower left are Syy(1). Sxx(1) is equal to the sum of the elements of Syy(1), or

$$Sxx(1) = r1 + 2\ r2 + r3 \tag{9}$$

or equivalently

$$Sxx(1) = r\ (1+r)^{**}2 \tag{10}$$

This result can also be obtained by calculating the expected value of the product of X with itself lagged one time period (one year). In doing this calculation, X is replaced by the sum of the two seasonal values in the year. It can be noted that Syy(0+1) is identical to what would be obtained for Syy if there were four seasons to the year. In a similar manner, the annual lag two covariance is determined to be

$$Sxx(2) = r^{**}3\ (1+r)^{**}2 \tag{11}$$

Next the annual serial correlation can be calculated by dividing the lagged covariances by the variance as given by (7). Denoting the annual correlation coefficients by R1 and R2,

$$R1 = r\ (1+r)\ /\ 2 \tag{12}$$

and

$$R2 = r^{**}3\ (1+r)\ /\ 2 \tag{13}$$

or

$$R2 = r^{**}2\ R1 \tag{14}$$

Similarly, this same calculation can be performed for higher lags. The correlation at lag k is given by

$$Rk = r^{**}2k\ R1 \tag{15}$$

Thus the correlogram is composed of a value for lag one (Equation 12) with an exponential reduction in values for higher lags (Equation 15). The reduction factor is equal to $r^{**}2$. This is exactly the form expected for an annual ARMA(1,1) model (see Salas et al., 1980; page 194). This alone is proof that a two season AR(1) model aggregates to an ARMA(1,1) annual model.

Although not an aggregation or disaggregation approach, persons interested in advanced study of multilevel models should read Moreau and Pyatt (1970). Their approach is similar in form to the extended approach of Mejia and Rousselle (1976) except that the current annual term is not included. This model, in the process of generating weekly values, preserves certain statistics of monthly flows. Moreau and Pyatt also noted the similarity between their model and the autoregressive models.

4. Disaggregation Modelling

Disaggregation has become a major technique for modelling hydrological time series. Multivariate applications and dissatisfaction with lower level models prompted the development of these applications. Harms and Campbell (1967) proposed essentially the first disaggregation approach. At that time, they termed their approach an extension of the Thomas-Fiering model (Thomas and Fiering, 1962) which was popular at that time. The Harms and Campbell model, while not having the mathematical sophistication or completeness of current disaggregation models, does possess most of the qualities desirable in disaggregations models. It did provide a method for generating the proportions needed to disaggregate an annual flow value. This model comes surprisingly close to producing the desired results, but it has never caught on because of its very obvious theoretic shortcomings. If disaggregation is viewed as a form of generating proportions which is not greatly unlike an interpolation scheme, it is not surprising that many very different approaches perform well. The concept itself is very robust with regard to the disaggregation model choice. The first well-accepted model was presented by Valencia and Schaake (1973). This model's introduction represented the true beginning of disaggregation as a technique for hydrological time series generation. Most importantly, because of its classic form, this approach provided a basis for almost all subsequent proposed disaggregation approaches. Mejia and Rousselle (1976) rated the Valencia and Schaake approach as "a significant benchmark in the literature of hydrological time series". Mejia and Rousselle proposed an expansion of the basic Valencia and Schaake model to overcome shortcomings of that model. Lane (1979), in developing a comprehensive computer package for the practical application of stochastic techniques to hydrological data, proposed several additional improvements. Procedures were added to allow for spatial disaggregation, and several methods were added for greatly reducing the number of required parameters. Later the package, named LAST, was improved by implementing a correction to the Mejia and Rousselle parameter estimation procedure (Lane, 1982b), and even later a staged seasonal disaggregation option was added (Frevert and Lane, 1985). As changes and additions were made to the computer package, the users manual was revised five times (Lane, 1980; Lane, 1982a; Lane, 1983; Lane and Frevert, 1985; Lane and Frevert, 1988). Salas et al. (1980) contains a thorough review of the disaggregation techniques available at that time. Santos and Salas (1983) proposed an improved disaggregation approach known as the STEP model.

Stedinger (1983) provided an excellent summary of areas of concern that required further improvement and study. Subsequently the areas of concern became the topics for a series of papers authored by Stedinger and various coauthors. One concern dealt with which persistence properties are most important in a multisite, multiseason disaggregation and what form models to do this should take on (Stedinger and Vogel, 1984). The second dealt with the need for keeping the number of parameters small (Stedinger et al.,

1985b; Grygier and Stedinger, 1988). Adjustments required to assure that the seasonal values would indeed add to give the original annual values received attention in Grygier and Stedinger (1988). The need for addressing and a method of treating parameter uncertainty was the fourth concern and this received attention in Stedinger et al. (1985b). Finally the fifth concern dealt with the need for a staged disaggregation approach (Stedinger et al., 1985a). Stedinger and Vogel (1984) also expanded the study of the original parameter estimation problem associated with the Mejia and Rousselle model (1976) and confirmed that the Lane (1982b) correction was a valid approach. Stedinger et al. (1985a) also made application of a long term persistence model for the upper level series. Grygier and Stedinger (1985) produced a software package entitled SPIGOT which embodies most of the procedures reported in the series of papers by Stedinger and his coauthors. While all of the concerns by Stedinger (1983) were recognized by Lane (1980), the SPIGOT programs actually implemented several items that the LAST model lacked. These included parameter uncertainty, ARMA modelling for the upper series, the use of an improved model to better preserve additivity, and better additivity corrections for the errors that remained.

The basic goal of any disaggregation model is to allow the preservation of statistical properties at more than one level. Disaggregation modelling has two additional attributes which make it attractive. First, it is a technique which allows for a reduction in the number of parameters with little or no corresponding loss of desirable properties in the generated data. Second, disaggregation allows for increased flexibility in the methods used for generation. For example, at the annual level, long-term persistence properties may be important. Disaggregation facilitates the use of long-term persistence models, particularly in multisite multiseason time series modelling.

On the other hand, disaggregation is not really well based physically or mathematically. Disaggregation is an approach which is an expedient way to generate very reasonable data samples with all of the "important" statistics preserved, with a minimum number of parameters, with errors in the process "hidden" in inconsequential locations within the time series, and with an easily followed and understood approach that is very versatile.

Disaggregation is an approach that grew not from a theoretic background but from a practical need. It uses, usually but not always, a simple annual model. The disaggregation model in effect only generates the proportions by which the already generated annual data are distributed within the seasonal time segments. Typically, the number of directly preserved properties are reduced as much as possible. Checks are made by examining the indirectly preserved properties and looking for agreement within expected sampling variation. An important concept that is not often stated in modelling is the concept of sweeping errors under the rug. That is, to spread errors to where they are neither noticeable nor important. This ties in with the relevance of the eventual use of the generated data. Errors or lack of completeness in the time series models may be either a real problem or inconsequential, depending on the final use of the data.

4.1. SIMPLIFICATIONS TO DISAGGREGATION

One simplification is using disjointed or uncoupled estimation. This allows for the deterministic parts of a model to be removed step by step with only a residual stochastic random term. For a long time this was the prevalent approach to times series modelling.

A second simplification is a staged disaggregation with multiple steps to reduce the number of parameters required, while keeping direct preservation of important statistical properties not only at the highest and lowest levels but also at intermediate levels. Of course, in doing so, more properties are relegated to indirect preservation.

A third simplification is using spatial disaggregation to reduce the number of parameters in multivariate cases. This is especially good where several tributaries in one basin are being generated and the aggregated or total basin flow can be generated first.

4.2. DESCRIPTION OF DISAGGREGATION MODELS

Disaggregation modelling is a process by which time series are generated dependent on a time series already available. Typically, the independent series is disaggregated, or broken apart, into subseries. Often, as in the case of seasonal disaggregation, the subseries (seasonal data) sum exactly to give the original series (annual data). The original series will be referred to here as the key series. The generation of the subseries dependent upon the key series is accomplished by using a linear model designed to preserve important statistical properties both between elements of the subseries and between the key and subseries elements. In this manner, statistical properties are preserved at both key and subseries level, and relationships between the two levels are maintained.

The two basic forms of disaggregation modelling are temporal and spatial. A temporal example is the disaggregation of an annual time series into a seasonal time series. Additional examples of temporal disaggregation include annual (key) to semiannual (sub), semiannual (key) to biweekly (sub), monthly (key) to biweekly (sub), annual (key) to monthly (sub), and annual (key) to biweekly (sub). An example of spatial disaggregation is the disaggregation of the total natural flow of a river basin into individual tributary flows. This is a classic use of spatial disaggregation.

4.3. GENERAL DISAGGREGATION MODEL

All disaggregation models can be reduced to a form which may be termed the linear dependence model. The linear dependence model is

$$Y = AX + BE \qquad (16)$$

where Y is the current observation of the series being generated (subseries or dependent series); Y is generated dependent on the current value of the X series (key series or independent series); E represents the current value from a completely random series (stochastic term); and A and B are the parameters (causal structure). Disaggregation models are a subset of the linear dependence model, as are autoregressive models. It should be noted that for these models, the X and Y series are transformed values that follow the normal distribution with a mean of zero. The assumption of zero mean is necessary with models in the form presented. It is possible to omit this assumption only if an additional column matrix is added to the basic equation. In terms of the linear dependence model (Equation 16), the form would now become

$$Y = AX + BE + C \qquad (17)$$

where C is parameter matrix with the same dimensions as Y. The values estimated for A and B would remain unchanged.

4.4. SINGLE SITE DISAGGREGATION MODELS

Three forms of basic single site temporal disaggregation are presented here: the basic model, an extended model and a condensed model. Only a single level of disaggregation will be presented, but it should be kept in mind that all forms can be cascaded or staged into a multilayer approach.
The basic model has the form (Valencia and Schaake, 1973)

$$Y = AX + BE \qquad (18)$$

For the case of disaggregating annual flows into seasonal flows, X is the annual flow value and Y is a column matrix containing the seasonal flow values which sum to X. Although X and Y are referred to as flow values, it should be remembered that they have zero means and are transformed values. The advantage of this model is its very basic clean form. This makes it the easiest of the three models considered here. The causal structure of this model is designed to preserve covariances between the annual value and its seasonal values and to preserve variance and covariances among the seasonal values. With n being the number of seasons, a total of n*n seasonal variance and covariance moments are preserved. One disadvantage is that the moments being preserved are not consistent. That is, the value for the last season of the year is generated preserving all covariances between itself and the preceding n-1 seasonal values, while the value for the first season is generated without preserving covariances between itself and any preceding seasonal values. A second disadvantage is that the number of parameters is large.
The extended model, developed by Mejia and Rousselle (1976), is an extension of the basic temporal model. An additional term is included in the model to preserve the seasonal covariances between the seasons of the current year and the seasons of the past year. The model takes the form

$$Y = AX + BE + CZ \qquad (19)$$

where Z is a column matrix containing as many seasonal values from the previous year as are desired and C is an additional parameter matrix. All other terms remain the same as for the basic model. Although this model is slightly more involved than the basic model, it is still quite straightforward and clean. Parameter estimation is considerably more complicated. The problem of an inconsistent causal structure is not corrected by the additional term, and the problem of an excessive number of parameters is made worse.
Lane (1979) has developed an approach which essentially sets to zero several parameters of the extended model which are not considered neces-

sary. This model here is referred to as a condensed model. The number of parameters to be estimated, and accordingly the number of moments preserved, are thus reduced drastically. The approach uses the extended model form, but on a one-season-at-a-time basis and with only one lagged season. The model equation can be written as

$$Y_s = As\ X + Bs\ E + Cs\ Y_{s-1} \qquad (20)$$

The small s denotes the current season being generated. Thus, if there are n seasons, there are n individual equations following the form of Equation 20. Also, there are n sets of parameters As, Bs and Cs. For the single site case, the parameter matrices are all single element matrices. This model is designed to preserve covariances between the annual value and its seasonal values, and to preserve variances and lag-one covariances among the seasonal values. The main advantage of this model is the reduction of the number of parameters. There are two major disadvantages. First, the model is not as clean and straightforward as the previous models. The second disadvantage is that since all seasons are not generated jointly, the seasonal data will not add exactly to give the annual time series. This problem can be dealt with in two ways. One is to assume the seasonal data are satisfactory and to recalculate the annual time series. The second and generally more preferable approach is to adjust the seasonal flows so that they add up exactly to the annual values. The benefit of the parameter reduction far outweighs these shortcomings. It should be noted that this same additivity problem is common to all of the approaches, basic, extended and condensed, if the data have undergone any transformations. Since seasonal data are seldom normally distributed, this nonpreservation of additivity is the rule rather than the exception. While Lane had recognized the need for adjusting the disaggregated data in order to preserve the highly desired additivity property, the magnitude of the problem was not well known. Later study and experience proved that the magnitude of the adjustments were often objectionably large. As a result work began to create models which would better address this problem. Both Santos and Salas (1983) with their STEP model and Stedinger et al. (1985b) with the SPIGOT model and Grugier and Stedinger (1985) have produced significant improvements in this area by adding one additional term to the disaggregation model.

4.5. MULTISITE DISAGGREGATION MODELS

Simply stated, the only difference between the multisite disaggregation and the single site methods just discussed, is that the matrices are larger. The multisite case does not require any new model forms, but only some minor complications because the moment matrices are much larger. The biggest effect is that the number of parameters and moments involved increase at a rate roughly proportional to the square of the number of sites.

Since the effect of going from single site to multisite applications is basically the same for all approaches, only the extended model will be examined. The model still has the form

$$Y = AX + BE + CZ \qquad (21)$$

Now, however, X is not a single value, but is a column vector which contains one annual value for each site. Likewise, the number of generated values in Y have increased so that Y now contains a complete set of seasonal data for each site. The number of stochastic terms contained in matrix E are also increased. If there are n seasons and m sites, and z seasons included in Matrix Z, then the dimensions are as follows:

Matrix Y, mn by one	Matrix A, mn by m
Matrix X, m by one	Matrix B, mn by mn
Matrix E, mn by one	Matrix C, mn by mz
Matrix Z, mz by one	

The advantage of a multisite approach over multiple applications of a single site approach is the preservation of some additional correlations. These are cross-correlations between the values at the various sites. The disadvantage is that in order to preserve these additional correlations, additional parameters are also required.

4.6. SPATIAL DISAGGREGATION MODEL

The form of the spatial disaggregation model used by Lane (1979) is

$$Y = AX + BE + CZ \hspace{3cm} (21)$$

where Y is a column matrix of current substation annual values being generated, X is a column matrix of current key station annual values (note that the model is not restricted to only one key station), and Z is a column matrix of the previous substation annual values. A, B and C are parameter matrices. This model is very similar in form to the Mejia and Rousselle temporal disaggregation model upon which it is based. As with all disaggregation approaches, the spatial disaggregation can be staged.

4.7. PARAMETER ESTIMATION FOR DISAGGREGATION MODELS

The parameters of disaggregation models have traditionally been estimated using the method of moments. This technique is rather straightforward in its application. It avoids trial and error, iterative procedures, and implicit solutions which are found in some other parameter estimation approaches. For this reason alone, it is a particularly appealing approach for the practitioner. This does not mean that it produces the "best" parameter estimates. More sophisticated techniques may produce better estimates in some senses, but these techniques have not yet been developed in the practical application of disaggregation methods. For the purposes of disaggregation, the method of moments estimates are probably more than adequate.

The method of moments estimates will be presented only for the basic disaggregation model. The estimates are presented without derivation. For other model parameter estimates and for the derivations of the estimates, the reader is urged to refer to Valencia and Schaake (1973), Mejia and Rousselle (1976), Salas et al. (1980), Lane and Frevert (1988), Grygier and Stedinger (1985), and Loucks et al. (1981).

4.8. PARAMETER ESTIMATION FOR THE BASIC DISAGGREGATION MODEL

The basic model has precisely the form of the linear dependence model. The estimates of the parameter Matrices A and B of the basic temporal disaggregation model (Equation 18) are given by

$$A = Syx \; Inv[Sxx] \tag{22}$$

$$BB' = Syy - Syx \; Inv[Sxx] \; Sxy \tag{23}$$

where Inv[ ] denotes matrix inverse. B is easily solved from BB' using any of several fairly standard methods. The parameter estimates are exactly the same as those for the linear dependence model. Now, however, meaning can be given to the three moment matrices involved. Syy is the matrix of covariances among the seasonal series, Syx is the matrix of covariances between the seasonal series and the annual series, and Sxx is the matrix of covariances among the annual series. The form of this model is designed to preserve Syy and Syx, while Sxx is already assumed to be preserved in the annual series. It can be noted that Syy completely specifies Sxx and Syx if no data transformation has been made.

5. Conclusions and Closing Remarks

Disaggregation models have existed for more than a decade as a very viable and practical approach to multisite multivariate modelling, especially when preservation of properties is desired or necessary at more than one horizon or level. Recent developments have improved abilities in the areas of parameter estimation, parameter uncertainty modelling and additivity adjustment. Additional improvements in these areas can also be expected in the future. In addition, the treatment of unequal length records in the multivariate case should receive additional attention in the future. Several computer packages or programmes have been developed to aid in the operational application of this type of modelling. Future computer improvements should be expected as the models are converted for use on micro computers. Joint estimation of parameters and the effect of regularity or smoothness conditions on parameter estimates can be fruitful areas for future research. Disaggregation modelling has major advantages over other modelling techniques. It is easily understood, easily applied and very flexible. There are opportunities to use long term memory models and to include parameter uncertainty.

Aggregation modelling is still very much a research area. Much work remains to be done before this modelling influences practical applications. The methodology shows great potential for aiding in model choice, parameter estimation, and for developing a strong theoretical background for other forms of modelling. Theoretic studies tend to show that most lower level models now popular aggregate to ARMA(1,1) models. Initial results with actual data seem to indicate that the low level seasonal models may need to be more complicated than the elementary AR or ARMA models. Similar approaches show promise in the study of other disciplines such as flood frequency analysis.

6. References

Bartolini, P., Salas, J.D. and Obeysekera, J.T.B. (1988) 'Multivariate periodic ARMA(1,1) processes', Water Resources Research, 24, 8, pp 1237-1246.

Frevert, D.K. and Lane, W.L. (1985) 'Development and application of a two level spatial disaggregation procedure for the LAST statistical hydrology package', Fourth International Hydrology Symposium, Fort Collin, Colorado.

Grygier, J.C. and Stedinger, J.R. (1985) 'SPIGOT: A synthetic glow generation software package, technical description', Cornell University, Ithaca, New York.

Grygier, J.C. and Stedinger, J.R. (1988) 'Condensed disaggregation procedures and conservation corrections for stochastic hydrology', Water Resources Research, 24,10, pp 1574-1584.

Harms, A.A. and Campbell, T.H. (1967) 'An extension to the Thomas-Fiering model for the sequential generation of streamflow', Water Resources Research, 3, 3, pp 653-661.

Kavvas, M.L., Cote, L.J. and Delleur, J.W. (1977) 'Time resolution of the hydrologic time-series models', Journal of Hydrology, 32, pp 347-361.

Lane, W.L. (1980) 'Applied stochastic techniques, user manual', Bureau of Reclamation, Denver, Colorado.

Lane, W.L. (1980) 'Applied stochastic techniques, user manual (first revision)', Water and Power Resources Service, Denver, Colorado.

Lane, W.L. (1982a) 'Applied stochastic techniques, user manual (second revision)', Bureau of Reclamation, Denver, Colorado.

Lane, W.L. (1982b) 'Corrected parameter estimates for disaggregation schemes, in statistical analysis of rainfall and runoff', V.P. Singh, editor, Proceedings of International Symposium on Rainfall-Runoff Modelling, May 18-21, 1981, Water Resources publications, Littleton, Colorado.

Lane, W.L. (1983) 'Applied stochastic techniques, user manual (third revision', Bureau of Reclamation, Denver, Colorado.

Lane, W.L. and Frevert, D.K. (1985) 'Applied stochastic techniques, user manual (fourth revision)', Bureau of Reclamation, Denver, Colorado.

Lane, W.L. and Frevert, D.K. (1988) 'Applied stochastic techniques, user manual (fifth revision)', Bureau of Reclamation, Denver, Colorado.

Loucks, D.P., Stedinger, J.R. and Haith, D.A. (1981) 'Water resource systems planning and analysis', Prentice-Hall, Englewood Cliffs, New Jersey.

Matalas, N.C. (1967) 'Mathematical assessment of synthetic hydrology', Water Resources Research, 3, 4, pp 937-945.

Mejia, J.M. and Rousselle, J. (1976) 'Disaggregation models in hydrology revisited', Water Resources Research, 12, 2, pp 185-186.

Moreau, D.H. and Pyatt, E.E. (1970) 'Weekly and monthly flows in synthetic hydrology', Water Resources Research, 6, 1, pp 53-61.

Obeysekera, J.T.B. and Salas, J.D (1982) 'On the aggregation and disaggregation of streamflow time series', EOS, 63, 18, p 321.

Obeysekera, J.T.B. and Salas, J.D. (1986) 'Modeling of aggregated hydrologic time series', Journal of Hydrology, 86, pp 197-219.

Rao, A.R., Rao, A.G. and Kashyap, R.L. (1985) 'Stochastic analysis of time-aggregated hydrologic data', Water Resources Bulletin, 21,5, pp 757-771.

Santos, E.G. and Salas, J.D. (1983) 'A parsimonious step disaggregation model for operational hydrology', EOS, 64, 45, p 706.

Salas, J.D., Delleur, J.W., Yevjevich, Y. and Lane W.L. (1980) 'Applied modeling of hydrologic time series', Water Resources Publications, Littleton, Colorado.

Stedinger, J.R. (1983) 'Advances in disaggregation modeling', Abstract for a paper presented at the American Geophysical Union Fall Meeting.

Stedinger, J.R., Lettenmaier, D.P. and Vogel, R.M. (1985a) 'Multisite ARMA(1,1) and disaggregation models for annual streamflow generation', Water Resources Research, 21, 4, pp 497-509.

Stedinger, J.R., Pei, D. and Cohn, T.A. (1985b) 'A condensed disaggregation model for incorporating parameter uncertainty into monthly reservoir simulations', Water Resources Research, 21, 5, pp 665-675.

Stedinger, J.R. and Vogel, R.M. (1984) 'Disaggregation procedures for generating serially correlated flow vectors', Water Resources Research, 20, 1, pp 47-56.

Thomas, H.A. and Fiering, M.B. (1962) 'Mathematical synthesis of streamflow sequences for the analysis of river basins by simulation', in Design of Water Resources Systems, A. Mass et al., pp 459-493, Harvard University Press, Cambridge, Massachusetts.

Valencia, D.R. and Schaake, J.C., Jr (1973) 'Disaggregation processes in stochastic hydrology', Water Resources Research, 9, 3, pp 580-585.

Vecchia, A.V., Obeysekera, J.T., Salas, J.D. and Boes, D.C. (1983) 'Aggregation and estimation for low-order periodic ARMA models', Water Resources Research, 19, 5, pp 1297-1306.

ARMAX AND TRANSFER FUNCTION MODELLING IN HYDROLOGY

J.B. MARCO
Catedratico de Ingenieria Hidraulica y Medio Ambiente
Universidad Politecnica de Valencia
Valencia
Spain

1. Introduction

Time series modelling has already taken place in hydrological technology. ARMA, AR models, etc. are devoted to preserving statistical properties from the stochastic process underlying a given sample, to generate long undistinguishable synthetic samples to provide for better analysis or derived processes. These models are characterized by the use of information from the analyzed series.

Of course, these models can be used to forecast. For instance, provided the situation and information at a given time, the model can forecast what is likely to occur within a few time intervals; the expected value as well as its possible variability. However, if an exogenous phenomenon is either linked to or is the cause of the process, it seems logical to introduce the information available on the phenomenon into the forecast. For instance, if we wish to forecast river flows and simultaneous rainfall data on the basin are available, it seems natural to use them for modelling.

Suppose we have two random variables X and Y. If it is desired to forecast Y as a function of X, the first idea would be a linear regression. In such case, a model would be written as

$$Y = a + b X + \epsilon \tag{1}$$

a and b being the model parameters and ϵ the model residual.

What would happen if X and Y were time series? The time evolution of X_t and Y_t would be linked to their previous and simultaneous values for both variables. X_t and Y_t can follow linear processes and a linear relationship can in fact be proposed between both time series. The residual itself, ϵ_t will also be a time series, likely to follow an ARMA model. This is how an ARMAX model is originated, as an extension of linear regression to the time series area. ARMAX stands for Autoregressive Moving Average Models with exogenous variables.

J.B. Marco et al. (eds.),
Stochastic Hydrology and its Use in Water Resources Systems Simulation and Optimization, 117–135.
© 1993 *Kluwer Academic Publishers.*

2. Background

ARMAX models were developed in 1970 from the Box and Jenkins (1970) systematization of linear modelling in the time interval domain. In continuous form and frequency domain there were similar techniques like cross-spectrum in wide use for systems and communication theory. In fact, Hino (1970) tried to use those techniques for rainfall-runoff modelling. Box and Jenkins called these models in time domain Transfer Function models of TFN - Transfer Function plus Noise models. Their work centres on the univariate modelling of a time series with information from a parallel series plus a noise.

In 1974, Young allowed for a more complete ARMA structure of the noise. He also proposed the ARMAX name for models with a moving average of white noise, keeping the TFN name for those models with a noise following a general ARMA model.

Kashyap and Rao (1976) systematized the multivariate modelling by making X_t and Y_t random vectors. Their work is largely based on Hannan's (1970) theory of multivariate time series and the canonical forms theory by Rissanen (1974). Young (1974) also developed the first sequential estimation algorithms. The interest of this type of techniques is obvious for real time forecasting.

The use of ARMAX models for intervention analysis was also developed in the mid-seventies by Box and Tiao (1975).

The most recent and comprehensive review of ARMAX techniques is attributed to Hannan and Kavalieris (1984). They established all aspects of this technique very clearly from a statistical point of view. All previously mentioned work deals with ARMAX models from a purely statistical point of view. Hydrological applications started in 1973 when Kashyap and Rao modelled daily runoff from the Wabash River with the parallel rainfall series over the basin. Simultaneous development of Kalman filter and Bayesian techniques involving considerable advantages for real-time problems, space-state equivalence of ARMAX models, have produced underdevelopment of ARMAX and TFN techniques from both statistical and hydrological points of view. Nowadays, when Kalman filtering capabilities have been largely explored, this state of things is changing. ARMAX models are really much more than a forecasting model; they represent the linear relationship among time series.

3. Univariate Modelling

Let Y_t be the time series to model or forecast as a function of the X_t time series of exogenous information. If a linear relationship is established between both series and the residuals, it can be written

$$Y_t = \sum_{i=1}^{p} \delta_i \, Y_{t-1} + \sum_{j=0}^{q} \omega_j \, X_{t-j} + \sum_{k=0}^{r} \theta_k \, \epsilon_{t-k} \qquad (2)$$

The model can be named ARMAX (p,q,r). In general Y_t and X_t are standardized stationary series. δ, ω, θ are the model parameters. ϵ_t is a white Gaussian noise of zero mean and unit variance, independent from

both X_t and Y_t series and, of course, itself.
   For instance, ARMAX(1,0,0) would be

$$Y_t = \delta_1 \, Y_{t-1} + \omega_0 \, X_t + \Theta_0 \, \epsilon_t \tag{3}$$

and ARMAX (1,1,1)

$$Y_t = \delta_1 \, Y_{t-1} + \omega_0 \, X_t - \omega_1 \, X_{t-1} + \Theta_0 \, \epsilon_t - \Theta_1 \, \epsilon_{t-1} \tag{4}$$

   Box and Jenkins used the notation $T(p,b,q)$ to indicate a model

$$Y_t = \sum_{i=1}^{p} \delta_i \, Y_{t-i} + \sum_{j=b}^{q} \omega \, X_{t-b-j} + \eta_t \tag{5}$$

where b is the time lag in the response of the system. η_t, the transfer function structure can have a more complex structure

$$\eta_t = \sum_{l=1}^{p'} \phi_1 \, \eta_{t-1} + \sum_{m=1}^{q'} \Theta_m \, \epsilon_{t-m} \tag{6}$$

   Such a model is called a TFN(p,b,q) model with ARMA(p',q') noise.
   If a backward shift operator is introduced, such that

$$B^a X_t = X_{t-a} \tag{7}$$

a more compact notation can be achieved

$$\delta(B)Y_t = \omega(B)X_t + \eta_t$$
$$\phi(B)\eta_t = \Theta(B)\epsilon_t \tag{8}$$

where

$$\begin{aligned}
\delta(B) &= 1 - \delta_1 B - \delta_2 B^2 - \ldots - \delta_p B^p \\
\omega(B) &= (\omega_0 - \omega_1 B - \omega_2 B^2 - \ldots - \omega_q B^q)B^b \\
\phi(B) &= 1 - \phi_1 B - \phi_2 B^2 - \ldots - \phi_p B^q \\
\Theta(B) &= \Theta_0 - \Theta_1 B - \Theta_2 B^2 - \ldots - \Theta_q B^{q'}
\end{aligned} \tag{9}$$

are polynomials in B.
   The transfer function model can be written, then, as

$$Y_t = \frac{\omega(B)}{\delta(B)} X_t + \frac{\Theta(B)}{\delta(B)\phi(B)} \epsilon_t \tag{10}$$

   ARMAX coefficients Θ_k can be computed by polynomial division $\Theta(B)/\phi(B)$, so a transfer function can be converted to ARMAX formulation.

3.1. IMPULSE-RESPONSE FUNCTION

To present the impulse-response function consider an ARMAX(2,1,0) model

$$Y_t = \delta_1 Y_{t-1} + \delta_2 Y_{t-2} + \omega_0 X_t - \omega_1 X_{t-1} + \Theta_0 \epsilon_t \tag{11}$$

This equation has to be satisfied for every t, so it can also be written for t-1

$$Y_{t-1} = \delta_1 Y_{t-2} + \delta_2 Y_{t-3} + \omega_0 X_{t-1} - \omega_1 X_{t-2} + \Theta_0 \epsilon_{t-1} \tag{12}$$

substituting this for the previous equation yields

$$Y_t = (\delta_2 + \delta_1{}^2)Y_{t-2} + \delta_1\delta_2 Y_{t-3} + \omega_0 X_t + (\delta_1\omega_0 - \omega_1)X_{t-1} + (\delta_1\omega_1)X_{t-2} + \epsilon_t + \delta_1\epsilon_{t-1} \tag{13}$$

Further substitutions will increase the number of terms in X_{t-i} while decreasing the Y_{t-1} terms. Coefficients affecting Y_{t-j} for $j>i$ are increasing powers of δ_1, which due to stability requirements tend to zero, so finally the model can be expressed as

$$Y_t = v_0 X_t + v_1 X_{t-1} + v_2 X_{t-2} + \ldots + \eta_t \tag{14}$$

with infinite terms $V_k X_{t-k}$.
Coefficients of V_k, $k=0,1,2,\ldots$, are known as the Impulse-Response function of the system since every V_k reflects the influence of a unit excitation at time t-k on the actual value of the output Y_t. If ω_j and δ_j are known, the V_k coefficients are easily computed by solving

$$v(B) = \frac{\omega(B)}{\delta(B)} \tag{15}$$

If this equation is expanded, the following system is obtained

$$
\begin{aligned}
v_i &= 0 & &; \quad i < b \\
v_i &= \delta_1 v_{i-1} + \delta_2 v_{i-2} + \ldots + \delta_p v_{i-p} + \omega_0 & &; \quad i = b \\
v_i &= \delta_1 v_{i-1} + \delta_2 v_{i-2} + \ldots + \delta_p v_{i-p} + \omega_{i-p} & &; \quad i = b+1, b+2, \ldots, b+p \\
v_i &= \delta_1 v_{i-1} + \delta_2 v_{i-2} + \ldots + \delta_p v_{i-p} & &; \quad i > b+p
\end{aligned}
\tag{16}
$$

which are the Yule-Walker equations for the T(p,b,q) model.
Note that no η_t related terms appear in these equations. Polynomial division can also be performed so,

$$\Omega(B) = \frac{\Theta(B)}{\delta(B) \; \varnothing(B)} \tag{17}$$

and an infinite set of terms in ϵ_{t-k} is obtained. So any transfer function model can be written as

$$Y_t = \sum_{k=0}^{\infty} v_k X_{t-k} + \sum_{k=0}^{\infty} \Omega_k \epsilon_{t-k} \tag{18}$$

If the η_t is retained, the model then is

$$Y_t = \sum_{k=0}^{\infty} v_k X_{t-k} + \eta_t \tag{19}$$

This could be viewed as a discretized convolution plus a perturbation term

$$Y_t = \int_0^{\infty} k(\tau) X(t-\tau) \, d\tau + \eta_t \tag{20}$$

So the impulse-response function is nothing but the kernel of a convolution with the excitation that yields the output. If we take X_t as the rainfall series and Y_t as the runoff, v_k are the unit hydrograph discrete values.

If this is so, what are the advantages of ARMAX or TFN formulation over traditional systems theory? The essential one stems from the fact that ARMAX or TFN models are parsimonious, that is to say, they have a minimum number of parameters. Impulse-response models have as many parameters as non-zero values of the kernel. Parsimony is fundamental for estimation and sampling properties of the parameters estimates (Hipel, 1986, Ledolter, 1981).

In fact, stability problems of unit hydrograph values induced Todini and Bouillot (1975) to develop their constrained least square method which can be considered as an early transfer function technique in hydrology.

The confidence interval around the parameters is narrower. Also, statistical tests are available to check the adequacy of the models. Since this type of modelling explicitly introduces ϵ_t, confidence limits can be computed around the forecasts.

A relationship can be derived for the impulse-response coefficients from the autocorrelation and crosscorrelation functions of the series. If Equation 19 is multiplied by X_{t-k} and an expectation operator applied, it yields

$$\rho_{yx}(K) = v_0 \, \rho_{xx}(K) + v_1 \, \rho_{xx}(K-1) + v_2 \, \rho_{xx}(K-2) + \ldots \tag{21}$$

This equation relates crosscorrelation functions of input and output to the impulse-response function and dependence structure of the input series. $\rho_{xx}(K)$ is the crosscorrelation function of Y_t and X_t and $\rho_{xx}(K)$ is the autocorrelation function from the X_t series.

If this equation is written for a sufficiently large K value so that v_{k+1}, $v_{k+2}, \ldots$ can be neglected, the following matrix equation is obtained

$$\begin{bmatrix} \rho_{yx}(0) \\ \rho_{yx}(1) \\ \rho_{yx}(2) \\ \cdot \\ \cdot \\ \cdot \\ \rho_{yx}(k) \end{bmatrix} = \begin{bmatrix} 1 & \rho_{xx}(1) & \rho_{xx}(2) & \cdots & \rho_{xx}(k) \\ \rho_{xx}(1) & 1 & \rho_{xx}(1) & \cdots & \rho_{xx}(k-1) \\ \rho_{xx}(2) & \rho_{xx}(1) & 1 & \cdots & \rho_{xx}(k-2) \\ \cdot & \cdot & \cdot & & \cdot \\ \cdot & \cdot & \cdot & & \cdot \\ \cdot & \cdot & \cdot & & \vdots \\ \rho_{xx}(k) & \rho_{xx}(k-1) & \rho_{xx}(k-2) & \cdots & 1 \end{bmatrix} \cdot \begin{bmatrix} v_0 \\ v_1 \\ v_2 \\ \cdot \\ \cdot \\ \cdot \\ v_k \end{bmatrix} \tag{22}$$

If ρ_{yx} and ρ_{xx} are known, v_k can be obtained by solving the system.

If the exogenous variable has no structure, that is to say, it is a white noise, then $\rho_{xx}(K)=0$; $k \neq 0$, and consequently

$$\rho_{yx}(K) = v_k \tag{23}$$

The impulse-response function is precisely the crosscorrelation between input and output. This property is very useful for model identification.

3.2. MODEL IDENTIFICATION

When ARMAX or TFN models are used, the traditional modelling sequence identification, parameter estimation and diagnostic checking must follow. Identification is a key part of the procedure since estimation methods are involved.

The Box and Jenkins method benefits from the crosscorrelation and impulse-response relationship. This method essentially stems from considering the response of the system to a white noise excitation. If a dependence structure for X_t is identified, then that structure is applied to both the Y_t and X_t series so the residuals are linked by the same transfer function. This procedure is called "prewhitening" the series.

Let the structure of X_t be an ARMA(p_x, q_x) model. Then

$$\delta_x(B)X_t = \omega_x(B) \, a_t \tag{24}$$

where a_t is a white Gaussian noise. If this structure is applied to the Y_t series, solving for the residuals

$$Y_{t'} = \frac{\omega_x(B)}{\delta_x(B)} \, Y_t \tag{25}$$

Since the system is linear, the transfer function will not change if the same linear transformation is applied to both input and output.

$$Y_{t'} = \sum_{k=0}^{\infty} v_k \, a_{t-k} + \eta_{t'} \tag{26}$$

But since a_t is white noise, v_k is obtained as

$$v_k = \rho_{y'a}(K) \tag{27}$$

Inspection of v_k yields the order of the model and, through the use of the Yule-Walker equations, a rough parameter estimation is performed. Then the η_t series is obtained and analyzed to fit an ARMA(p', q') model. This procedure yields an idea of what the orders of the model could be and an initial parameter estimation. Nevertheless, Akaike's information criterion will provide the final model discrimination among competing models.

$$AIC = 2 \ln L + 2K \tag{28} \quad \text{(Akaike, 1974)}$$

where L is the maximum likelihood function and K is the number of free parameters in the model. The order of the model will not depart very much from the prewhitening method results. This method has been used by Thompstone (1985) as proposed by Hipel and McLeod (1985) for TFN model identification.

Miller (1981) applied multiple regression and significance tests for regression coefficients to determine the order of the model.

Cooper and Wood (1982a) present a multivariate identification method based on canonical analysis. The model is set as a relationship between the future, represented by a vector $\{Y_t, Y_{t+1}, Y_{t+2}, \ldots\} = \underline{Y}$, and the past, $\underline{X} = \{X_t, Y_{t-1}, X_{t-1}, Y_{t-2}, X_{t-2}, \ldots\}$. Canonical correlations are obtained between these two vectors. These are the eigenvalues of the eigensystem

$$| S_{yx} \, S_{xx}^{-1} \, S_{xy} - \lambda S_{yy} | = 0 \tag{29}$$

where S_{xx}, S_{yx}, S_{xy}, S_{yy} are the sample covariances of the Y and X vectors. The $\underline{X}$ vector is set $\underline{X} = \{X_t, Y_{t-1}, X_{t-1}, \ldots, X_{t-k}, Y_{t-k}\}$, where K is larger than the maximum expected order of the model.

The future vector is then considered $\underline{Y} = \{Y_t, Y_{t+1}, \ldots, Y_{t+1}\}$ and canonical correlations are obtained. If the minimum canonical correlation is significantly different from zero, then the vector is expanded and the procedure continues until the order of the model is obtained. Initial prewhitening of the series is also required.

3.3. PARAMETER ESTIMATION

If a model is formulated as

$$Y_t = \sum_{i=1}^{p} \delta_i \, Y_{t-i} + \sum_{j=0}^{q} \omega_j \, X_{t-j} + \eta_t \tag{30}$$

where η_t is serially uncorrelated, the least-square estimation is consistent and asymptotically efficient. Nevertheless, if η_t has a structure, Y_t will be correlated with η_t and least-squares estimates become inconsistent. If η_t is not a white noise, the estimation is complex.

Estimation methods have been thoroughly discussed by O'Connell and Clarke (1981) and lately by Patry and Marino (1984).

Box and Jenkins point out that if initial values X_0, Y_0, are available for a given set β of parameters, ϵ_t can be computed. In this situation, the conditional least-square function can be minimized.

$$\min_{\beta} \, \sum_{t=1}^{n} \, \epsilon_t^2 \, (\beta | X_0, Y_0, \epsilon_0) \tag{31}$$

yielding consistent and asymptotically efficient estimates, very close to maximum likelihood. Non-linear least-square methods, using Newton-Raphson or any other gradient-search techniques, are usually needed.

Young (1974) introduced the instrumental variable approach by decomposing the problem into two coupled estimations. The Y_t values are substituted by $\hat{Y}_t$, recursively estimating the parameters until convergence and then the residual series is modelled.

Cooper and Wood (1982b) developed and applied maximum likelihood methods for parameter estimation. Experience shows that maximum likelihood dramatically improves model results.

Whatever the estimation method, Budzianowski and Stupczewski (1981) clearly warn about the risk of decoupling the problem of estimating separately the transfer function and noise parameters. Joint estimates of the whole parameter set must be obtained.

All the preceeding techniques were originally developed for off-line estimation. If the models are to be used for real-time forecasting, sequential estimation is needed. Among sequential methods, recursive ordinary least squares (ROLS) and recursive instrumental variables (RIV) were developed by Young (1974 and 1975). Both models are very convenient if η_t has no structure. With ARMAX structure, recursive maximum likelihood estimates were developed and applied in hydrology by Cooper and Wood (1982b). Recursive parameter estimation is frequently achieved within the Kalman filter framework. Any TFN or ARMAX model can be cast in space-state form. Cooper and Wood (1982a) explore this relationship in depth.

3.4. DIAGNOSTIC AND CHECKING

Global adequacy of the model must be shown by overall tests. Akaike's information criterion is the most frequently used discriminating technique among different competing models as already explained.

Patry and Marino (1984) state that model order underestimation severely reduces the model performance, much more than overestimation does.

For basic hypotheses concerning ϵ_t, the model residuals must also be checked. Since ϵ_t is assumed to be serially independent and also independent from X_t and Y_t, not only must the residual autocorrelation function be tested for independence, but also the crosscorrelations functions $\rho_{\gamma\epsilon}$ and $\rho_{\chi\epsilon}$.

Finally, residuals are assumed normal so this must also be checked. Violating the assumption does not have as strong an effect on the transfer functions as it does with common ARMA modelling (Thompstone et al., 1985).

4. Forecasting with TFN and ARMAX Models

Assume that a Y_t forecast is sought with L intervals time lead. X_t and Y_t values are known up to time t. If this forecast is denoted $Y_{t+L|t}$, in order to have minimal quadratic error, it must be minimized.

$$\min_{Y_t, X_t} E[Y_{t+L} - \hat{Y}_{t+L|t}]^2 \tag{32}$$

This is equivalent to forecasting with

$$Y_{t+L|t} = \sum_{i=1}^{p} \delta_i Y_{t+L-i|t} + \sum_{j=0}^{q} \omega_j X_{t+L-j|t} + \sum_{k=0}^{r} \theta_k \epsilon_{t+L-k|t} \tag{33}$$

by using the expected values at time α, $X_{\alpha|\beta}$, $Y_{\alpha|\beta}$, $\epsilon_{\alpha|\beta}$, conditioned to the observed values at time β. Rules for obtaining these values are simple: if the value has already occurred, the observed value will be used. If the value is in the future, expected values will be used for $X_{\alpha|\beta}$, $Y_{\alpha|\beta}$ and zero for $\epsilon_{\alpha|\beta}$. A similar method can be used with the TFN formulation.

All TFN and ARMAX models explicitly yield forecasting variances.

If the model for Xt series is

$$\delta_I(B)X_t = \omega_I(B)a_t \tag{34}$$

Then obtaining X_t and introducing it into the TFN formulation

$$Y_t = \frac{\omega(B)\;\omega_I(B)}{\delta(B)\;\delta_I(B)}\;a_t + \frac{\Theta(B)}{\delta(B)\;\phi(B)}\;\epsilon_t \tag{35}$$

Performing both polynomial divisions,

$$V(B) = \frac{\omega(B)\;\omega_I(B)}{\delta(B)\;\delta_I(B)} \tag{36}$$

$$\Omega(B) = \frac{\Theta(B)}{\delta(B)\;\phi(B)}$$

the Y_t series is expressed as a moving average of two white Gaussian noises a_t and ϵ_t of unit variance.

$$Y_t = V(B)a_t + \Omega(B)\;\epsilon_t \tag{37}$$

the forecasting variance will be,

$$\mathrm{var}\;(\hat{Y}_{t+L|t}) = \sum_{j=0}^{L-1} V j^2 + \sum_{j=0}^{L-1} \Omega_j^2 \tag{38}$$

Obtaining confidence limits of around $\hat{Y}(t+L|t)$ is straightforward.

5. Multivariate Modelling

Univariate models, like those described above, can be extended to multivariate models when X_t, Y_t and ϵ_t are random vectors. In this situation an ARMAX model can be written as

$$\underline{Y}_t = \sum_{i=1}^{p} \underline{\delta}_i\; \underline{Y}_{t-i} + \sum_{j=0}^{q} \underline{\omega}_j\; \underline{X}_{t-j} + \sum_{k=0}^{r} \underline{\Theta}_k\; \underline{\epsilon}_{t-k} \tag{39}$$

where $\underline{Y}_t$ is a random vector of M dimension, $\underline{X}_t$ is a random vector of N dimension, $\underline{\epsilon}$ is a random vector of zero mean and M dimension and covariance matrix $\underline{\sigma}^2 . \underline{\Theta}_k$ and $\underline{\delta}_j$ are [MxM] parameter matrices, and ω_j are [MxN] parameter matrices.

If the TFN model is generalized to multivariate situations, the noise model will also be a multivariate structure

$$\mathfrak{n}_t = \sum_{l=1}^{p'} \emptyset_l \ \mathfrak{n}_{t-l} + \sum_{m=1}^{q'} \underline{\Theta}_\blacksquare \ \epsilon_{t-\blacksquare} \tag{40}$$

where $\emptyset_l$ and $\underline{\Theta}_\blacksquare$ are [MxM] parameter matrices. With Box and Jenkins notation, $\underline{\delta}(B)$, $\underline{\omega}(B)$, $\emptyset(B)$ and $\underline{\Theta}(B)$ will become matrix polynomials.

As happens with multivariate time series models with an order increase, the number of parameters increases quadratically. This is a common drawback for multivariate models. If multivariate modelling is necessary, the order of the model must be very low. Otherwise, the sampling properties of the parameter estimates will deteriorate.

With multivariate structure there is redundant information. In fact, given an ARMAX model there are infinite transformations yielding the same results. It is then necessary to define canonical forms in order to achieve uniqueness and the best estimating properties. Kashyap and Rao (1976) established three canonical forms for ARMAX models.

Canonical form I: $\delta(B)$ Lower Triangular, $\delta(0) = I$
        $\omega(B)$, $\Theta(B)$, σ^2 Arbitrary
Canonical form II: $\Theta(B)$ Diagonal, $\delta(0) = 1$
        $\delta(B)$, $\omega(B)$, σ^2 Arbitrary
Canonical form III: $\Theta(B)$, σ^2 Diagonal, $\delta(0)$ Lower triangular
        $\delta(B)$, $\omega(B)$ Arbitrary

Canonical forms with $\Theta(B)$ diagonal allow the individual noise to be separated from each vector component. This is a clear advantage for model identification since the problem centres on the transference. On the other hand, canonical form I is usually the best for parameter estimation since it has the minimum number of parameters.

From the estimation point of view, it seems that conditional maximum likelihood seems to be the most favourable approach whether off-line as proposed by Kashyap and Rao (1976) (CHL), or on-line, as developed by Cooper and Wood (1982a).

Confidence intervals for the forecast can also be generalized.

If matrix polynomials $V(B)$ and $\Omega(B)$ are defined with [MxN] and [MxM] dimensions and obtained from the equations

$$\delta_I(B) \ \delta(B) \ V(B) = \omega(B) \ \omega_I(B)$$
$$\emptyset(B) \ \delta(B) \ \Omega(B) = \Theta(B) \tag{41}$$

the covariance matrix of the forecasts will then be

$$\text{Cov} \ [\underline{Y}(t+L|t) \ \underline{Y}^T(t+L|t) = \sum_{j=0}^{L-1} V_j \sigma_a^2 \ V_j^T + \sum_{j=0}^{L-1} \Omega_j \sigma^2 \ \omega_j^T \tag{42}$$

and the joint confidence intervals around the forecast values are easily computed.

6. ARMAX and TFN Applications to Rainfall-Runoff Modelling

6.1. FIRST ATTEMPTS

Hydrological applications of these models have been scarce in comparison with their potential. The most common area has been, of course, rainfall-runoff modelling. Kashyap and Rao were the first, in 1973, to apply explicitly a TFN model for daily river flow forecast using rainfall data.

In 1975, Todini introduced his CLS (constrained least square) method and applied it to the Arno and Ombrone Rivers in Italy (Natale and Todini, 1974, Todini and Bouillot, 1975, Todini, 1975). This model is not parsimonious, so it can be considered as a hybrid between a TFN and a unit hydrograph method. His method of handling the nonlinearity of rainfall-runoff processes must be pointed out (Todini and Wallis, 1977).

The UK Natural Environment Research Council as far back as 1975 recommended the use of transfer functions for river flow forecasting.

In Belgium, Lorent (1975) used ETP values to build up surrogates of effective rainfall and introduce them as the exogenous information for transfer.

At that time, Kalman filter developments greatly diverted hydrologists in that direction. During the 1975-81 period several authors developed different recursive estimation techniques and used transfer functions and ARMAX models within the space-state approach. However, from the hydrological point of view, only after 1980 was any progress made.

Anselmo and Ubertini (1979, 1981) applying the TFN on daily and hourly scales, introduced the complete TFN model.

Demareé (1981), modifying Lorent's method, also applied the TFN model to the River Dijle in Belgium.

6.2. ANALYSIS OF HYDROLOGICAL NONLINEARITY

If Y_t is the surface water runoff and X_t is the effective rainfall, linear statistical models would be completely justified for modelling. But as O'Connell and Clarke (1981) point out, neither the effective rainfall, nor the base flow are known. So, this nonlinear hydrological behaviour has to be taken into account in some way. There are several methods for dealing with this behaviour. Moore (1981) discusses several of them at length.

First of all, a moisture content index can be introduced as an additional exogenous variable. That is the proposal by the UK Natural Environment Research Council using API5 (Antecedent Precipitation Index-5 days) and moisture content deficit SMD to define m_t

$$API_t = K \ API_{t-1} + P_{t-1}$$
$$m_t = API5_t - SMD \tag{43}$$

Instead of using m_t as an additional variable, Whithead and Young (1975) use it to scale the precipitation.

$$P_t{}^* = m_t P_t \tag{44}$$

using $P_t{}^*$ as the exogenous series X_t.

Todini's method (Todini and Wallis, 1977) has been employed by several authors. A threshold T is defined on a moisture-related variable on the basin, namely m_t. Then, two precipitation series are generated

$$m_t > T \qquad P_{1t} = P_t \qquad P_{2t} = 0$$
$$m_t \leq T \qquad P_{1t} = 0 \qquad P_{2t} = P_t \tag{45}$$

which are used as two independent inputs for the transfer.

Belgian hydrologists favour a more physical-conceptual approach. Demareé (1981) substracts ETP_t from rainfall. The resulting r_t series is filtered as

$$S_t = S_{t-1} + 1/T_s \ (r_t - S_{t-1}) \tag{46}$$

in order to account for the soil drying process. T_s is a time constant. Finally, an input to the model X_t is obtained as a modulation of the r_t series.

$$X_t = r_t \ s_t \tag{47}$$

Anselmo and Ubertini (1981) tried as input series

$$X_t = \sum_{k=0}^{M} c^k(1-c) \ P_{t-k} \tag{48}$$

where c is obtained by minimizing the difference between Y_t and X_t, but it did not introduce a major improvement in the results.

Nonlinearity also appears in the output. For modelling rivers with an important base flow, a threshold for the Y_t series may be needed as Moore (1981) suggests.

In 1979, Ganendra introduced the perlog variable

$$pl(t) = \frac{d}{dt} (\ln X(t)) = \frac{1}{X(t)} \frac{d}{dt} (X(t)) \tag{49}$$

Perlog can be generated on both input and output. The rationale behind its use is the fact that it amplifies the time rate of flow increase. This endeavours to correct a classical problem with linear forecasting models. Forecasts are much better when the system is near its mean state than with extreme values.

6.3. MULTIVARIATE MODELLING

All the applications discussed consider univariate output. A single variable is forecast with the ARMAX or TFN model. However, for several model uses, joint forecast of a set of flow values may be needed. This approach has received less attention by rainfall-runoff specialists than other hydrological areas like water quality.

In fact, the first application of this type was presented by Cooper and

Wood (1981) as an example of their multivariate models.

6.4. LARGE TIME SCALES

Previously presented applications deal with time series on daily or hourly
time scales.   Short-term forecast is the fundamental objective and,
consequently, on-line schemes are essential, but problems like seasonality
remain secondary.   By increasing the time scale, for instance weekly or
monthly, the forecasting problem takes on a different scope. Medium-term
water management is the objective. Less data are available, but more time
for computation can be provided. Off-line methods become more competitive.
On the other hand, measurement errors are smaller.   The space-state
approach becomes less dominant.
   In 1985, Oliver and Marco proposed a weekly flow forecasting model.
Seasonality was taken into account through standardization

$$X_t = \frac{p_{t,\tau}-\mu p_\tau}{\sigma p_\tau} \qquad ; \qquad Y_t = \frac{q_{p,\tau}-\mu q_\tau}{\sigma q_\tau} \qquad (50)$$

Fourier analysis in the sense proposed by Yevjevich (1972) was performed,
selecting the same number of significant harmonics on both input and
output.
   Thompstone et al. (1986) forecasted quarter-monthly river flows for
hydropower purposes with a complete TFN model. The results show how a
simple model like this can predict better and cheaper than complex
conceptual models can.   Seasonality was also taken into account by using
periodic parameters.

6.5. FLOOD ROUTING

ARMAX and TFN models can clearly be used for flood-routing.   In this
situation, upstream flow would be the input and downstream flow the output.
Noise would take into account lateral inflows and model adequacy. In fact,
it is easy to show that the Muskingum method and other similar methods can
be formulated as transfer functions, and fitting the model can be used for
estimating its parameters (Young, 1986). Nevertheless, there are very few
references to it in the literature. it seems clear that practitioners in
this area prefer deterministic methods.

7.   Some Other Areas of Hydrological Application

Among TFN and ARMAX applications outside the rainfall-runoff analysis,
three areas must be recalled: snowmelt, water quality and groundwater.

7.1. SNOWMELT

ARMAX modelling has proved to be an efficient tool for snowmelt analysis
since it can introduce several exogenous variables. Damsleth in 1978 first
employed a TFN model using rainfall and temperature as exogenous
information.

A second study in Canada (Burns and McBean, 1985) used only degree days as the exogenous variable.

A combined model, based on Sugawara's tank model, was developed (Mizumura and Chin, 1985). It tries to model separately snowmelt, surface and groundwater flows using ARMAX techniques plus a Kalman filter.

Haltiner and Salas (1988) present an ARMAX model which is shown to be similar to the Martinec-Rango deterministic model. A combined snowmelt-rainfall index is used as the exogenous input

$$X_t = (T_t - T_b) S_t + P_t \tag{51}$$

where $T_t - T_b$ are temperature series and threshold, S_t areal cover of snow and P_t rainfall.

7.2. WATER QUALITY

ARMAX and TFN models for water quality modelling have been tools used widely from the start. The first applications in this area were due to Whitehead and Young (1975). They applied a TFN model to BOD and BO forecasting in the Bedford-Ouse River system, using as input a host of hydrological variables like river flows and radiation, etc.

An essential characteristic for these applications is the multivariate approach. Young's work has been centred on this area for 15 years.

7.3. GROUNDWATER STUDIES

The first model in this area was developed by Marco and Yevjevich (1981). They formulated a multivariate transfer function model, using rainfall data as input to predict a vector of moisture content at different depths from a soil profile with a 15-day interval. Seasonality was taken into account through Fourier analysis of means and standard deviations. The model was used to generate synthetically natural groundwater recharge samples.

Estrela and Sahuquillo (1985) used transfer functions to identify a response model for a karstic spring.

8. Potential Uses

TFN and ARMAX models are not bound to hydrological forecasting. In this area, they have to compete with the space-state approach which has the advantage of handling measurement errors. Real-time forecasting with Bayesian methods needs an "a priori" knowledge of parameters, structure and covariances. This is not a problem when real-time data acquisition systems are available to provide hundreds of data. But when the discretization interval increases, the number of data decrease, and "a priori" ignorance of ARMAX modelling becomes an advantage.

ARMAX and TFN models were developed as a generalization of regression analysis to the time series domain. Their explanatory value becomes the main asset. In this sense, there are many unexplored possibilities for hydrological use. For instance, filling-in of missing data. River flows can be completed using rainfall and previous flows information thus preserving all statistical relationships.

Another very promising area seems to be intervention analysis; Whitfield and Woods (1984) presented an ARMAX-based intervention analysis for environmental impact evaluation after a reservoir completion. The model was

$$Y_t - \bar{Y} = X_t + \sum_{i=1}^{12} \omega_{0i} (\xi_t - \bar{\xi}_i) + \frac{\epsilon_t}{1 - \phi_i B} \tag{52}$$

where X_t is the pre-intervention periodic mean and ξ_i the intervened variable periodic mean having 1 or 0 value for reservoir construction. This was proposed by Hipel (1977). In general, ARMAX and TFN models can be used to search for causality in hydrology.

This type of model can also be used for sample generation, for instance, when the available sample is very small but a larger sample exists from an X_t forecasting series. This approach can be convenient if the transfer explains most of the Y_t variance, as is the case when the causality is clear. This was the approach used by Lawrence and Kottegoda (1977) for river flow generation or by Marco and Yevjevich (1983) for synthetic moisture content series generation.

9. Some Unexplored Questions

ARMAX and TFN modelling in hydrology is in its infancy. It seems that underlying statistical techniques are consistent and provide a solid backing, but many unexplored capabilities remain for hydrological analysis.

For instance, no theory is available to handle periodic series. Like ARMA models, ARMAX can be generalized to periodic. A PARMAX model family, as suggested by Haltiner (1985) having the structure

$$Y_{t,\tau} = \sum_{i=1}^{p} \delta_{i,\tau} \ Y_{t,\tau-i} + \sum_{j=0}^{q} \omega_{j,\tau} \ X_{t,\tau-j} + \sum_{k=0}^{r} \theta_{k,\tau} \ \epsilon_{t,\tau-k} \tag{53}$$

where t is the year and τ the season, could be developed.

Very few applications have been developed with multivariate output outside the water quality domain. Previous multivariate analysis of variables has not been researched. Imagine a multivariate rainfall-runoff model like

$$\underline{q}_t = \underline{\underline{\delta}}_1 \ \underline{q}_{t-1} + \underline{\underline{\omega}}_0 \ \underline{p}_t + \underline{\underline{\theta}}_0 \ \underline{\epsilon}_t \tag{54}$$

What would happen if a previous principal components analysis is performed on q_t and p_t vectors? Furthermore, the previous hydrological treatment of the exogenous data has not been analyzed in depth. For instance, if a rainfall-runoff transfer model is sought, a p_t rainfall vector can be used, but Thiessen averaging on the basin or even krigging techniques could previously have yielded a unique exogenous series.

Parsimony for multivariate transfer models can be achieved using canonical forms. But, based on physical knowledge, many parameters can be set to zero previously. If a basin, as in Figure 1, is to be modelled with an ARMAX(1,1,0), the following simplified model seems to be physically reasonable.

$$
\begin{bmatrix} q_1 \\ q_2 \\ q_3 \end{bmatrix}_t = \begin{bmatrix} \delta_{11} & 0 & 0 \\ 0 & \delta_{22} & 0 \\ \delta_{31} & \delta_{32} & \delta_{33} \end{bmatrix} \begin{bmatrix} q_1 \\ q_2 \\ q_3 \end{bmatrix}_{t-1} + \begin{bmatrix} \omega_{11} & \omega_{12} & 0 & 0 & 0 \\ 0 & 0 & \omega_{23} & 0 & 0 \\ 0 & 0 & 0 & \omega_{34} & \omega_{35} \end{bmatrix} \cdot [p_1 p_2 p_3 p_4 p_5]_t^T +
$$

$$
+ \begin{bmatrix} \epsilon_1 \\ \epsilon_2 \\ \epsilon_3 \end{bmatrix}_t
$$

$$(55)$$

10. Conclusion

ARMAX and TFN modelling in hydrology are emerging techniques. If the statistical theory seems sufficiently developed, their hydrological application remains largely unexplored. But, to exploit their capabilities to the full it must be understood that these models are more than forecasting methods.

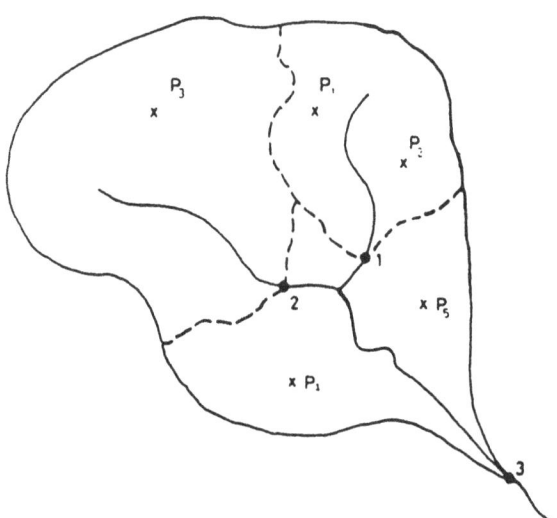

Figure 1. A basin with three flow measurement points and five raingauges.

11. References

Akaike, H. (1974) 'A new look at the statistical model identification', IEEE Transactions on Automatic Control, AC-19, pp 716-723.

Anselmo, V. and Ubertini, L. (1979) 'Transfer function-noise model applied to flow forecasting', Hydro. Sci. Bull, 24, 3, pp 353-359.

Anselmo, V., Melone, F. and Ubertini, L. (1981) 'Application of multiple stochastic models to rainfall-flow relationship of the Toce River', Proc. Ins. Con. Rainfall-Runoff Modeling, V. Singh, ed.

Box, G.E.P. and Jenkins, G.M. (1970) 'Time series analysis forecasting and control', Holden Day, Inc., San Francisco, California.

Box, G.E.P. and Tiao, G.C. (1975) 'Intervention analysis with applications to economic and environmental problems', Journal Amer. Statis. Assoc. 70, pp 70-79.

Box, G.E.P. and Tiao, G.C. (1977) 'A canonical analysis of multiple time series', Biometrika, 64, 2, pp 355-365.

Budzianowski, R.J. and Strupczewski, W.G. (1981) 'On the structure of the linear stochastic forecasting models', Proc. Int. Conf. Rainfall-Runoff Modeling, V. Singh, ed.

Burn, D.H. and McBean, E.A. (1985) 'River flow forecasting model for Sturgeon River', ASCE J. of Hyd. Eng. Vol. III, No 2.

Cluckie, I.E. (1980) Hydrological Forecasting, IAHS Publication NǪ 129.

Cooper, D.M. and Wood, E.F. (1982a) 'Identification of multivariate time series and multivariate input-output models', Water Resour. Res. 18 (4), pp 937-946.

Cooper, D.M. and Wood, E.F. (1982b) 'Parameter estimation of multiple input-output time series models: application to rainfall-runoff processes', Water Resour. Res. (18(5), pp 1352-1364.

Damsleth, E. (1978) 'Analysis of hydrologic data with linear transfer models', Publ. 602 Norwegian Computing Centre, Oslo, Norway.

Demareé, G. (1981) 'Hybrid conceptual-stochastic modeling of rainfall-runoff processes applied to the Dijle catchment', Proc. Int. Conf. Rainfall-Runoff Modeling, V. Singh, Ed.

Estrela, T. and Sahuquillo, A. (1985) 'Modeling the response hydrograph of subsurface flow', in Multivariate Analysis of Hydrologic Processes, M.W. Shen et al., Eds. Fort Collins, pp 141-152.

Ganendra, T. (1979) 'Real-time forecasting and control in the operation of water resources systems', Ph D. Thesis, University of London.

Haltiner, J.P. (1985) 'Stochastic modeling of season and daily streamflow', Ph D. Dissertation, Colorado State University, Fort Collins, Colorado.

Haltiner, J.P. and Salas J.D. (1988) 'Short-term forecasting of snowmelt runoff using ARMAX models', Water Res. Bull., 24(5), pp 1083-1089.

Hannan, E.J. (1970) Multiple time series, J. Wiley, New York.

Hannan, E.J. and Kavalieris, L. (1984) 'Multivariate linear time series models', Adv. Appl. Prob. 16, pp 292-561.

Hino, M. (1970) 'Runoff forecast by linear predictive filter', ASCE H. Hydraulics, 96 (HY3), pp 681-707.

Hipel, K.W., McLeod, A.I. and McBean (1977) 'Stochastic modelling of the effects of reservoir operation', J. of Hydrol. 31, pp 97-113.

Hipel, K.W. and McLeod, A.I. (1985) Time Series Modelling for Water Resources and Environmental Engineers, Elsevier, Amsterdam.

Hipel, K.W. (1985) 'Time series analysis in perspective', Water Res. Bull. 21(4), pp 609-624.

Kashyap, R.L. and Rao, A.R. (1973) 'Real time recursive prediction of river flows', Automatica 9, pp 179-183.

Kashyap, R.L. and Rao, A.R. (1976) 'Dynamic-stochastic models from empirical data', Academic Press, New York.

Katayama, T. (1976) 'Application of maximum likelihood identification to river flow prediction', IIASA Workshop on real time forecasting/control of water resources systems, E. Wood (Ed.), Pergamon Press.

Lawrence, A.J. and Kottegoda, N.T. (1977) 'Stochastic modeling of river-flow time series', J.R. Statist. Soc. Ser. A, 140, pp 1-27.

Ledolter, J. and Abraham, B. (1981) 'Parsimony and its importance in times series forecasting', Technometrics 23(4), pp 411-414.

Lorent, B. (1975) 'Test of different river flow predictors', in Modelling and Simulation of Water Resources Systems, G.C. Vansteenkiste (Ed.), North Holland, Amsterdam.

Marco, J. and Yevjevich, V. (1985) 'Stochastic modelling of ground-water recharge', 21st Congress IAHR, Proceedings, Melbourne, Australia.

Miller, R.B., Bell, W., Ferreiro, O and Wang R.Y. (1981) 'Modeling daily river flows with precipitation input', Water Resour. Res. 17(1), pp 209-215.

Mizumura, K. and Chiu, C.L. (1985) 'Prediction of combined snowmelt and rainfall-runoff', ASCE J. of Hyd. Eng. 2, pp 179-193.

Moore, R.J. (1981) 'Transfer functions, noise predictors and the forecasting of flood events in real time', Prec. Int. Conf. Rainfall-Runoff Modelling.

Natale, L. and Todini, E. (1976) 'A stable estimator for linear models', Water Resour. Res. 12(4), pp 667-676.

Natural Environment Research Council, UK (1975) Flood Studies Report, Vol. 1, chap. 7, pp 513-531, London.

O'Connell, P.E. and Clarke, R.T. (1981) 'Adaptative hydrological forecasting, a Review', Hydrol. Sci. Bull., 26(2), pp 179-205.

Oliver, J. and Marco, J.B. (1985) 'Real time management of an irrigation water resources system', Multivariate Analysis of Hydrologic Processes, Shen et al., Eds., Colorado State University, Fort Collins, pp 703-715.

Patry, G.G. and Mariño, M.A. (1984) 'Parameter identification of time varying noise difference equations for real-time urban runoff forecasting', J. of Hydrol. 72, pp 25-55.

Piccolo, D. and Ubertini, L. (1979) 'Flood forecasting by intervention transfer stochastic models', Proc. of IAHR 18th Congress, Vol. 5, pp 319-326.

Rissanen, J. (1974) 'Basis of invariants and canonical forms for linear dynamic systems', Automatica 10, pp 175-182.

Thompstone, R.M., Hipel, K.W. and McLeod, A.I. (1985) 'Forecasting quarter-monthly riverflow', Water Res. Bull., 21(5), pp 731-741.

Todini, E. and Bouillot, D. (1975) 'A rainfall-runoff Kalman filter model', in: System Simulation in Water Resources, G.C. Vansteenkiste, Ed. North Holland, Amsterdam.

Todini, E. (1975) 'The Arno River model. Problems, methodologies and techniques', in: Modelling and Simulation of Water Resources Systems, G.C. Vansteenkiste, Ed. North Holland, Amsterdam.

Todini, E. and Wallis, J.R. (1977) 'Using CLS for daily or longer period rainfall-runoff modelling', in Mathematical Models in Surface Water Hydrology, (Ed. by Ciriani, Maiore and Wallis), pp 148-168, J. Wiley, London.

Whitehead, P.G. and Young, P.C. (1975) 'A dynamic stochastic model for water quality in part of the Bedford-Ouse River System', in: Computer Simulation of Water Resources Systems, G.C. Vansteenkiste, Ed. North Holland, Amsterdam.

Whitfield, P.H. and Woods, P.F. (1984) 'Intervention analysis of water quality records', Water Res. Bull., 20(5), pp 657-667.

Yevjevich, V. (1972) 'Stochastic processes in hydrology', Water Resources Publications, Fort Collins, Colorado.

Young, P.C. (1974) 'Recursive approaches to time series analysis', Bull Inst. Math. Appl. 10, pp 209-224.

Young, P.C. (1986) 'Time series methods and recursive estimation in hydrological systems analysis', in: River Flow Modelling and Forecasting, D.A. Kraijenholl and J.R. Moll, Eds. Reide, The Netherlands.

STOCHASTIC MODELS OF TEMPORAL RAINFALL: REPRODUCIBILITY, ESTIMATION AND PREDICTION OF EXTREME EVENTS

P. BURLANDO and R. ROSSO
Institute of Hydraulics, Politecnico di Milano
Piazza Leonardo da Vinci, 32
20133 Milan
Italy

ABSTRACT. Temporal rainfall models based on the point-process theory are reviewed. First the compound Poisson approach is examined. Theoretical analysis concerned with the extreme properties is presented for the Independent Poisson Marks model, and an example of application to the real world is discussed. The clustered Poisson approach with rectangular pulses is then analyzed in terms of both the Neyman–Scott model, and the Bartlett-Lewis one. The available theoretical work is reviewed, and some problems arising in statistical inference are further analyzed. Criteria of model fitting and parameter estimation are also presented in the light of applications to real-world hydrology. Finally, an extensive field data analysis is presented, in order to assess the capability of the mathematical models investigated to represent the actual behaviour of the natural process.

1. Introduction

Water resources systems and planning require relatively long records of rainfall and runoff. Unfortunately, direct observations of historical series, when available, are often too short for the purpose. In other circumstances these series are available with too coarse a detail as dictated by the time scale required for the analysis. Finally, the lack of direct observations is a common problem in hydrological practice. Therefore, these must be replaced with data generated by means of appropriate mathematical techniques. In this light, precipitation can be seen as the major issue of diverse hydrological investigations, since it forms the basic input of hydrological systems.

Recent trends in hydrological research reveal an increasing need for systematic and accurate analyses of precipitation fields. The solution to many hydrological problems is intimately related to a better knowledge of the rainfall process, such as the development of reliable streamflow and groundwater models, the prediction of floods and droughts, the design of data collection networks, among others.

These problems require the mathematical representation of the rainfall process over a wide range of time and space scales. Such a requirement can

137

J.B. Marco et al. (eds.),
Stochastic Hydrology and its Use in Water Resources Systems Simulation and Optimization, 137–173.
© 1993 *Kluwer Academic Publishers.*

be met either by using different models at any scale of interest or a comprehensive mathematical description capable of preserving the fluctuations of the investigated process over this range of scales. Although the first approach is straightforward, some fairly substantial problems arise from aggregation and disaggregation due to the intermittency and erratic fluctuations displayed by rainfall fields at short time scales. Continuous space-time modelling of rainfall fields can overcome these problems, but numerous difficulties arise from the mathematical representations needed to describe the space-time process, and from the availability of adequate data for the analysis. Accordingly, most of the theoretical work has been developed with respect to the temporal process of rainfall at a point in space. Although this cannot provide an exhaustive representation of the precipitation space-time structure, a number of theoretical results have been derived which can be viewed perspectively for application to real-world hydrology.

In the present paper some major results stemming from the development of the point process theory are briefly reviewed and discussed; some problems concerned with statistical inference are further analysed, and some examples of application to real-world data are presented.

2.   Space-time Variability of Rainfall Fields

Precipitation is caused by the cooling of moist air masses resulting from evaporation where tropical oceans are the major source of water vapour. Cooling can occur through different processes which always lead to a situation characterized by high non-stationarity, time-space variability and fluctuations.   These could be interpreted as a symptom not of deterministic or stochastic, but chaotic behaviour, which any attempt to model the rainfall fields must keep in mind.   However, a stochastic approach can currently provide the only effective route towards an engineering description of precipitation in the absence of satisfactory mathematical and physical representations of the laws governing its complexity.

Rainfall fluctuations are also embedded in the structure of clouds which has been observed to be non uniform.   A rough image of rain-producing clouds can be depicted as a random conglomeration of 'cells' of random size and 'mean life-time'; these move about in a random manner relative to the mean motion of the whole precipitation field.   Systematic studies on precipitation patterns have been carried out by several researchers from the meteorological point of view.   The results obtained by radar telemetering (see e.g. Austin and Houze, 1972; Harrold and Austin, 1974) showed high variability of rain intensity to occur over space intervals of less than one kilometer, and time intervals of less than one minute. Radar observations also confirmed the highly complex, and apparently random characteristics of clouds, because in all types of storms precipitation areas different in size, shape and intensity occur simultaneously. These can be categorized according to their horizontal extension in order to understand better those major patterns which must be taken into account in the modelling of rainfall fields.

As introduced by Austin and Houze (1972), the experimental surveys indicate that three major ranges of extension are exhibited by rainfall fields, namely:

(i) areas equal to, or larger than 10^4 km^2, known as 'synoptic areas', are typically associated with cyclonic storms. Their 'life span' lasts generally from one to several days;

(ii) a 'mesoscale area' is defined for ranges extending from $0.5 \cdot 10^2$ to 10^4 km^2. Within this range, large sub-synoptic precipitation areas (10^3 + 10^4) are referred to as the 'large mesoscale areas (LMSA)'. These generate and dissipate within a synoptic area in a variable number generally ranging from one to six, their life span being several hours. The LMSA move in relation to a moving synoptic region, sometimes moving even faster within it. Observations of rainfall intensity showed this to be always higher in an LMSA than in the region surrounding it. Smaller areas can also be identified within a mesoscale region; these are referred to as 'small mesoscale precipitation areas (SMSA)' ranging from 10^2 to 10^3 km^2, where rainfall intensity is higher than that in the LMSA. The average duration of SMSA is in the order of one hour;

(iii) regions of cumulus convective precipitation, commonly known as 'convective cells (CC)', which represent the smallest 'rainfall' structure that can be identified at present. These range from 10 to 30 km^2 in size depending upon the storm type. Convective cells generate and dissipate within an LMSA occurring generally in the form of clusters due to the tendency for new cells to generate in the proximity of existing cells. The latter fact has been explained by the action of the cold air spread out from existing cells, the downdraft of which gives an upward impulse to the neighbouring warm air resulting in the formation of a new cell. It has been observed that CC develop into three stages, namely the cumulus stage, the mature stage and the dissipation stage (Amorocho and Wu, 1977), along a life span ranging from several minutes to about half an hour. Due to their short duration, the movement of the CC takes on the movement of the LMSA, while rainfall intensity in the CC has been observed to exceed the one in the region surrounding them.

The above mentioned patterns, which are sketched in Figure 1, provide quite a general framework, despite these were clearly observed for cyclonic storms only. Similar features are indeed exhibited by other types of storms. Air mass thunderstorms are much less extended, and these obviously do not present synoptic or mesoscale regions, but still they occur in the form of cells, or cell clusters. The same consideration applies to frontal bands, which do not cover synoptic areas; however, these can be identified as LMSA containing several SMSA, as well as cells, and cell clusters.

One can observe that the consistent occurrence of precipitation areas at the sub-synoptic scale, which displays similarities both in pattern and in behaviour throughout the life span of the field, should allow for some description and modelling of the distribution of rainfall. As it will be shown in the following sections, several attempts have been carried out, mainly based on a stochastic approach, which allow for a satisfactory reproduction of rainfall occurrences and subsequent cell clustering in time. The basic idea stems from describing storm occurrences as a random point on the time axis, the arrival of which is represented by a counting process. The compound point-process theory can then be used to describe the random storm rainfall intensity generated by each arrival, and its duration. The rainfall structure within a storm can be described by cell arrivals clustering after each storm occurrence.

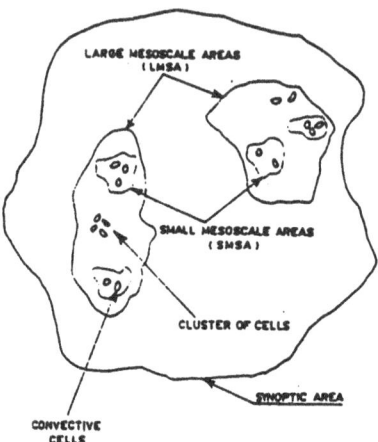

Figure 1. Precipitation patterns.

3. Stochastic Models of Temporal Rainfall Based on Poisson Arrivals

3.1. COMPOUND POISSON MODELS

As introduced above, the patterns of space-time variability previously outlined suggest that rainfall occurrences at a point in space can be viewed as a random collection of points along the time axis. Many mathematical tools available for the purpose have been proposed in the scientific literature, as pointed out by the excellent review by Waymire and Gupta (1981). Nevertheless, the adequacy of many of these fails with rainfall variability, particularly with respect to the non-linearity and non-normality displayed by precipitation series. Moreover, the cluster dependence, which is clearly exhibited by precipitation both in the space and time domains, requires extending the analysis beyond the second-order properties, in order to achieve a closer insight into the physical process involved. The approach which best accounts for these requirements is probably the point-process theory (see, e.g., Cox and Isham, 1980), as has frequently been indicated in the scientific literature (see, e.g., Waymire and Gupta, 1981). Within this framework it is indeed possible to derive a number of properties which allow for quite a complete description of the precipitation process. Moreover, the point process theory can be developed as easily in higher dimensional spaces as over the time domain, so that it allows for easy moving from the temporal analysis to the space-time one, as outlined by recently presented space-time models (see, e.g., Rodriguez-Iturbe et al., 1986; Rodriguez-Iturbe and Eagleson, 1987).

The random point process approach characterizes storm occurrences as

highly localized events distributed randomly along the time continuum.
Various models can be used for the counting process, but the Poisson
process seems to provide the best compromise between the conflicting needs
of simplicity and generality. Moreover, most of the properties of the
homogeneous Poisson process, which is strictly stationary, can be extended
to the non-homogeneous case, i.e. the time-dependent behaviour, without any
restriction. Therefore, one can handle the problem of modelling temporal
rainfall by considering the counting process to be stationary rather than
periodic, and further by analyzing the periodicity in the context of non-
homogeneity (see, e.g. Snyder, p. 39, 1975).

Different levels of complexity can be used to represent the above
mentioned precipitation features. First, each storm can be described as
a Poisson arrival to which a random depth of rainfall is associated in the
form of a random mark. This model, which does not provide any description
of the duration of the storm, has been referred to as the 'Poisson White
Noise Model' by Rodriguez-Iturbe et al. (1984). Eagleson (1972) proposed
a compound Poisson model where two random variables, representing the
average intensity and the duration of the storm respectively, are
associated with each point occurrence. Although it is recognized that this
model may not be entirely satisfactory in all circumstances and that more
complex formulations might be appropriate (see, e.g., Eagleson, 1978, 1981,
and following section 4), it has been extensively adopted in hydrological
practice, e.g. to derive flood frequency distribution from climate and
basin characteristics (see, e.g. Carlson and Fox, 1976; Caroni and Rosso,
1977; Chan and Bras, 1979; Hebson and Wood, 1982; Cordova and Rodriguez-
Iturbe, 1983; Diaz-Granados et al., 1984). This model is analyzed in the
following text together with more complex formulations. The present
analysis is first aimed at evaluating the capability of matching the
second-order statistical properties of temporal rainfall at different
scales of aggregation. However, the model performance should also be
investigated in terms of parsimonious model parametrization, a crucial
issue for reliable application to real-world data. Major emphasis is also
given to the model capability of reproducing the extreme properties of
rainfall records over a range of scales, another crucial issue in the light
of model application to prediction and design problems.

Compound Poisson models can be built either in terms of marked point
processes or in the form of continuous point processes. The marks involved
in the first approach include a random variable, i.e. the storm duration,
which somehow interacts with the timing of the process. Therefore, the
second approach is more correct than the first one, from a theoretical
viewpoint. However, this allows straightforward analytical results to be
achieved, on the one hand, and it has been extensively used in hydrological
practice due to its simplicity and physical meaningfulness, on the other.
Moreover, because the order of size point interarrivals is generally much
larger than the duration of a storm, such an interaction is generally weak.
Both these approaches are analyzed here, and their performances with
respect to extreme values are compared.

3.2. INDEPENDENT POISSON MARKS MODEL

In the case of the Independent Poisson Marks (IPM) model (Eagleson, 1972)
the occurrence of storm events are represented by Poisson arrivals at rate

λ; this is expressed by the probability mass function (pmf) of the counting process

$$P[N(0,t)=n] = \frac{(\lambda)^n \exp(-\lambda t)}{n!} \tag{1}$$

where $N(0,t)$ is the number of arrivals during a period of length t. Accordingly the expected value, and the variance of storm occurrences are given by expressions such as

$$E[N(0,t)] = \lambda t \tag{2}$$

$$Var[N(0,t)] = \lambda t \tag{3}$$

To each arrival we associate a random vector mark of two variables: the average storm intensity, i_s, and the storm duration, t_s. As previously mentioned, the values taken by the latter are further assumed to be negligible as compared with the interarrival time, w_s, of the storm occurrences, which is exponentially distributed with parameter λ, that is

$$f_{W_s}(w_s) = \lambda \exp(-\lambda w_s) \tag{4}$$

Consequently, this model represents the temporal precipitation process as a sequence of non-overlapping rectangular pulses, as shown in Figure 2. By specifying the probability density function (pdf) of i_s and t_s respectively the model is entirely characterized.

Various pdf's can be used to describe storm duration and intensity. Eagleson (1972) assumed both duration and intensity to be independent, identically distributed (iid), and to follow exponential distributions. This assumption allows for easy derivation of the properties of the modelled precipitation process, inclusive of its extreme behaviour. This aspect was investigated by Bacchi et al. (1987) who derived the analytical formulation of the cumulative distribution function (cdf) of the maximum storm intensity for both the instantaneous intensity process, $X(t)$, and the aggregated one, $X_T(t)$, which is obtained by locally integrating $X(t)$ over a moving interval of size T, namely

$$X_T(t) = 1/T \int_{t-T/2}^{t+T/2} X(u) \, du \tag{5}$$

The cdf of maximum storm intensity within a period of length τ is found to be

$$L_\tau(x) = \exp\{[-\lambda\tau (1-F_{i_s}(x))]\}, \qquad 0 < t \le \tau \tag{6}$$

for the instantaneous process, and

$$L_\tau(x_T) = \exp[-\lambda\tau \iint_{R(X_T)} f_{i_s,t_s}(v,u) \, dv \, du] \tag{7}$$

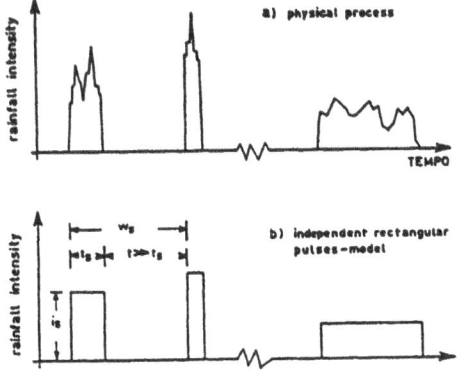

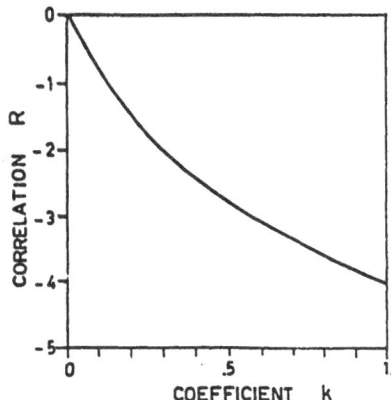

W_g = interarrival time of events (pulses)
t_g = pulse duration
i_g = pulse intensity

Figure 2. The temporal rainfall process and its representation according to the Independent Poisson Marks model.

Figure 3. Bivariate Exponential Distribution. Parameter k as a function of the correlation coefficient R (Gumbel, 1960).

for the aggregated process, with $Fi_g(.)$, $fi_g,t_g(.,.)$ denoting the storm intensity cdf, and the joint pdf of i_g and t_g, respectively, while $R(X_T)$ the region within which the local average exceeds threshold x_T.

By specifying $fi_g,t_g(.,.)$ one can thus evaluate the extreme value distribution of the instantaneous and locally averaged intensity using (5) and (6), respectively. A particular solution is obtained under the assumption of mutual independence for i_g and t_g, that is

$$fi_g,t_g(v,u) = fi_g(v) \ ft_g(u) \qquad (8)$$

which can be substituted in Equation 6 to give

$$L_T(x_T) = \exp \left[-\lambda\tau \{1 - Fi_g(x) [1 - Ft_g(T))] - \int_0^T ft_g(u) \ Fi_g(xT/u) \ du\} \right] (9)$$

where $Fi_g(.)$ and $Ft_g(.)$ denote the marginal cdf's of storm intensity and duration, respectively. Equations 6 and 9 can be put in dimensionless form, through rescaling storm duration according to its expectation, $E[t_g]$, and storm intensity according to its expectation, $E[i_g]$. This approach was followed by Bacchi et al. (1989) who recently investigated three different alternatives in the choice of a probability model for the joint random variables of storm duration and intensity. They assumed i_s and t_s to be (i) iid according to exponential pdf's (Eagleson, 1978), (ii) iid according to gamma pdf's (Hebson and Wood, 1982), (iii) negatively correlated according to a bivariate exponential pdf in the form introduced by Gumbel (1960).

Model (iii) has been shown to give the most interesting results, as will be outlined in the following Section 6.1, due to the accounting for negative correlation between storm duration and its average intensity. This is taken into the model by assuming

$$F_{i_s,t_s}(v,u) = 1 - \exp(-v/\mu_I) - \exp(-\delta u) + \exp(-v/\mu_I - \delta u - \delta kuv/\mu_I) \qquad (10)$$

where marginals $f_{i_s}(.)$ and $f_{t_s}(.)$ are exponential with parameter $1/\mu_I$ and δ respectively, and k is a parameter describing correlation $R(i_s,t_s)$ between i_s and t_s, which takes values between 0 and 1. Since it holds (Gumbel, 1960)

$$R(i_s,t_s) = \int_0^\infty (1+ky)^{-1} \exp(-y) \, dy \qquad (11)$$

this model can reproduce negative correlation between i_s and t_s, with correlation coefficients from -0.40365 to 0, as shown in Figure 3. It can also be seen that the independent exponential model will therefore correspond to the particular case for k=0. The corresponding pdf is given by

$$f_{i_s,t_s}(v,u) = (\delta/\mu_I) \left[(1+kv/\mu_I)(1+\delta ku)-k\right] \cdot \exp(-v/\mu_I - \delta u - \delta kuv/\mu_I) \qquad (12)$$

which can be used in (7) to obtain

$$L_{\gamma t}(x_{\gamma t}) = \exp\{-M[\exp(-x^* - T^* - kx^*T^*) + \\ + \int_0^{T^*} (1+kx^*T^*/u) \exp(-u - kx^*T^* - x^*T^*/u) \, du]\} \qquad (13)$$

with $x^* = x/\mu_I$, $T^* = \delta T$, and $\tau^* = \beta\tau$. Also for model (iii) $L_{\gamma t}(x^*)$ will still be obtained by Equation 16, that is

$$L_{\gamma t}(x^*) = \exp[-M \exp(-x^*)] \qquad (14)$$

It can be observed that for T tending to zero, Equation 13 reduces to Equation 14, that is, the locally averaged extreme value tends to the instantaneous one, which is distributed according to the EV1 distribution.

The above models have been tested in Bacchi et al. (1989) with respect to model capability in reproducing average storm intensities in the intensity-duration-frequency domain. A summary of the results is presented in the following Section 6.1, while the reader is referred to the paper by Bacchi et al. (1989) for further theoretical details.

3.3. POISSON RECTANGULAR PULSES MODEL

In addition to the IPM model, the Poisson Rectangular Pulses (PRP) model describes the temporal evolution of the rainfall process as a sequence of rectangular pulses associated with a Poisson process of arrival rate λ. The Poisson arrivals denote the occurrence of rainfall events, but the word "event" only has a suggestive connotation in the present context. In fact, different storm occurrences may overlap, in the same way that they do not need separation to identify independent rainstorms. The events are characterized by a rectangular pulse which starts at the time of occurrence of the Poisson point process. Each pulse is characterized by a random height, i_r, representing the average rainfall intensity, and a random duration, t_r, as sketched in Figure 4. It is assumed that the event characteristics are independent of the occurrence time and, furthermore, they are independent and identically distributed (iid) random variables.

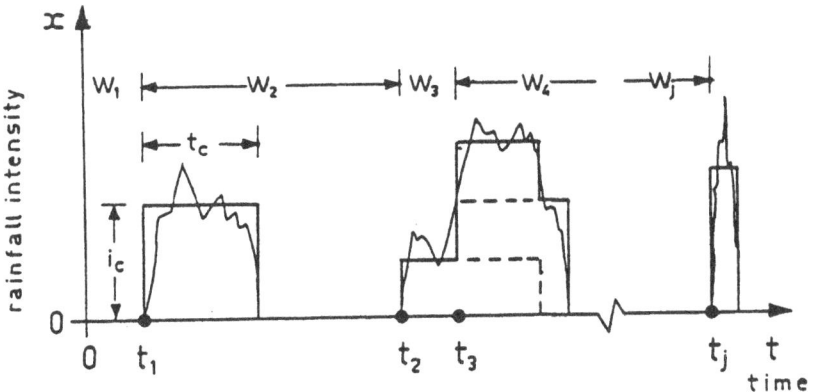

w_j = interarrival time of events (pulses)
t_c = pulse duration
i_c = pulse intensity

Figure 4. The temporal rainfall process as represented by the Poisson Rectangular Pulses model.

In order to achieve analytical results in a simple form, it is usually assumed that i_r and t_r follow the exponential distribution. That is

$$f_{i_r}(i_r) = 1/\mu_x \ \exp(-i_r/\mu_x) \tag{15}$$

$$f_{t_r}(t_r) = \delta \ \exp(-\delta t_r) \tag{16}$$

The stochastic properties of this model were studied in detail by Rodriguez-Iturbe et al. (1987b). Second-order analysis yields to

$$m \quad = E[X(t)] = \lambda\mu_x\delta^{-1} \tag{17}$$

$$\sigma^2 \quad = \text{Var} \ [X(t)] = 2\lambda\mu_I{}^2\delta^{-1} \tag{18}$$

$$R(s) = \text{Corr}[X(t), X(t+s)] = \exp(-\delta s) \qquad s \geq 0 \tag{19}$$

The scale of fluctuation, θ, of the process $X(t)$ provides a global measure of the correlation structure of the process, which allows assessment of the model's capability of reproducing the dependence structure of the physical process. Moreover, it gives the time interval required to obtain stable (low variance) estimates of the mean of the fluctuating process of rainfall intensity. This can be obtained either as the zero-frequency ordinate of the one-sided unit-area spectral density function (Taylor, 1921), or by integrating the autocorrelation function over the whole real-time axis (see Vanmarcke, p. 187, 1983), namely

$$\theta = 2 \int_0^{+\infty} r(t) \ dt \tag{20}$$

Straightforward application of (19) and (20) gives the PRP model

$$\theta = 2\delta^{-1} \tag{21}$$

The crossing properties of $X(t)$ were studied by Rodriguez-Iturbe and Cox (1986) and Rodriguez-Iturbe et al. (1987b), to whom the reader can refer for further theoretical details. Accordingly, under the Poisson assumption for the crossing process, the cdf of the maximum storm intensity (instantaneous) within a period of length τ can be written as

$$L_\tau(x) = \exp\left[-\lambda\tau \ \exp(-x/\mu_I-\lambda/\delta) \ I_0[2\sqrt{\lambda}x/(\mu_I\delta)]\right] \tag{22}$$

where $I_0[.]$ is the modified Bessel function of zero order. By substituting dimensionless formulation of the parameters, namely

$$\lambda = \lambda*\delta \qquad\qquad x = x*\mu_I \qquad\qquad \tau = \tau*\delta^{-1} \tag{23}$$

for λ, x and τ in Equation 22 one obtains

$$L_{\tau*}(x*) = \exp\left[-\lambda*\tau* \ \exp(-x*-\lambda*) \ I_0(2\sqrt{\lambda}*x*)\right] \tag{24}$$

One can observe that for small $\lambda*$, $I_0(2\sqrt{\lambda}*x*)$ is approximatively unitary, and Equation 24 tends to overlap with Equation 6, that is the PRP model performs similarly to the IPM.

3.4. PARAMETER ESTIMATION

Both IPM and PRP models require the estimation of λ, μ_I and δ. This can be worked out from the observed data which are either in the form of measured depth over fixed duration or are already available in the event-aggregated form. In the first case Equations 17, 18 and 19 can be used to estimate the PRP model for the iid case. These can also be used for the IPM model in the (i) and (ii), according to what has been explained in Section 3.2.

Alternatively, the parameter λ can be estimated by evaluating the mean number of storms within a reference period. This can be identified either as a calendar reference period, say one year or one month, or as a physically homogeneous period, for instance a rainy season. The arrival rate λ can also be evaluated by fitting the exponential distribution to the observed series of interarrival times. Eagleson (1978) observed that these two methods of estimation often produce different values of λ. The parameters μ_{τ} and δ are then estimated by fitting the distributions which are assumed to be representative to the observed series of intensity and duration of storm events.

The estimation procedure, which is relatively simple and straightforward, has its crucial point in the identification of storm events. This is often a subjective task, since no standard methods for identifying individual storms are adopted in hydrological practice. As pointed out by Bonta and Rao (1988), two main factors among others affect the subjective separation of independent storms, namely the sampling interval of the precipitation data and the method of computing the "critical duration" which separates storm periods (on average) from each other. An "objective procedure" (see, e.g., Bonta and Rao, 1988), therefore, should be adopted to identify storm events, in order to reduce the effect of arbitrary separation techniques on the model performance.

A special procedure can be derived to fit the model for the purpose of reproducing the extreme precipitation process (Bacchi et al., 1989). In this case the estimation of the parameters of the models has been performed by means of an approximate procedure which relates the EV1 parameters, as obtained by an intensity-duration-frequency analysis of the storm data, to the parameters of the model. This procedure relies on the availability of the extreme rainfall observations for a number k of selected durations, and on the preservation of the extreme properties for a given range of integration intervals. Since for time intervals lower than the average storm duration, the instantaneous and averaged process are characterized by the same extreme cdf, the data of maximum averaged intensity over the lower available duration, T_L, can be taken to be representative of the instantaneous ones. Accordingly the three parameters μ_{τ}, λ, and δ^{-1} can be expressed as a function of the scale and location parameters of the EV1 cdf fitted for T_L and T_U, that is

$$\mu_{\tau} = a_L \tag{25}$$

$$M = \lambda\tau = \exp(b_L/a_L) \tag{26}$$

$$\delta \approx -(y^*_U/T_U) \ln(x_U/\mu_{\tau}y^*_U) \tag{27}$$

where (27) is an approximate solution of Equation 23 valid for y^*_U close to 5, and x_U and y^*_U are obtained for a fixed quantile L, by means of

$$x_U = b_U + a_U(-\ln(-\ln(L))) \tag{28}$$

and

$$y^*_U = \ln M - \ln(-\ln(L)) \tag{29}$$

4. Stochastic Models of Temporal Rainfall Based on Clustered Poisson Arrivals

As shown by Rodriguez-Iturbe (1986), the compound Poisson models can give reasonably flexible models for the temporal rainfall process at a particular level of aggregation, but they can be very limited when a range of time scales is contemplated. Models based upon clustered point processes have therefore been developed for the purpose of aggregating and disaggregating temporal rainfall records through a continuum of scales (e.g. Kavvas and Delleur, 1981; Rodriguez-Iturbe et al., 1984; Rodriguez and Bras, 1985; Rodriguez-Iturbe, 1986; Rodriguez-Iturbe et al., 1987b). All these models are based upon the Poisson process to represent storm arrivals but they differ in the storm structure. Storm origins still occur in a point process, but each point is replaced by a cluster of points, where the original point is regarded as the cluster centre. As for the PRP model, with each point of the clustered point process we associate a rectangular pulse, the physical meaning of which is taken as the occurrence of a storm cell with a given intensity and duration.

Two families of clustered processes can be adopted for this purpose, the Neyman-Scott (NS) process and the Bartlett-Lewis (BL) process, the properties of which have been extensively studied (see, for example, Cox and Isham, Sec. 3.4, 1980). The difference between the two concerns the way in which the cells are distributed in time, but it is very unlikely that empirical analysis of data can be used to choose between these. Both processes are based on a Poisson process of storm origins. To each origin is associated a random number of cells. In the NS process the positions of these cells in the time axis are given by a set of iid random variables, where the location of the distribution is determined by the time origin. In the BL process, the interarrival times of successive cells are iid random variables.

4.1. NEYMAN-SCOTT CLUSTERING OF RECTANGULAR PULSES

Storm origins occur in a Poisson process of rate λ, and a random number C of cell origins is associated with each storm, C being not less than one. The cell origins are independently displaced from the storm origin, so that the waiting time of each cell origin from the storm origin is an iid random variable, which is assumed to be exponentially distributed. No cell origin will therefore be located at the cluster centre, i.e. at the origin of the storm. As before, a rectangular pulse is independently associated with each cell origin, its duration and intensity being independent random variables which are assumed to be exponentially distributed. Independence among clusters of cells associated with different storms is also assumed. A schematic depiction of the model based on this process is shown in Figure 5.

The second-order properties of this process have been studied in detail by Rodriguez-Iturbe and Cox (1986). The cells are assumed to be geometrically distributed, that is,

$$P[C=c] = \frac{(1 - 1/\mu_c)^{c-1}}{\mu_c} \tag{30}$$

where μ_c is the mean number of cells in a cluster, and the origins of cells are displaced from the storm origin according to an exponential distribution, namely

$$f(t_d) = \beta \exp(-\beta t_d) \qquad\qquad \beta > 0 \qquad\qquad (31)$$

where β^{-1} is the mean displacement of a cell from the storm origin. An exponential distribution is also assumed to characterize the intensity and duration of a cell, namely

$$f(i_c) = 1/\mu_I \exp(-1/\mu_I i_c) \qquad\qquad \mu_I > 0 \qquad\qquad (32)$$

$$f(t_c) = \delta \exp(-\delta t_c) \qquad\qquad \delta > 0 \qquad\qquad (33)$$

where μ_I and δ^{-1} are the mean intensity and the mean duration respectively. Under these hypotheses the second-order properties of the instantaneous intensity process $X(t)$ are given by

$$E[X(t)] = \lambda \mu_c \mu_I / \delta \qquad\qquad (34)$$

$$\mathrm{Var}[X(t)] = \frac{\lambda \mu_c \mu_I^2}{\delta} \left[2 + \frac{\beta(\mu_c - 1)}{(\delta + \beta)} \right] \qquad\qquad (35)$$

$$\mathrm{Cov}[X(t), X(t+u)] = \frac{\lambda \mu_c \mu_I^2}{\delta} \left[2 - \frac{\beta^2(\mu_c - 1)}{(\delta^2 - \beta^2)} \right] e^{-\delta u} + \frac{\lambda \mu_c (\mu_c - 1) \mu_I^2 \beta}{(\delta^2 - \beta^2)} e^{-\beta u} \qquad (36)$$

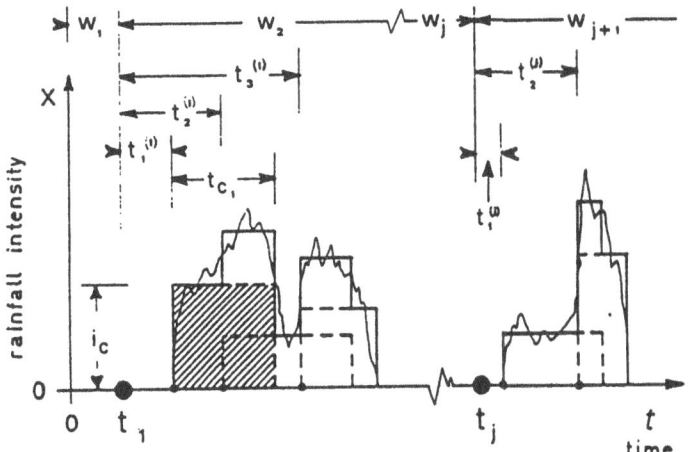

$t_i^{(j)}$ = displacement of the j-th cell from the cluster center
w_j = interarrival time of events (pulses)
t_{ci} = duration of the i-th pulse
i_{ci} = intensity of the i-th pulse

Figure 5. The temporal rainfall process as represented by the Neyman-Scott clustering of rectangular pulses.

Similarly, the same properties are derived for the aggregated process $X_T(t)$ in the form

$$E[X_T(t)] = T\lambda\mu_c\mu_x / \delta \tag{37}$$

$$Var[X_T(t)] = \frac{\lambda\mu_c\mu_x^2}{\delta^3} [2- \frac{\beta^2(\mu_c-1)}{(\delta^2-\beta^2)}](\delta T-1+e^{-\delta T}) + \frac{\lambda\mu_c(\mu_c-1)\mu_x^2}{\beta(\delta^2-\beta^2)} (\beta T-1+e^{-\beta T}) \tag{38}$$

$$Cov[X_T^i, X_T^{i+k}] = \frac{\lambda\mu_c\mu_x^2}{\delta^3} [2- \frac{\beta^2(\mu_c-1)}{(\delta^2-\beta^2)}] (1-e^{-\delta T})^2 e^{-\delta(k-1)T} +$$
$$+ \frac{\lambda\mu_c(\mu_c-1)\mu_x^2}{\beta(\delta^2-\beta^2)} (1-e^{-\beta T})^2 e^{-\beta(k-1)T} \tag{39}$$

Following Vanmarcke (1983), the scale of fluctuation of the process $X(t)$ - see Equation 20 - is obtained as

$$\theta_{NS} = 2\delta^{-1} \frac{(\mu_c+1)}{2+\dfrac{(\mu_c-1)\beta}{(\delta+\beta)}} \tag{40}$$

4.2. BARTLETT-LEWIS CLUSTERING OF RECTANGULAR PULSES

As before, storm origins arise in a Poisson process of rate λ. Each origin is followed by a Poisson process of rate ϵ, of cell origins. Therefore, the rain cells are not placed randomly from a cluster centre which is not itself a rain cell, as with the NSRP model, but the storm begins with the arrival of the cluster centre which is a cell in itself. Then, the occurrence of cells in time is characterized by an exponential distribution, namely

$$f(t_i = \epsilon \exp(-\epsilon t_i) \qquad \epsilon > 0 \tag{41}$$

which governs the time lapse between subsequent cells, as shown in Figure 6. After a time, exponentially distributed with rate α the process terminates. All other assumptions made for the NSRP model remain the same for the BLRP models, that is exponential iid cell intensity with rate μ_x which is independent of the duration of cells, also taken as exponential iid random variables with mean δ^{-1}.

The number of cells per storm thus has a geometrical distribution the mean of which is given by

$$m_c = 1 + \epsilon/\alpha \tag{42}$$

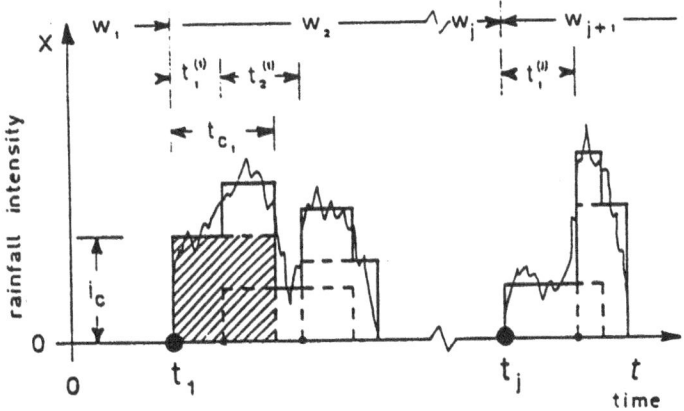

$t_i^{(j)}$ = interarrival time between cells
w_j = interarrival time of events (pulses)
t_{ci} = duration of the i-th pulse
i_{ci} = intensity of the i-th pulse

Figure 6. The temporal rainfall process as represented by the Bartlett-Lewis clustering of rectangular pulses.

The mean and variance of the rainfall intensity process, $X(t)$, are given by Rodriguez-Iturbe et. al. (1987b) as follows:

$$E[X(t)] = \lambda \mu_c \mu_x / \delta \tag{43}$$

$$Var[X(t)] = \frac{\lambda \mu_c \mu_x^2}{\delta} \left[2 + \frac{\beta(\mu_c-1)}{\beta+\delta(\mu_c-1)} \right] \tag{44}$$

$$Cov[X(t), X(t+u)] = \frac{\lambda \mu_c \mu_x^2}{\delta} \left[2 + \frac{\beta^2}{\beta^2-\delta^2(\mu_c-1)^2} \right] e^{-\delta u} -$$

$$- \frac{\lambda \mu_c \mu_x^2}{\delta} \left[2 + \frac{\beta \delta (\mu_c-1)^2}{\beta^2-\delta^2(\mu_c-1)^2} \right] e^{-\alpha u} \tag{45}$$

Similarly, the aggregated process properties are written as

$$E[X_T(t)] = T \lambda \mu_c \mu_x / \delta \tag{46}$$

$$Var[X_T(t)] = \frac{2 \lambda \mu_c \mu_x^2 T}{\delta^2} \left[2 + \frac{\epsilon}{\alpha} \right] + 2 \lambda \mu_c \mu_x^2 \epsilon \frac{(1-e^{-\alpha T})}{\alpha^2(\alpha^2-\delta^2)} -$$

$$- \frac{2 \lambda \mu_c \mu_x^2}{\delta^2} \left[2 + \frac{\epsilon \alpha}{\alpha^2-\delta^2} \right] (1-e^{-\delta T}) \tag{47}$$

$$Cov[X_T^i, X_T^{i+k}] = \frac{\lambda\mu_c\mu_I^2}{\delta^3} [2+ \frac{\epsilon\alpha}{\alpha^2-\delta^2}] (1-e^{-\delta T})^2 e^{-\delta(k-1)T} -$$

$$- \lambda\mu_c\mu_I^2\epsilon \frac{(1-e^{-\alpha T})}{\alpha^2(\alpha^2-\delta^2)} e^{-\alpha(k-1)T}$$

(48)

Similarly to NSRP, a global measure of the correlation structure is still given by the scale of fluctuation, which is obtained according to Vanmarcke (1983) as

$$\theta_{BL} = 2\delta^{-1} \frac{\mu_c+2}{2+\frac{\epsilon\mu_c}{\delta(\mu_c+1)+\epsilon}}$$

(49)

An easier accessibility of the Bartlett-Lewis process to mathematical analysis also allowed the theoretical derivation of various properties which are detailed by Rodriguez-Iturbe et al. (1987b). Some of these can be of major interest for a more practical application. Among these, which are left to the original derivation of Rodriguez-Iturbe et al. (1987b), the probability of zero rainfall (i.e. a dry period) can be used for estimation purposes, while the mean duration of an event and the mean duration of a dry period can be used for testing the performances of the model.

4.3. PARAMETER ESTIMATION VIA OPTIMIZATION

The general problem of parameter estimation within the point process framework has been widely approached in statistical literature (see, e.g., Kutoyants, 1984). Nevertheless the application of the available methods often results in more complex problems than the theoretical approach involved.

The fitting of temporal rainfall models based on point processes can be obtained either by the method of moments or by that of the maximum likelihood, provided that the features of aggregated rainfall over a certain interval of time are available from the sample data set. On the one hand the likelihood function is often difficult to evaluate analytic-ally for point processes and for large data set (see, e.g. Foufoula-Georgiu and Guttorp, 1987; Smith and Karr, 1985a,b). Nevertheless, in view of its higher efficiency, it should be used any time the small data set and the point process involved allow for its evaluation.

On the other hand, the method of moments often yields simpler expressions for the estimate. The expressions for theoretical moments have to be equated to estimates of these moments as obtained from the sample data set, using as many equations as the unknown parameters. Due to the fact that high-order moments often yield to cumbersome relationships and are difficult to estimate accurately, low-order moments and their combinations at different levels of aggregation are used for the estimation (e.g., mean, variance and lag-one autocorrelation at one hour, and variance and lag-one autocorrelation at twenty-four hours, for the NSRP model). Table 1 shows an extended set of combined equations for the purpose.

TABLE 1. Equation sets for parameter estimation using rainfall data at different levels of aggregation for NSRP and BLRP models

a) NSRP model

Equation No. Set	30'	1 hr	3 hr	6 hr	12 hr	24 hr
1		$\mu,\sigma^2,r(1)$		$r(1),\sigma^2$		
2		$\mu,\sigma^2,r(1)$			$r(1),\sigma^2$	
3		$\mu,\sigma^2,r(1)$				$r(1),\sigma^2$
4			$\mu,\sigma^2,r(1)$	$r(1),\sigma^2$		
5	$\mu,\sigma^2,r(1)$			$r(1),\sigma^2$		
6	$\mu,\sigma^2,r(1)$		$r(1),\sigma^2$			
7	$\mu,\sigma^2,r(1)$				$r(1),\sigma^2$	
8	$\mu,\sigma^2,r(1)$					$r(1),\sigma^2$
9				$\mu,\sigma^2,r(1)$		$r(1),\sigma^2$

b) BLRP model

Equation No. Set	30'	1 hr	3 hr	6 hr	12 hr	24 hr
1		$\mu,\sigma^2,r(1),\pi_0$		$r(1)$		
2		$\mu,\sigma^2,r(1)$				$r(1),\pi_0$
3	$\mu,\sigma^2,r(1)$					$r(1),\pi_0$
4	$\mu,\sigma^2,r(1)$			$r(1),\pi_0$		
5	$\mu,\sigma^2,r(1)$	$r(1),\pi_0$				
6				$\mu,\sigma^2,r(1)$		$r(1),\pi_0$
7	$\mu,\sigma^2,r(1)$		$r(1),\pi_0$			
8		$\mu,\sigma^2,r(1)$		$r(1),\pi_0$		
9		$\mu,\sigma^2,r(1)$			$r(1),\pi_0$	

The non-linear system of equations which results from the method of moments can be solved via optimization by minimizing a linear combination of the involved equations (Burlando, 1989). In fact, if one assumes that $Z = (z_1, \ldots, z_j)$ is the parameter vector and $f_j(Z)$ is the j-th moment equation, the non-linear system can be written as

$$f_1(Z) = w_1$$
.
.   (50)
.
$$f_j(Z) = w_j$$

where $W = (w_1, \ldots, w_j)$ denotes the vector containing the statistical properties of the sample data set. The system is solved, that is the parameter vector Z is obtained, by minimizing the objective function expressed as

$$F = \sum_{1}^{j} f_i*(Z) \tag{51}$$

where the right-side of the equation is written as

$$f_i*(Z) = [\frac{f_i(Z)}{w_i} - 1]^2 \tag{52}$$

The estimation procedure is therefore reduced to an unconstrained minimization problem which can be solved by any proper method. In the applications hereafter referred to it has been solved by means of the Rosenbrock's algorithm (Rosenbrock, 1960). This choice, stemming in fact from a direct search method, although very sensitive to initial conditions, seems to help in finding the "absolute" minimum of the objective function, which is characterized by a large number of relative minima.

4.4. CAN THE EMBEDDED STRUCTURE OF CLUSTERING BE IDENTIFIED?

The structural differences between the NS and the BL process are not sufficient to discriminate between them, thereby indicating what approach can best fit the empirical data. Moreover, data analysis does not appear to provide any determining information to this end. Nevertheless, analysis of the second-order properties of these two models can provide some useful arguments for meaningful comparison. The variance and the scale of fluctuation of the process will therefore be used to develop a tentative objective criterion for the choice of the model.

The difference between the two models stems from the way in which cells are distributed in time. Looking at the physical meaning of the parameters and variables involved, it is clear that these, namely λ, the storm arrival rate which depends on the meteorological mechanism, δ and μ_x, the cell describing parameters, and μ_c, the mean number of cell production in a storm, from a purely physical point of view, should take the same value in both formulations. Accordingly, the mean of the precipitation intensity process proves to be the same as shown by Equations 34 and 43. Under this hypothesis, some considerations can be issued on the scale of fluctuation of the two processes, aimed at model comparison.

By introducing the following dimensionless parameters

$$\lambda* = \lambda/\delta; \qquad \beta* = \beta/\delta; \qquad \tau* = \tau/\delta \tag{53}$$

the variance and the scale of fluctuation can be expressed as

$$\sigma^2_{NS*} = \frac{\sigma^2_{NS}}{\lambda*\mu_c\mu_x} = [2 + (\mu_c-1)\frac{\beta*}{1+\beta*}] \tag{54}$$

$$\Theta_{NS*} = 2 \; \frac{\mu_c + 1}{2 + \dfrac{(\mu_c - 1)\beta*}{1+\beta*}} \tag{55}$$

for the NS process, and

$$\sigma^2_{BL*} = \frac{\sigma^2_{BL}}{\lambda * \mu_c \mu_I} = [\; 2 + \frac{\epsilon * \mu_c}{1+\mu_c+\epsilon*}\;] \tag{56}$$

$$\Theta_{BL*} = 2 \; \frac{\mu_c + 2}{2 + \dfrac{\epsilon * \mu_c}{1+\mu_c+\epsilon*}} \tag{57}$$

for the BL process. If we assume that the two models should have the same variance, a relationship between the key parameters of cell time distribution, β and ϵ, can be obtained in the form

$$K = (\mu_c - 1) \; \frac{\beta*}{1+\beta*} = \frac{\epsilon * \mu_c}{1+\mu_c+\epsilon*} \tag{58}$$

This yields the expression of Θ_{BL} for a model which preserves the mean and variance of the NS process, namely

$$\Theta_{BL*=NS*} = 2\;[\; 2 + \frac{\epsilon * \mu_c}{1+\mu_c+\epsilon*}\;]^{-1} + \Theta_{NS*} \tag{59}$$

where the first of the right-hand members can be indicated as K'.

In the case of the BLRP and NSRP models presenting the same mean and variance of the continuous process, Equation 59 shows that the BLRP's scale of fluctuation always exceeds that of the NSRP. According to the definition of the scale of fluctuation reported by Equation 20, the BLRP model therefore displays a much richer correlation structure than the NSRP.

Moreover, following the definition given by Vanmarcke (1983), that is

$$\theta = \lim_{T \to \infty} T \; g(T) \tag{60}$$

where T is the level of temporal aggregation of the process $X(t)$, and $g(T)$ is the variance function of the same process, which measures the reduction of the point variance under local averaging, it could be argued that a higher scale of fluctuation denotes a limit of the model in reproducing the short-time averaged characteristic of the process, namely when T is relatively small. Therefore, it can be seen that the scale of fluctuation gives a measure of the lower limit of the model's sensitivity to represent the short-time fluctuations of the process. From this point of view, the NS approach should therefore provide a better representation of low-band fluctuations of the continuous rainfall intensity process, as well as of the extreme one. Further details will be introduced in the following section, when the results from the simulation runs will be discussed. Implications of these theoretical results will also influence the reproducibility of the extreme process, as it is discussed ahead.

5. Continuous Simulation of Temporal Rainfall

As first pointed out by Eagleson (1978, 1981) and further observed by Rodriguez-Iturbe et al. (1984) and Rodriguez-Iturbe (1986), the compound Poisson models do not provide an adequate description of the features of the natural process through a continuum of scales. Their use, therefore, is more confined to applications where a simple mathematical structure is requested, in order to derive analytically the properties of the processes which are mainly driven by rainfall, than to simulation problems. In fact, their ability to preserve the historical characteristics of the rainfall data is generally confined to the scale at which the model (i.e. the parameters) has been estimated. Moreover, the structure of the compound Poisson models do not attempt any description of the internal structure of the storms, although the PRP allows for overlapping of pulses, so that a satisfactory description of the historical process can be expected for rainfall data at aggregated periods which are of the order of a typical storm event. For these reasons only results obtained by clustered point processes are presented with respect to the continuous simulation of temporal rainfall.

The higher flexibility of the cluster-based approach is initially evident from the structure of the models, which explicitly accounts for the internal description of storms. Subsequently, the results obtained from extensive application of the NSRP and BLRP models (see, e.g., Rodriguez-Iturbe et al., 1987a; Burlando, 1989) show that good results can be obtained through a continuum of scales in preserving the characteristics of the historical records. Nevertheless, an important limitation is still evident in disaggregating data below time scales corresponding to the mean life time of a rain cell, which is the basic element of the model structure.

Both the NS and BL approaches provide a satisfactory reproduction of the natural process, but the structural difference between the two processes does not allow for understanding which can be the proper descriptor based on the empirical analysis of the historical rainfall data. The extensive application of the two models, NSRP and BLRP, to several sets of data which are characterized by different climatic characteristics, length and quality of data span is therefore needed in order to assess the practical feasibility of temporal modelling by these approaches.

The NSRP and BLRP have been investigated in depth with respect to the simulation of the continuous process, by analyzing their ability to preserve the historical patterns of rainfall at stations which are different in terms of hydrological regime and length of data set (Burlando, 1989). In the following text some results are reported for the Ximeniano Station, which is located in the urban area of Florence in the Arno River basin of central Italy. A set of twenty-five years of five minute data was available there. Some reference is also made to applications carried out by the Centro Station, which is located in northern Italy in the sub-alpine catchment of the Rio della Gallina river, where a good set of only three years of five minute data was available for model estimation. The results from this application allow some comments to be made about the robustness of the models.

First it can be seen that the estimated parameters of the models are characterized by two types of variability, as clearly shown in Table 2. The first one, according to the estimation procedure described in the previous Section 4.3, depends on the scale of temporal aggregation of first and second-order moments used in the estimation. This dependence is higher for the BLRP than for the NSRP (Burlando, 1989), although the estimated values are shown to be meaningful with respect to the related physical quantities which they represent. The second variability, which is strictly related to climatic and seasonal non-homogeneities, reveals the need for identification of physically homogeneous reference periods (e.g. seasons or months) in order to obtain stable and reliable estimates of the parameters, and to perform a correct fit of the model to the historical rainfall patterns. Accordingly, a sensitivity analysis of the effect of climatic and seasonal non-homogeneities was performed by applying both NSRP and BLRP on different scales, namely on the seasonal scale for the Centro Station, and on the annual and monthly scale for the Florence Station.

The NSRP model is shown to be capable of preserving quite satisfactorily the first and second-order historical moments while also the lag-one to lag-three autocorrelations are correctly reproduced, both in the case of yearly and monthly scales of simulation. Similarly, the historical and simulated probabilities of dry and small rainfall periods concur, but the model shows a general tendency to overestimate the historical pattern, which is larger for small scales of aggregations and for the percentage of dry intervals. Nevertheless, the concurrence is quite satisfactory, as summarized in Figure 7 for the month of May at the Florence Station. This kind of behaviour is generally also valid for the other months, as well as for the model applied on the yearly scale. As was expected, the performance of the model can vary according to the size of the historical data set. In fact, the results obtained for the Florence Station, where a large record was available, are better than the ones obtained for the Centro Station, where only a three year record was available. However, even in this case, the historical pattern is preserved and the tendency to errors is generally found not to exceed 10%, showing that the NSRP approach can be considered sufficiently robust.

The results obtained by the BLRP model are shown to be close to the ones produced by the NSRP approach, as shown in Figure 7 for the Florence Station. The model is therefore capable of reproducing the historical first and second-order statistics, although the errors are somewhat larger than for the NSRP. Some irregularities of behaviour and discrepancies can be detected in the autocorrelation values and decay patterns. This occurred especially on the yearly scale because of the larger influence of non-homogeneities than on the seasonal one, and for the Centro Station due to the shortage of the historical record available for model estimation. Similarly, probabilities of zero and small rainfall periods generally match historical patterns, even though the low-scale values obtained by the NSRP model are generally closer to these than the results provided by BLRP. The discrepancies generally increase as the scale of aggregations are smaller and closer to the mean life-time of a cell. Accordingly, the BLRP model appears to be less flexible than the NSRP along a continuum of scales.

TABLE 2

a) Station: XIMENIANO, FLORENCE. Parameters for yearly simulations of NSRP and BLRP models estimated using different sets of equations

SET of EQ.^{ns}	NSRP λ [hr^{-1}]	μ_c [-]	δ^{-1} [hr]	β^{-1} [hr]	μ_x [mm/hr]	BLRP λ [hr^{-1}]	μ_c [-]	δ^{-1} [hr]	ϵ^{-1} [hr]	μ_x [mm/hr]
1	0.01041	6.89	0.327	4.980	3.706	-	-	-	-	-
2	0.00936	6.71	0.380	6.905	3.636	0.00998	4.60	0.526	3.914	3.593
3	0.00850	6.21	0.458	11.050	3.590	0.01034	5.47	0.334	2.732	4.596
4	-	-	-	-	-	0.00646	8.59	0.339	3.167	4.613
5	0.01043	7.69	0.244	4.610	4.440	-	-	-	-	-
6	0.01240	7.51	0.212	2.995	4.405	0.00888	3.24	1.665	9.368	1.813
7	0.00927	7.83	0.268	6.395	4.471	-	-	-	-	-
8	0.00834	7.64	0.301	10.067	4.528	-	-	-	-	-
9	-	-	-	-	-	0.01034	4.47	0.522	3.721	3.599

b) Station: XIMENIANO, FLORENCE. Estimated parameters for monthly simulations of NSRP and BLRP models

MONTH	NSRP λ [hr^{-1}]	μ_c [-]	δ^{-1} [hr]	β^{-1} [hr]	μ_x [mm/hr]	BLRP λ [hr^{-1}]	μ_c [-]	δ^{-1} [hr]	ϵ^{-1} [hr]	μ_x [mm/hr]
JAN	0.01277	18.53	0.184	4.834	1.838	0.01572	8.08	0.452	0.776	1.393
FEB	0.01712	9.97	0.466	4.224	1.171	0.01960	5.54	0.679	1.121	1.264
MAR	0.01365	7.54	0.560	7.450	1.598	0.01423	6.08	0.649	1.810	1.642
APR	0.01397	8.51	0.343	5.038	1.782	0.01223	7.06	0.446	1.213	1.754
MAY	0.01269	6.96	0.243	4.093	3.845	0.01302	4.42	0.425	2.470	3.380
JUN	0.01024	2.92	0.456	7.585	4.351	0.00902	3.32	0.453	3.926	4.360
JUL	0.00654	2.30	0.240	3.520	10.420	0.00541	3.04	0.201	1.845	11.430
AUG	0.00928	2.96	0.220	3.520	14.811	0.01130	2.12	0.279	3.619	13.369
SET	0.00865	5.46	0.259	3.053	6.387	0.00963	4.07	0.393	2.785	5.897
OCT	0.01193	4.73	0.596	6.842	2.834	0.01088	5.90	0.533	1.278	2.792
NOV	0.01140	10.39	0.432	6.626	2.956	0.01631	4.24	0.703	2.883	3.114
DEC	0.01234	7.74	0.762	9.774	1.390	0.01566	6.06	0.780	1.470	1.378

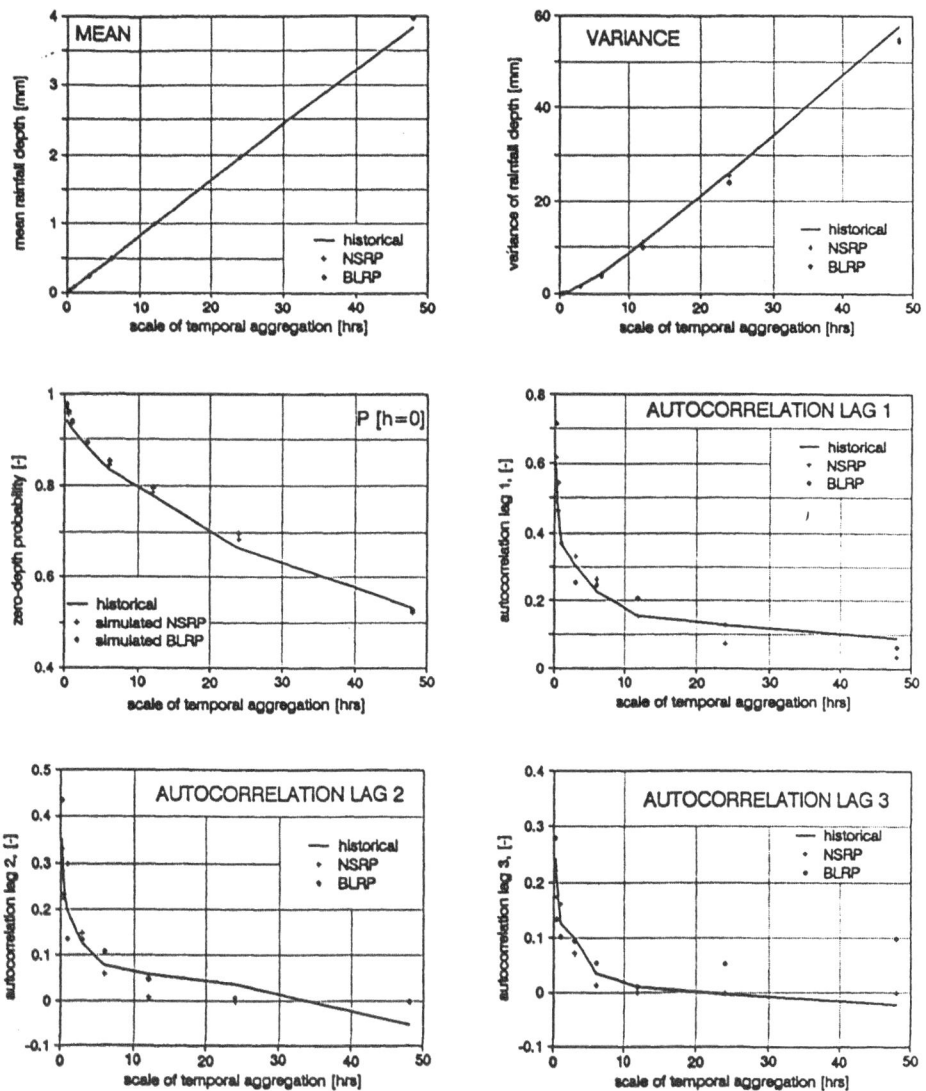

Figure 7. Station: XIMENIANO, FLORENCE. Statistical characteristics on different levels of aggregation. Historical and fitted values are shown for both the NSRP and BLRP model on the monthly scale of simulation (month of May).

5.1. SOME REMARKS ON THE CONTINUOUS SIMULATION VIA NSRP AND BLRP CLUSTERED MODELS

The results just referred to were to some extent expected, following the considerations expressed in the preceding Section 4.4 about the scale of fluctuation of the BLRP process, which has been shown theoretically to be always higher than the one of the equivalent NSRP. In order to verify this expectation and the theoretical result, a comparison was made of the scale of fluctuation computed for both NSRP and BLRP, with the one estimated from the data observed.

The values of the scale estimated from historical data, θ, were computed for this purpose by means of the estimator proposed by Vanmarcke (1983), namely

$$\theta = \frac{\hat{\Gamma}(T)T}{1-\hat{\Gamma}(T)} \qquad (61)$$

where $\hat{\Gamma}(T)$ is the estimated variance function for the temporal level of aggregation T. It must be noted that, despite the fact that the use of (61) is only recommended if T is much higher than θ, it provided reasonably stable estimates when these were obtained using three different values of T, namely 12, 24, and 48 hours. The estimates provided by Equation 61 can therefore be assumed as valid reference values.

Some interesting observations can be taken up from Table 3, where a summary of the comparison is shown. First, it can be observed that both the θ values computed according to Equations 55 and 57 generally underestimate the historical values, but are still shown to follow the same trend throughout the year. Moreover, it can be observed that the NS values are always lower than the BL ones, prevalently ranging from 70 to 85% of the latter. The ratio between θ_{NS} and θ_{BL} shows small changes, even when θ_{BL} is computed as a sum of θ_{NS} and the factor K', which is a function of the model parameter, i.e., according to Equation 59, either using the relationship based on the NSRP parameters for K', or that based on the BLRP ones, respectively indicated as K'$_{NS}$ and K'$_{BL}$ in Table 3. Accordingly, it is not unrealistic to assume that the four parameters, λ, μ_c, μ_x and δ, should have the same value for both models, so that the criterion proposed in Section 4.4 for model comparison can be reasonably accepted and supported.

Moreover, it should be remembered that the scale of fluctuation gives the time interval required to obtain stable (low variance) estimates of the mean of the fluctuating process, i.e., the rainfall intensity (Taylor, 1921; Vanmarcke, 1983). As stated above, following the definition given by Vanmarcke (1983), the scale of fluctuation to some extent represents the lower limit of the model's sensitivity to reproduce the short-time fluctuations of the rainfall intensity process. Accordingly, the higher the value of θ, the larger the level of aggregation, T, at which the model is able to give reliable estimates. Both the theoretical analysis and the results from the simulations proved that the BLRP model always presents higher values of θ than the NSRP, thereby indicating that the BLRP should be characterized by poorer performances at low scales of aggregation than the NSRP. This is correctly what the simulation of the continuous process showed, such behaviour being emphasized by the length of the historical record used for the model estimation.

Therefore, on the basis of the results presented (see Burlando, 1989, for a more complete and detailed analysis), it could be observed that both the approaches are suitable for simulation purposes. The lower temporal limit for practical purposes still remains the reproduction of physical patterns for time scales which are below the mean life time of a rain cell, according to the actual structure of the models. Nevertheless, the most interesting result, with respect to the continuous simulation of the natural process, is represented by the dependence of the performance of the model on the length of data sets used for estimation. This has been found to be surprisingly small if compared with the great difference of historical record length between the two sites considered. In view of this, temporal modelling can be appealing for a variety of hydrological problems where long rainfall records are not available but are still needed for planning and investigation.

6. Reproducibility of Extreme Events

An important aspect for the evaluation of model performances is the ability to reproduce the extreme properties of the natural process. Many planning problems for which temporal rainfall modelling may be useful deal in fact with the extreme behaviour of the process. The above-illustrated models must therefore be tested with respect to their extreme behaviour. Moreover, this problem has not yet been studied in enough depth, as it has also been pointed out by Rodriguez-Iturbe et al. (1987a). In the following text some considerations are given on the analytical formulation of the maximum storm intensity by the compound Poisson approach. Solutions of the theoretical equations for physically reasonable cases and an application to real world cases are presented. The clustered point process approach is also investigated. Due to the lack of theoretical work available on the extreme properties of clustered models of point processes, simulation runs were carried out at two different sites using both NSRP and BLRP models. In order to test the flexibility of each model, the statistical similarities of model extremes to historical ones are evaluated within the intensity-duration-frequency (IDF) domain by comparing the IDF curves obtained from the historical data with the ones obtained from the simulated values.

6.1. MODELS BASED ON POISSON ARRIVALS

6.1.1. *Independent Poisson Marks (IPM) Model.* The reproducibility of extreme events by means of the Independent Poisson Marks model was investigated through both controlled experiments and application to real world situations.

First, the *exponential model* was analyzed for $M=50$. Equation 6 is shown in Figure 8, where we also show the different solutions of Equation 7 which were worked out by numerical integration for the different values of T^*, i.e., the dimensionless interval used for averaging. It can be observed that the extreme value distribution of the local averages still follows the EV1 distribution very closely for any length of the averaging interval less than the mean duration of a storm. Moreover, the extreme value distribution of the local average is found to be poorly sensitive to the length of the averaging interval within this range; for instance, if T* takes values as low as 0.2 + 0.1, the local average and the instantaneous values practically overlap at the same frequency level. On the contrary, the effect of increasing the averaging intervals becomes relevant for intervals greater than the mean duration of a storm. This effect is well demonstrated by plotting the values of x* versus T* for a given frequency level to obtain a diagram similar to the traditional intensity-duration-frequency curves, as shown in Figure 9.

TABLE 3. Scale of fluctuation of the rainfall process at the Ximeniano Station, Florence. Comparison of estimated and fitted values

Month	estimated from variance function for T=24 hrs	θ_{NS}	θ_{BL}	θ_{NS}/θ_{BL}	$\theta_{BL}' = \theta_{NS}+K_{NS}'$	θ_{NS}/θ_{BL}'	$\theta_{BL}'' = \theta_{NS}+K_{BL}'$	$\theta_{NS}/\theta_{BL}''$
JAN	6.056	2.719	3.662	0.743	3.061	0.888	3.083	0.882
FEB	6.210	3.535	4.148	0.852	4.005	0.883	4.085	0.865
MAR	5.344	3.892	4.572	0.851	4.420	0.880	4.458	0.873
APR	5.079	2.632	3.502	0.752	2.992	0.880	3.018	0.872
MAY	3.775	1.657	2.555	0.649	2.022	0.820	2.055	0.806
JUN	2.665	1.695	2.311	0.733	2.124	0.798	2.129	0.796
JUL	1.843	0.761	0.973	0.782	0.954	0.798	0.954	0.797
AUG	2.220	0.823	1.122	0.734	1.087	0.757	1.095	0.752
SEP	3.604	1.424	2.261	0.630	1.759	0.810	1.796	0.793
OCT	4.283	2.952	3.603	0.819	3.412	0.865	3.408	0.866
NOV	5.854	3.822	4.006	0.954	4.368	0.875	4.465	0.856
DEC	6.499	5.354	5.189	1.032	5.981	0.895	5.998	0.893

It must also be observed that for a length of the averaging interval exceeding the mean storm duration, i.e., T*>1, the form of the cdf tends to depart from the EV1 behaviour. Moreover, the cdf of the extreme value averaged over large intervals, say T*=10, displays a higher skewness than the EV1 distribution, which satisfactorily fits the cdf of extreme values locally averaged over short intervals, let alone the instantaneous ones. According to this model, therefore, the extreme cdf value should display increasing skewness at increasing duration of the averaging window, as is often supported by the analysis of rainfall records.

The form of the dimensionless IDF curves, shown in Figure 9 for the three models and a return period of 100 years, can be compared with the ones resulting from the statistical analysis of rainfall data. These are generally plotted as straight lines on the double logarithmic plane (log x, log T) within a certain (limited) range of durations, their slope ranging from -0.3 to -0.5, according to the geographical and climatic conditions of the site. It can be seen that the exponential model can provide satisfactory predictions of the extreme storm intensities for a narrow band of durations only, as will be outlined in the following section where IDF curves computed by this approach are compared with the ones fitted from observed storm data.

The results obtained from the *gamma model* taking the shape parameters of

the Gamma distribution of the intensity, n, and of the duration, r, equal
to 1 and 0.5 respectively, and M equal to 50, do not show any improvement.
This can easily be detected from the plot of the 100-year IDF curve which
is compared to the ones from the other models in Figure 9.

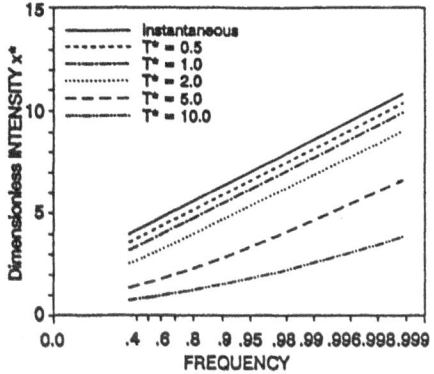 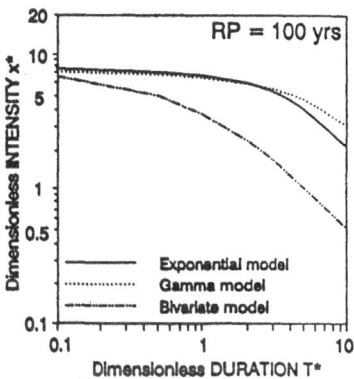

Figure 8. Reliability function of
storm intensity for different
durations. Independent Exponential
Model, M=50.

Figure 9. Dimensionless IDF.
Exponential, Gamma and Bivariate
models for M=50 and for a return
period of 100 years.

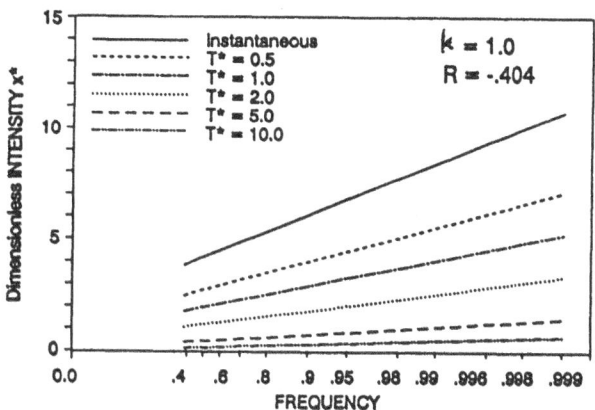

Figure 10. Reliability storm intensity function for different durations.
Bivariate Exponential Model, k=1, M=50.

Finally, the extreme behaviour of the *bivariate exponential model* was
investigated. Figure 10 reports on the reliability functions obtained from
Equations 13 and 14 by numerical integration for different durations and
for k=1 and M=50. As compared with the ones resulting from the Independent

Exponential model, it can be observed that the EV1 cdf behaviour is displayed for a large band of durations. The effect of assuming correlated intensity and duration reduces the overlapping effect which was typically shown by independent models for small values of T*, and tends to reduce the departure from EV1 cdf behaviour. Such a behaviour is also displayed by the 100-year IDF curve shown in Figure 9. This shows, for dimensionless durations larger than 0.5, slopes close to those which normally characterize the IDF curves which are statistically inferred from storm intensity data. Nevertheless, the departure from this behaviour is still sensitive to small values of T*. It is also interesting to note that the sensitivity of the model to variations in the values of the parameter k, i.e., in the correlation coefficient $R(i_s, t_s)$, is not at all negligible. Although this result is not reported here, this appears clearer by keeping in mind that for k=0 the model coincides with the Independent Exponential Model.

6.1.2. *Application of the IPM Model.* The IPM model has been further applied to a real case of intensity-duration-frequency analysis for the urban area of Milan, Italy. The results provided by the models for the Palazzo Marino gauge station (Milan City Hall) are compared in terms of both intensity-duration-frequency curves and reliability functions to the curves available from a previous fitting analysis of storm data (Moisello, 1976).

The special estimation procedure shown in Section 3.4 was applied to the Palazzo Marino Station. Data were available for the lower duration of T=15 min and for the upper one of T=360 min. The following values were obtained:

$$\alpha^{-1} = 19.942 \text{ mm/hr}; \qquad \beta^{-1} = 0.815 \text{ hr}; \qquad M = 25.91$$

A further analysis of the available depth and duration data allowed an estimate to be made of the parameters which characterize the storm distribution and which are given in Table 4.

TABLE 4. IPM parameters estimated for different characterizations of storm intensity and duration.

DISTRIBUTION		PARAMETERS				
Intensity	Duration	Intensity			Duration	
		α	n	δ	β	r
Expon.	Expon.	0.05	1.0	0	1.23	1.00
Expon.	Gamma	0.05	1.0	0	1.23	0.62
Bivar. Expon.		0.05	1.0	1	0.13	1.00

The reliability function and the IDF curves were worked out by numerical integration of Equation 7 for different values of the dimensionless duration T*. The results of the integration are given in Figure 11 where the numerical solutions of the IDF curves are compared with the ones fitted to the storm intensity data for a return period (RP) of 100 years. On the one hand it can be observed that neither the Independent Exponential model nor the Independent Gamma-Exponential one are capable of reproducing the fitted curves, being too sensitive in terms of the averaging interval. On the other hand, the Bivariate Exponential model shows a fairly satisfactory performance in reproducing the characteristics of extreme values, at least for a duration range which can be of technical interest. Nevertheless, the effect of the averaging interval is still sensitive to long durations as well as the departure from the EV1 behaviour for short durations. This behaviour is confirmed for different return periods, for which the IDF plots are not reported here.

Similarly, Figure 12 shows the reliability function obtained from the numerical integration as compared to the statistically inferred ones for the dimensionless duration T*=1. The performances of the Independent Exponential and the Gamma Independent Models yield the conclusion that they are not suitable for reproducing extreme properties of the rainfall process. The Bivariate Exponential Model shows greater suitability for the purpose, at least for values of the dimensionless duration T* larger than 0.5÷1, where at this level the correlation effect is still capable of reproducing the storm characteristics rather than the more complex effects of the internal structure properties.

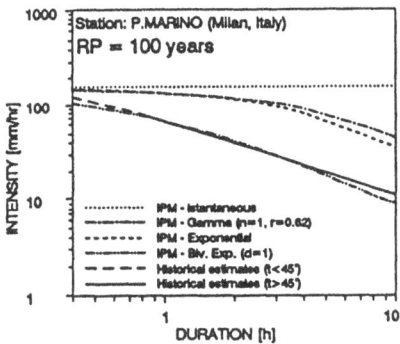
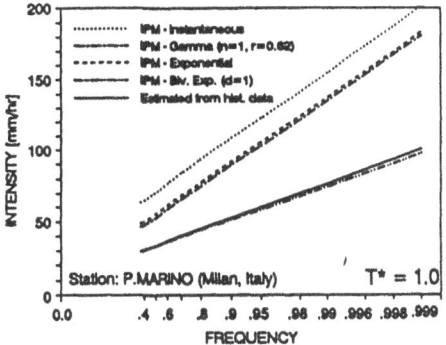

Figure 11. IPM model. Comparison of fitted and modelled IDF curves for return period RP=100 years at the Palazzo Marino Station (Milan, Italy).

Figure 12. IPM model. Comparison of fitted and modelled reliability function of storm intensity for dimensionless duration T*=1 at the Palazzo Marino Station (Milan, Italy).

6.1.3. *Poisson* Rectangular Pulses (PRP) Model. The PRP model was investigated by means of numerical experiments performed to analyze the extreme properties of the averaged process, $X_\tau(t)$, although no comparison was made with respect to the data observed. These experiments were carried out by continuous simulation runs of the time evolution of dimensionless intensity, x*, for 1000 sampling periods of length τ*, and with given values of λ*. The results obtained for τ*=5000, and λ*=0.01 are shown in Figure 13. It can be seen that the instantaneous extreme value distribution (24) applies in practice for all durations of the averaging interval less than the mean duration of one pulse. In addition to the

166

instantaneous values, the local average is found to follow approximately
the extreme properties of the IPM model, such approximation increasing in
accuracy with decreasing values of parameter λ^*. An example is given in
Table 5, where the estimate of the 99% quantile, i.e., a return period of
100 years, of dimensionless storm intensity, x*, which are obtained from
simulation runs with different values of λ^* are compared. No significant
improvement in the extreme reproducibility with respect to the Poisson
marks approach is obtained by allowing overlapping of the storm, for the
sake of accuracy in reproducing the storm profile. Therefore, no gain
seems to exist in using PRP instead of IPM, given also that the more
complex theoretical structure of PRP than IPM does not allow for a
numerical solution of the cdf of extremes. The simulation route therefore
must be followed for this purpose, and the availability of continuous data
is required for parameter estimation.

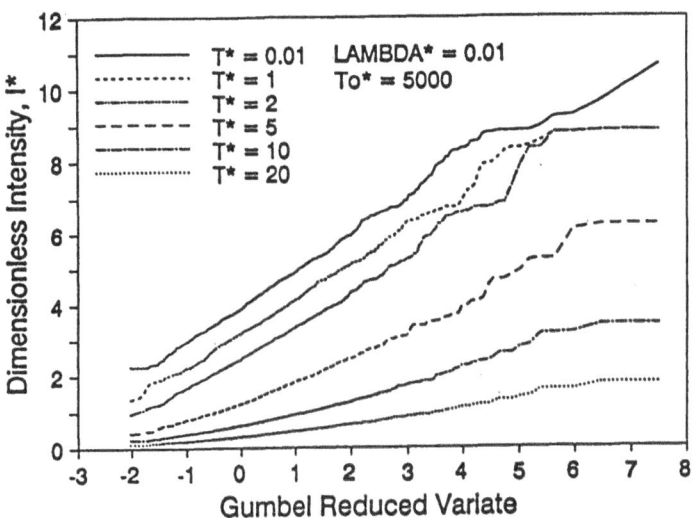

Figure 13. Traces of generated extreme values of the PRP model plotted for
different dimensionless durations.

TABLE 5. Estimates of the 99% quantile of the dimensionless maximum storm
intensity from simulation of 1000 periods with M = $\lambda\tau$=50 points per period
(Compound Poisson models).

Model	λ^*	Dim.less Duration -->	0	.1	1	5	10
IPM	∞		8.40	8.38	7.63	4.55	2.53
PRP	.1		9.15	9.03	7.63	4.58	2.72
PRP	.01		8.78	8.41	7.75	4.53	2.53
PRP	.001		8.14	7.94	7.33	4.91	2.57

6.2. MODELS BASED ON CLUSTERED POISSON ARRIVALS

6.2.1. *Extremes of the Neyman-Scott Clustering Process.* Maximum intensities for the NSRP model were estimated for different levels of aggregation from 200 reference periods – years, seasons, months – of generated data (for a detailed analysis see Burlando, 1989). The historical behaviour was generally matched by the simulated values, and a linear trend of the EV1 distribution in Gumbel plot respected, although some discrepancies for the highest return periods and for the smallest scales of aggregation were to be observed. Such behaviour was also observed for the simulated extremes at Denver by Rodriguez-Iturbe et al. (1987a), who justified that result as a small-sample effect determined by the dependence structure in the storm occurrence process incorporated in the cluster-based models. The fact that similar trends may also occur in the actual process has yet to be observed.

The IDF curves derived from the simulated extremes – expressed in the form $i(t) = at^{n-1}$, where a and n are estimated independent of the return period – once more reveal the satisfactory agreement between simulated and historical values. The curves derived from the historical data at the Florence station are generally quite well preserved on any scale of application. In fact, good coherence was also obtained on the yearly scale of application, although a homogeneous trend to underestimate the maximum intensity was observed. This was not dependent on the return period, but was shown to be larger for small scales of aggregation. Such lack of performance was overcome by the monthly model which reproduced very well the observed patterns on any of the examined scales of aggregation and independently of the return period, as shown in Figure 14.

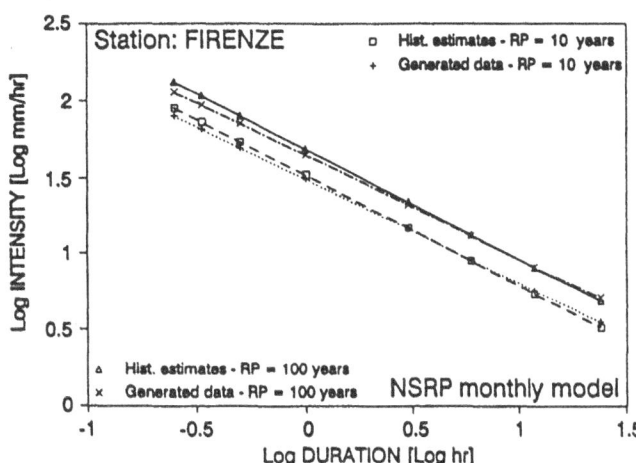

Figure 14. NSRP model, monthly simulations. Comparison of fitted and modelled IDF curves for return periods RP=10 and RP=100 years at the Ximeniano station (Florence, Italy).

Good results, albeit only briefly reported here, were also obtained for the Centro station where the model was applied on the seasonal scale. Small overestimations and underestimations of the IDF curves on the lower and upper scales of aggregation were revealed, although the error should indeed be considered negligible if analyzed with respect to the short historical record available for model estimation.

Two major considerations arise then from the application to real world situations of the NSRP model, namely (i) the need for a homogeneous reference period for simulation, and (ii) the robustness of the model, which is capable of preserving historical characteristics and of giving satisfactory results even where only a short historical record is available to be more controlled by climatic non-stationarities, which differ from season to season, than by the length of the estimation record.

6.2.2. *Extremes of the Bartlett-Lewis Clustering Process.* The same schedule of simulation runs which was performed for the NSRP model was used to analyze the extreme properties of the BLRP one, in order to compare the results and behaviour of the two models.

The maximum intensities simulated for the Florence station follow the EV1 behaviour, with some tendency on the part of the values corresponding to higher return periods to curve upwards, this characteristic being found to be stronger than in the case of NSRP, both on the yearly and monthly scale of application. This is probably due to some lack of performance also shown by the model in reproducing the statistical patterns of the continuous process.

Slightly different considerations apply to the derived IDF's. In fact, these match the historical patterns considerably well, showing the same pattern of the NSRP behaviour. Therefore, the monthly simulations allow a perfect reproduction to be made of the historical IDF's, while the values from the yearly simulations generate underestimated IDF's. Again, it can be observed that the performances are practically independent from the return period, so that the result presented in Figure 15 for 10 and 100 years return period can also be extended to higher frequencies.

A further remark can be made with respect to the IDF estimated for the Centro station. The relatively short record available for the estimation is in fact capable of highlighting some of the differences between the two models, as these were introduced in the previous Section 4.4. Compared to the historically derived ones, the "simulated" IDF provided underestimation on low scales of aggregation and overestimation on high scales, this being the opposite of what was obtained from the NSRP, as shown in Figure 16. A possible reason for that stems from the theoretical analysis which was introduced above about the scale of fluctuation of the process. In fact, following Equation 20, the higher scale of fluctuation for the BLRP than for the NSRP means that the former is characterized by a richer correlation structure. This implies that the BLRP should have a larger memory of the process than the NSRP. Accordingly, the low intensity values which generally characterize the rising event, could have a strong effect on the values aggregated on low temporal scales, producing underestimation of the simulated values. Similarly, such memory could turn into a large effect of the peak values of the intensity process on the tail of the event, so generating overestimation when the process is aggregated on large scales.

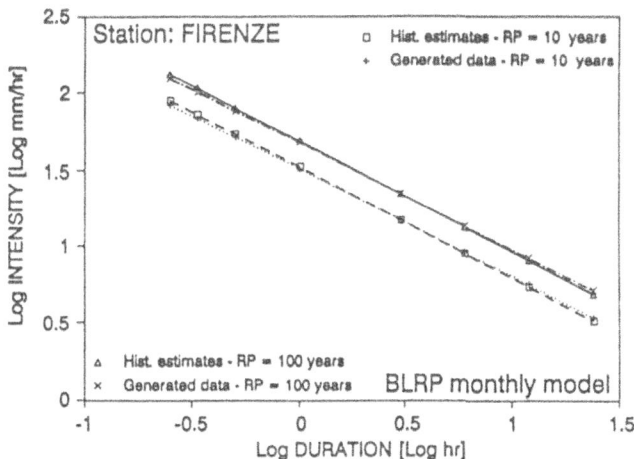

Figure 15. BLRP model, monthly simulations. Comparison of fitted and modelled IDF curves for return periods RP=10 and RP=100 years at the Ximeniano station (Florence, Italy).

6.2.3. *Some Further Remarks on the Extremes of the Clustered Poisson Models.* Both the NSRP and BLRP models are demonstrated to be suitable for reproducing the process of the extremes. The results which have been presented show that both of the approaches are robust, that is, estimation can also be successful based on a few years of data, let alone the need for a homogeneous scale of application. Data must indeed be characterized by good time resolution – say, at least, ten-minute data, or less – and by high quality of recording. Unfortunately, such data are not frequently available in practice. The BLRP model has been shown to be more sensitive to data quality and availability than the NSRP, revealing a larger performance gap between cases of large and scarce availability of data. It must be observed that, although differences can be detected, these are not likely to represent a possible standard for model choice with regard to extreme reproducibility. It can further be observed that both the models displayed poor results in reproducing the extremes for levels of aggregation which are lower than the mean life time of a cell.

The latter result, which can be considered as the major limitation of the approach, offers a possible argument to support a theoretically based model guideline. It is in fact usually very important for the model to be capable of reproducing the low duration extreme intensities which are of critical importance for many projects. In this light, the NSRP denoted a stronger robustness in reproducing non-stationarities which are typical of the short averaged characteristic of the process, while a greater lack of performance was exhibited by the BLRP. That is, the experimental evidence supports the theoretical reason for which the NS process should perform better than the BL one on low levels of aggregation, due to its lower scale of fluctuation. It must however be recognized that this argument still requires further research.

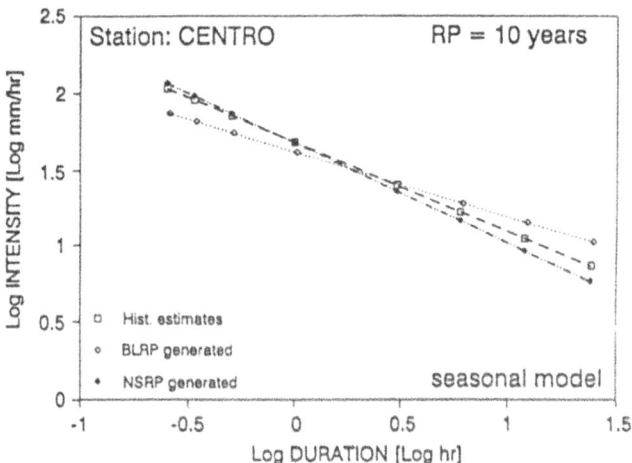

Figure 16. Comparison of estimated, and BLRP and NSRP derived IDF curves for return period RP=10 years at Centro station (Rio della Gallina, Italy).

7. Conclusions

The analysis presented above reveals that the theoretical results which have been made available in scientific literature lead to model formulations that match the natural process of temporal precipitation considerably well, let alone the limits of simple models and the cumbersome formulations of high-descriptive approaches.

For the purpose of reproducing the continuous process, the clustered models (NSRP and BLRP) show satisfactory feasibility, although fine resolution of data is required for model fitting on a scale which is characterized by the homogeneity and stationarity of meteorological and climatic patterns. Both cluster models examined can accomplish aggregation and disaggregation of rainfall throughout a wide range of temporal scales, the fitting being controlled on only one temporal scale. The lower level of aggregation which can be satisfactorily reproduced is controlled by the mean duration of a cell, which is likely to be of the order of 15 to 30 minutes. In this aspect the NSRP model generally performs better than the BLRP. This concurs with the fact that the fluctuation scale of the NSRP model has been found, both theoretically and from the simulations, to be smaller than that of the BLRP model equivalent in the value of the mean and the variance. Accordingly, the NS process is characterized by a lower limit in representing the short-time fluctuations of the natural process.

Interesting considerations also arise from an analysis of the extreme properties. The theoretical analysis presented for the IPM model and the subsequent application to real data showed that its use should be confined to problems where mathematical tractability is required, such as derivation of flood frequency distributions and urban drainage analysis. Conversely, the cluster models revealed a higher robustness than the compound

Poisson approach. In fact, both the NSRP and the BLRP models were able to reproduce the extreme behaviour of the historical process, even in the case of poor availability of historical records for estimation. In these circumstances the NSRP model showed a larger flexibility due to the lower sensitivity to natural fluctuations and non-homogeneities which does not yet affect the ability of the model to reproduce the historical IDF's satisfactorily. Therefore, both the NSRP and the BLRP can be perspectively useful in practical hydrology, for the purpose of either planning or prediction, regardless of the difficulties lying in the choice of which model could best fit the natural process. Consequently, to the question of whether a purely temporal approach to rainfall modelling at a point in space can provide an exhaustive representation of the precipitation process in view of the real-world hydrology, a satisfactory answer could come from the analysis presented, although a more realistic rainfall modelling effort should account for space-time interactions.

Nevertheless, some issues still remain open in order to reduce the uncertainties of this modelling approach. It does not seem that the temporal lower limit to model performance, i.e., the mean cell lifetime, can be overcome within the cluster framework. The author's idea therefore is that a dependence between the number of cell arrivals and their distribution in time should be included. This could accomplish a better representation of the climatic variations which the analysis clearly showed to bear a strong influence on the variability parameters. In this light, a better representation of the process and an implicit accounting for meteorological fluctuations could also be achieved by randomizing the temporal process of either the storm arrival or the cell chronology. Further research is then presently directed towards these modifications. This could in fact allow for a better representation of the probabilities of small rainfall, which has been shown to be perhaps the major limitation to temporal modelling by cluster models.

8. References

Amorocho, J. and Wu, B. (1977) 'Mathematical models for the simulation of cyclonic storm sequences and precipitation fields', Journal of Hydrology, 32, pp 329-345.

Austin, P.M. and Houze, R.A. Jr (1972) 'Analysis of the structure of the precipitation patterns in New England', J. of Appl. Meteorol., 11, pp 926-934.

Bacchi, B., La Barbera, P. and Rosso, R. (1987) 'Storm characterization by Poisson models of temporal rainfall', Proc. XXII IAHR Congress, Vol. Urban Drainage Hydraulics and Hydrology, pp 35-40, Lausanne, September.

Bacchi, B., Burlando, P. and Rosso, R. (1989) 'Extreme value analysis of stochastic models of point rainfall', 3rd Scientific Assembly of IAHS, Poster Session, Baltimore, May 10-19.

Bonta, J.V. and Rao, A.R. (1988) 'Factors affecting the identification of independent rainstorm events', J. Hydrol., 98, pp 275-293.

Burlando, P. (1989) 'Modelli stocastici per la simulazione e la previsione della precipitazione nel tempo', Ph. D. Thesis, Institute of Hydraulics, Politecnico di Milano, Milan, Italy.

Carlson, R.F. and Fox, P. (1976) 'A northern snowmelt-flood frequency model', Water Resources Research, 12(4), pp 786-...

Chan, S.O. and Bras, R.L. (1979) 'Urban storm management: distribution of flood volumes', Water Resources Research, 15(2), pp 371-382.

Cordova, J.R. and Rodriguez-Iturbe, I. (1983) 'Geomorphoclimatic estimation of extreme flow probabilities', J. Hydrol., 65, pp 159-173.

Cox, D.R. and Isham, V. (1980) Point Processes, Chapman e Hall, London.

Diaz-Granados, M.A., Valdes, J.B. and Bras, R.L. (1984) 'A physically based flood frequency distribution', Water Resources Research, 20(7), pp 995-1002.

Eagleson, P.S. (1972) 'Dynamics of flood frequency', Water Resources Research 8(4), pp 878-898.

Eagleson, P.S. (1978) 'Climate, soil and vegetation, 2. The distribution of annual precipitation derived form observed storm sequences', Water Resources Research, 14(5), pp 713-721.

Eagleson, P.S. (1981) 'Some limiting forms of the Poisson distribution of annual station precipitation', Water Resources Research 17(3), pp 752-757.

Foufoula-Georgiu, E. and Guttorp, P (1987) 'Assessment of a class of Neyman-Scott models for temporal rainfall', J. of Geophys. Res., 92 (D8), pp 9679-9682.

Gumbel, E.J. (1960) 'Bivariate exponential distributions', J. Am. Stat. Assoc. 55, pp 698-707.

Harrold, T.W. and Austin, P.M. (1974) 'The structure of precipitation systems. A review', J. de Rech. Atmosph. 8 (1-2), pp 41-57.

Hebson, C. and Wood, E.F. (1982) 'A derived flood frequency distribution using Horton ratios', Water Resources Research 18(5), pp 1509-1518.

Kavvas, L. and Delleur, J.W. (1981) 'A stochastic cluster model of daily rainfall sequences', Water Resources Research 17(4), pp 1151-1160.

Kutoyants, Yu. A. (1984) Parameter Estimation for Stochastic Processes, Heldermann, Berlin.

Moisello, U. (1986) 'Il regime delle piogge intense di Milano', Ingegneria Ambientale 5, pp 545-561.

Rodriguez-Iturbe, I., Gupta, V.K. and Waymire, E. (1984) 'Scale considerations in the modelling of temporal rainfall', Water Resources Research 20(11), pp 1611-1619.

Rodriguez-Iturbe, I., Cox, D.R. and Eagleson, P.S. (1986) 'Spatial modelling of total storm rainfall', Proc. R. Soc. London, Ser. A, 403, pp 27-50.

Rodriguez-Iturbe, I. and Cox, D.R. (1986) 'Aggregation of temporal rainfall records through a continuum of scales: part I. Theoretical models', unpublished manuscript.

Rodriguez-Iturbe, I. (1986) 'Scale of fluctuations of rainfall models', Water Resources Research 22(9), pp 15S-37S, August.

Rodriguez-Iturbe, I. and Eagleson, P.S. (1987) 'Mathematical models of rainstorm events in space and time', Water Resources Research 23(1), pp 181-190.

Rodriguez-Iturbe, I., Febres de Power, B. and Valdes, J.B. (1987a) 'Rectangular pulses point process models for rainfall: analysis of empirical data', Jour. of Geophys. Res., 92 (D8), pp 9645-9656.

Rodriguez-Iturbe, I., Cox, D.R. and Isham, V. (1987b) 'Some models for rainfall based on stochastic point processes', Proc. R. Soc. London, Ser. A, 410, pp 269-288.

Rosenbrock, H.H. (1960) 'An automatic method of finding the greatest or the least value of a function', Computer Journal, 3.

Rosso, R. and Caroni, E. (1977) 'Storm sewer capacity design under risk', Proc. XVII Congr. International Association for Hydraulic Research (IAHR), Vol. 4, pp 537-542, Baden-Baden, 15-19 August.

Smith, J.A. and Karr, A.F. (1985a) 'Statistical inference for point process models of rainfall', Water Resources Research, 21(1), pp 73-79.

Smith, J.A. and Karr, A.F. (1985b) 'Parameter estimation for a model of space-time rainfall', Water Resources Research, 21(8), pp 1251-1257.

Snyder, D.L. (1975) Random Point Processes, New York, J. Wiley & Sons.

Taylor, G.I. (1921) 'Diffusion by continuous movement', Proc. London Math. Soc., Vol. 20, pp 196-211.

Vanmarcke, E. (1983) Random Fields: Analysis and Synthesis, MIT Press, Cambridge, Massachusetts.

Waymire, E. and Gupta, V.K. (1981) 'The mathematical structure of rainfall representations', Water Resources Research, 17(5), pp 1261-1294.

PRACTICAL RESULTS IN SIMULATION AND FORECASTING

K.W. HIPEL
*Departments of Systems Design Engineering and
Statistics and Actuarial Science
University of Waterloo
Waterloo, Ontario
Canada N2L 3G1*

ABSTRACT. Experimental results are presented for a range of situations in which ARMA (autoregressive-moving average) models are used for simulation and forecasting. After presenting an approach for obtaining unbiased simulated sequences, simulation is employed for demonstrating how ARMA models statistically preserve important historical statistics that are used in defining the Hurst phenomenon. Forecasting experiments with both nonseasonal and seasonal hydrological time series show that ARMA-type models forecast well when compared to their competitors.

1. Introduction

The first step to be taken when fitting time series or stochastic models to a data set is to decide upon which families of models should be considered. For example, when modelling an average monthly riverflow time series one could entertain using periodic models such as periodic autoregressive (PAR) and periodic autoregressive-moving average (PARMA) models. Another class of models that could be employed is deseasonalized models whereby an autoregressive-moving average (ARMA) model is fitted to the deseasonalized monthly riverflows. Whatever the case, after specifying the class or classes of models to consider, one can select the overall best model by following the identification, estimation, and diagnostic check stages of model construction (see, for example, Box and Jenkins, 1976; Salas et al., 1980; and Hipel and McLeod, 1993). By building a model in an iterative fashion within the model construction stages, one can obtain a model which provides a good statistical fit to the data using a minimum number of parameters.

A particular stochastic model fitted to a given time series can serve a number of useful purposes. For instance, the calibrated model provides an economic means for encoding the basic statistical properties of the time series into a few model parameters. In the process of carrying out the model building procedure, one obtains a better understanding about the underlying statistical properties of the data set. In addition to the insights which are always gained when iteratively fitting a model to a time

175

J.B. Marco et al. (eds.),
Stochastic Hydrology and its Use in Water Resources Systems Simulation and Optimization, 175–188.
© 1993 *Kluwer Academic Publishers.*

series, there are two important types of applications of time series models which are in widespread use by practitioners. These application areas are forecasting and simulation. The main objective of this paper is to present a survey of simulation and forecasting results for hydrological time series. For the experiments reported in this paper, the author was always a member of the research teams.

Simulation and forecasting experiments can be used for model discrimination purposes. More specifically, simulation can be utilized to test whether or not a class of time series models statistically preserves important historical statistics of the data sets to which the set of models is fitted. One would, quite naturally, like to employ models which can account for the key statistical characteristics of hydrological time series. Forecasting experiments can be utilized for determining which classes of models furnish the most accurate forecasts when fitted to hydrological as well as other types of series. As will be shown in the upcoming sections on simulation and forecasting, ARMA-type models perform very well when both simulating and forecasting hydrological time series.

2. Simulation

2.1. INTRODUCTION

The main objective of forecasting is to use the time series model fitted to a data set to obtain the most accurate estimate or prediction of future unknown observations. The goal of simulation is to employ the fitted model to generate a set of stochastically equivalent sequences of observations which could possibly occur in the future. These simulated sequences are often referred to as synthetic data by hydrologists because they are only possible realizations of what could take place. As a matter of fact, the overall science of fitting stochastic models to hydrologic data and using these models for simulation purposes is often called synthetic hydrology. Other titles for this field include stochastic and operational hydrology.

Simulation is now a widely accepted technique to aid in both the design and operation of water resources systems. Vogel and Stedinger (1988), for instance, demonstrate that using synthetic data generated by stochastic streamflow models can lead to improvements in the prediction of reservoir design capacity estimates. Besides the design and operation of large-scale water engineering systems (see for example Hipel et al., 1979), another main use of simulation is to investigate the theoretical properties of stochastic models. Often it is analytically impossible to derive certain theoretical characteristics of a given type of time series model. However, by using simulation one can determine these theoretical properties to any desired level of accuracy. In Section 2.3, simulation is used with ARMA models to study theoretical problems related to what is called the Hurst phenomenon.

When simulating with an ARMA or other type of model, one can avoid the introduction of bias into a simulated sequence by using starting values that are randomly generated from the underlying stochastic process. McLeod and Hipel (1978b) present two procedures for obtaining unbiased starting values and simulated sequences. WASIM1 (Waterloo Simulation Procedure 1) consists of using the moving average (MA) form of an ARMA model to simulate the starting values and then employing the original ARMA model to simulate the remaining synthetic data. The WASIM2 procedure is based upon a knowledge of the theoretical autocovariance function for an ARMA process and is described in the next subsection. Finally, one can employ the WASIM3 procedure for incorporating parameter uncertainty into a simulation study.

Using simulation experiments, it is shown in Section 2.3.3 that ARMA models fitted to average annual geophysical time series statistically preserve statistics related to the Hurst phenomenon (Hipel and McLeod, 1978a). For the case of average monthly riverflow series, it is pointed out in Section 2.4 that PAR models statistically retain various critical period statistics (Thompstone et al., 1987).

2.2. WATERLOO SIMULATION PROCEDURE 2

Let w_t, $t=1,2,\ldots,n$, be a stationary nonseasonal time series to which an ARMA model is fitted to produce the ARMA(p,q) model

$$\phi(B)w_t = \theta(B)a_t \tag{1}$$

where

$$\phi(B) = 1 - \phi_1 B - \phi_2 B^2 - \ldots - \phi_p B^q$$

is the nonseasonal autoregressive (AR) operator or polynomial of order p such that the roots of the characteristic equation $\phi(B)=0$ lie outside the unit circle for nonseasonal stationarity and the ϕ_i, $i=1,2,\ldots,p$, are the nonseasonal AR parameters;

$$\theta(B) = 1 - \theta_1 B - \theta_2 B^2 - \ldots - \theta_p B^p$$

is the nonseasonal moving average (MA) operator or polynomial order q such that the roots of $\theta(B)=0$ lie outside the unit circle for invertibility and θ_i, $i=1,2,\ldots,q$, are the nonseasonal MA parameters; the a_t's are identically independently distributed innovations with mean 0 and variance $\sigma_a^2[IID(o,\sigma_a^2)]$ and often the disturbances are assumed to be normally independently distributed $[NID(0,\sigma_a^2)]$.

Prior to fitting a model to a series one may wish to take a Box-Cox power transformation (Box and Cox, 1964) in order to alleviate problems with non-normality and/or heteroscedasticity of the model residuals. In addition, nonstationarity can be first eliminated using differencing. Models with a non-zero mean (or any other type of deterministic component) are simulated by first generating the corresponding zero-mean process and then adding on the mean component.

Suppose that it is required to generate k terms of an ARMA(p,q) model with innovations that are $NID(0,\sigma_a^2)$. The following simulation procedure is exact to simulate w_1, w_2, $\ldots$, w_k, for all stationary ARMA(p,q) models.

1. Obtain the theoretical autocovariance function γ_j for $j = 0,1,\ldots,p-1$, by using the algorithm of McLeod (1975) with $\sigma_a^2 = 1$.

2. Expand the ARMA(p,q) model in Equation 1 as a pure MA model in order to obtain the pure MA coefficients ψ_j for $j=1,2,\ldots,q-1$.

3. Form the covariance matrix $\underline{\Delta}\sigma_a^2$ of w_p, $w_{p-1},\ldots,w_1,a_p,a_{p-1},\ldots,a_{p-q+1}$.

$$\underline{\Delta} = \begin{bmatrix} [\gamma_{i-j}]_{pxp} & [\psi_{j-i}]_{pxq} \\ [\psi_{i-j}]_{qxp} & [\delta_{i,j}]_{qxq} \end{bmatrix}_{(p+q)x(p+q)} \tag{2}$$

In Equation 2, the (i,j) element and dimension of each partitioned matrix are indicated. The values of $\delta_{i,j}$ are 1 or 0 according to whether $i=j$ or $i \neq j$, respectively. When $i-j<0$, then $\gamma_{i-j}=\gamma_{j-i}$ and $\psi_{i-j}=0$.

4. Determine the lower triangular matrix $\underline{M}$ by Cholesky decomposition (Ralston, 1965, p. 410) such that

$$\underline{\Delta} = \underline{M}\,\underline{M}' \tag{3}$$

5. Generate $e_1, e_2, \ldots, e_{p+q}$, and $a_{p+1}, a_{p+2}, \ldots, a_k$, where the e_t and a_t sequences are $NID(0, \sigma_a^2)$.

6. Calculate $w_1, w_2, \ldots, w_p$, from

$$w_{p+1-t} = \sum_{j=1}^{t} m_{t,j}e_j \qquad t = 1, 2, \ldots, p \tag{4}$$

where $m_{t,j}$ is the t,j entry in matrix $\underline{M}$.

7. Determine $a_{p-q+1}, a_{p-q+2}, \ldots, a_p$, from

$$a_{p+1-t} = \sum_{j=1}^{p+t} m_{t+p,j}e_j \qquad t = 1, 2, \ldots, q \tag{5}$$

8. Obtain $w_{p+1}, w_{p+2}, \ldots, w_k$, using

$$w_t = \phi_1 w_{t-1} + \phi_2 w_{t-2} + \ldots + \phi_p w_{t-p} + a_t - \theta_1 a_{t-1} - \theta_2 a_{t-2} - \ldots - \theta_q a_{t-q}$$
$$t = p + 1, p + 2, \ldots, k \tag{6}$$

9. If another series of length k is required, then return to Step 5.

For a particular ARMA model, it is only necessary to calculate the matrix $\underline{M}$ once, no matter how many simulated series are synthesized. Therefore, WASIM2 is economical with respect to computer time required, especially when many time series of the same length are generated.

Often the white noise disturbances can be assumed to be $NID(0, \sigma_a^2)$ and it is desirable to have as much accuracy as possible in order to eliminate bias. For this situation, McLeod and Hipel (1978b) recommend using WASIM2 for a pure AR model or an ARMA process. When simulating a pure MA process with innovations that are $NID(0, \sigma_a^2)$, the WASIM1 and WASIM2 procedures are identical.

WASIM2 is based upon a knowledge of the theoretical autocovariance or autocorrelation function (ACF). This general type of approach can also be used for simulating using other types of stochastic models. For example, Hipel (1975) and McLeod and Hipel (1978a) employ the theoretical ACF of the fractional Gaussian noise model for developing an exact simulation technique for that model.

2.3. THE HURST PHENOMENON

2.3.1. *Background*. The Hurst phenomenon created one of the most interesting, controversial and long-lasting scientific debates ever to arise in the field of hydrology. The genesis of the Hurst phenomenon took place over forty years ago in Egypt. Just after World War II, a British scientist by the name of H.E. Hurst became deeply involved in studying how the Nile River could be optimally controlled and utilized for the benefit of both Egypt and Sudan. As Director-General of the Physical Department in the Ministry of Public Works in Cairo, Egypt, Hurst was particularly interested in the long-term storage requirements of the Nile River. In addition to annual riverflow series, Hurst analyzed a wide variety of other yearly geophysical time series in order to examine the statistical properties of some specific statistics that are closely related to long term storage. These statistical studies led Hurst to develop an empirical law upon which the definition of the Hurst phenomenon is based.

The fact that the Hurst phenomenon arose from scientific work carried out in Egypt provided the controversy with an aura of mystery and intrigue. Was the Hurst phenomenon more difficult to solve than the riddle of the Sphinx? A range of explanations have been put forward for solving the Hurst phenomenon. Furthermore, in the process of studying the Hurst phenomenon, many original contributions have been made to the fields of hydrology and statistics. For example, fractional Gaussian noise (Mandelbrot and Wallis, 1969) and other kinds of long memory models were developed. In addition, the shifting level processes were defined by Boes and Salas (1978) and extended by Salas and Boes (1980) as well as Ballerini and Boes (1985). Interpretations and descriptions of the Hurst phenomenon are provided by authors including Hipel (1975), McLeod and Hipel (1978a), Klemes (1974) and Salas et al. (1979).

The solution to Hurst's riddle originally put forward by Hipel (1975) and Hipel and McLeod (1978a) is outlined in Section 2.3.3 after the Hurst phenomenon is defined in the next subsection. Simulation experiments demonstrate that when the most appropriate ARMA models are fitted to a wide variety of annual geophysical time series, a statistic called the Hurst coefficient is "statistically preserved" by the models. Therefore, although the Hurst coefficient and other related statistics are not directly incorporated as model parameters in the design of an ARMA model, these statistics can still be indirectly accounted for or modelled by ARMA models.

2.3.2. *Definition*. Consider a time series $w_1, w_2, \ldots, w_n$. Define the kth adjusted partial sum S^*_k as

$$S^*_k = S^*_{k-1} + (w_k - \bar{w}_n) = \sum_{i=1}^{k} w_i - k\bar{w}_n \quad k = 1, 2, \ldots, n \qquad (7)$$

where $S^*_0 = 0$, $S^*_n = 0$, $\bar{w}_n = \sum_{i=1}^{n} w_i/n$ is the mean of the first n terms of series. The adjusted range is defined as

$$R^*_n = M^*_n - m^*_n \qquad (8)$$

where $M^*_n = \max (0, S^*_1, S^*_2, \ldots, S^*_n)$ is the adjusted surplus, and $m^*_n = \min (0, S^*_1, S^*_2, \ldots, S^*_n)$ is the adjusted deficit. Finally, the rescaled adjusted range (RAR) is

$$\bar{R}^*_n = R^*_n / D^*_n \tag{9}$$

where $D^*_n = n^{-1/2} \left(\sum_{i=1}^{n} (w_i - \bar{w}_n)^2 \right)^{1/2}$ is the sample standard deviation. The

above and other related statistics are very useful in reservoir design.

Hurst (1951, 1956) stimulated interest in the RAR statistic by his studies of long-term storage requirements on the Nile River. On the basis of a study of 690 annual time series comprising streamflow, river and lake levels, precipitation, temperature, pressure, tree ring, mud varve, sunspot and wheat price records, Hurst implied that $\bar{R}^*_n$ varies with n as

$$\bar{R}^*_n \alpha n^h \tag{10}$$

where h is the generalized Hurst coefficient. The above equation can be written in the general form

$$\bar{R}^*_n = an^h \tag{11}$$

where a is a coefficient that is not a function of n. It should be noted that Hurst did not explicitly state the generalized Hurst law of Equation 11 in his research papers. However, by choosing the coefficient a to have a value of $(1/2)^h$, Hurst in effect estimated h by the Hurst coefficient K in the empirical equation

$$\bar{R}^*_n = (n/2)^K \tag{12}$$

By taking logarithms of Equation 12, an explicit relationship for K is then

$$K = \frac{\log \bar{R}^*_n}{\log n - \log 2} = \frac{\log R^*_n - \log D^*_n}{\log n - \log 2} \tag{13}$$

Employing series that varied in length from 30 to 2000 years, Hurst found K to range from 0.46 to 0.96 with a mean of 0.73 and a standard deviation of 0.09.

Assuming a normally independently distributed (NID) process, Hurst (1951) utilized some coin-tossing experiments to develop a theoretical asymptotic relationship for the expected value of the adjusted range. Using the theory of Brownian motion, Feller (1951) rigorously established Hurst's formula for any sequence of identically independently distributed (IID) random variables possessing finite variance. It follows from a standard convergence theorem in probability theory (Rao, 1973, p. 122) that for large n,

$$E(\bar{R}^*_n) = 1.2533 n^{1/2} \tag{14}$$

Even though Hurst studied the RAR for small n and not for the adjusted range, the form of Equation 14 prompted him to use K in Equation 13 as an estimate of h and also to assume K to be constant over time. However, for 690 geophysical time series, Hurst found K to have an average of 0.73, while the asymptotic, or limiting, value of K given by Equation 14 is 0.5. This discrepancy is referred to as the Hurst phenomenon. The search for a reasonable explanation of the Hurst phenomenon and the need for methods whereby the statistics related to Hurst's work can be incorporated into mathematical models have intrigued researchers for decades.

2.3.3. *Simulation Experiments.* A major challenge in stochastic hydrology is to determine models that preserve important historical statistics such as the rescaled adjusted range (RAR), or equivalently the Hurst coefficient K. The major finding of Hipel (1975) and Hipel and McLeod (1978a) which is also reported in this section, is that ARMA models do statistically preserve the historical RAR statistics or equivalently the Hurst coefficients denoted using K's. This interesting scientific result is what solves the riddle of the Hurst phenomenon.

In their research, Hipel (1975) and Hipel and McLeod (1978a) fit ARMA models to 6 annual riverflow, 1 mud varve, 1 temperature, 1 precipitation, 1 sunspot, 1 minimum annual flow, and 12 tree ring time series. Simulation studies are then carried out to determine the small sample empirical cumulative distribution function (ECDF) of the RAR or K for various ARMA models. The ECDF for each of these statistics is shown to be a function of the time series length n and the parameter values of the specific ARMA process being considered. Furthermore, it is possible to determine as accurately as desired the distribution of the RAR or K. Then it is shown by utilizing simulation results and a given statistical test that ARMA models do preserve the observed RAR or K of the 23 geophysical time series.

Suppose that the determination of the exact distribution of the RAR or K is required. The expected value of $\bar{R}^*_n$ is now known theoretically for both an independent and a symmetrically correlated Gaussian process (Anis and Lloyd, 1976). At present, the cumulative distribution function (CDF) or $\bar{R}^*_n$ for a white noise process and, in general, any ARMA model is analytically intractable. However, by simulation it is possible to determine as accurately as is desired for practical purposes the CDF for $\bar{R}^*_n$. Because both $\bar{R}^*_n$ and K are functions of n, their CDF's are defined for a particular length of series n. The CDF for $\bar{R}^*_n$ is

$$F = F(r;n,\phi,\theta) = Pr(\bar{R}^*_n \leq r) \tag{15}$$

where n is the length of each individual time series; ϕ is the set of known AR parameters; θ is the set of known MA parameters; and r is any possible value of $\bar{R}^*_n$.

When simulating a series of length n, WASIM1 is used for ARMA(0,q) models while WASIM2 (see Section 2.2) is used for ARMA(p,0) and ARMA(p,q) models. Because the RAR or K is independent of the variance of the innovations, any value of σ_a^2 may be used. Consequently, it is simplest to set $\sigma_a^2=1$ and, hence, to assume that the residuals are NID(0,1).

Suppose that $\bar{n}$ simulations of length n are generated for a specific ARMA model and the $\bar{n}$ RAR's given by $\bar{R}^*_{n1}$, $\bar{R}^*_{n2}$, ..., $\bar{R}^*_{n\bar{n}}$, are calculated for each of the simulated series. If the sample of the RAR is recorded such that

$\bar{R}^*_{n(1)} \leq \bar{R}^*_{n(2)} \leq \ldots \leq \bar{R}^*_{n(\bar{n})}$, it is known that the maximum likelihood estimate of F is given by the ECDF (Gnedenko, 1968, pp 444-451):

$$F_{\bar{n}} = F_{\bar{n}}(r;n,\phi,\theta) = 0 \quad r \leq \bar{R}^*_{n(1)}$$

$$F_{\bar{n}} = F_{\bar{n}}(r;n,\phi,\theta) = k/n \quad \bar{R}^*_{n(k)} < r \leq \bar{R}^*_{n(k+1)} \tag{16}$$

$$F_{\bar{n}} = F_{\bar{n}}(r;n,\phi,\theta) = 1 \quad r > \bar{R}^*_{n(\bar{n})}$$

By employing the ECDF of the RAR or K in conjunction with a specified statistical test, it is now shown that ARMA models do preserve the historically observed Hurst statistics. Because the Hurst coefficient K is widely cited in the literature, the research results for this statistic are described. However, K and $\bar{R}^*_n$ are connected by the simple transformation given in Equation 12, and, therefore, preservation of either statistic automatically implies retention of the other by an ARMA model.

The ARMA models fitted to 23 geophysical time series ranging in length from n=96 to n=1164 are listed by Hipel and McLeod (1978a, Table 3). For exactly the same n as the historical data, 10^4 simulations are done for each model to determine the ECDF of K, or equivalently $\bar{R}^*_n$. The probability p_i of having K for the ith model greater than the K calculated for the ith historical series is determined from the ith ECDF as

$$Pr(K>K_i^{obs}|model) = p_i \tag{17}$$

where K_i^{obs} is the K value calculated for the ith observed historical time series. If the chosen ARMA model is correct, then by definition, p_i would be uniformly distributed on (0,1). For k time series it can be shown (Fisher, 1970, p 99) that

$$-2 \sum_{i=1}^{k} \ln p_i \approx \chi_{2k}^2 \tag{18}$$

Significance testing can be done by using Equation 18 to determine whether the observed Hurst coefficient or the RAR is preserved by ARMA models. The test could fail if the incorrect model were fitted to the data or if ARMA models do not retain the Hurst K. Careful model selection was done, thereby largely eliminating the former reason for test failure. If it is thought (as was suggested by Mandelbrot and Wallis, 1968) that the observed K is larger than that implied by an appropriate Brownian domain model, then a one-tailed rather than a two-tailed test may be performed.

The results of the χ^2 test in Equation 18 for the 23 geophysical phenomena confirm that there is no evidence that the observed K, or equivalently the RAR, is not adequately preserved by the ARMA models. Hipel and McLeod (1978a, Table 4) summarize the information used in the test. The calculated χ^2 value from Equation 18 is not significant at the 5% level of significance for the 23 time series for either a one-sided or a two-sided test. Therefore, on the basis of the given information, ARMA models do preserve K or the RAR when considering all the time series. Furthermore, when the set of annual riverflows, miscellaneous data, and tree ring indices are inspected individually, it can be seen that ARMA models preserve the historical Hurst statistics for all three cases (Hipel and McLeod, 1978a).

2.4. PRESERVATION OF CRITICAL PERIOD STATISTICS

In an important paper on stochastic hydrology, Hall et al. (1969) describe problems related to the design and operation of a reservoir when water shortages must be considered. They define the critical period as the period of time during which a given riverflow time series is most critical with respect to meeting water demands. After defining a number of statistics connected to critical period, Hall et al. (1969) conclude using simulation experiments that the time series models they consider do not adequately preserve the critical period statistics. In a more comprehensive study, Askew et al. (1971) find that a wide variety of stochastic models are not capable of retaining critical period statistics.

Thompstone et al. (1987) employ split sample experiments to demonstrate that PAR models do statistically preserve the critical period statistics. More specifically, the most appropriate PAR models are constructed for modelling three average monthly hydrological time series where the last 20 years of data for each series is omitted during model calibration. These last 20 years of data are used to estimate the critical period statistics. Next, 1000 synthetic monthly sequences of 40 years each are generated for each of the three models and the first 20 years of each sample are dropped to provide 1000 effectively independent sequences equal in length to the series used to calculate the historical critical period statistics. Similar to the procedure described in Section 2.3.3 for the Hurst statistics, the ECDF for each critical period statistic and each model can be utilized for obtaining the p_i values used in Equation 18. By employing Equation 18, for each critical period statistic across the three models a χ^2 test can be carried out. For the three types of critical period statistics studied, Thompstone et al. (1987) show that PAR models statistically preserve these statistics.

3. Forecasting Experiments

3.1. INTRODUCTION

The general purpose of forecasting is to provide the best possible estimates of what will happen at specified points in time in the future. By following the three stages of model construction, one can fit the most appropriate time series model to a given time series. Based upon the calibrated model and the most recent observations, one can obtain minimum mean square error forecasts of future observations. Forecasts for a riverflow series, for example, can be used for deciding upon the operating rules of a reservoir. Forecasting can also be used for model discrimination purposes. When models from a variety of different classes of models are fitted to a time series, one can select the model which provides the most accurate forecasts. In order to determine this, one can carry out statistical tests to ascertain if one model forecasts significantly better than another.

In practical applications, one-step-ahead forecasts are often required for effectively operating a large-scale engineering project such as a system of reservoirs. When a new observation becomes available, the next one-step-ahead forecast can be made for deciding upon the operating rules in the subsequent time period. Furthermore, a theoretical advantage of one-step-ahead forecasts is that they are statistically independent. This property allows one to develop statistical tests to determine if one model forecasts significantly better than another. Three tests that can be used for comparing one-step-ahead forecast errors from two models to determine if one model forecasts significantly better than another are the Wilcoxon signed rank (Wilcoxon, 1945), likelihood ratio (Noakes et al., 1988) and the Pitman (1939) tests.

In the upcoming subsections, interesting conclusions are reported for a variety of forecasting experiments carried out by the author and his colleagues during the past few years. One finding is that ARMA type models perform very well when compared to their competitors for forecasting one-step-ahead forecasts for both nonseasonal and seasonal hydrological time series. Another result is that time series models that are developed by following the three stages of model construction forecast better than models that are not built using this approach. If a good model can be found for modelling a specific time series, forecasts cannot be improved by combining it with forecasts from models which are less parsimonious or less adequate. Finally, ARMA-type models often forecast better than more complicated conceptual models.

3.2. FORECASTING STUDIES WITH ANNUAL GEOPHYSICAL TIME SERIES

Noakes et al. (1988) carry out forecasting experiments with two sets of annual geophysical time series. The first set of time series consists of four average annual riverflow series and one annual temperature series while the second set is made up of nine tree ring indices series. The nonseasonal models considered in the study are ARMA (Box and Jenkins, 1976), fractional Gaussian noise (see McLeod and Hipel (1978a) and Hipel and McLeod (1978b) for a maximum likelihood procedure for estimating the model parameters), fractional ARMA (Li and McLeod, 1986), fractional differencing (a special case of the fractional ARMA model) (Granger and Joyeux, 1980; Hosking, 1981, 1984), first order Markov chain, and nonparametric regression (Yakowitz, 1985a,b) models. To execute split sample experiments, the nonseasonal models are fitted to the first part of each series and then used to obtain one-step-ahead forecasts for the last 30 years of the data. The forecast errors are compared using the root mean squared errors in each experiment. Pitman's (1939) test is employed to test pairwise between models for statistically significant differences in the root mean squared errors of the forecasts. The new likelihood ratio test of Noakes et al. (1988) is also used.

Overall, the ARMA and nonparametric regression models forecast the best in the two sets of experiments. The fractional Gaussian noise and fractional differencing models do not perform well.

3.3. FORECASTING EXPERIMENTS WITH MONTHLY RIVERFLOW SERIES

Noakes et al. (1985) use mean monthly riverflow from 30 rivers in North and South America to test the short-term forecasting ability of seasonal ARIMA (autoregressive integrated moving average), deseasonalized ARMA and PAR models. Each of the time series is transformed using natural logarithms and all of the seasonal models are fitted to the first part of the series which does not contain the last 36 months of data. In total, seasonal models, consisting of one seasonal ARIMA, two types of deseasonalized ARMA and six kinds of PAR models, are fitted to the first section of the series. Each model is then used to generate one-step-ahead forecasts for the last 36 months of data. The forecasts, when simply the mean of the logarithmic flows are used, are also computed. The logarithmic forecast errors associated with each of the forecasting models are then compared using various performance measures including root mean squared errors and mean absolute derivation. Along with Fisher's (1970, p. 99) method for combining significance levels, the Wilcoxon signed rank (Wilcoxon, 1945) is used to test for statistically significant differences in the forecasting ability of the various procedures.

The overall finding of this rather extensive forecasting study is that PAR models identified using the periodic partial autocorrelation function forecast the best. This is not surprising since hydrologists realize that riverflow time series exhibit an autocorrelation structure that depends not only on the time lag between observed flows but also on the season of the year (Moss and Bryson, 1974). For example, if snowmelt is an important factor in runoff that might occur in March or April, the correlation between observed riverflows may be negative for these months whereas at other times of the year it is usually positive.

3.4. COMBINING HYDROLOGICAL FORECASTS

McLeod et al. (1987) employ two case studies to investigate the usefulness of combining forecasts from different seasonal models in order to improve the accuracy of the forecasts. These authors also define a new technique for calculating the combining weights across two models when seasonal data are being forecasted.

In the first study, a transfer function-noise (Box and Jenkins, 1976) model, a PAR model, and a conceptual model are employed to forecast quarter-monthly riverflows. These models all approach the modelling and forecasting problem from three different perspectives, and each has its own particular strengths and weaknesses. The forecasts generated by the individual models are combined in an effort to exploit the strengths of each model. The results of this case study indicate that significantly better forecasts can be obtained when forecasts from different types of models are combined. In the second study, PAR models and seasonal ARIMA models are used to forecast 30 monthly riverflows. Combining the individual forecasts from these two types of time series models does not result in significantly better forecasts. This is because on its own the seasonal ARIMA model does not forecast seasonal riverflow series as well as PAR models.

Thompstone et al. (1985) also compare quarter-monthly riverflow forecasts for transfer function-noise, PAR and conceptual models, while Thompstone et al. (1983) show how transfer function-noise models can be used for powerhouse inflow forecasting. However, they do not examine the problem of combining forecasts across models. Finally, using a Kalman filter framework, Jimenez et al. (1989) develop a computationally efficient method for obtaining maximum likelihood estimates for the parameters in a periodic ARMA model. They also present a small forecasting experiment with these models.

4. Conclusions

ARMA-type models work remarkably well for simulating both nonseasonal and seasonal hydrological time series. As demonstrated by the simulation experiments outlined in Section 2.3.3, ARMA models statistically preserve the Hurst statistics and thereby provide a clear answer to the riddle of the Hurst phenomenon. Furthermore, as pointed out in Section 2.4 PAR models statistically preserve the critical period statistics for monthly riverflow time series.

When forecasting annual geophysical time series, ARMA models and the non-parametric regression model of Yakowitz (1985a,b) produce more accurate forecasts then their other competitors (Section 3.2). For the case of average monthly riverflows, PAR models identified using the periodic partial autocorrelation function provide accurate forecasts. As explained in Section 3.4, combining forecasts from different models can produce more accurate forecasts when the individual models are quite different in design and both models produce reasonably accurate forecasts. However, because seasonal ARIMA models do not forecast seasonal riverflow data nearly as well as PAR models, combining forecasts across these two models produces forecasts that are less accurate than the PAR forecasts on their own.

When developing a time series model for describing a given time series, experience has shown that better models can be developed by following the identification, estimation and diagnostic check stages of model construction. Only a properly designed and calibrated model has the potential to work well in simulation and forecasting.

5. Acknowledgements

The research on simulation and forecasting that is referenced and summarized in this paper is based upon collaborative work of the author with Prof. A. Ian McLeod, Dr. Donald J. Noakes, Dr. Robert M. Thompstone, Dr. Carlos Jimenez and Prof. Sidney Yakowitz. The author greatly appreciates the fine contributions of these colleagues and having the opportunity to carry out joint research with them.

6. References

Anis, A.A. and Lloyd, E.H. (1976) 'The expected value of the adjusted rescaled Hurst range of independent normal summands', Biometrika 63, pp 111-116.

Askew, A.J., Yeh, W.W.G. and Hall, W.A. (1971) 'A comparative study of critical drought simulation', Water Resources Research 7, pp 52-62.

Ballerini, R. and Boes, D.C. (1985) 'Hurst behaviour of shifting level process', Water Resources Research 12, 11, pp 1642-1648.

Boes, D.C. and Salas, J.D. (1978) 'Nonstationarity of the mean and the Hurst phenomenon', Water Resources Research 14, 1, pp 135-143.

Box, G.E.P. and Cox, D.R. (1964) 'An analysis of transformations', Journal of the Royal Statistical Society, Series B 26, pp 211-252.

Box, G.E.P. and Jenkins, G.M. (1976) Time Series Analysis: Forecasting and Control, Second Edition, Holden-Day, San Francisco.

Feller, W. (1951) 'The asymptotic distribution of the range of sums of independent random variables', Annals of Mathematical Statistics 22, pp 427-432.

Fisher, R.A. (1970) Statistical Methods for Research Workers, Oliver and Boyd, Edinburg, UK.

Gnedenko, B.V. (1968) 'Theory of Probability', Chelsea, New York.

Granger, C.W.J. and Joyeux, R. (1980) 'An introduction to long-memory time series models and fractional differencing', Journal of Time Series Analysis 1, pp 15-29.

Hall, W.A., Askew, A.J. and Yeh, W.W.G. (1969) 'Use of the critical period in reservoir analysis', Water Resources Research 5, 6, pp 1205-1215.

Hipel, K.W. (1975) 'Contemporary Box-Jenkins modelling in water resources', Ph.D. Thesis, University of Waterloo, Waterloo, Ontario.

Hipel, K.W. and McLeod, A.I. (1978a) 'Preservation of the rescaled adjusted range, 2, simulation studies using Box-Jenkins models', Water Resources Research 14, 3, pp 509-516.

Hipel, K.W. and McLeod, A.I. (1978b) 'Preservation of the rescaled adjusted range, 3, fractional Gaussian noise algorithms', Water Resources Research 14, 3, pp 517-518.

Hipel, K.W. and McLeod, A.I. (1993) Time Series Modelling of Water Resources and Environmental Systems, Elsevier, Amsterdam.

Hipel, K.W., McBean, E.A. and McLeod, A.I. (1979) 'Hydrologic generating model selection', Journal of the Water Resources Planning and Management Division, ASCE 105, WR2, pp 223-242.

Hosking, J.R.M. (1981) 'Fractional differencing', Biometrika 68, pp 165-176.

Hosking, J.R.M. (1984) 'Modeling persistence in hydrological time series using fractional differencing', Water Resources Research 20, 12, pp 1898-1908.

Hurst, H.E. (1951) 'Long-term storage capacity of reservoirs', Transactions of the American Society of Civil Engineers 116, pp 770-808.

Hurst, H.E. (1956) 'Methods of using long-term storage in reservoirs', Proceedings of the Institute of Civil Engineers 1, pp 519-543.

Jimenez, C., McLeod, A.I. and Hipel, K.W. (1989) 'Kalman filter estimation for periodic autoregressive-moving average models', Stochastic Hydrology and Hydraulics 3, 3, pp 229-242.

Klemes, V. (1974) 'The Hurst phenomenon: a puzzle?', Water Resources Research 10, 4, pp 675-688.

Li, W.K. and McLeod, A.I. (1986) 'Fractional time series modelling', Biometrika 73, 1, pp 217-221.

Mandelbrot, B.B. and Wallis, J.R. (1968) 'Noah, Joseph and operational hydrology', Water Resources Research, 5,5, pp 909-918.

Mandelbrot, B.B. and Wallis, J.R. (1969) 'Computer experiments with fractional Gaussian noises, Parts 1 to 3', Water Resources Research 5, pp 228-267.

McLeod, A.I. (1975) 'Derivation of the theoretical autocovariance function of autoregressive-moving average time series', Journal of the Royal Statistical Society, Series C (Applied Statistics) 24, 2, pp 255-256.

McLeod, A.I. and Hipel, K.W. (1978a) 'Preservation of the rescaled adjusted range, 1, a reassessment of the Hurst phenomenon', Water Resources Research 14, 3, pp 491-508.

McLeod, A.I. and Hipel, K.W. (1978b) 'Simulation procedures for Box-Jenkins models', Water Resources Research 14, 5, pp 969-975.

McLeod, A.I., Noakes, D.J., Hipel, K.W. and Thompstone, R.M. (1987) 'Combining hydrologic forecasts', Journal of the Water Resources Planning and Management Division, ASCE, 113, 1, pp 29-41.

Moss, M.E. and Bryson, M.C. (1974) 'Autocorrelation structure of monthly streamflows', Water Resources Research 10, pp 737-744.

Noakes, D.J., McLeod, A.I. and Hipel, K.W. (1985) 'Forecasting monthly riverflow time series', International Journal of Forecasting 1, pp 179-190.

Noakes, D.J., Hipel, K.W., McLeod, A.I., Jimenez, J. and Yakowitz, S. (1988) 'Forecasting annual geophysical time series', International Journal of Forecasting 4, pp 103-115.

Pitman, E.J.G. (1939) 'A note on normal correlation', Biometrika 31, pp 9-12.

Ralston, A. (1965) A First Course in Numerical Analysis, McGraw-Hill, New York.

Rao, C.R. (1973) Linear Statistical Inference and Its Applications, Second Edition, John Wiley, New York.

Salas, J.D. and Boes, D.C. (1980) 'Shifting level modelling of hydrologic series', Advances in Water Resources 3, pp 59-63.

Salas, J.D., Boes, D.C., Yevjevich, V. and Pegram, G.G.S. (1979) 'Hurst phenomenon as a pre-asymptotic behaviour', Journal of Hydrology 44, pp 1-15.

Salas, J.D., Delleur, J.W., Yevjevich, V. and Lane, W.L. (1980) Applied Modelling of Hydrologic Series', Water Resources Publications, Littleton, Colorado.

Thompstone, R.M., Hipel, K.W. and McLeod, A.I. (1983) 'Transfer function-noise modelling for powerhouse inflow forecasting', INFOR 21, PP 258-269.

Thompstone, R.M., Hipel, K.W. and McLeod, A.I. (1985) 'Forecasting quarter-monthly riverflows', Water Resources Bulletin 21, 5, pp 731-741.

Thompstone, R.M., Hipel, K.W. and McLeod, A.I. (1987) 'Simulation of monthly hydrological time series', in I.B. MacNeill and G.J. Umphrey (eds), A.I. McLeod (assoc.ed.), Advances in Statistical Sciences, Festschrift in Honor of Professor V.M. Joshi's 70th Birthday, Vol. IV, Stochastic Hydrology, D. Reidel Publishing Co., Dordrecht, The Netherlands, pp 57-71.

Vogel, R.M. and Stedinger, J.R. (1988) 'The value of stochastic streamflow models in over-year reservoir design applications', Water Resources Research 24, 9, pp 1483-1490.

Wilcoxon, F. (1945) 'Individual comparisons by ranking methods', Biometrics 1, pp 80-83.

Yakowitz, S.J. (1985a) 'Nonparametric density estimation and prediction for Markov sequences', Journal of the American Statistical Association 80, pp 215-221.

Yakowitz, S.J. (1985b) 'Markov flow models and the flood warning problem', Water Resources Research 21, pp 81-88.

USE AND MISUSE OF STOCHASTIC HYDROLOGY IN WATER RESOURCES

VUJICA YEVJEVICH
Professor Emeritus of Civil Engineering
Colorado State University
Fort Collins, Colorado 80523
USA

ABSTRACT. Proper use of stochastic modeling and generation of samples in utilitarian hydrology significantly outweighs misuse and unwarranted expectations. A major use of this stochastic technology is in the solution of water resources problems for which time horizons of solutions and the number of spatial points needing generation of samples are relatively large. This technology then produces multiple solutions, permitting the drawing up of the frequency distribution and the economic optimization or conventional decision making of the unique solution to be provided by the system. It further enables investigators: to experimentally derive the sampling frequency distributions; to check the analytically derived exact, asymptotic and approximate solutions of stochastic hydrology; to assess the characteristics of inferred or assumed models; to solve the complex problems of the regional water resources character; to develop the optimization-based operational rules for the water resources systems; to study the effects of uncertainty in the models and in their estimated parameters; to determine the confidence limit solutions; to investigate the sensitivity of solutions to various changes in models; to condense information; and to solve several other problems for which the modeling/simulation technology gives more accurate results than any other method available. This technology is good for the extraction and transfer of information contained in observational data. It is not a technology that could increase information beyond that contained in the data, except when the physical information available is combined with the statistical information.

1. Introductory Statements

It frequently happens that both proper use and misuse of most new technologies develop, at least in the first place where they are applied. It is therefore to be expected that the most recent developments in the application of stochastic hydrology to water resources would consist of both correct use, but also misuse of this technology. It is, however, important that the proper use significantly outweighs the misuse.

189

J.B. Marco et al. (eds.),
Stochastic Hydrology and its Use in Water Resources Systems Simulation and Optimization, 189–208.
© 1993 *Kluwer Academic Publishers.*

The introduction of innovations into the solving of water resources problems, especially when it requires complex mathematical techniques, is often very slow. Most current leaders in the water resources field obtained their basic education in water resources disciplines, heavily based on advanced mathematics, a couple of decades ago. They are usually very well versed in the ongoing standard methods of solving water resources problems. However, when new technologies are proposed, these leaders are correct in asking the simple question of what can these innovations do better, more accurately and advantageously than the existing technology. Innovators along with practitioners are therefore responsible for proving the advantages of new technologies. The responsibility for the introduction and proper use of new methods of stochastic hydrology into the practice of water resources planning, design and operation rests then with both innovators and practitioners. Otherwise, proper use may be replaced by eventual misuse or an unwarranted expectation.

Because inputs and outputs of water resources systems are random variables and often involve complex stochastic processes, solutions of water resources problems are also random variables, or they vary from one sample or subsample to another. These samples or subsamples are then conceived as time horizons for which solutions are sought. These multiple solutions permit either drawing up of empirical frequency distributions or the derivation of probability distribution functions (PDF's) of these solutions. Then, these distributions become input for the economic analysis and/or conventional decision making in the form of which solution value of each random variable should be provided by a system. Solutions of classical water resources problems are most often given as unique values of the random variable (such values as the fixed capacity of a plant, diameter of a tunnel, benefit, cost, risk, etc.). Each unique value then divides this probability or frequency distribution into two parts, the range of variable values with the satisfactory performance of the relatively high probability of non-exceedence, and the range of variable values with the unsatisfactory performance of the relatively small probability of exceedence or risk. Stochastic hydrology then represents new technologies which produce solutions even when the standard or conventional methods give only highly unreliable results.

The two terms, simulation and generation of samples, are considered in this text as synonyms and will be used interchangeably. Other terms used elsewhere as synonyms are: Monte Carlo method, synthetic hydrology, experimental stochastic method or stochastic simulation method. However, these latter terms will not be used in this text except in the historical context.

2. Solving Problems by Generating Samples

2.1. SOLUTION OF PROBLEMS INVOLVING LONG TIME HORIZONS

The length of historic samples of input and output of water resources structures and systems may often be sufficient for the solution of their current problems provided the size of these samples is not too small (say, at least 15 years of data) and the time horizon of searching for the solution is also relatively small in comparison with the available length of the historic sample. In this case the division of the sample into subsamples, with each subsample equal by its size to the time horizon of solutions, will produce as many results as there are subsamples.

A simple example would then be the search for the storage capacity needing to be built in order to provide the daily peaking hydropower or the daily peaking water supply of a city. The day then becomes the subsample or the time horizon which would produce as many daily needed storage capacities as there were days in the sample. So, ten years of experience with peaking power or water supply would represent the 3650 potential solutions to the problem. In this case there is no need for generation of new samples or subsamples since there will be a sufficient number of solutions to produce a reliable frequency distribution of the needed storage capacities. The optimization method or conventional decision making will determine which storage size should be built and the risk of not having enough storage capacity to satisfy demand.

Similarly, for the weekly cycle of determining the storage capacity to be provided for varying weekly hydropower or water supply, there will be 520 solutions for ten years of experience or data on flow regulation. This information would then enable a reliable frequency distribution to be drawn up of the needed storage capacity and the corresponding decision what storage capacity should be provided for the given or optimized risk of not being able to satisfy demand.

With the time horizon increased, such as the seasonal or within-the-year flow regulation, the subsample or horizon in this case is equal to the year, with the number of years providing the number of solutions of the needed storage capacity for seasonal regulation. The condition in this case is that the firm or guaranteed dry season water releases are only a small fraction of the mean of water input, which will not require any water carryover from year to year, except under exceptional conditions. Then 15 years of data would give 15 solutions of the needed storage capacity leading to their approximate frequency distribution and the unique capacity size with the corresponding risk of not being sufficient.

When the time horizon starts to increase beyond the year, say for the over-the-year flow regulation with a longer and longer time period of the average filling-emptying cycle of regulation (the time from one empty state of the reservoir to the next is a random variable), the number of subsamples in the historic sample decreases as the horizon increases. Then there would be a smaller and smaller number of solutions of needed storage capacity as the horizon increases, thus making the drawing up of the frequency distribution highly unreliable, or there would be no reliable solution for the time horizon which exceeds the historic sample size.

The basic criterion for a meaningful application of the method of generation of samples by using models of stochastic hydrology in order to obtain more accurate solutions of problems with the long time horizons than with the other methods could be stated simply as

$$m = N/n \tag{1}$$

where N = available sample size; n = subsample or the time horizon for which a solution is sought; and m = number of solutions enabling the drawing up of the frequency distribution needed for the final economic analysis and/or conventional decision making. When m is very small, the solution using the historic sample is either highly unreliable, or one does not theoretically exist even when m < 1. The extraction of information from the historic sample in the form of a hydrologic stochastic model therefore permits the generation of as many samples of the size equal to the time horizon as the accuracy of the solution may require, or of a very long sample if this approach would produce the best solution.

The generation of samples in hydrology has two basic cases: (1) the search for solutions or properties of population of a particular property of the process in which case the size of only one generated long sample is determined by the desired accuracy of the solution and the economy of generating the sample; and (2) the search for properties of the particular sample size(s), that are the characteristics of results of simulation associated with a finite time horizon, in which case a set of m samples of that size is generated to enable the corresponding frequency distribution to be drawn up.

For problems involving the finite time horizons, the present technology of stochastic hydrology permits the simulation of as many pairs of potential future samples of input and output of a water resources system, and at as many points in space as needed by the problem posed which the accuracy of the solution to a problem may require. The number of generated pairs of samples of input and output at all the spatial points is the major factor in the accuracy of solutions along with errors in models and sampling errors in their estimated parameters. The basic conditions to satisfy in this approach are the proper reproduction by modeling and simulation of: (1) the time structure and temporal dependence of input and output; (2) the dependence between the stochastic input and output components at each point; (3) the spatial distribution of the estimated parameters; and (4) the spatial dependence of the stochastic components of both, input and output, at each pair of points.

Routine techniques as well as software for many computers are already available to conduct modeling and generation. This approach to the solution of water resources problems may be properly called the experimental numerical stochastic method. In some way, it is analogous to physical experimentation in hydraulics and hydrology, or to numerical experimentation in fluid mechanics and other water-related disciplines.

2.2. EXPERIMENTAL ESTIMATION OF SAMPLING DISTRIBUTIONS

Statistical inferences require sampling distributions of the parameters estimated, of the parameters needed for testing the hypotheses, of specific values of distributions such as percentiles, of nonparametric values, and of special characteristics of hydrologic stochastic processes such as the estimated parameters of range, runs and the other water-resources related stochastic variables. Most analytically obtained sampling distributions are approximations as a result of simplifying assumptions in their derivation, such as independence, normality, stationarity, ergodicity, etc. Often, one is not sure how well these derived probability distributions of tested or testing parameters approximate the true sampling distributions. When most parameters are periodic, time series intermittent, and stochastic temporal and spatial dependencies relatively complex, one may suspect that the analytically derived sampling distributions could significantly depart from the true distributions.

Mathematical modeling of hydrologic stochastic processes and methods of simulation or generation of samples may be a very effective tool for experimentally finding the sampling frequency distributions of tested or testing parameters and of various other estimates. Two useful benefits may result from this stochastic technology: (1) if relatively reliable analytical sampling probability distribution functions (PDF) are available, or claimed to be good approximations to the true distributions by their developers or advocates, modeling and simulation of a sufficient number of samples of given size would produce as many estimates of tested or testing parameters or of nonparametric values as there are samples generated; and (2) if there is no reliable, or claimed as reliable, analytically developed sampling PDF, modeling and simulation of that stochastic process would produce sampling frequency distributions as the closest approximations to the true sampling probability distributions. The comparison in the first case between the analytically obtained sampling PDF and the experimentally obtained sampling frequency distribution will determine whether to use, in that particular case and in future cases, the PDF or the experimentally derived frequency distribution. In complex cases one may be inclined to routinely use the experimental technology of mathematical modeling and the generation of samples.

Figure 1 shows a case for which the sampling frequency distribution was derived by the modeling/simulation technology. Definitely, the reliability of that distribution depends on the reliability of models, estimates of their parameters, simulation technique and the number of simulated samples. This latter factor is usually an economic problem whenever the true cost of the time of modelers, simulation professionals and of computer operators were the controlling factors.

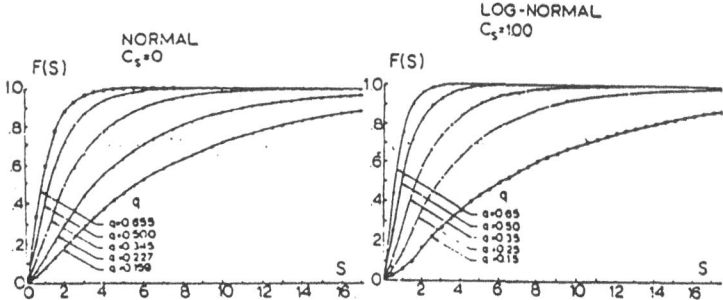

Figure 1. Frequency distributions of positive run-sums for the normal (C_s = 0) and lognormal (C_s = 1) independent stationary stochastic processes.

2.3. CHECKING ANALYTICAL DEVELOPMENTS BASED ON HYPOTHESES

The classical application of simulation, called the Monte Carlo method by probabilists and specialists on stochastic processes, was to check the analytically developed expressions and theorems by simulating random numbers of given probability distribution and stochastic properties. In this case, a large number of samples of finite horizons or a very long sample for populations permitted computations which have served as checks of accuracy of the expressions derived. The first probabilists would use their students to throw dice or coins, draw cards or observe roulettes in gambling casinos, or else do it by themselves, in order to obtain a large number of occurrences of studied complex random events, and to check their mathematical expressions usually found analytically or based on intuitive hypotheses.

Similar practice is useful in stochastic hydrology in two areas of analytical attempts to derive the exact solutions of current problems. These developments may be related either to exact, asymptotic or approximate probability distributions or related stochastic processes, or to values or relationships of their exact, asymptotic or approximate parameters to some variable such as the sample size. These two areas are: (1) expressions obtained analytically, with simulation of many samples of given size or a very long sample serving to check whether the experimentally obtained results closely followed those of these expressions, and if they did not, how much they differed one from the other; and (2) investigators advanced a hypothesis which led to the mathematical expression as the solution, but which was then checked by simulation of sample(s). The larger and larger the number of these samples and their sizes, the smaller and smaller would become the expected differences on the average as either the sample size or the number of generated samples increased. If differences between the results obtained by derived expressions and by simulation continuously decreased, either with the increase of the sample size or with the increase of the number of generated samples, the expectation and the assumption would be that for the infinite sample size or the infinite number of finite samples these differences would become zeros.

An example of the approximate analytical expression is the expected range as a function of the sample size n for the first-order autoregressive normal process. The exact expression for the expected range of independent normal process as function of the sample size was available (Annis and Lloyd, 1953) as

$$E(R_n) = \sqrt{2/\pi} \sum_{i=1}^{n} i^{-1/2} \tag{2}$$

but not for the dependent normal processes. The complexity of mathematics involved in the case of the first-order autoregressive normal process did not permit a development of general, exact analytical solutions for the expected range of any sample size. It was hypothesized (Yevjevich, 1967) that the approximation to the exact expected range may have the form of

$$E(R_n) = \sqrt{2/\pi} \sum_{i=1}^{n} i^{-1} (\text{Var } S_i)^{-1/2} \tag{3}$$

with $\text{var}(S_i)$ = variance of accumulated first-order autoregressive normal process up to the i-th value instead of independent process as in Equation 2. This approximation then was

$$E(R_n) = \sqrt{2/\pi} \ (1-\rho^2)^{-1/2} \ \sum_{i=1}^{n} i^{-1/2} \left[\frac{1+\rho}{1-\rho} - \frac{2\rho(1-\rho^i)}{i(1-\rho)^2} \right]^{1/2} \tag{4}$$

where ρ = first autocorrelation coefficient and n = sample size. For the exact expected range of the first-order autoregressive model for small n of 3 and 4 only (Boes and Salas, 1973), it was found (Salas, 1974; Troutman, 1974) that the approximate values of Equation 4 departed from these exact expected values only by 0.1 and 0.2 per cent, respectively, for n=3 and n=4. The method of sample generation has shown that the computed means were very close to the mean values obtained by Equation 4 for a wide range of n. However, it would have represented great difficulties to detect such minute differences, except for a very large number of generated samples. Therefore, one should be careful not to mistake the approximate analytical solutions for the exact values, even when differences were so small as in this example.

The classical statistical methods of solving problems followed often the approach of first searching for exact solutions. If they were not easy to obtain, then attempts would be made to find the asymptotic solutions. If neither this approach was successful, the last resort became a development of approximate solutions. In all three approaches, simulation of samples as the experimental method would serve as the check whether the results were acceptable or not. If these attempts were made for finding PDF's, and were not successful, then investigators would attempt to find the exact, asymptotic and approximate values of parameters of these PDF's, in that order of preference. Figure 2 shows an example of the application of this latter approach to the expected range of normal independent variable, by comparing the results of derived expressions for the mean range as functions of sample size n (the exact of Equation 2, asymptotic and approximate expression) and the results of simulation.

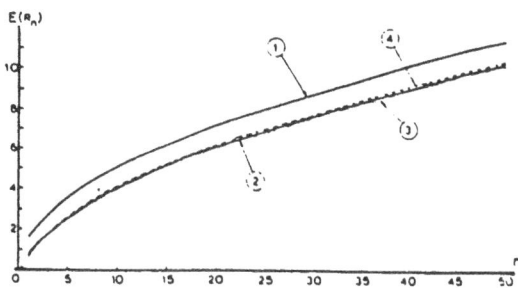

Figure 2. Comparison of means of range: (1) mean of asymptotic distribution, $E(r_n) = 1.6 \sqrt{n}$; (2) exact means, Equation 2; (3) means determined by generation of samples, R_n; and (4) an approximation given by $E(R_n) = 1.6 \sqrt{n-1}$; with the results of the latter three cases practically overlapping (Colorado State University Hydrology paper No. 10).

2.4. GENERAL PROPERTIES OF ASSUMED OR INFERRED MODELS

It is well known that probability density functions most currently used in hydrology, such as normal, lognormal, gamma, beta, double exponential, bounded exponential, etc., cannot be integrated in closed forms for finite integration limits. Therefore, they are presented in statistical tables, or can be obtained from these tables because of relationships between some of these variables. If one tries to find the exact distributions of various properties of these functions, say of their independent or simple dependent stochastic processes, in the form of probability distributions or relationships of their parameters to sample size, such properties as the range, adjusted range, conditional range, rescaled range, maximum deficit (depletion or maximum drawdown of a reservoir for given sample size), adjusted deficit, maximum run-length of a sample (maximum drought duration), maximum run-sum of a sample (maximum drought deficit), and similar derived random variables of these stochastic processes, the number of available exact, asymptotic and approximate analytical expressions is relatively limited (Pegram et al, 1980). To compensate for the lack of these expressions, the experimental approach permits the determination of approximate frequency distributions or of values of parameters, and as accurately as the cost of the large number of generated samples and the number of their sizes would be justified by the increased accuracy.

It is sufficient to infer that a PDF of a variable and its stochastic model were well fitted to many cases in a region, or the fits consistently passed the rigorous test, for this PDF and the model to be repeatedly used in that region. In the absence of exact, asymptotic and approximate PDF's of major derived variables of significant application to water resources systems, one may be tempted to develop experimentally, especially for large or important projects, the frequency distributions or the values of their parameters, or even some other major characteristics of these processes. Figure 3 shows an example of frequency distributions of the three drought describing variables for the monthly precipitation of the Upper Great Plains (USA). Figure 4 gives the estimated parameters and extremes of these drought variables in function of the probability of drought defining truncation level for the monthly precipitation of these plains. The relative smoothness of the fitted functions or curves, respectively for Figures 3 and 4, reveals that these results should be very close to the unknown true functions.

Instead of using the inferred PDF's and the corresponding stochastic models, one can assume these functions, study how their derived properties behaved and use these derived properties in various investigations. The closer these assumed models are to the real cases in practice, the more beneficial would be the experimentally developed frequency distributions and values of their parameters.

2.5. EXPERIMENTAL SOLUTIONS IN CASE OF COMPLEX PROBLEMS

Whenever complexity exceeds a given level, even when the time horizons of looking for solutions were relatively small, generation of samples may be either a useful tool or a necessary check of results obtained by another method. Solution may come out to be a serendipity, namely one tried to find solution to a problem but found this simulation to be more useful for solving another problem.

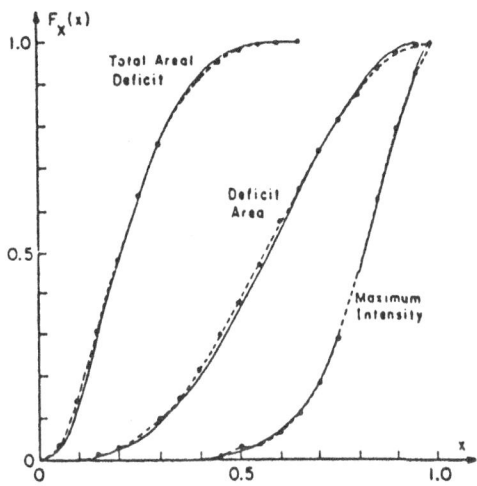

Figure 3. Frequency distributions (broken lines), obtained by modeling/simulation method, and fitted Beta probability distribution functions (solid lines) for the relative total drought areal deficit, relative drought deficit area and relative drought maximum indensity, for the Upper Great Plains of the United States of America (Colorado State University Hydrology Paper No. 87)

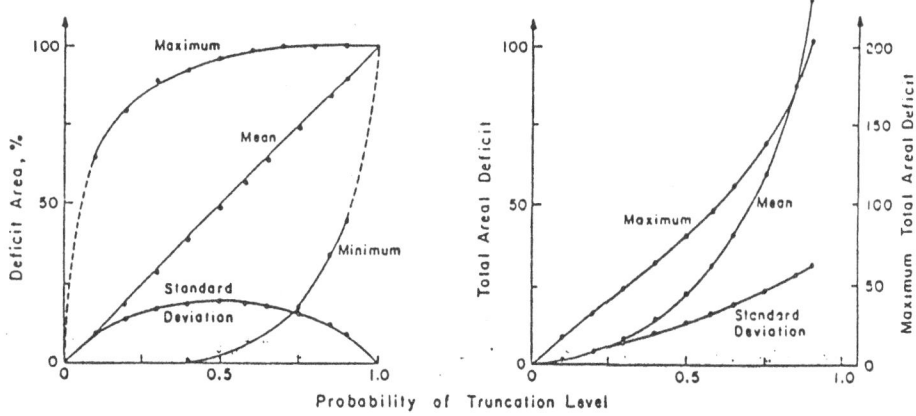

Figure 4. Variation of basic statistics of drought deficit area and drought total areal deficit as functions of the probability of truncation level of the monthly precipitation which defines droughts for the sample size N=600 (the Upper Great Plains of USA).

An example of such a case was the simulation of the net 10-days inputs of water into the Great Lakes of Canada and the United States of America, carried out for the US Corps of Engineers, Chicago Division (Colorado State University Hydrology Paper No. 72; Yevjevich, 1975). The purpose of simulation was to eventually improve the long standing operation of water releases through weirs at the connecting rivers between the lakes for the lake outflow regulation. It came out that simulation could not significantly improve the operational rules for these lakes, because the long experience with the operation and continuous historical improvement of these rules required little change. However, serendipity occurred due to the feasibility of assigning probabilities of exceedence to the highest levels of these lakes, as well as probabilities of non-exceedence to their lowest levels. Because the highest and lowest levels of these lakes have significant economic and safety impacts on inhabitants at and around the lake shores, the assessment of risk of these highest levels being exceeded, and of the risk of these lowest levels not being exceeded, was the major result of simulation.

The problem of the probability of exceedence or non-exceedence of the high and low levels of bodies of water can be extended from lakes with free outflow, like in the case of the Great Lakes, to lakes with subsurface outflow only (such cases as the outflow from regularly inundated and enclosed karst poljes), to lakes with both surface and subsurface outflows, but particularly to terminal lakes with no outflow except by evaporation. These latter lakes often have developments of communities and businesses around them. Long sequences of wet or dry years of relatively large return periods often produced unusually high or low lake levels, thus flooding large occupied areas or leaving developments far above the lowest levels.

A typical example of the problems with terminal lakes is the Great Salt Lake in Utah, or several salt lakes in Turkey and elsewhere. To reliably find the 100-year or any other return period of lowest and highest levels of these terminal (or even non-terminal) lakes, one must simulate a long sample of all the terms of the lake water budget equation, such terms as surface inflow, precipitation on the lake, evaporation, bank inflows and outflows, any artificial diversion in or out the lake, etc. Then the budget equation would give the net changes in water volume along the selected time intervals, and the corresponding volume-lake level relationship would provide the long time series of lake levels. The longer the simulated samples, the more accurate would be the estimated 100-year or any other return period, highest and lowest levels of the lake.

2.6. EXPERIMENTAL METHOD FOR OPERATIONAL OPTIMIZATION DECISIONS

The operation of water resources structures or systems can often be improved by the generation of relatively short samples, along with the use of systems state variables and any forecast of input or output. At a given time, one may well know the values of state variables of the system as well as the forecast of incoming input and output. Then the operational decisions in the form of water releases or deliveries become the conditional random variables, with conditions being these states and forecasts.

If one developed mathematical models of stochastic processes of these conditional variables, they would permit simulation of a number of samples of expected input and output in the immediate future. Figure 5 schematically presents such a case. For each generated sample, say in the case of seasonal or over-the-year regulating reservoirs, of the size of 12-36 months or so, there would be an optimal release in the immediate time unit, in this case the forthcoming month. Then one may decide to use the average or median release value out of all optimal values of generated samples, or by using any other decision making method.

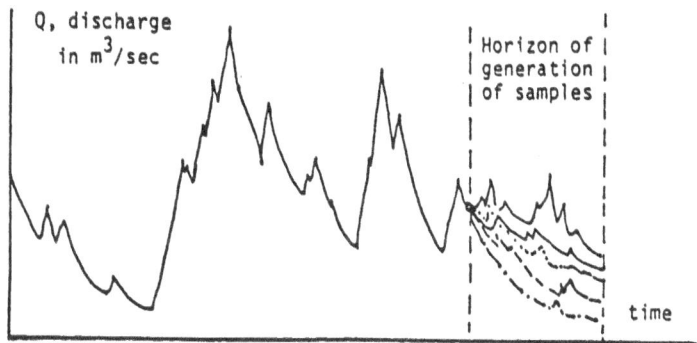

Figure 5. Schematic representation of using the samples generation method of the conditional variable (conditioned on systems state and forecast variables) to optimize the reservoir releases immediately in the forthcoming months.

2.7. MISUSE OF MODELING/SIMULATION APPROACH

The basic misuse of the generation methods would lie in applying them when they were either unnecessary or redundant. This is particularly true in the case when the time horizon of solutions is very small, so that the historic samples gave a sufficient number of results for subsamples equal to horizons in order to derive reliable frequency distributions of these solutions. In that case, the use of the method practically reveals that the investigator was well acquainted with, and can apply, the advanced stochastic modeling and simulation rather than improve on the solution of the problem posed.

Misuse of the method is current whenever there is no clear objective and good justification for the application of stochastic models and simulations. Also, misuse may be found in cases when either too long a sample was generated, or too many finite samples were simulated, namely the use of computer and labor time was an economic waste because the quality of the original data, modeling, estimation of parameters, and the generation itself did not justify the search for the high accuracy which is beyond the limits imposed by the entire process of data analysis, modeling and simulation.

3. Condensation of Hydrologic Information

3.1. DILEMMA BETWEEN STORAGE OF INFORMATION IN THE FORM OF DATA OR IN THE FORM OF MODELS

Stochastic modeling in hydrology seems not to have progressed sufficiently enough for a finite step of condensed information, namely that models replace the bulk of historic data in data-banks. For all practical purposes, models with their estimated parameters are not likely to soon replace the original data as it concerns information. As already discussed in the introductory write-up on the application of stochastic hydrology to water resources, major difficulties come from complexities in modeling short-interval hydrologic time series, such as the hourly or daily series. In these cases the nonparametric methods of treating periodicity in parameters required an avalanche of estimated parameters, which by the statistical principle of parsimony in the number of parameters could not all be very accurately estimated, because of large sampling variations, as Figure 6 demonstrates for the case of estimated daily precipitation means by the nonparametric method at Austin, Texas. When the parametric method of fitting the Fourier series to the estimated parameters over the intervals of the cycle(s) was used to describe the periodicity (such as in the case of Figure 6), the problem of proper inference on the minimum number of harmonics in the Fourier series fits often presents several difficulties. They are mainly due to the effects of periodicity in parameters involving higher order statistical moments and to complexities in stochastic dependence models, often also with periodic autocorrelation and autoregression coefficients.

Information of stochastic hydrologic models must first be shown to be identical, or very close, to information contained in the historic sample in order to even become a candidate for an eventual replacement of a historic sample. To advance in this direction of condensing information, two substantial progresses must first occur: (1) modeling of short-interval time series, such as the daily river flow series or the hourly precipitation series must be significantly advanced and widely tested in order that modeling becomes generally accepted as the full extraction of information contained in the historic sample(s); and (2) technology must be further developed for measuring how much information is contained in the time series of the shortest time interval available for data, say by the use of the entropy concept and measures, as well as how much information percentage-wise the model of that series has extracted. Therefore, the attractive concept of condensing hydrologic information through replacing large sequences of raw data in data banks by models and the set of estimated parameters on data storage disks, may have to wait for further developments in hydrology and in measuring the available and the extracted information of hydrologic time series.

3.2. BRIDGING PAST AND FUTURE SAMPLES THROUGH MODELING

Of all the geophysical sciences, hydrology and its processes are most affected by human activities and various disruptions in nature. Entire river basins exist in the world which have their water regimes completely changed, both in terms of quantity and quality. These changes are dynamic.

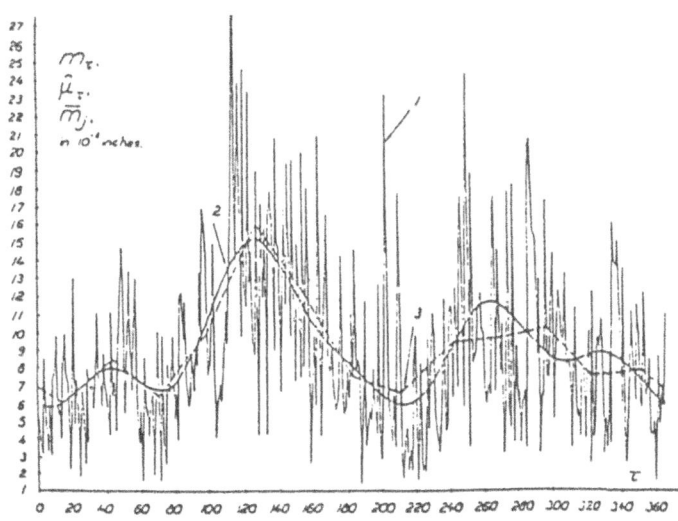

Figure 6. Sampling variation of 365 estimated means and the periodicity of daily precipitation means at Austin, Texas: (1) 365 estimated means, m_τ; (2) fitted periodic function μ_τ by the Fourier series analysis to the 365 values m_τ, with five significant harmonics and (3) daily means, $\bar{m}_j$, averaged over the 28-day intervals (only 13 average daily means for the year).

This dynamism accelerates as economies of countries progress. Water is an essential factor in these developments. As a consequence of these changes, hydrologic time series are often subject to various nonhomogeneities, thus their parameters continuously evolve. Therefore, trends and slippages in parameters of past time series will rarely repeat themselves in the same form in the future. Similarly, coefficients of periodic functions of parameters may be different in the future from what they were in the past, basically due to changes in degrees of water regulation and other developments.

The best method for taking into account these repeated modifications in parameters of the series would be by the identification of all changes in basic parameters during the past observations, particularly of changes in the form of trends, jumps and amplitudes and phases of periodicity. By removing or correcting them, homogeneous series will be produced. Then modeling becomes feasible. Inferences on future predominant changes in parameters will represent an important step in the further modification of the models and their parameters. These new model forms and new functions for parameters would serve the generation of potential future samples, and as close to reality as the reliability of projected changes permit. The more complex the past and future nonhomogeneities in hydrologic time series, the more useful modeling and simulation become.

3.3. CONDENSING INFORMATION FOR EASIER INPUT INTO COMPUTERS

The easy input of hydrologic information into computers in the form of a set of equations, with numerical parameters, may be advantageous in many cases. This factor would further favour developments in modeling and in bridging the psychological barrier of users between the presentation of information in the form of long tables of historic data and in the form of models of an equivalent information content.

4. Solving Problems with Large Numbers of Spatial Points

4.1. TYPES OF PROBLEMS NEEDING GENERATION OF SAMPLES AT LARGE NUMBERS OF SPATIAL POINTS

Many water resources problems become regional with time, requiring hydrologic information at a larger and larger numbers of points. Historic information may be scattered at points that are far from a uniform distribution along a line, across an area or over a volume. Among many potential examples of required information on a large number of points, only three will be described here to demonstrate both the need for such information and the potential for better solutions of water resources problems by spatial-temporal modeling of processes and simulation of samples than is the case without modeling and simulation.

When many reservoirs are built in a river basin or in a region, the increased connections in distance and density of water or power distribution networks make all the reservoirs operationally interconnected, even if physically they may be independent. With time dependence in outputs (demand) and inputs (supply), if not also with dependence among inputs and outputs of these reservoirs, the need may arise in the planning and development of the operational hierarchical rules for the generation of samples of the same sizes at all reservoirs. Historic data may not exist at all for the reservoirs, or even if they do, samples may be of different sizes and may not coincide completely in time at all points.

Large droughts cover vast areas. The longer and deeper a drought, the larger the region it usually covered. When regions are homogeneous in production, these large droughts may be economically and socially devastating. The large drought in the Great Plains of the United States of America in the 1930's is an example of how devastating droughts may be. Similar potential, or even historically experienced, droughts may or may have involved the other bread baskets of the world, such as the Canadian Prairies, Argentinian Pampa, Australian wheat regions, Plains of India, Central Europe, European Plains of USSR, and many other large food producing areas of the world. To determine the chances (probabilities or risks) for the occurrence of these exceptionally large droughts in these vast regions, as well as to find their main characteristics, the only approach is to simulate large samples of drought controlling hydrologic and hydrometeorologic variables at many points over these regions.

The water quality of reservoirs, lakes and large aquifers has become one of the major problems in water resources planning and operation, not to mention their eventual cleaning after long neglect. Information on quality variables, especially on toxic materials, may be needed at many points of these bodies of water. The most attractive approach then becomes the modeling of these processes at a grid of spatial points, and eventually simulating their samples by preserving spatial dependence, particularly the stratification by density, mixing, deposits of pollutants, etc.

Similar examples can be found in the investigation of aerial coverage by large storms with exceptional floods, risks that droughts may cover two areas of unique major crops at the same time for given distances along their centers, and similar regional problems. All of them will best be solved by the modeling and simulation of samples. Whenever the number of spatial points exceeds a small critical number, say four to five, modeling and simulation should use the principal components approach in preserving spatial dependence among the time series of these points.

4.2. GENERATION OF SAMPLES AT REGULAR GRIDS OF SPATIAL POINTS

Whenever there is a need to generate samples at a set of spatial points, the dilemma becomes whether to generate samples at irregular grids of observational points or at a new grid of regular points, specially selected to better and more easily solve the problems posed than if irregular grids are used. As already briefly stated, the historic time series at irregular observational networks may have different sample sizes and samples may not completely coincide in time. Therefore, information may not be of the same accuracy at most points. Furthermore, more weight is given in solutions to those parts of the space (line, area, volume) which have a denser coverage by points and longer time series.

Spatial modeling provides isolines of parameters. If hydrologic parameters slowly vary over the space in the case of hydrologicly homogeneous regions, such as the large plains or lightly undulating terrains, isolines of parameters may be replaced by mathematical expressions

$$\bar{m}_{t,i} = 2.4114 - 0.0991Y - 0.1203X + 0.0026Y^2 - 0.0046X^2 + 0.0105XY \qquad (5)$$

as Equation 5 demonstrates for the mean monthly precipitation over the upper part of the Great Plains, USA, as function of the latitude X and longitude Y. Then, it becomes simple to determine the values of all the model parameters at a regular grid of points, assuming that the model structure and type of equations do not change from point to point while the parameters continuously varied. Figure 7 presents the observed spatial points with isolines of the standard deviation of monthly precipitation, while Figure 8 gives both the observational points and the 80-point regular grid for the Upper Great Plains of the United States of America used for the generation of samples in the study of its droughts (CSU Hydrology Paper No. 87; Tase, 1976).

To study important regional water resources problems by using the regular grid of points, the only method presently available is to generalize regional models, compute all their model parameters at points of this regular network of points, and to simulate their samples. Either a very long sample at each point is generated when the population properties of the solution are sought, or a set of samples of different sample sizes are simulated, if the sample size is a critical aspect of the solution.

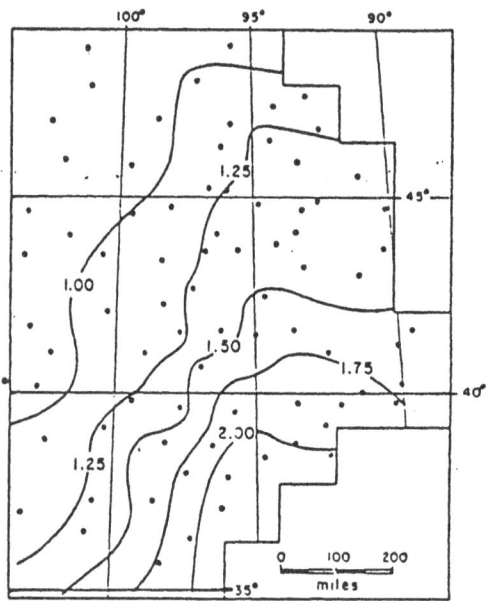

Figure 7. Isolines of the 30-year general standard deviation of monthly precipitation in the Upper Great Plains (USA), with points (dots) of observed time series.

5. Solutions of Problems Under Uncertainties

5.1. FINDING CONFIDENCE LIMIT SOLUTIONS TO PROBLEMS

Any water resources solution based on hydrologic statistical or stochastic information is subject to various uncertainties in the information. One of the most important uncertainties are the sampling errors in estimates. All estimates of model parameters are subject to this uncertainty. As sampling uncertainty is measured by the confidence limits to these estimates, the best method to find the confidence limits to solutions of water resources problems is to produce the joint probability or frequency sampling distributions of these estimates, then draw the confidence limits to major model parameters. If the investigation refers to the design of the needed storage capacities, the three major parameters of the model would be the mean, the standard deviation particularly, and the measure of stochastic dependence.

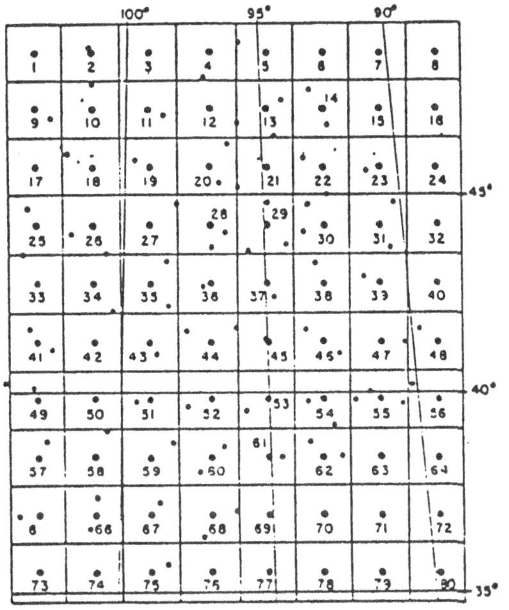

Figure 8. Grid system (big dots) of points of generated samples for drought study, with the 100-mile grid interval over the Upper Great Plains (USA) and the grid of observed stations (small dots) with their series used for the objectives of modeling the spatial-temporal monthly precipitation process.

The important factor in solving uncertainty dilemmas is the mutual dependence of the estimates of these three parameters, thus complicating the selection of the confidence limits. Furthermore, complication arises from the fact that confidence limits of these three parameters are the points of a spatial surface, say the surface close to an ellipsoid. There are too many potential three-number combinations of these parameters. Therefore, some selection rule must be applied to limit these combinations to a manageable number. In this case the model parameters are replaced by the confidence limit parameters, samples are generated and solutions to the problem found. These new results permit conclusions on the following water resources aspects: (1) how sensitive the solutions are to the sampling uncertainty; (2) how would the system or structure perform in the future if the lower or higher confidence limits of the parameters are closer to the true population values than the estimated parameters; and (3) what would be the minimal benefit of the system if the lower confidence limits came out to be the expected parameter values of the system's inputs and outputs, with this benefit determining the pricing policy for the product of the system.

If all parameters of the model are to be changed to their joint confidence limits, which is a difficult task if not decided on the basis of some simplified assumptions, the only solution method would be the generation of samples. Both the drawing of confidence limits in using experimentally obtained joint frequency distributions of all estimated parameters of the model, and the simulation of samples of these models with the selected confidence limits of all parameters, would require a ubstantial number of generated samples.

5.2. MODIFICATION OR ALTERNATIVES TO INFERRED MODELS DUE TO UNCERTAINTY

Uncertainty may also be in models. To take care of this uncertainty, either minor modifications to model composition and expressions may be used, or the alternative models may be investigated. In both cases the simulation of samples by using these modified or alternative models would produce water resources solutions, which in turn would permit conclusions of how sensitive these solutions are to uncertainty in models themselves.

One has to be always aware that modeling and simulation in utilitarian hydrology are not the objectives of study by themselves, but are applied exclusively to solve practical problems. Therefore, investigations of models with confidence limits of parameters, and the use of modified or alternative models, should serve the decision making of how sensitive to these uncertainties the solutions of those problems were.

6. Simulation Versus Information

6.1. SIMULATION AS METHOD FOR FULL EXTRACTION OF INFORMATION

It is often impossible to directly extract the full and specific information from historic samples. The reason may be either the shortness of the observed time series, the small number of observational points, or the lack of standard methods for such an extraction. By modeling these temporal-spatial processes, even if data are limited in size, and by simulating a sufficient number of time series of given size or of a long sample at the selected set of spatial points, frequency distributions can be found for any property of this process. This is then equivalent to the relatively full extraction of information for that property. This is especially important in the case of complex processes. Figure 9 gives an example of this use of the extraction of information for a simple dependent normal process in the form of frequency distributions of the range for several values of the sample size.

Similarly, models and sample generation can be used to transfer information in space from the points at which the observations were made to points where information was needed. This is discussed in the previous text on the generation of samples at a regular new grid of spatial points.

6.2. MISCONCEPTION ABOUT CREATION OF INFORMATION

Whoever has taught a course in stochastic hydrology to graduate students from developing countries must have faced the problem of explaining in detail that the modeling/simulation methods were a technology for the extraction or transfer of information, and can in no way create information. It is understandable that the scarcity of observational data in many developing countries would generate the idea of compensating this lack of data by the samples simulation approach.

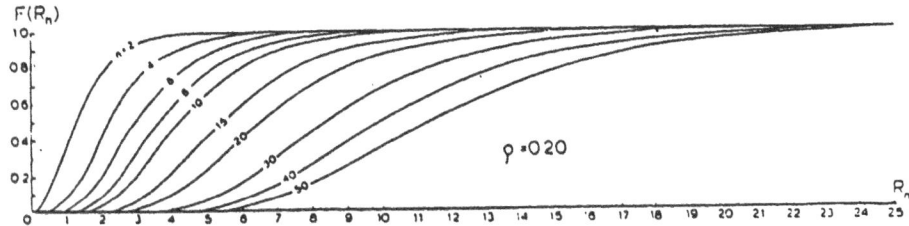

Figure 9. Distribution of range R_n, of standardized normal variable for various sample sizes n, in the case of a first-order autoregressive model with $\rho = 0.20$ (average world-wide first autocorrelation coefficient of annual river flows), obtained by the samples generation method.

However, extracting the full information from a limited amount of data is crucial and this technology should always help. The shorter the sample of observed data, the more powerful the methods of information extraction used should be. This is the opposite to many prevailing concepts in international consulting circles, namely that the shorter the sample available, the simpler the methods used should be. If one had 70 years of data and used some simple method of information extraction or transfer, and thus extracted or transferred only the equivalent of 60 years of data, it would have less impact on decision making that if 15 years of data were available and only an equivalent of ten years of data were extracted or transferred.

7. Conclusions

In assessing the proper use and eventual misuse or unwarranted expectations in the application of stochastic modeling/simulation technology to water resources, the following conclusions are drawn:

(1) The modeling of processes and the generation of samples should be considered a problem solving technology of the utilitarian hydrology and water resources decision making.

(2) The appropriate use of this problem solving technology is for cases involving long time horizons of solutions and the need for information at a large number of spatial points. The misuse occurs when this technology is either redundant or has no advantage in comparison with simple and less expensive methods.

(3) The generation of a long sample is required when the estimates of population characteristics are needed, while the generation of a large number of samples of given sizes is required when the characteristics of sample properties are needed.

(4) The stochastic modeling of processes and simulation of samples may also be conceived as the technology of extraction and transfer of information, especially in cases of complex processes and needs for specific

information, for which this technology becomes more appropriate than the other statistical methods of information extraction and transfer.

(5) To enhance the power of modeling/generation technology in the extraction and transfer of information, methods should be further developed for measuring the contents of hydrologic stochastic information in observed data, as also what degree of this information each statistical or stochastic method used has extracted or transferred.

(6) The assessment of the effects of uncertainties in models and in the estimates of model parameters on water resources solutions could be best investigated by generating samples for modified or alternative models and for confidence limit values of the estimated model parameters.

(7) The modeling/simulation technology cannot under any circumstances increase information beyond that contained in the observational data, except by combining any relevant additional physical information with the already available stochastic information.

(8) Models and their estimated parameters represent condensed information which however cannot yet fully replace data from which the models are derived, because of the complexities of modeling short-interval time series in utilitarian hydrology, though this goal is worthy of further investigations.

(9) As the complexity of water resources problems increases with time, especially on a regional basis, so also will the usefulness of the application of the modeling/simulation technology.

(10) As in the general statistics and probability theory, samples generating techniques will continue to be useful tools for the eventual checking of analytically derived exact, asymptotic and approximate solutions of utilitarian stochastic hydrology.

8. References

Annis, A.A. and Lloyd E.H. (1953) 'On the range of partial sums of a finite number of independent normal variates, Biometrica, 40, pp 35-42.

Boes, D.C. and Salas, J.D. (1973) 'On the expected range and expected adjusted range of partial sums of exchangeable random variables', Journal of Applied Probability, 10, pp 671-677.

Pegram, G.G.S., Salas, J.D., Boes, D.C. and Yevjevich, V. (1980) 'Stochastic Properties of Water Storage, Colorado State University Hydrology' Paper No. 100 (48 pages), Fort Collins, Colorado.

Salas, J.D. (1974) 'Range of cumulative sums, I. Exact and approximate expected values', Journal of Hydrology, 2, 3, pp 39-66.

Tase Norio (1976) 'Area-deficit-intensity characteristics of droughts', Colorado State University, Hydrology Paper No. 87 (40 pages), Fort Collins, Colorado.

Troutman, B.M. (1974) 'Expected range of partial sums', M.S. Thesis, Colorado State University, Fort Collins, Colorado.

Yevjevich, V. (1967) 'Mean range of linearly dependent normal variables with application to storage problems', Water Resources Research, Vol. 3., pp 633-671.

Yevjevich, V. (1975) 'Generation of hydrologic samples, case study of the Great Lakes', Colorado State University, Hydrology Paper No. 72 (40 p), Fort Collins, Colorado.

SOME THOUGHTS ABOUT THE STATE-OF-THE-ART IN STOCHASTIC HYDROLOGY AND STREAMFLOW FORECASTING

D.P. LETTENMAIER
NATO Advanced Study Institute

Over the last four days, we have heard papers on two related subareas of stochastic hydrology: Simulation and Forecasting. I shall attempt to give a brief summary of what I believe is the state of the art based on what I have heard and my own experience, and perhaps to clarify a few issues. My general objective will be to identify what we can do reasonably well now, what we cannot do, and where I think the best possibilities for advances lie.

First, I shall consider the area of stochastic streamflow simulation, also known as synthetic hydrology and streamflow synthesis. What this field attempts to do is to approximate the joint distribution, in space and time, of streamflows in such a manner as to allow what amounts to derived distribution applications via simulation. For instance, if we want to know the reliability of a reservoir system, it could, in theory, be derived mathematically if we knew the probability distribution of all the inflows. In practice, this is not possible, so we try to estimate the reliability by simulating lengthy (or multiple) sequences of reservoir inflows, and routing them through some kind of model of the reservoir system operation. Incidentally, little has been said about estimating reservoir inflows, which we usually need, as opposed to gauged flows, which are usually what we have.

It seems to me that we can do a reasonable job at simulating annual streamflows at multiple (correlated) sites, so long as the annual flow distributions are continuous and are not too variable (e.g., coefficient of variation less than one or so, and skew coefficients less than two or so). Clearly, the case that Juan Marco talked about does not fall within this range, and we are in trouble there. Under the above constraints, though, we have a range of models that we can use. For practical purposes, we are probably talking about low order AR or ARMA(1,1) models. For multiple sites, if ARMA(1,1) models are used, my preference is to use univariate techniques such as maximum likelihood (which are reasonably well developed) to fit ARMA(1,1) models, or the special case AR(1) thereof, at the individual sites. The phi and theta matrices can then be taken to be diagonal, and the lag zero cross-correlations preserved through the noise covariance matrix.

For simulating seasonal flows, the most practical approach seems to be

209

J.B. Marco et al. (eds.),
Stochastic Hydrology and its Use in Water Resources Systems Simulation and Optimization, 209–215.
© 1993 *Kluwer Academic Publishers.*

disaggregation, using the methods Bill Lane described. At present, I do not think aggregation models are a viable alternative unless you can live with some distortion at the aggregate (e.g., annual) scale. For the reason I mention below, I doubt we will completely resolve the deficiencies of the aggregation approach without looking at precipitation. Useful alternatives to multi-site annual modelling and disaggregation in time are modelling of spatially summed flows with subsequent spatial and/or temporal disaggregation. We need to keep in mind that there are some approximations in what Bill Lane refers to as the condensed approach. In particular, we end up with some correlations that are not explicitly preserved, and the choice of approaches depends on what it is best to ignore. We also have to deal with the problem of adjustments to preserve additivity over sites or time, resulting both from the form of the condensed models, and from the non-linearity of the transformations. Again, from a practical viewpoint, we need to transform both annual and seasonal flows to the normal domain, and this leads to the first serious limitation of existing methods.

When either the seasonal or annual flows cannot be transformed to be at least approximately normal, we are in trouble. The case mentioned by Marco is certainly one example; I am aware of another, which is the Salt River, Arizona in the US, which is in an extremely arid climate where streamflow is basically the result of a mixture of meteorological processes, one of which controls almost always and results in little or no precipitation. While we would not have any trouble fitting, say, a log normal mixture to the empirical distribution of flows using maximum likelihood or even graphical methods, we do not have the mathematics that allow us to use this kind of distribution (which is not additive) in our streamflow synthesis models.

Another significant problem arises in arid climates where streamflow may be intermittent (e.g., ephemeral streams). In this area, there is some ongoing work by Salas and Boes, and perhaps others, but to my knowledge we still lack tools for application. Of course, there are some "patches" that could be used if the problem applies only to a few seasons (e.g., seasons of variable length could be used to eliminate the zeroes), but such approaches remain to be demonstrated.

A couple of final comments regarding streamflow synthesis. The first probably should have been raised at the beginning, and this is the obvious point that there is nothing we accomplish in the modelling and simulation process that creates information. If we have 20 years of historic data, the fact that we generate many sequences of length, say, 100 years each through a reservoir operation model, does not create any more information than is in the 20 years of historic data. As I indicated earlier, we are just trying to approximate the joint distribution of streamflows, and we are not going to do a very good job with such a short record length. Of course, this is reflected in the answers we get, so long as we incorporate the parameter uncertainty in our simulation. As we have heard, there is at least one package (Jery Stedinger and Jan Grygier's SPIGOT) that does this, and I am glad that we have seen some examples that incorporate this source of uncertainty.

Another point, which is related, concerns the value of the whole streamflow synthesis approach. This, I believe, is record length-dependent. If the historic record is too short, estimating stochastic models is going to be overkill. You could argue that including parameter uncertainty will at least account for short record length difficulties, but incorporation of parameter uncertainty itself requires some assumptions that are record length-dependent. So, we could view the value of the streamflow synthesis approach as increasing with the length of record available for parameter estimation. However, if we have a sufficiently long record, we could just do our analysis using the historic record, and we would not be too far off. For instance, Professor Yevjevich has used in his paper an example of the estimation of dry weather water supply from a run-of-the river or small (much less than mean annual flow) reservoir with 15 years of data, and notes that the 15 resulting values would give you a reasonable basis for estimating system reliability, especially if you did not need to go too far out on the tails of the distribution. Likewise, if we needed to estimate the "firm field" of a water supply reservoir for a ten year operating horizon and had 100 years of data, the ten nearly-independent sub-sequences of the record might be enough. Vogel and Stedinger, in a recent ASCE paper (I believe) address this tradeoff.

Finally, back to the issue of rainfall modelling, aggregation, and so on. Over the last decade or so, a lot of researchers in hydrology have decided they would like to know more than just the joint probability distribution of streamflows; they would like to understand more about runoff generation mechanisms. This in itself is fine, but is perhaps not of much direct use to the practitioner. The related problem which is, or should be, of practical interest, is as follows. Quite often, we simply do not have the 50+/- years of streamflow data, and/or we do not have it where we really need it, for proper implementation of streamflow synthesis methods. We often do have point precipitation for some period of record, and off in the future somewhere, if the powers that be in agencies like NOAA's National Climatic Data Center in the US think things through properly, we may be able to obtain archival spatial rainfall data in the form of radar records (the storage of which will be a nontrivial issue because of the tremendous magnitude of data generated by radar).

If we were able to characterize the precipitation process at points compatible with a rainfall-runoff model based on relatively short records (for instance, ten years or so), and if a rainfall-runoff model could be calibrated from a similarly "short" coincident record of rainfall and runoff, we then might be able to simulate lengthy sequences of rainfall, and the corresponding runoff. Some encouragement for such an approach is given in the previously mentioned work of Eagleson, which suggests that if you know about the properties of short term (daily, as I recall) precipitation for a few years (e.g., five in Eagleson's work), you can do about as well in estimating the probability distribution of annual precipitation as through a "brute force" approach using in the order of 50 years of annual precipitation. Further, we (and others) have found that in the use of conceptual simulation models such as NWSRFS, HSPF, and so on, you do not improve your parameter estimates much once you get three or so years of coincident rainfall and runoff data, so long as the years you have represent a reasonable range of conditions so that the various modes of the model are excited. I will not say more about stochastic precipitation models, because I believe Juan Valdez will talk about them. I think, though, that some of the problems found in streamflow aggregation models, notably, the incompatibility between correlation structures on the seasonal and annual scale, will need to be addressed through modelling of event-scale (e.g., hourly to daily) precipitation and the rainfall to runoff transformation, rather than through attempting what I view as a surrogate approach of more complex short-term stochastic runoff models.

I would like to turn now to runoff forecasting, which is the other major topic we have heard discussed. First, I want to impose on you my definition of forecasting. To me, forecasting means estimation at a specific future time. Prediction is the same, with elimination of the specific time. Therefore, we <u>forecast</u> the stage at a given stream gauge ten hours from now, or the inflow to a reservoir next month, but we <u>predict</u> the 100-year flood. This is somewhat standard terminology, except that we have some contradictions like the US National Weather Service's <u>Extended Streamflow Prediction</u> approach (ESP), which is in fact a long range hydrological forecasting method.

The first thing to consider in forecasting is what we are really trying to do. We have heard about ARMAX, Kalman filters, conceptual models (which have been criticized rather severely by certain of my colleagues, probably not with complete justification), but we have not heard much about why you would want to forecast and how the forecast would be used. I can understand that this seems obvious; unfortunately in my experience this is not always the case. Often, water managers are fairly adamant that they need better forecasts, but do not really know what they could do with them. I hope that this problem will receive more attention next week when the discussion shifts to system operation, but for now, let me say that we are wasting our time reducing forecast error if it does not somehow lead to improved system operation, in terms of reduced loss of life or property damage for flood forecasts, or improved net economic benefits, improved environmental attributes, or other gain for society for longer term forecasts.

If we want to forecast runoff, the first thing we need to do is to figure out how far in advance we need forecasts, and where we can best obtain the appropriate forecast lead time. A reasonable conceptualization is to view the runoff generation process as consisting of three components: atmospheric circulation, which leads to precipitation at the land surface, hydrological response, which leads to channel runoff, and channel routing, which leads to runoff (more importantly, river stage) at a point of interest along the stream channel. Therefore, we have three processes we can hope to forecast, each of which has some associated response time and forecasting error:

1) the atmosphere itself (meteorological forecast);
2) the hydrological system (hydrological response forecast); and
3) channel routing.

As we proceed from channel runoff back to the atmospheric process, we can consider the potential accuracy to decrease; that is, given the upstream flows, we can make reasonably accurate downstream forecasts based on our knowledge of the hydraulics of open channel flow; given the precipitation inputs, we can be moderately successful in forecasting channel contributions and, finally, our colleagues in the atmospheric sciences still have a rather difficult time with quantitative precipitation forecasting. It is only fair to note, though, that the accuracy of meteorological forecasts has improved rather considerably along with breakthroughs in computer technology and data acquisition over the last 40 years, while the accuracy of hydrological forecasts has not improved much at all, at least in the US.

What this all suggests is that if the channel routing time is long enough and the local inflows are small enough, it is best just to measure upstream flows and do some routing. If that does not give us enough lead time, we need a hydrologic model, and if that still is not enough, then we need to do something about forecast period precipitation. In this context, it is not a real surprise that the stage of the Mississippi River at Vicksburg can be forecast quite accurately a week or two in advance using some fairly simple methods; on the other hand, the error in a forecast of seasonal runoff from mountainous (snow-fed) rivers in the western US is largely attributable to lack of knowledge of precipitation (and temperature) in the forecast period.

I would like to make a few comments about forecast approaches. But first I need to distinguish between different types of forecasts based on their lead time. I shall consider short-term forecasts as having lead time up to 96 hours, which happens to be the longest forecast period for the US National Meteorological Center's Quantitative Precipitation Forecasts (QPF's) in the US. Intermediate forecasts will be from 96 hours to a month, and long term forecasts greater than a month.

One key point that applies to forecasts of all three durations is the need to update. There is almost always some information in the error of past forecasts (that is, the difference between what was forecasted and what actually happened) that can help us to improve future forecasts. Eric Wood described some quantitative approaches to performing updating for short term forecasts. I think it is important to note that some structural deficiencies in the forecast model (e.g., the use of a simple linear versus a nonlinear model) may not be as important for forecasting as for simulation, since the updating keeps you close to the true state. Of course, if you can reduce the forecast error variance through whatever means, that is never going to hurt. The problem of how to update medium and long term forecasts can be a bit more difficult than updating short term forecasts since, with a longer time step, it is not always relevant to consider information from the same time step in the past. For instance, if we are making a runoff forecast of lead time four months, knowing the error in the previous four month's forecast may not be much help, we may be better off using the error in a forecast for the week or two immediately past.

For shorter term runoff forecasts, one unfortunate tendency of research papers has been to emphasize forecasts with relatively short lead times (which results in smaller forecast errors, hence implies more utility for the forecasting model being advocated) rather than considering the forecast use. For flood forecasts, undoubtedly the greatest value in forecasting is to prevent loss of life; you can move people out of the way of the flood if there is enough time, but there is not always a lot you can do about property loss in the short term. This is the reason that, in the US, the emphasis in flood forecasting has shifted from flash floods. It is important to keep in mind, though, that you need forecast lead times long enough to move people. If you were forecasting for two hours in advance, but it takes six hours for an emergency response system to get people out of the way, it will not be any big help. Another problem, which we will have to leave to our colleagues in psychology, is to make sure that people really heed the warnings and move.

A final comment regarding forecasting, which is applicable to all lead times, is that provision of forecast error information is essential. As I have noted above, for flood forecasting it is far more important that we provide forecasts of the best possible accuracy for lead times comparable to emergency response time, rather than more accurate forecasts at lead times that are too short. What decision makers need (or need to be trained to use) is information about the uncertainty of the forecasts. For instance, if we have a 50% probability that a flood peak will cause a potential loss of life, that information needs to be provided to the emergency response people (FEMA in the US). If the probability is 10%, it is less obvious what the response should be, but it is still clear that we need to provide that information.

For medium to long-term forecasts, a similar consideration applies. In 1987, Eric Wood and I were asked to develop a runoff forecasting model for the Seattle Water Department in fairly short order (three weeks as I recall between project initiation and issuance of our first forecast). The Department faced a potential crisis, since it had been an extremely dry summer and reservoirs were nearing critical levels. Nonetheless, we found (and they already knew!) that there would be no problem given normal rainfall during the fall months. In fact, our forecasting model, which was of the autoregressive type, gave forecasts that rather rapidly decayed to the long-term mean, which implied that there would be no problem under the "best" forecast. What actually happened was that the fall rains were late in arriving, and the actual forecast period runoff for both one week and end-of-season forecasts (twelve weeks length for the first forecast, declining as we progressed through the fall) were all quite close to the lower fifth percentile of the forecast error distribution. The information about forecast errors was therefore essential, and the "best" forecast was given relatively little consideration by the Department's water managers.

I shall offer one final thought regarding forecast models, which mostly applies to the medium and long term, then I shall let Juan Valdez continue with some thoughts on precipitation modelling. There has been some discussion over the last few days about the relative merits of linear and nonlinear models, both for simulation and forecasting. I shall emphasize again that, to the practitioner, the only real concern is the "bottom line". If he (or she) can do as well with a relatively simple linear model, for which there is a lot of well developed theory for model identification, parameter estimation and forecasting, as for a nonlinear model, there is no need to bother with the nonlinear model, even if there is no particularly sound physical or theoretical justification. That view is somewhat similar to that of the technical stock analyst, who essentially asks for the market to tell him what it is doing, rather than trying to figure out on the basis of fundamentals what is going on. For short term forecasting, where we tend to be data rich, it is possible to design some relatively straightforward comparisons to evaluate model performance. For long-term forecasting, it is a bit more difficult, since our archival data base usually has statistically "small" sample sizes.

The problem with linear models for medium and long-term forecasting is that they tend to do well near the centre of the distribution, but the greatest payoff for forecasts is during the extremes. Unfortunately, it is difficult because of the relatively small number of extremes during most historic records, to make such a claim hold up statistically. At this point, some anecdotal information will have to suffice. In 1977, which was the driest year on record for a number of Californian rivers, the traditional long term runoff forecasts, which mostly are based on regression of runoff on observed snow course data and past precipitation, resulted in forecasts of negative flows for several rivers. In addition, we did some comparisons of forecast errors using the Lettenmaier/ Wood model of the Seattle system I mentioned previously, with retrospective forecasts derived from the Nanjing model Eric Wood mentioned previously. Although not complete, our results seem to indicate that the overall runoff error distribution for the two models is not much different, but in the two most extreme years of record, the nonlinear model performed better. I certainly will not claim that this is conclusive evidence, but I will offer it as a caution to those who want to use linear models for runoff forecasting during droughts.

ISSUES IN THE MODELLING OF PRECIPITATION

J.B. VALDES
Department of Civil Engineering
Texas A&M University
College Station, Texas 77843-3136
USA

In this part of the panel presentations I shall attempt to summarize the recent developments in the modelling of precipitation, both at a point and in space, and to discuss the major strengths and weaknesses of these models. In doing so and following Dennis Lettenmaier's path, I shall report on what I have read about and what are my own thoughts in this matter.

To start with, everybody realizes that precipitation is a highly variable process, both in space and in time. Also, precipitation is a process which has an embedded structure in space and time. These two, somewhat conflicting statements have been known for some time. Some examples are Amorocho and Wu (1973, 1977); Duckstein et al. (1968); Corotis (1975); Sorman and Wallace (1973) among many others. Even though there was general agreement on this, precipitation modelling has not in the past been the most advanced part of hydrology. One reason is the feeling that the catchment, acting as a filter for the high frequencies of rainfall, makes detailed modelling of precipitation unnecessary. The complexity of modelling the dynamics of precipitation has also made it more attractive to postulate simpler stochastic models and spatial interpolation techniques took care of the spatial variability of rainfall. However, there are several examples, both in controlled experiments and the real world, where the above statement is not correct.

One traditional approach in hydrology, both for the modelling of point and spatial precipitation, has been to fit probability distributions to significant components of the precipitation process. The major problem is then the estimation of the parameters of the particular PDF's selected to model the precipitation process.

As an example, one stochastic model frequently used to represent point precipitation is the Poisson process, from the very simple Instantaneous Poisson Process model (e.g., Eagleson, 1972) to the Rectangular Pulses Poison model (e.g., Eagleson, 1972; Rodriguez-Iturbe, 1984). Several limitations weaken this kind of approach. From the parameter estimation point of view, the problem was the isolation of independent storm events since the parameter estimation techniques dealt with estimation of the parameters of the instantaneous process whereas the rainfall records were aggregated in time. Restrepo and Eagleson (1978) proposed an ingenious ad-hoc procedure to determine the minimum time between successive storms but

217

J.B. Marco et al. (eds.),
Stochastic Hydrology and its Use in Water Resources Systems Simulation and Optimization, 217–220.
© 1993 *Kluwer Academic Publishers.*

still the problem remains.

In the last decade new approaches to modelling point precipitation have been presented in the literature. In order to capture the embedded structure, in this case only in time, models based on point process theory have been proposed. The Barlett-Lewis Rectangular Pulses and the Neyman-Scott Rectangular Pulses models have been proposed in the last few years. The major advantage of these models over the previous ones is the explicit recognition of the time structure of precipitation, the "clustering effect" usually observed in rainfall records. The increase in the number of parameters is not significant and provides more realistic representations of the precipitation process. One additional problem has been dealt with in the papers presenting these models. To overcome the problem of the isolation of independent storm events and the fact that precipitation records are usually recorded in a time agreement manner, Rodriguez-Iturbe et al. (1984; 1988) have analytically derived the first and second order moments of the time aggregated process of the rainfall intensity; i.e.

$$X_T(t) = \int_0^T X(t) \, dt$$

where $X_T(t)$ is the time aggregated point rainfall depth for a time period of duration T and $X(t)$ is the instantaneous point precipitation. This aggregated first and second order moments, have the advantage that no storm separation is required to estimate the model parameters. Bill Lane and Levant Kavvas have spoken in this Workshop on the issues of aggregation versus disaggregation in hydrological records. Their comments also apply here.

Some models only perform well on the scale of temporal aggregation for which their parameters were estimated (e.g., hourly) but are unable to preserve the statistics at other levels of aggregation (e.g., daily). Thus one criteria to select appropriate point models may also be the ability of the model to preserve the statistical properties of the precipitation (i.e. the first and second-order moments) at several levels of temporal aggregation. Rodriguez-Iturbe et al. (1978) analyzed the ability of Neyman-Scott and Barlett-Lewis to represent precipitation records at Denver. Entekhabi et al. (1989) have discussed the parameter estimation and simulation of the modified Barlett-Lewis method. In this workshop Burlando and Rosso have discussed at length the capabilities of point-process based models for the simulation of precipitation traces and the derivation on the intensity-duration-frequency curves and the structural differences between the two most promising models: Barlett-Lewis and Neyman-Scott. They have also discussed the seasonality in the parameter estimation. Thus I will stop here on the description of these kind of models.

Multidimensional Models

Several stochastic models that incorporate physical features of precipitation have been proposed. Some of them are:

i) Mesoscale precipitation models: the WGR model (Waymire and Gupta, 1981; Waymire et al., 1984; Wood and Silvapalan, 1987), that represents the most common features observed in mesoscale precipitation.

ii) RCE model (Rodriguez-Iturbe et al., 1986) that represents the spatial representation of total rainfall without considering its temporal structure. Rodriguez-Iturbe and Eagleson (1987) addressed the temporal and spatial distribution of precipitation and derived the first and second-order moments of the ground rainfall intensity.

Again, one of the major problems is the estimation of the model parameters. The WGR model, even when reduced to its minimum parameter requirements, has nine parameters that need to be evaluated. The estimation of the parameters through the analysis of radar images is a time consuming task and not very practical for most hydrological applications. There is however a major advantage of the models mentioned above and other precipitation models. The advantage being that in the above models analytical expressions for the first and second-order moments of instantaneous point rainfall intensity have been found. These analytical expressions are a function of the model parameters. Thus, the general approach followed for the estimation of the parameters of point-precipitation models may also be seen here. Islam et al. (1988) have developed a procedure to estimate the parameters of the WGR model from the sample moments of the time-aggregated records of rainfall at raingauges. He applied the approach to a dense network in Walnut Gulch with very satisfactory results. Koeppsell and Valdes (1989) applied the same methodology to a more sparse network in Texas with also satisfactory results. Valdes et al. (1989) developed an approach using areal averages (as obtained by radar) rather than time averages and applied their approach to the GATE (GARP Atlantic Research Program) oceanic rainfall records also with satisfactory results.

These models, from a simple space-time AR(1) model, similar to the ones described in this Workshop to the WGR model, have been used in estimating the sampling error of ground and space-borne sensors of precipitation, both active and passive, in determining climatological estimates of precipitation. North and Nakamoto (1989) have presented a mechanism in which the square of the sampling error is a function of the spectrum of the instantaneous precipitation field and a design filter which is a function of the sampling characteristics of the sensor. There exists a large interest in these estimations and they are crucial to the planning of future missions like TRMM (Tropical Rainfall Measurement Mission) and TRAMAR (Tropical Rainfall Areal Mapping Radar) which are being studied by NASA.

Applications of Precipitation Models in Water Resources

The models described above have numerous applications in addition to rainfall-runoff modelling. Rosso and Burlando have mentioned the advantage of short-term rainfall forecasting to increase the lead time in flood predictions.

Some of those possible applications are in detecting the influence of precipitation in water demand: Maidment and co-workers at the University of Texas-Austin (1984-1989) have done extensive research in determining the impact that precipitation has on the decrease of urban water demand. Sastri and Valdes (1989) have also proposed a procedure to evaluate this impact.

Water demand for agriculture is also impacted by precipitation and the determination of optimal operating policies following the stochastic optimization techniques to be presented next week in this Institute may be applied to solve the so-called irrigation scheduling problem. Researchers at MIT have used alternative point precipitation models to determine the optimal policies (Bras and Cordova, 1980; Rhenals and Bras, 1982; Ramirez and Bras, 1982, 1985).

PART II

CONTRIBUTED PAPERS ON STOCHASTIC HYDROLOGY

SOME APPLICATIONS OF TIME SERIES MODELS TO CLIMATE STUDY (SOUTHERN OSCILLATION): MODELLING, FORECASTING AND PREDICTABILITY

PAO-SHIN CHU
Department of Meteorology
University of Hawaii
Honolulu, Hawaii
USA

The Southern Oscillation (SO) is the most dominant mode in short-term climate variations over the globe, particularly in the tropics and midlatitudes, accounting for a significant portion of the variance in the global climate system. In a compact expression, the SO is a large-scale phenomenon in which atmospheric masses are exchanged between centres in the Pacific and Indian Oceans. Climate variability and hydrological cycle in many parts of the world (e.g. southern Europe) are intimately related to the SO. The state of the SO is generally described by an index (SOI), which is the normalized sea level pressure difference between Tahiti and Darwin, Australia.

Using a time-domain approach, autoregressive-moving average (ARMA) processes are applied to model and predict the SOI on a monthly and seasonal basis. Based on the sample autocorrelation function, sample partial autocorrelation function, and the Bayesian Information Criterion, an ARMA(1,7;1) process is found to fit the monthly SOI and an AR(3) process to fit the seasonal SOI adequately (Chu and Katz, 1985). Figure 1a shows the theoretical autocorrelation function (broken curve) for an ARMA (1,7;1) process and the sample autocorrelation function of the monthly SOI. Figure 1b shows the theoretical autocorrelation function (broken curve) for an AR(3) process and the sample autocorrelation function of the seasonal SOI.

Regarding prediction, Figure 2 displays the one-season-ahead forecast values of the seasonal SOI for the AR(3) model, along with the actual observations from Summer 1982 to Winter 1984. Over this period, unprecedented large anomalies occurred between Summer 1982 - Spring 1983. With the exception of Summer 1982, a close correspondence between the observed and predicted series is readily seen, both in the direction of variation and in magnitude.

Besides their use as diagnostic and forecasting tools, time series models have other potential uses. For instance, Chu and Katz (1987) developed a method to provide a measure of actual, rather than potential predictability of a climate variable on the basis of its past history alone. This method explicitly estimates how well such a time series model can forecast future behaviour. Specifically, the proportion of variance explained by the ℓ-step ahead forecast is:

J.B. Marco et al. (eds.),
Stochastic Hydrology and its Use in Water Resources Systems Simulation and Optimization, 223–227.
© 1993 *Kluwer Academic Publishers.*

$$\lambda_\ell = 1 - \frac{v(\ell)}{\sigma^2} \qquad\qquad \ell = 1, 2, \ldots$$

where $v(\ell)$ denotes the variance of the ℓ-step ahead forecast error, and σ^2 the variance of the X_t process as

$$X_t = \sum_{i=1}^{p} \phi_i X_{t-i} + a_t - \sum_{j=1}^{q} \theta_j a_{t-j}$$

Note that $0 \leq \lambda_\ell \leq 1$, with $\lambda_\ell = 1$ for the case of perfect predictability and $\lambda_\ell = 0$ for the case of no predictability.

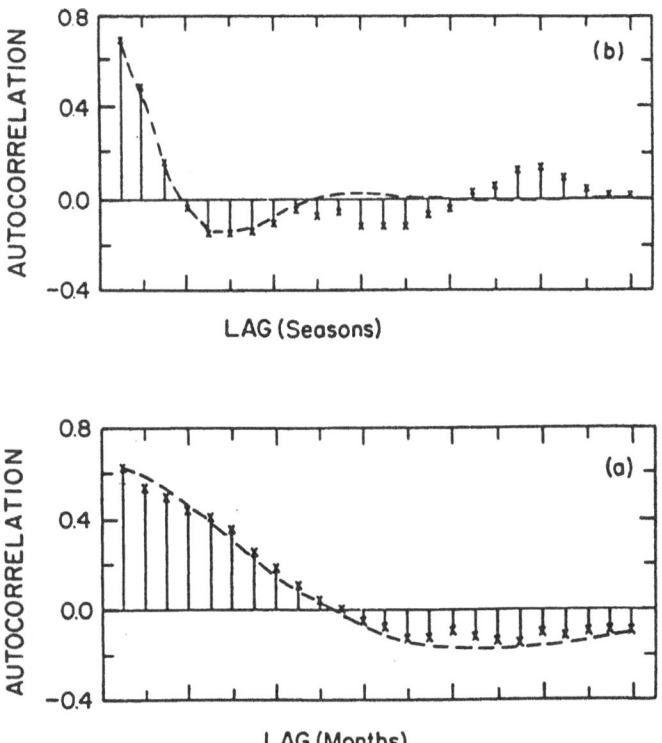

Figure 1. (a) Sample autocorrelation function of the monthly SOI and the theoretical autocorrelation function (broken line) for the ARMA (1,7;1) process. (b) Sample autocorrelation function of the seasonal SOI and the theoretical autocorrelation (broken line) for the AR(3) process.

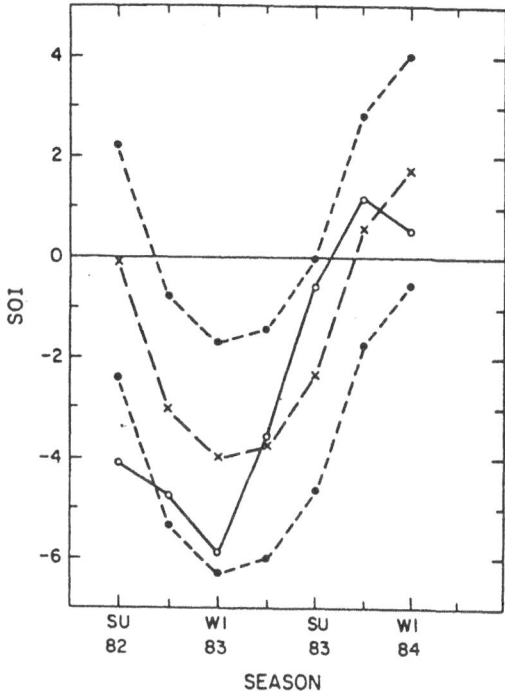

Figure 2. Time series plots of one-season-ahead forecast (long broken line) and the observed (solid line) values of the seasonal SOI from Summer 1982 (SU 82) through to Winter 1984 (WI 84). The 95% prediction limits for the forecast values are shown by short broken lines. The period of the analyzed model is Spring 1935 through Spring 1982.

Expressions for the theoretical predictability of the AR(p) and ARMA (1,1) processes are derived. Using an ARMA(1,1) process to represent the monthly SOI fluctuations, up to 44% of the variation in SO can be theoretically predicted one month ahead and 35% two months ahead (Fig. 3). On a seasonal time scale, the theoretical predictability of the SO is 53% one season ahead and 31% two seasons ahead (Fig. 4).

Another feature of time series models is that they are capable of producing "pseudo periodic" behaviour, meaning that this approach allows for the detection of quasi-periodicities that many climate variables (e.g., precipitation, drought index) have been known to possess. Figure 5 gives the power spectra determined from the ARMA processes identified earlier. A large portion of the variance is accounted for by periods in the neighbourhood of 12 to 20 seasons (three to five years), a feature consistent with that obtained from the conventional frequency-domain method (Chu and Katz, 1989). Hence, the spectrum derived from time series models provides another criterion by which to assess the adequacy of the models

226

in representing the temporal behaviour of a climate system, recalling that these models were originally selected on the basis of a different time-domain criterion.

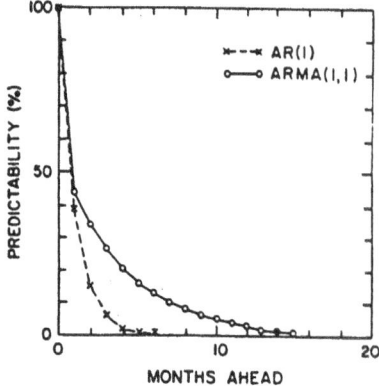

Figure 3. Estimated theoretical predictability of the monthly SOI based on the AR(1) and ARMA(1,1) parameter estimates obtained using data from January 1935 through November 1984.

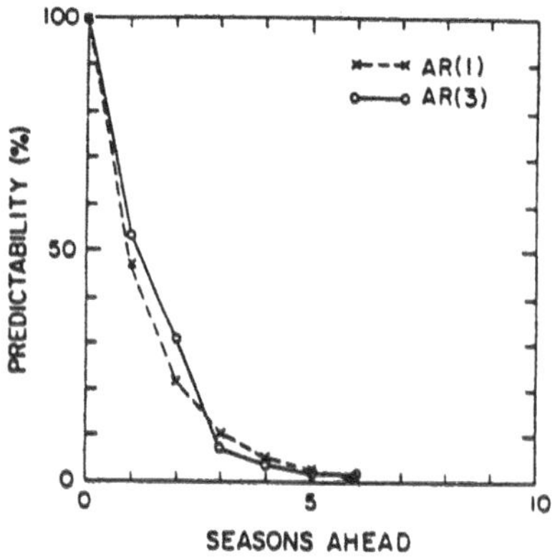

Figure 4. Estimated theoretical predictability of the seasonal SOI based on the AR(1) and AR(3) parameter estimates obtained using data from Spring 1935 through Fall 1984.

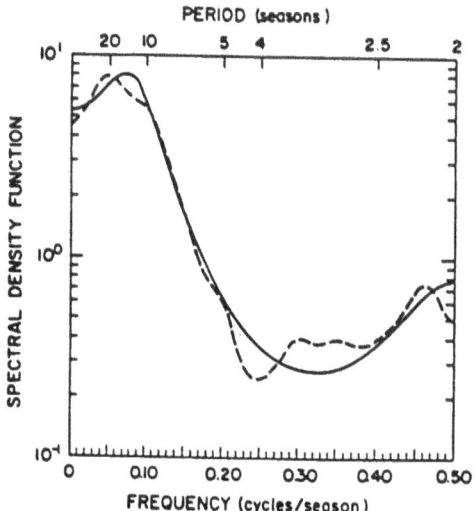

Figure 5. Spectral density function of the seasonal SOI from the frequency-domain approach (broken curve) and time-domain approach (solid curve). The analysis period is from Spring 1935 to Summer 1983. The Parzen lag window is used with a maximum lag of 30 seasons (broken curve). Note that the scale on the ordinate is logarithmic, whereas the scale on the abscissa is linear.

References

Chu, P.-S. and Katz, R.W. (1985) 'Modeling and forecasting the Southern Oscillation: a time-domain approach', Monthly Weather Review, 113, pp 1876-1888.

Chu, P.-S. and Katz, R.W. (1987) 'Measures of predictability with applications to the Southern Oscillation', Monthly Weather Review, 115, pp 1542-1549.

Chu, P.-S. and Katz, R.W. (1989) 'Spectral estimation from time series models with relevance to the Southern Oscillation', Journal of Climate, 2, pp 86-90.

CONCEPTUAL BASIS OF STOCHASTIC MODELS OF MONTHLY STREAMFLOWS

P. CLAPS
Dept. Environm. Eng. & Phys.
University of Basilicata
Potenza
Italy

1. Summary

Annual and monthly streamflow time series are mostly modelled using linear models of the ARMA class. Model identification is usually performed through statistical procedures (e.g. Noakes et al., 1985) or, sometimes, by describing the processes as linear conceptual models whose equations can be rearranged in order to assume AR or ARMA representation (e.g. Salas and Smith (1981), Moss and Bryson (1974).
   The aim of the present work is to reproduce streamflow time series at both annual and monthly scales by mean of ARMA models whose order is identified through simple conceptual models of the processes. Explicit correspondences between conceptual and stochastic parameters result from the identification procedure.

2. Conceptual Model of Annual Streamflows

Efficient simulation of the annual streamflow process is provided by the Thomas-Fiering conceptual model. In this model the precipitation x_t in the year t is assumed to reach the stream in the amount $(1-c_1-c_2)x_t$ (surface runoff) where $c_1 x_t$ is the amount of infiltration and $c_2 x_t$ is the amount of evaporation. The annual streamflow is the sum of the surface runoff and of the groundwater contribution $c_3 V_{t-1}$, where V_t is the storage volume at the end of the year t.
   This conceptual model has been demonstrated (Salas and Smith, 1981) to lead to an ARMA(1,1) stochastic model, which seems to be the most adequate to describe the annual streamflow process.
   However, this model, in the form derived in the above mentioned paper, does not provide explicit relationships between conceptual and stochastic parameters.
   In order to reconcile this lack of agreement, a modification of the conceptual model with regard to the system input is proposed. The watershed is supposed to be fed by the *effective* rainfall rather than by the *total* rainfall. In such a way, the evapotranspiration process disappears from the mass balance equations, reducing the conceptual para-
229

J.B. Marco et al. (eds.),
Stochastic Hydrology and its Use in Water Resources Systems Simulation and Optimization, 229–235.
© 1993 *Kluwer Academic Publishers.*

meters to the number of two.

The streamflow and the groundwater storage mass-balance equations are:

$$D_t = c_K V_{t-1} + a(1-r_K)I_t + (1-a)I_t$$

$$V_t = (1-c_K) V_{t-1} + a r_K I_t$$

In the above equations, D_t is the streamflow, V_t is the storage volume, I_t is the effective rainfall (or net input) and $(1-c_K)$ is the *recession coefficient of the storage volume*.

Conceptual parameters are: the infiltration coefficient a, and the storage coefficient K ($K = -[\ln(1-c_K)]^{-1}$).

The feasibility of accounting for the within-year effective rainfall distribution is also added through a predetermined coefficient r_K, which is called the *recession coefficient of the within-period infiltration* and represents the rate of the infiltration volume reaching the stream at the end of the year. The coefficient r_K depends on the aquifer storage coefficient K and on the shape of the within-year effective rainfall curve (see Moss and Bryson (1974) for the case of concentrated input).

If an analytical representation is given to the effective rainfall curve i(t), the r_K - K relationship can be obtained by solving the expression:

$$(1-r_K) \int_0^1 i(z)dz = \int_0^1 e^{-z/K} / K \cdot \int_0^z i(m)e^{-m/K} dm\, dz$$

The proposed conceptual model is shown in Figure 1.

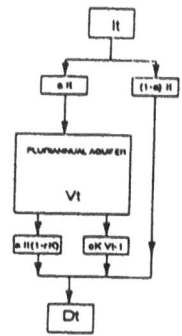

Figure 1.   Conceptual model of annual flows.

3. Annual Flows: Relationships Between Conceptual and Stochastic Parameters

By combining the above mass balance equations, the following ARMA(1,1) model is obtained with parameters a and K:

$$D_t - (1 - c_K) D_{t-1} = (1-a\, r_K) I_t - [(1 - c_K) - a\, r_K] I_{t-1}$$

On the other hand, a general ARMA(1,1) stochastic model can be written as:

$$D_t - \Phi\ D_{t-1} = I^*_t - \Theta\ I^*_{t-1}$$

with parameters: Φ = autoregression coefficient, Θ = moving average coefficient. The parameters a and K can be expressed in terms of Φ and Θ as follows:

$$a = (\Phi - \Theta)\ /\ (1 - \Theta) \qquad K = -[\ln\ (\Phi)]^{-1}$$

and the relationship between the residual I^*_t and the net input (effective rainfall) I_t is:

$$I^*_t = I_t\ (1 - a\ r_R)$$

4. Conceptual Model of Monthly Flows

In Italian basins with two distinct climatic seasons and no snow melt runoff, the presence of at least two different groundwater contributions can be observed during the recession phase. The first one is due to the so-called *deep aquifer* and its presence is evident at the end of the dry season. It has a pluriannual decay, which implies that the storage coefficient is greater than one year.

The second one is due to the so-called *seasonal aquifer*, which can be observed at the end of the rainfall season, where an exponential decay occurs below flood peaks. This aquifer has a seasonal response, because it fills and empties within the year.

Those considerations suggest a conceptual model with two linear reservoirs in parallel plus a diversion, with no lag, representing the direct runoff (Fig. 2).

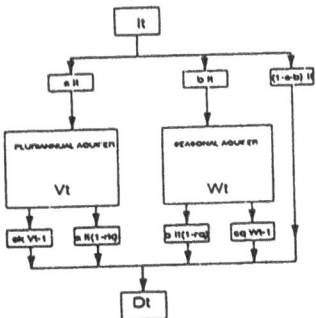

Figure 2. Scheme of the proposed conceptual model of monthly flows.

The model described therefore has four conceptual parameters: two storage coefficients k and q (respectively for the deep and for the seasonal aquifer) both expressed in months, and the two infiltration coefficients a (which is the same as for the conceptual model of annual streamflows) and b.

5. Monthly Flows: Relationships Between Conceptual and Stochastic Parameters

Streamflow and groundwater storage mass balance equations can be written as:

$$D_t = c_k V_{t-1} + a(1-r_k)I_t + c_q W_{t-1} + b(1-r_q)I_t + (1-a-b)I_t$$

$$V_t = (1-c_k) V_{t-1} + a r_k I_t$$

$$W_t = (1-c_q) W_{t-1} + b r_q I_t$$

With regard to the seasonal aquifer, W_t is the storage volume and $(1-c_q)$ is the recession coefficient of the storage volume.

Rearranging, an ARMA(2,2) stochastic model is obtained:

$$D_t - \Phi_1 D_{t-1} - \Phi_2 D_{t-2} = I*_t - \Theta_1 I*_{t-1} - \Theta_2 I*_{t-2}$$

The coefficients of the above canonical representation have conceptual expressions:

$$\Phi_1 = (1-c_k) + (1-c_q) \qquad\qquad \Theta_1, \Theta_2 = f(a,b,c_q,c_k,r_q,r_k)$$

$$\Phi_2 = -(1-c_k) \cdot (1-c_q)$$

And the residual – net input relationship is:

$$I*_t = I_t (1 - a r_k - br_q)$$

Given that the net input is a seasonal process, the stochastic model is an ARMA(2,2) with pseudo-periodic residual.

6. Estimation of ARMA(2,2) Model Parameters

The model was applied to streamflow time series recorded in Italy.

Parameters estimation, performed with classical methods, did not give satisfactory results in our judgment. There are two reasons for this.

First, the storage coefficient k of the deep aquifer assumes very high values with respect to the time scale. In fact, the acceptance zone of the stochastic AR parameters is determined by the constraints:

$$\Phi_2 + \Phi_1 < 1 \; ; \; \Phi_2 - \Phi_1 < 1 \; ; \; -1 < \Phi_2 < 1$$

and when k increases, the coefficient $(1-c_k)$ approaches unity and the sum $\Phi_2 + \Phi_1$ moves toward the limit value 1.

Second, the seasonal aquifer runoff and the total runoff have similar periodicities. For that reason, any deseasonalisation of the streamflow series tends to eliminate the presence of the seasonal aquifer contribution.

The aforementioned drawbacks seem to prove the impossibility of simultaneously estimating the parameters of the two aquifers.

A way to overcome the former problem is to estimate deep and seasonal aquifer parameters independently, using separate models in two different time scales.

The latter point suggests not deseasonalizing the time series under consideration before the estimation stage.

7. Two-stage Model of Monthly Flows

A combined model is proposed which takes advantage of the conceptual and stochastic features of the streamflow process at both annual and monthly time scales.

At the annual time scale, pluriannual aquifer parameters and effective annual rainfall series are estimated.

Then, the deep aquifer runoff is calculated at the monthly time scale. The estimation is performed by disaggregating the series of annual estimated effective input and putting the resulting monthly series into the deep aquifer component. The disaggregation is actually a 'seasonalization' of the mean annual net input and consists of constraining the net input monthly series to have the same periodicity as the monthly flow series.

At that time, a decomposition of the flow is possible at the monthly scale, in order to separate the flow consisting of direct and seasonal aquifer components (which can be considered a 'subprocess' D'_t of the process D_t) from the deep aquifer flow (see Fig. 3).

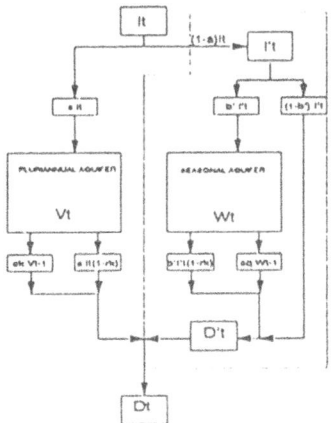

Figure 3. Scheme of the proposed two-stage conceptual model of monthly flows.

The conceptual model of the "direct plus seasonal aquifer" flow leads to an ARMA(1,1) stochastic model with pseudo-periodic residual that arises from the equation of a system composed by one linear reservoir plus

234

diversion with no lag, as in the case of annual streamflows, fed by the pseudo-periodic process $(1-a)I_t$.

The model estimation structure consists of the following steps:

```
TWO-STAGE MODEL OF MONTHLY FLOWS: ESTIMATION PROCEDURE

ESTIMATION OF THE DEEP AQUIFER CONCEPTUAL PARAMETERS
VIA STOCHASTIC MODEL OF ANNUAL FLOWS

DISAGGREGATION OF THE ANNUAL EFFECTIVE INPUT UNDER
THE PERIODICITY HYPOTHESIS

EVALUATION OF THE DEEP AQUIFER MONTHLY RUNOFF
ORIGINATED BY THE ABOVE INPUT

CALCULATION OF THE 'SUBPROCESS' SERIES D'_t

ESTIMATION OF THE SEASONAL AQUIFER CONCEPTUAL PARAMETERS
VIA STOCHASTIC MODEL OF THE 'SUBPROCESS' FLOWS D'_t

EVALUATION OF THE EFFECTIVE INPUT MONTHLY SERIES I_t

EVALUATION OF THE DEEP AND SEASONAL AQUIFER RUNOFFS
FROM THE INPUT I_t
```

8. Applications and Conclusions

The combined model described has been applied with satisfactory results to some river time series in southern Italy. As shown in Figure 4, deep and seasonal aquifer flow reconstruction seems reasonable, as well as the values of conceptual parameters. Nevertheless, further studies are needed in order to improve estimation efficiency, which is stressed by the periodicity of the residuals.

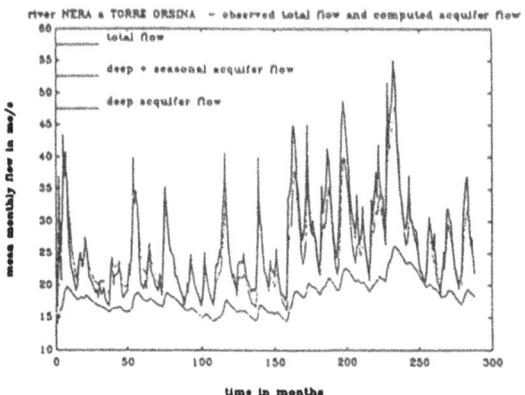

Figure 4. Reconstruction of the deep and seasonal aquifer runoffs. The conceptual parameters, calculated from stochastic parameter estimates, are: a=0.71, K=4.13 years, b=0.252, q=2.93 months.

9. References

Moss, M.E. and Bryson, M.C. (1974) 'Autocorrelation structure of monthly streamflows', Water Resources Research, 10(4), pp 737-744.

Noakes, D.J., McLeod, A.I. and Hipel, K.W. (1985) 'Forecasting monthly riverflow time series', Int. J. Forecast. 1, pp 179-190.

Salas, J.D. and Smith, R.A. (1981) 'Physical basis of stochastic models of annual flows', Water Resources Research, 17(2), pp 428-430.

A NONLINEAR, NONSTATIONARY STOCHASTIC MODEL FOR THE PREDICTION OF STORAGE IN THE EDWARDS AQUIFER, TEXAS

FIDEL A. SAENZ DE ORMIJANA
Ph.D. Agroman
Hurtado de Mendoza, 4
28036 Madrid
Spain

1. Introduction

The Edwards aquifer is a carbonate formation that underlies an area of some 8,200 square km in south-central Texas, with a total capacity estimated at 20 to 40 cubic km (Kelly, 1985). At least nine cities and towns with a combined population in excess of one million, among them the city of San Antonio, rely exclusively on the aquifer for their water supply. Springflow from the aquifer feeds a number of important streams that supply water to downstream users as well as fresh water to coastal estuaries in the Gulf of Mexico. With pumping from the aquifer increasing over the last fifty years from 15% to almost 90% of the annual recharge, there is growing concern about the possibility of important reductions in storage or about the capacity of the aquifer to stand dry periods at ever increasing levels of pumping.

Although linear stationary models (such as ARMA or transfer function techniques) are extremely popular in Stochastic Hydrology, there is growing interest in stochastic models that can accommodate the general *nonlinear and nonstationary* effects which occur frequently in actual hydrological processes. There is also growing interest in stochastic models that include some *physical* information, as opposed to purely stochastic models. This paper presents a very simple stochastic model that predicts the evolution of the annual storage in the Edwards aquifer, including the three characteristics mentioned above: the model is nonlinear because of the *threshold* effect - springflow ceases when storage falls below a certain level - and nonstationary because of the long-term increases in pumping; physical information is brought in through the continuity equation, which is used to compute the characteristics of the storage process from the combination of the different flow components in the aquifer. Finally, the model presented here is *lumped* because it treats the aquifer as a single control volume. In the proposed approach, the characteristics of the storage process are computed without resorting to *Monte Carlo* simulation. Instead, the discrete-time storage series is approximated by an equivalent continuous-time process. One can then solve the *Fokker-Planck* equation which controls the evolution of the cumulative distribution function (CDF) of this continuous-time storage process.

237

J.B. Marco et al. (eds.),
Stochastic Hydrology and its Use in Water Resources Systems Simulation and Optimization, 237–246.
© 1993 *Kluwer Academic Publishers.*

2. Netflow Components

As a control volume, the Edwards aquifer is subject to the three processes shown in Figure 1: recharge, pumping, and springflow. The combination of these three processes constitutes the aquifer *netflow*,

$$N_t = I_t - W_t - Q_t \tag{1}$$

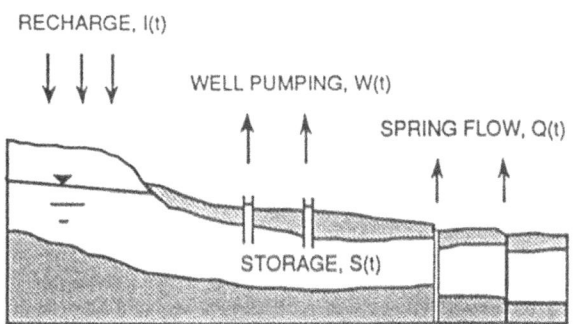

Figure 1. Schematic representation of the Edwards aquifer.

The purpose of stochastic storage analysis is to derive the properties of the storage (and outflow) process, from the properties of the netflow N_t, using the continuity equation

$$S_t = S_{t-1} + N_t \tag{2}$$

together with the netflow-storage relationship. Historic records for the annual netflow components, together with the storage, are shown in Figure 2.

The annual recharge series is approximately random Gaussian. As shown in Figure 2a, the average recharge was higher in the second half of the period of record than in the first half, as a result of the higher precipitation experienced in the 1956-1987 period. However, since it is not possible to offer a physical explanation for this difference in precipitation between the two periods, the recharge series was considered stationary, and its variation considered purely stochastic. On the other hand, as shown in Figure 2b, pumping constitutes a nonstationary process: it follows a quadratic trend with superimposed random variation. If the present trend continues, pumping will exceed the mean annual recharge around the year 2000.

The actual storage in the aquifer cannot be known exactly: the storage shown in Figure 2d corresponds to the cumulative storage change since 1933, which is why negative storage values appear in the graph. The dependence of the springflow is evident if one compares the corresponding historic series shown in Figures 2c and 2d. This dependence is approximately linear, so that the Edwards aquifer can, in principle, be modelled as a *linear reservoir*. The linear reservoir, that is, a storage system where

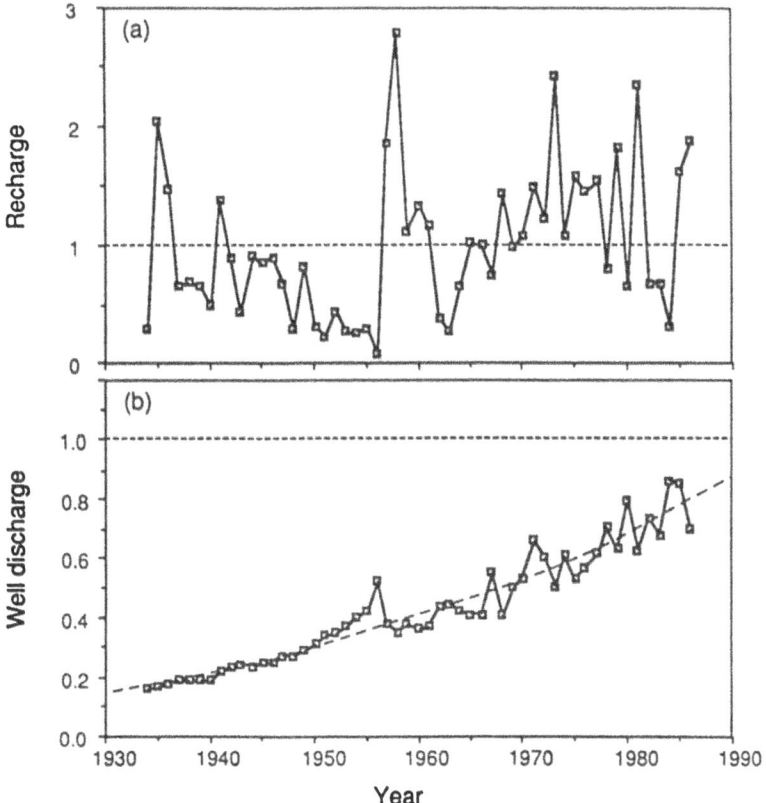

Figure 2. Annual netflow components and storage in the Edwards aquifer: (a) recharge, and (b) well discharge. All the series are given in units of mean annual recharge, equal to 758.5 Hm^3/yr (1 Hm^3 = 10^6 m^3).

outflow depends linearly on storage, has been widely applied in the past to treat stochastic storage problems (see, for example, Langbein, 1958; Klemes, 1973, 1974, 1978; Klemes and Boruvka, 1975). If one regards the Edwards aquifer as a single control volume, however, the linear-reservoir analogy cannot be applicable over the whole storage range. Springflow must cease when storage falls below a certain level, otherwise springflow would become negative for a low enough storage level. This *threshold* effect represents a very simple type of nonlinearity, which corresponds to *linear threshold* processes. These are processes which behave piecewise linearly, depending on the state of the system, are a natural extension of linear processes, and constitute a well-known device to approximate nonlinear systems (Tong, 1983). In this case, the springflow-storage relationship was described using the *linear single-threshold* regression model

$$Q_t = \left[\begin{array}{ll} \beta_0 + \beta_1 I_t + \beta_2(\theta S_t + (1 - \theta) S_{t-1}) + \bar{Q}_t \ , & \text{if } S_{t-1} > S^{(0)} \\ 0 & , \quad \text{if } S_{t-1} \leq S^{(0)} \end{array} \right. \quad (3)$$

where $S^{(0)}$ is the *threshold* storage below which springflow is zero. The

240

value of $S^{(0)}$ was determined with the condition that $E[Q_t]$ must be zero for $S_{t-1} = S^{(0)}$. This yields $S^{(0)} = -3,781$ Hm³. Equation 3 is just an approximation, since, in reality, the start-of-year storage does not uniquely determine whether springflow during that year is zero or not. In Equation 3, $\bar{Q}_t$ is a zero-mean stochastic disturbance of a general ARMA type, which in the present case proved to be purely random, and β_i, $i = 0,1,2$ and $\theta (0 \leq \theta \leq 1)$ are coefficients. Maximum likelihood estimates for these coefficients were obtained using the historic data as shown in Table 1. One-year-ahead springflow forecasts calculated using the springflow-storage relationship in Equation 3 are compared with the historic values in Figure 3.

TABLE 1. Maximum likelihood estimates for the coefficients of the annual springflow-storage relationship. The standard errors are shown in brackets below the estimated value.

Model	β_0 10^6 m³yr^{-1}	β_1	β_2 yr^{-1}	θ		σ_ϵ 10^6 m³yr^{-1}	R^2
I	327.36 (17.38)	.1966 (0.0175)	.126 (.0109)	0		59.57	.822

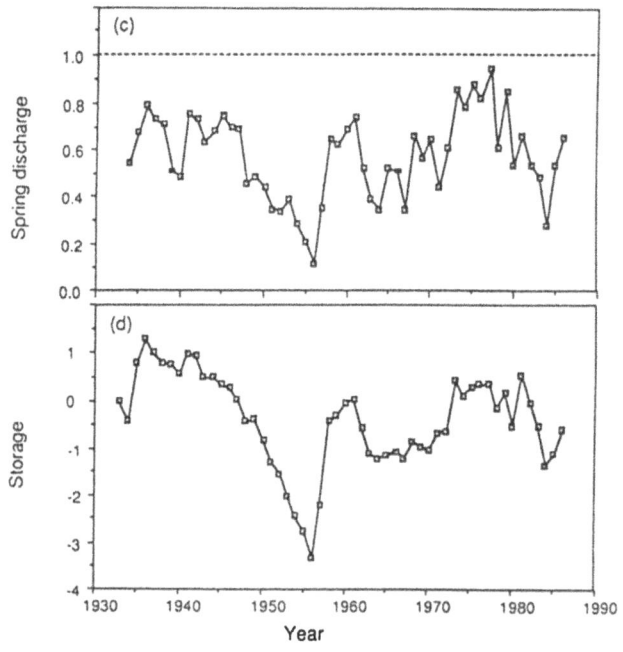

Figure 2 (continued). Annual netflow components and storage in the Edwards aquifer: (c) springflow, and (d) storage. All the series are given in units of mean annual recharge, equal to 758.5 Hm³/yr. The storage corresponds to the cumulative storage change since 1933.

3. Methodology

Once stochastic models have been established for all the netflow components, the properties of the storage process can be deduced (using the continuity equation) from those of the netflow process, which is conditioned on the start-of-year storage,

$$
N_t = \left[\begin{array}{ll}
I_t - W_t - Q_t & \text{, if } S_{t-1} > S^{(0)} \\
I_t - W_t & \text{, if } S_{t-1} \leq S^{(0)}
\end{array} \right. \tag{4}
$$

since springflow is zero when storage falls below the threshold $S^{(0)}$. As shown above, the parameters of each stochastic component in this relationship (recharge, pumping, and springflow) are estimated separately. Because the stochastic disturbances of the component models (for I_t, W_t, and Q_t) are random, the netflow process can be written, conditioned on the start-of-year storage, as

$$
N_t | S_{t-1} = E[N_t | S_{t-1}] + V[N_t | S_{t-1}] z_t \tag{5}
$$

where z_t is a random Gaussian series.

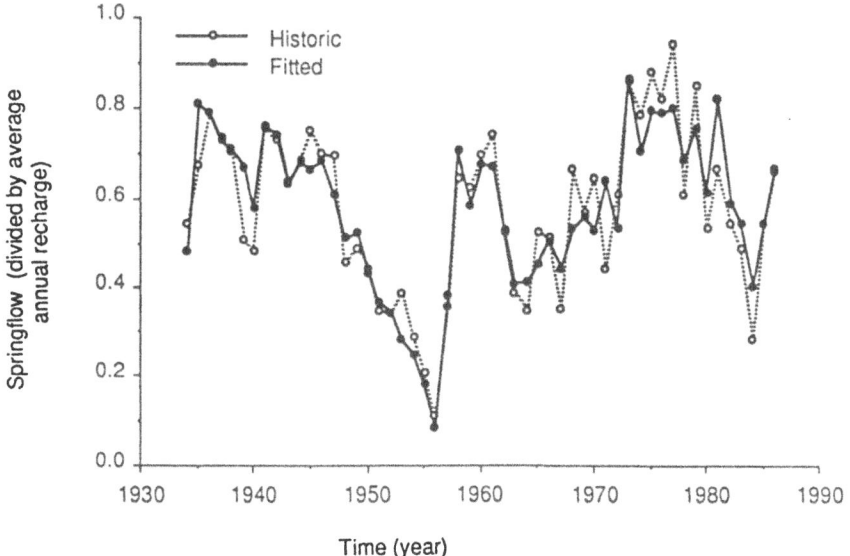

Figure 3. Comparison of historic springflow with one-year-ahead predictions using the springflow-storage relationship. All values are given in units of mean annual recharge, equal to 785.5 Hm^3/yr.

The essence of the approach proposed here is to work in continuous time, substituting the random series z_t in the previous equation by a continuous-time white noise process $z(t)$. This means that the equivalent storage process is no longer defined by the discrete-time continuity equation shown in Equation 2, but by a stochastic differential equation

$$d_t S_t = N_t \tag{6}$$
$$= E[N_t|S_{t-1}] + V[N_t|S_{t-1}]\, z(t)$$

where d_t indicates derivative with respect to time. The conditions under which the substitution of the discrete-time storage process by a continuous-time one is acceptable have been examined in a previous study (Saenz de Ormijana and Maidment, 1989). In general, this substitution is acceptable, although in certain cases it may be necessary to increase the continuous-time variance for the discrete- and continuous-time results to agree. The storage process S_t defined by the stochastic differential equation above is a continuous-time Gaussian Markov process, whose transient cumulative distribution function for storage x at time t, $F(x,t)$, is controlled by a *Fokker-Planck* equation (Saenz de Ormijana and Maidment, 1989)

$$\partial_t F(x,t) = -u(x,t)\partial_x F(x,t) + \tfrac{1}{2}\, v(x,t)\partial_x(v(x,t)\partial_x F(x,t)) \tag{7}$$

This form of the Fokker-Planck equation corresponds to the Stratonovich interpretation of stochastic calculus, which defines white noise as the ideal limit of actual, coloured (i.e. with memory) noises (Risken, 1989). In Equation 7, $u(x,t)$ is the *drift* coefficient which represents the average rate of change of storage

$$u(x,t) = E[N_t|S_{t-1}] = E[I_t] - E[W_t] - E[Q_t|S_{t-1}]$$

$$= \left[\begin{array}{ll} E[I_t] - E[W_t] & ,\; S_{t-1} \leq S^{(0)} \\ -\beta_0 + (1-\beta_1)E[I_t] - E[W_t] - \beta_2 S_{t-1} & ,\; S_{t-1} > S^{(0)} \end{array} \right.$$

$$= \left[\begin{array}{ll} a_t & ,\; S_{t-1} \leq S^{(0)} \\ a_t + b(S_{t-1} - S^{(0)}) & ,\; S_{t-1} > S^{(0)} \end{array} \right. \tag{8}$$

where a_t is a quadratic function of time due to the quadratic pumping trend

$$a_t = 627.3 - 5.024(t-1934) - 0.074(t-1934)^2$$

in units of Hm^3/yr, and $b = -\beta_2 = 0.126\ yr^{-1}$. Because of the linear threshold model assumed for the springflow-storage relationship, the drift coefficient is a piecewise linear function of the storage. Because of the pumping trend, the drift is also time-varying. The coefficient $v(x,t)$ in the Fokker-Planck equation (Equation 7) corresponds to the square root of the *diffusion* coefficient which represents the rate of change of the mean square change in storage, that is

$$v(x,t)^2 = V[N_t|S_{t-1}] = V[I_t - W_t] - Q_t|S_{t-1}$$

$$= \left[\begin{array}{ll} V[I_t - W_t] & ,\; S_{t-1} \leq S^{(0)} \\ V[(1-\beta_1)I_t - W_t - Q_t] & ,\; S_{t-1} > S^{(0)} \end{array} \right.$$

$$= \left[\begin{array}{ll} c^2_1 & ,\; S_{t-1} \leq S^{(0)} \\ c^2_2 & ,\; S_{t-1} > S^{(0)} \end{array} \right. \tag{9}$$

where $c_1 = 488.8$ $\text{Hm}^{9}\text{yr}^{-\frac{1}{2}}$, and $c_2 = 397.4$ $\text{Hm}^{9}\text{yr}^{-\frac{1}{2}}$. Equation 9 is a piecewise constant function of the storage. In general, an analytical solution of the Fokker-Planck equation is not possible when the drift coefficient varies with time and depends nonlinearly on the storage. In this case, the equation was solved numerically using an accurate high-order finite-difference method (Leonard, 1984).

4. Storage Results

Once the Fokker-Planck equation is solved, and the transient probability distribution $F(x,t)$ is known, one may calculate any other aspects of the distribution required. For example, Figure 4 shows the probability of storage falling below the threshold, which corresponds approximately to the probability of zero springflow. Figure 4 shows the asymptotic probabilities once the effect of the initial storage has died out, and the transient probabilities conditioned on the 1986 storage level. Eventually, all forecasts must converge to the asymptotic curve, as shown in Figure 4. For purposes of comparison, Figure 4 also shows the probability of storage falling below the threshold (conditioned on the 1986 storage level) if one ignores the effect of the threshold, that is, assuming that the simpler linear model applies over the whole storage range - in this case springflow will become negative for low enough storage. The storage level in 1986 was fairly large, so the effect of the threshold is small and the two curves (linear and threshold) are very close at first. Once pumping exceeds the mean annual recharge in the year 2000, however, two opposite effects occur. On the one hand, springflow may be negative for the linear system. This is equivalent to an excess recharge and determines higher storage levels and lower probability of below-threshold storage in the linear system. On the other hand, the diffusion coefficient for the threshold model is larger below the threshold ($c_1 > c_2$). This means that, when storage is very often below the threshold, more upcrossings of the threshold will occur in the threshold system than in the linear one, which only considers the smaller positive-storage value of the diffusion coefficient. That is why the probability of storage falling below the threshold proves to be larger for the linear system after the year 2000.

Similarly, Figure 5 shows a storage forecast (with the 95% confidence limits) from the low-storage conditions in 1956. The dotted line corresponds to the asymptotic mean, that is, once the effect of the initial conditions has died out. As pointed out before, all forecasts converge asymptotically to this line. The forecast curve shows that, on average, the aquifer takes a long time to recover from a severe drawdown; the annual variation of the recharge, however, is fairly large, so that the aquifer may recover much faster, as shown by the 95% confidence limits in Figure 5.

Finally, Figure 6 shows the agreement between the empirical cumulative distribution function (CDF) and the CDF predicted by the theory. Although the storage process is nonstationary, this was not taken into account for this figure. The empirical CDF was obtained in the usual manner, as if the storage values were identically distributed, which is not true because of the pumping trend. The theoretical CDF shown in Figure 6 corresponds to the average pumping over the period of record (which corresponds approximately to the 1962 pumping level). The coherence between theoretical and empirical CDFs is quite good, considering that this is not an exercise in curve fitting, but that the theoretical CDF was obtained indirectly using netflow data.

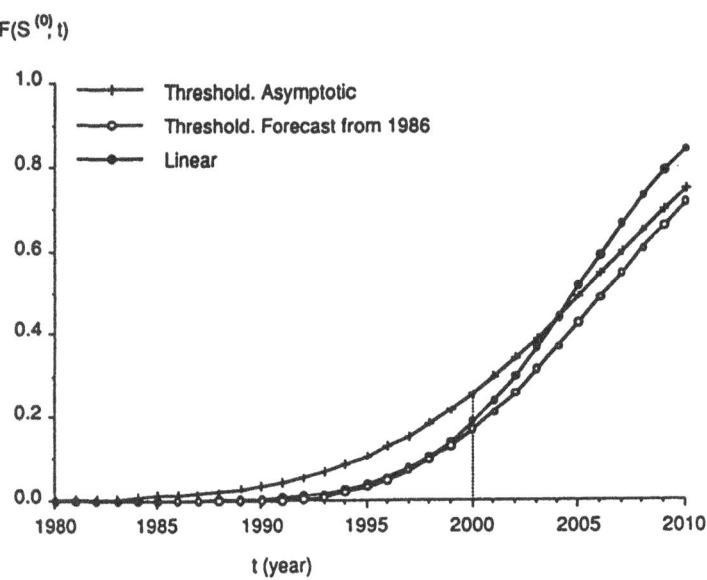

Figure 4. Probability of storage falling below the threshold (which corresponds approximately to the probability of zero annual springflow).

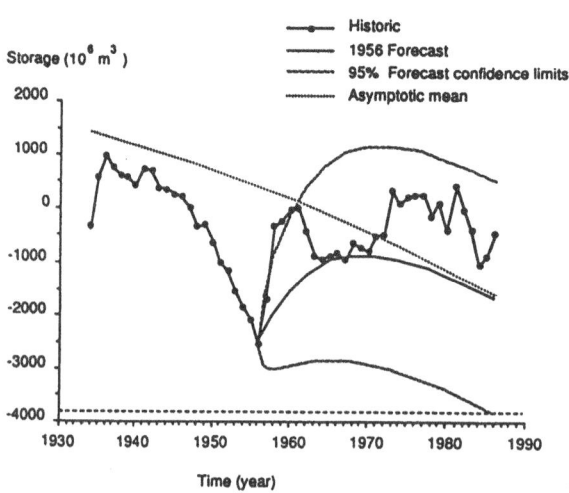

Figure 5. Storage forecast from 1956.

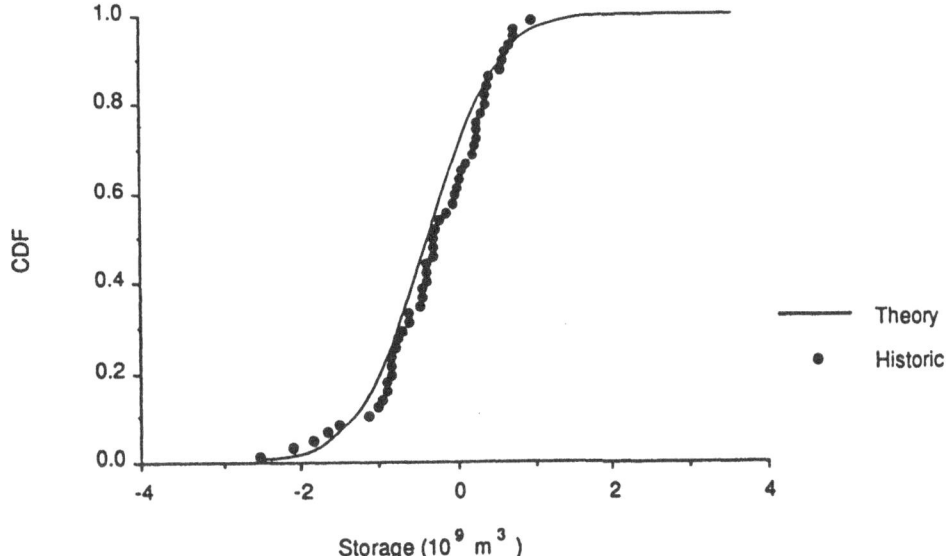

Figure 6. Comparison of the empirical and theoretical CDFs. The theoretical CDF corresponds to average conditions for the period of record (1934-1987).

5. References

Gardiner, C.W. (1985) Handbook of Stochastic Methods for Physics, Chemistry, and the Natural Sciences, 2nd ed, 442 p, Springer-Verlag, Berlin.

Kelly, M.Q. (1985) 'The Edwards underground water district', Issues in Groundwater Management, ed. by E.T. Smerdon and W.R. Jordan, pp 237-241, Center for Research in Water Resources, The University of Texas at Austin, Austin, Texas.

Klemes, V. (1973) 'Watershed as semi-infinite storage reservoir', Journal Irrigation and Drainage Division ASCE, 99, pp 477-491.

Klemes, V. (1974) 'Probability distribution of outflow from a linear reservoir', Journal of Hydrology, 21, pp 305-314.

Klemes, V. (1978) 'Physically based stochastic storage analysis', Advances in Hydroscience, 11, ed by V.T. Chow, Academic Press, New York.

Klemes, V. and Boruvka, L. (1975) 'Output from a cascade of discrete linear reservoirs with stochastic input', Journal of Hydrology, 27, pp 1-13.

Langbein, W.B. (1958) 'Queueing theory and water storage', Journal Hydraulic Division ASCE, Paper 1811, 84, pp 1-24.

Leonard, B.P. (1984) 'Third order upwinding as a rational basis for computational fluid dynamics', Computational Techniques and Applications, CTAC-83, ed. by J. Noye and C. Fletcher, pp 106-120, Elsevier, The Netherlands.

Risken, H. (1989) The Fokker-Planck Equation: Methods of Solution and Applications, 2nd edition, 472 p, Springer-Verlag, Berlin.

Saenz de Ormijana, F. and Maidment, D.R. (1989) 'Stochastic analysis of nonlinear, nonstationary water storage systems in continuous and discrete time', CRWR Report No. 228, The University of Texas at Austin, Austin, Texas.

Tong, H. (1983) 'Threshold models in non-linear time series analysis', Lecture Notes in Statistics, 21, 323 p, Springer-Verlag, New York.

APPROPRIATE TIME RESOLUTION FOR STOCHASTIC DROUGHT ANALYSIS

R. RODRIGUES[1], M.A. SANTOS[2], F.N. CORREIA[3]
Directorate General of Natural Resources (DGRN)
Av. Almirante Gago Coutinho 30
1000 Lisbon
Portugal

ABSTRACT. The continuous use and improvement of a rationale for the stochastic characterization of regional droughts led the authors to determine how short the characterization time unit can be in order to provide a better description of such drought characteristics as duration, magnitude and severity in a system analysis context without losing the macroview of the drought phenomenon given by an annual characterization.

This paper is a first approach to this time resolution tuning. River flow is the drought determinant factor chosen and a river basin in southern Portugal is used as case study. Monthly, seasonal, six-monthly and annual time intervals of integration of the river flows at four streamgauges are used as discretization time units. The distribution functions best fitting the data were derived for each time unit in order to estimate the percentiles to be used as truncation levels. Areas contributing to the stream gauges are the possible drought-affected areas. The screening of the historic truncated time series and the risk analysis of the outputs from a stochastic drought model were the validation criteria used.

Preliminary results from 26 years of recorded data presented the seasonal approach to drought characterization as the most promising one.

1. Introduction

A common attitude when facing drought problems is to stress the absence of standards and common terminology due to the specific nature and complexity of the drought phenomena. Although recognized as regional phenomena, no regionalization approach for drought studies is known as typical or basic. Also recognized as being long-lasting phenomena, no fundamental time unit of analysis is proposed as such, although the month is specified as the lowest discretization unit in order to distinguish drought analysis from low-flow analysis. Known as being related to a water deficit, it is

----
[1] Head of the Surface Water Division
[2] Head of Hydrology Department
[3] Director-General

J.B. Marco et al. (eds.),
Stochastic Hydrology and its Use in Water Resources Systems Simulation and Optimization, 247–265.
© 1993 *Kluwer Academic Publishers.*

difficult to establish the nature of the physical factor or factors that determine the shortage of water: precipitation, streamflow, soil moisture, groundwater levels, reservoir storage or water supply. Most of the time the type of analysis (climatic, hydrological, agricultural, hydropower, social and/or economic) and undoubtedly the data availability might help in that selection.

As a consequence of this conditional nature, conclusions from a particular analysis must be carefully generalized. As for the methodology proposed for any drought analysis model, the concepts must be very synthetic and the techniques very flexible in order to make that model a very portable one.

Bearing this in mind, the authors presented in a previous work (Correia et. al., 1987) a methodology for the characterization of regional droughts in terms of engineering risk. With the methodology proposed a very flexible model was built allowing for different definitions of drought and drought recovery, making it applicable to different climatic conditions and regions. Reading the work referred to is strongly recommended for a better understanding of the present paper.

One thing that was not clear after the first run of the model was the discretization time of the data to be used in the model. The time resolution of models is a concept constantly used in deterministic models but, although familiar to stochastic hydrology, its reference in stochastic modelling is not so frequent. In an attempt to contribute to this problem, the authors analyzed a case study.

2. Problem Formulation

In a previous work (Santos, 1983), the author introduces a precise definition of regional drought and presents a methodology for drought characterization that grasps the long-lasting and regional nature of the drought phenomenon allowing for its stochastic nature. This work is mainly oriented to a climatic characterization where such concepts as drought duration, deficit area, or drought intensity and/or severity are its main concern. The regional drought characterization methodology is based upon the basic concepts of point-drought and drought-affected area, the former being related to a threshold (truncation level or critical level) below which a water shortage exists, whereas the later concept is related to another threshold (critical area) above which the integration of the various areas affected by point-droughts represent a significant portion of the whole region under study (Fig. 1). The analysis of the time series through the statistical theory of runs as proposed by Yevjevich (1967) provides the means to characterize point-droughts through three characteristic variables: the deficit (measure of drought severity), defined as the cumulative deficiency below the truncation level, the duration (measure of drought persistence), defined as the lapse of time during which deficiencies occur below the truncation level; and the intensity (measure of drought magnitude), defined as the average deficiency below the truncation level throughout each duration. In the work by Santos, rainfall data recorded in 98 raingauges spread over the Portuguese continental area were used as a drought-determinant simulator. The year was chosen as the discretization unit.

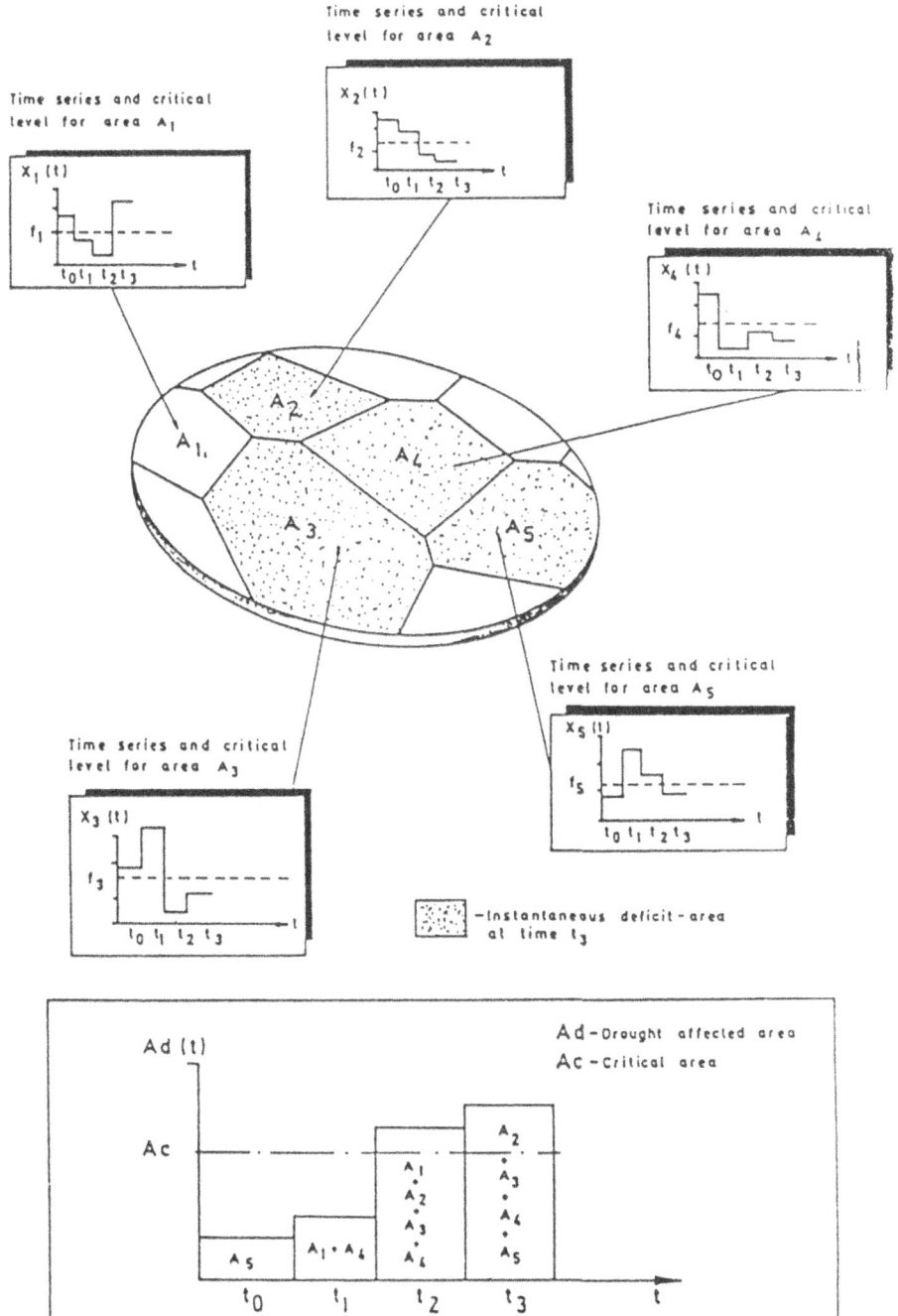

Figure 1. Schematic representation of the proposed methodology for regional drought characterization.

From an engineering point of view there is a need to orient this methodology to problems of planning and operating drought-control measures. A formulation of such methodology in order to incorporate such concepts as "risk of failure", e.g. risk of drought, and "consequences of failure", e.g. drought consequences was performed later (Correia et al., 1987). The new methodology relates the risk with drought occurrences and reliability with non-occurrences. Although this procedure is somewhat strict because it does not include any consideration of severity, that is taken care of under other concepts of the methodology. The mission risk concept considers only those drought occurrences in a pre-defined within-year period (mission period). As to resilience, since it measures the recovery time of a system, it can be immediately formulated with drought duration, especially in yearly terms. But if mission periods are used, it is reasonable, from a physical point of view, to consider that the recovery is accomplished only when an excess of moisture replenishes the exhausted resources above a recovery level - which may equal the truncation level or more realistically be taken as average conditions. The vulnerability concept, which expresses the severity of a system failure in terms of its consequences, can be equated either with total areal deficits, drought intensities or a loss function where the vulnerability variable may depend on any combination of drought characteristics.

Since more than a descriptive approach of historic regional drought events was intended, this work uses a multivariate stochastic model to generate a synthetic series of 1,000 years of monthly precipitation values at six distinctive regions defined within a river basin. Thirty-three years of monthly data were used to calibrate the parameters of the model that reproduced two important characteristics of the sample: the log-normality nature of the standardized monthly precipitation and its serial independence. It was important to use a discretization unit of less than one year in order to study the mission reliability and risk numerically, and to investigate its influence on the regional drought characterization. In fact, it is very common for drought problems to be relevant only during a given period of the year and this was illustrated in the study with the critical crop growing period.

Although the use of more detailed information might introduce some serial dependency among the drought characteristic variables, the methodological concepts themselves prevented this. In terms of the number of droughts, grouping of related occurrences was taken into account once the risk assessment was made on a yearly basis and a dry year was considered as a year in which at least one dry period occurs. As to the drought duration, one way of preventing a long drought period from being interrupted and each sub-division being considered a drought was to establish that a drought does not end until a portion of the accumulated water deficit (recovery rate) was recovered and the recovery level was set equal to the mean value on each discretization period (in the example, the mean monthly precipitation). The standardization of the severity or magnitude of droughts defined for different discretization units was based on the same principles as in the drought durations: while the drought is not recovered (full drought period) all the deficits occurring separately are summed up to give a global idea of the severity and this is also true for the regional intensity once the global deficit is divided by the summation of all drought durations occurring within the full drought period.

The macro-flowchart of the computer program in Figure 2 presents the basic concepts of the risk assessment methodology under a structured approach while Figure 3 sums up the typical outputs of the model. Since a regional drought is defined by means of two distinct variables, namely a truncation level and a percentage of drought-affected area, it is possible to study the relationship between one of these variables and the drought return period while the other variable is set constant. This is why the critical area loop and the truncation level loop in Figure 2 are used as enveloping loops of the whole structure. The consideration of different percentages of a same region area (for instance 50% and 100%) as critical areas allows the analysis of such features of the drought-affected region as its size, shape and areal spreading, while the use of several possible percentiles as truncation levels (usually from 1% to 20%) are used to define such drought characteristics as its duration, magnitude and severity.

The descriptive statistics performed with the output from the risk assessment drought model numerically materialized the expected risk, mission risk, resilience and vulnerability functions with such general shapes as presented in Figure 3. As can be seen from the risk functions, there is a region where the definition of the critical area is important as well as the percentile used as truncation level. In the authors' experience, this happens with return periods of ten years, if no mission period is used, or 20 years when mission risk analysis is performed. This means that very severe droughts are likely to affect large continental areas instead of regions on a watershed scale. The distribution functions used to study resilience present, for each duration, an increase in non-exceedance probability for higher values of truncation levels - in the authors' experience, an increase is also observed for larger critical areas. If a more exigent definition of drought recovery is adopted with the consideration of a recovery level greater than the truncation level, larger recovery times and slower trends in distribution functions are obtained (the same applies for greater recovery rates). As to the vulnerability functions, it is observable that, for the same return period, an increase in drought magnitude is related to an increase in the truncation level. At the same time, to that increase there usually corresponds a decrease in the critical area. However, the authors have observed some exceptions where the larger the persistence of drought, the smaller the drought-affected areas. Those exceptions, when incurred, were attached to the highest truncations level used (the 20% percentile) that defines droughts in a less exigent way, e.g. related to shorter return periods.

The proposed methodology proved to be very adequate for the characterization of regional droughts in terms of engineering risk. Also the use of different discretization units, especially those smaller than one year, showed that the model is sensitive to the time series discretization. For example, the model with a monthly time resolution failed to reproduce multiyear droughts. At first, the monthly generation process used was taken as the cause for such failure and an alternate yearly generation process with disaggregation capabilities was considered. However, the run of the 33-year monthly recorded data through the model showed that the problem was not in the stochastic generation process but in the inability of the monthly data to reproduce multiyear droughts in a

252

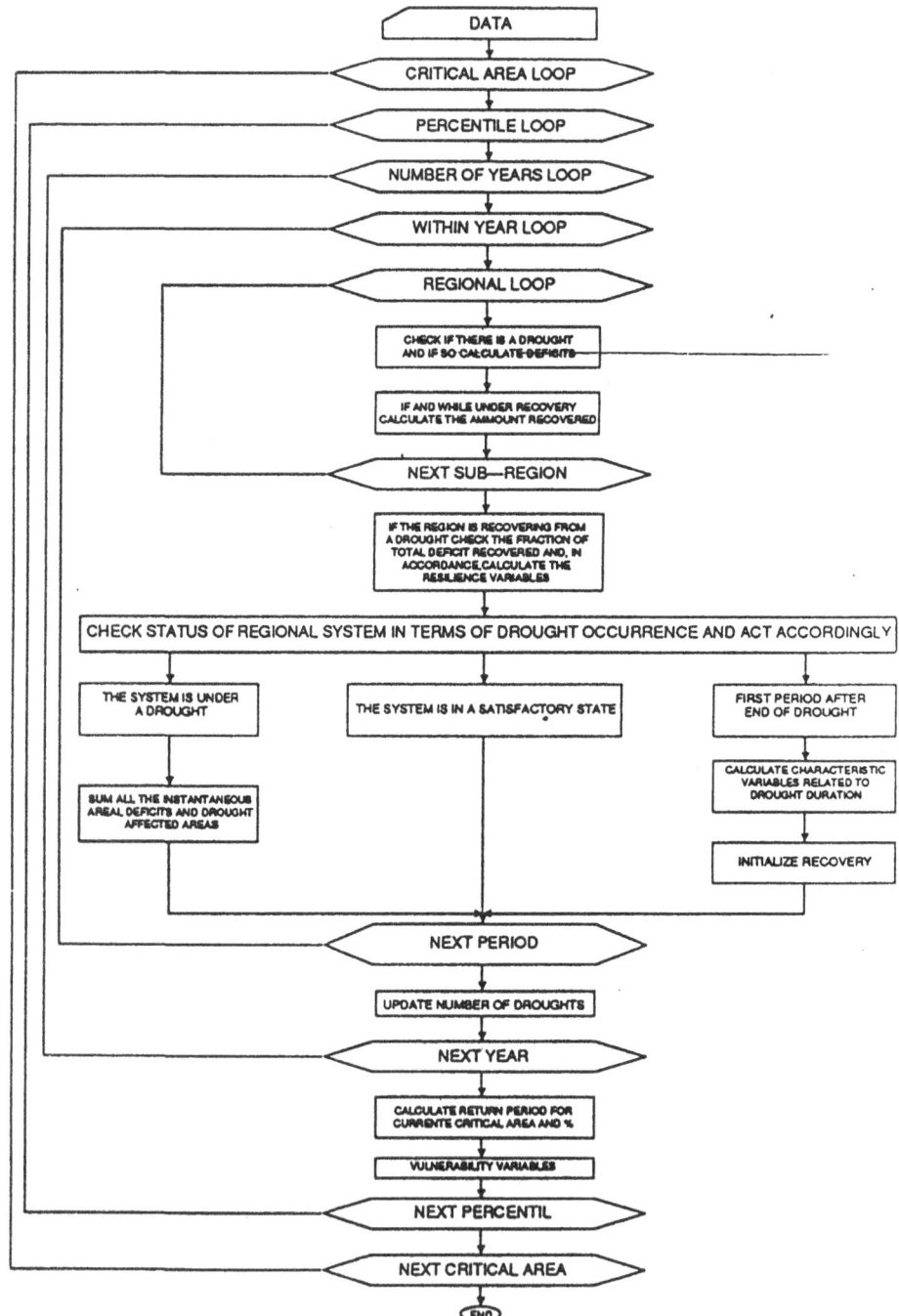

Figure 2. Macro-flowchart of computer program R-DROUGHT.

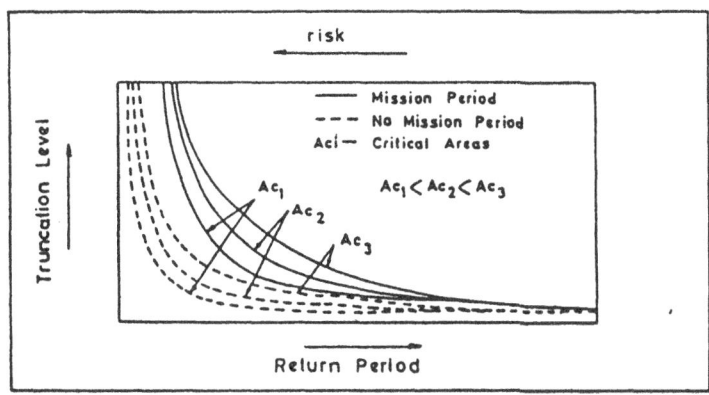

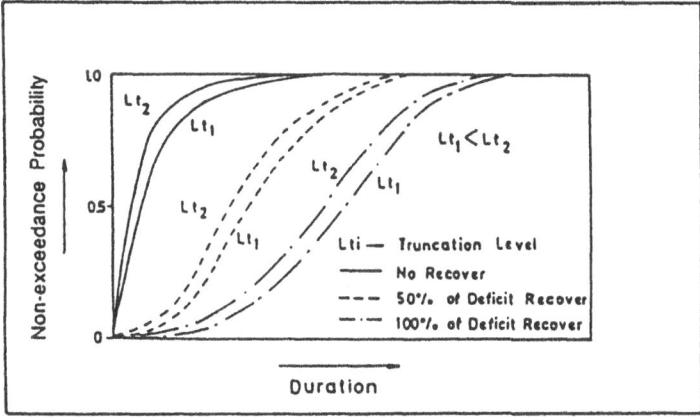

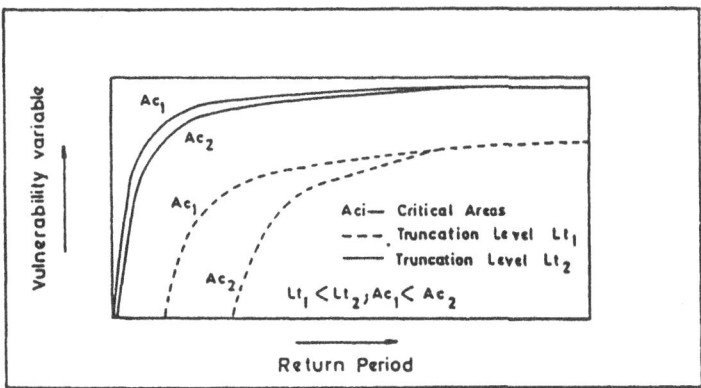

Figure 3. Theoretical functions for the risk and mission risk, resilience and vulnerability of droughts.

threshold abstraction environment. What happened was that the random oscillation of the monthly values over a periodic trend most of the time upcrossed the periodic truncation level (defined as a non-exceedance probability of the monthly values), interrupting the drought duration in yearly terms. Some of these interruptions occurred in such a way that the magnitude of the non-dry months was able to promote the recovery of the whole accumulated water deficit (Fig. 4) making almost no distinction between simple drought duration and drought duration until the system recovers. This is why no drought durations greater than nine months were observed. These observations showed that some research for determining the adequate time unit to be used in drought assessment is needed. This paper aims to contribute to the solution of this problem.

3. Methodology

Maintaining all the basic concepts and criteria for the regional drought definition used previously, it was decided not to verify the time unit constraint of the risk assessment drought model other than to change its standard parameters and simply allow monthly data to characterize multiyear droughts (one easy way to do it is to increase the recovery level). In this study river flow is the variable selected as drought-determinant factor in order to test the effects of serial dependency on the model for short time units of discretization. This way, a better determination of the time resolution of the model can be achieved. Although the title of this paper might seem too generic, it is only the time resolution of the present methodology that is being tested.

The validation of the appropriate discretization units to be used in the model to reproduce both multiyear trends and detailed quantification of drought characteristics is done in two steps: a joint screening of the different discretized historic time series superimposed on the respective truncation levels and the analysis of the risk, resilience and vulnerability of the regional droughts observed. The 20 percentile from the adjusted distribution functions to each discretized data period is used as a truncation level in order to identify a larger number of possible droughts. In order to reflect a regional approach in the analysis and, at the same time, reduce the number of point-series to be visually inspected, one of the streamgauges to be used in the case study will have to drain an area greater than 50% of the whole study area.

4. Case Study

4.1. BASIC DATA

The proposed methodology was applied to the River Sado basin that drains an area of about 7,700 km², the largest river basin in the Portuguese territory.

The River Sado basin is located in southern Portugal and has common boundaries with two international river basins from the Iberian peninsula: River Tagus basin to the North and the River Guadiana basin to the East (Fig. 5).

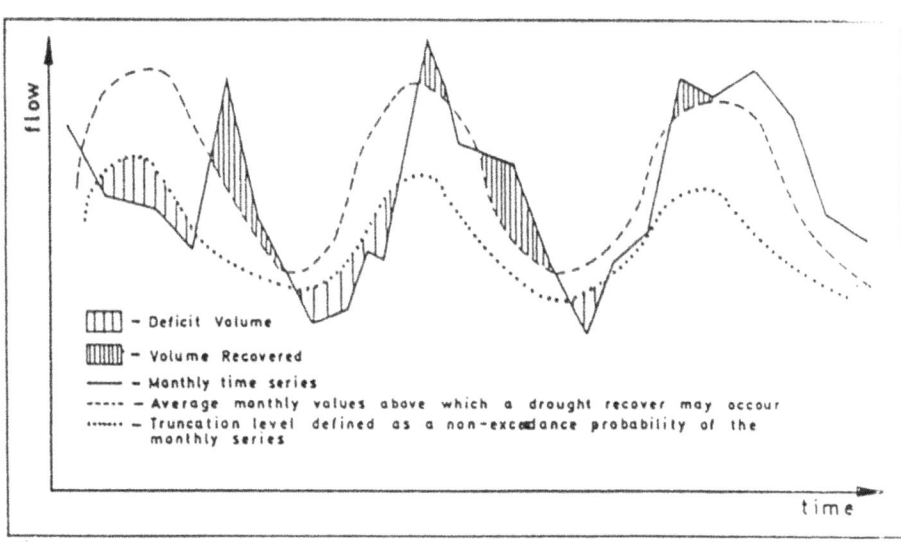

Figure 4. Different types of drought interruptions for a monthly time series.

Figure 5. Location of the River Sado basin.

The climate within the watershed can be considered moderate with exceptions of a short wet coastal area in the Southwest and South headwaters, with annual precipitations higher than 800 mm, and a dry area in the East where the annual precipitation is below 500 mm.

One of the main hydrological problems in the river basin is the within-year partition of the rainfall amounts and its large spatial distribution. In fact, the average number of rainy days in the several raingauges covering the basin ranges from 60 (SE and E) to 100 (W, SW and N). These numbers are reduced, respectively, to 15 and 30 rainy days for daily rainfall amounts above 10 mm.

The streams in the basin reflect this unbalanced partition of water within the year. Curiously some of their names retain as a prefix a modified spelling of the Arab word for the northern African rivers, the Oueds, known as having torrential responses to precipitation phenomena and also long periods of low flows when rivers might even dry completely. As an example one can mention the River Odivelas located precisely in one of the driest parts of the basin and presenting an average period of 3.4 months of null flows. Another example of this unbalanced seasonal behaviour in the encircling region is the Old Anas River from the Romans that became the Oued-ana to the Arabs, and finally Guadiana.

In order to introduce some regulation in the flow regimes determined by the intermittent rainfall amounts, a great number of reservoirs for irrigation purposes have been built in the basin since 1949. The first two aimed only at the irrigation of a small area along the Sado valley but, since 1963, a big irrigation plan that accounts for inter-basin transfers of water is being implemented. Figure 6 locates the main reservoirs in the basin, shows the regulation volumes and the year activity commenced for each reservoir.

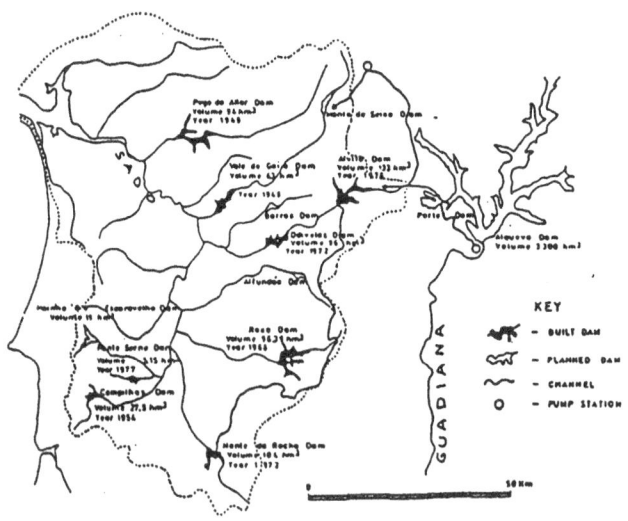

Figure 6. Location of reservoirs built within the River Sado Basin.

The geomorphology of the basin is centred on the filling during the late Terciary of the ancient gulf known as Tagus-Sado which now constitutes the surrounding terrains of the Sado and Tagus stream beds. The deep faults and consecutive collapses created a non-homogeneous hydrogeological system in the basin.

Since the stream gauging started in 1933 in the basin to account for future irrigation studies and since most of the man-made changes to the streams started in the late sixties, it is possible to select 25-year series of unregulated flows.

The large areal size and different precipitation regimes coupled with the flow data availability and the economic interests of the region makes the River Sado basin the ideal hydrogeological system to test the proposed methodology.

In this study, data from four streamgauges are used, covering approximately 50% of the whole river basin area (Fig. 7). A common period of 26 years of unregulated flows (1934/35 to 1959/60) is used to illustrate the methodology proposed in the previous section.

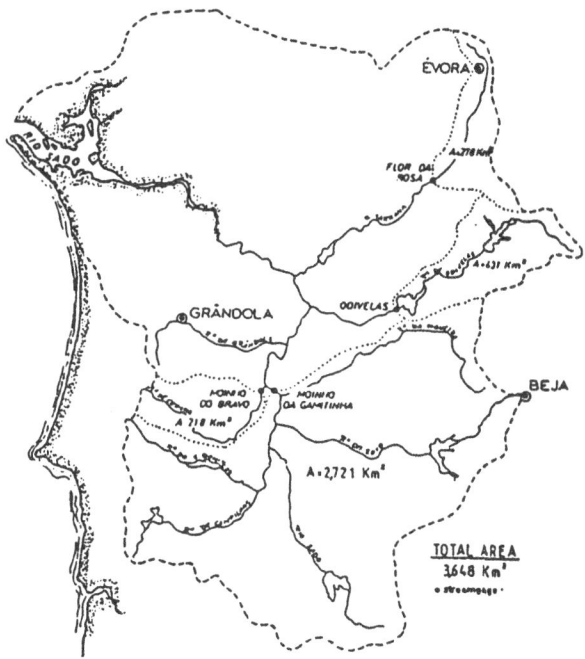

Figure 7. Location of streamgauges used in the study.

4.2. SCREENING OF TIME SERIES TRUNCATION

As mentioned earlier, a screening of the time series from the River Sado sub-basin that drains an area equal to 35.6% of the whole river basin and 74.6% of the study area (Fig. 7) was performed on annual, six-monthly, seasonal (three-monthly) and monthly bases. The objective was to identify in each series and for the same truncation level (the 20% percentile), the drought duration and drought deficits and to cross-reference this information in order to make some qualitative judgement about detail improvement versus short-term perspectives.

Figure 8 overlays all the relevant common periods thus enabling a time-history study. From its analysis the following conclusions are drawn:
- the year is the most adequate time of discretization for quick identification of multiyear drought periods and recovery volumes. The only two drawbacks are related to the precision of recovery periods – since a full year is used as a minimum sensitive unit – and the inability to reproduce within-year mission periods;
- there are no significant differences in the analysis when using six-monthly instead of annual discretizations besides the possibility of locating droughts within-year periods thus enabling mission risk analysis. Note that in Figure 8 all the annual drought periods were equally identified on the six-monthly approach, except for the 1952/53 drought, which was confined to the first six-months of that year thus adding some accuracy to drought durations;
- with the seasonal time of discretization, the plot of the whole series became difficult to manage. That is why making use of the previous annual screening, emphasis was laid on the analysis of the relevant seasonal periods in terms of multiyear drought trends, identified as the 43-45 period and the 56-58 period. Figure 8 shows some interruptions in the multiyear drought periods that occurred in the seasonal series although only one of the two-year drought periods was affected by these interruptions that lasted three months each. As to the true drought duration in a resilience context, not only does the non-recovery of drought deficits during interrupted seasons ensure the multiyear drought trends maintenance, but also defines its duration in a clearer way;
- when using the monthly time series, more and larger drought interruptions (six months in 1944) were detected in the same two relevant drought periods. Furthermore, the drought volumes began to be unmatchable with the corresponding seasonal ones, especially in the 56-58 period where the larger drought periods corresponded to insignificant deficit enhancements below the minimum average monthly value (Fig. 8). Still in this period, one can observe how the interruptions of drought periods in the wet months yielded less deficits. This happened not only due to greater fluctuations in the series, but also because of the lack of monotony in the behaviour of the truncation level. Since the truncation levels are extracted from the 20% percentiles of the best-fitted distribution functions derived for each monthly flow series, different types of distributions are to be expected and consequently different shapes in the average and truncation level functions. Although a smoothing of the truncation level function could be performed in order to avoid constant interruptions of long drought periods, that would introduce some rearrangements into the existing standard methodology which is beyond the scope of this paper. Therefore, it seemed better to skip the monthly time series in the next sensitivity analysis since only 26 years of data were being used to derive the twelve different distribution functions.

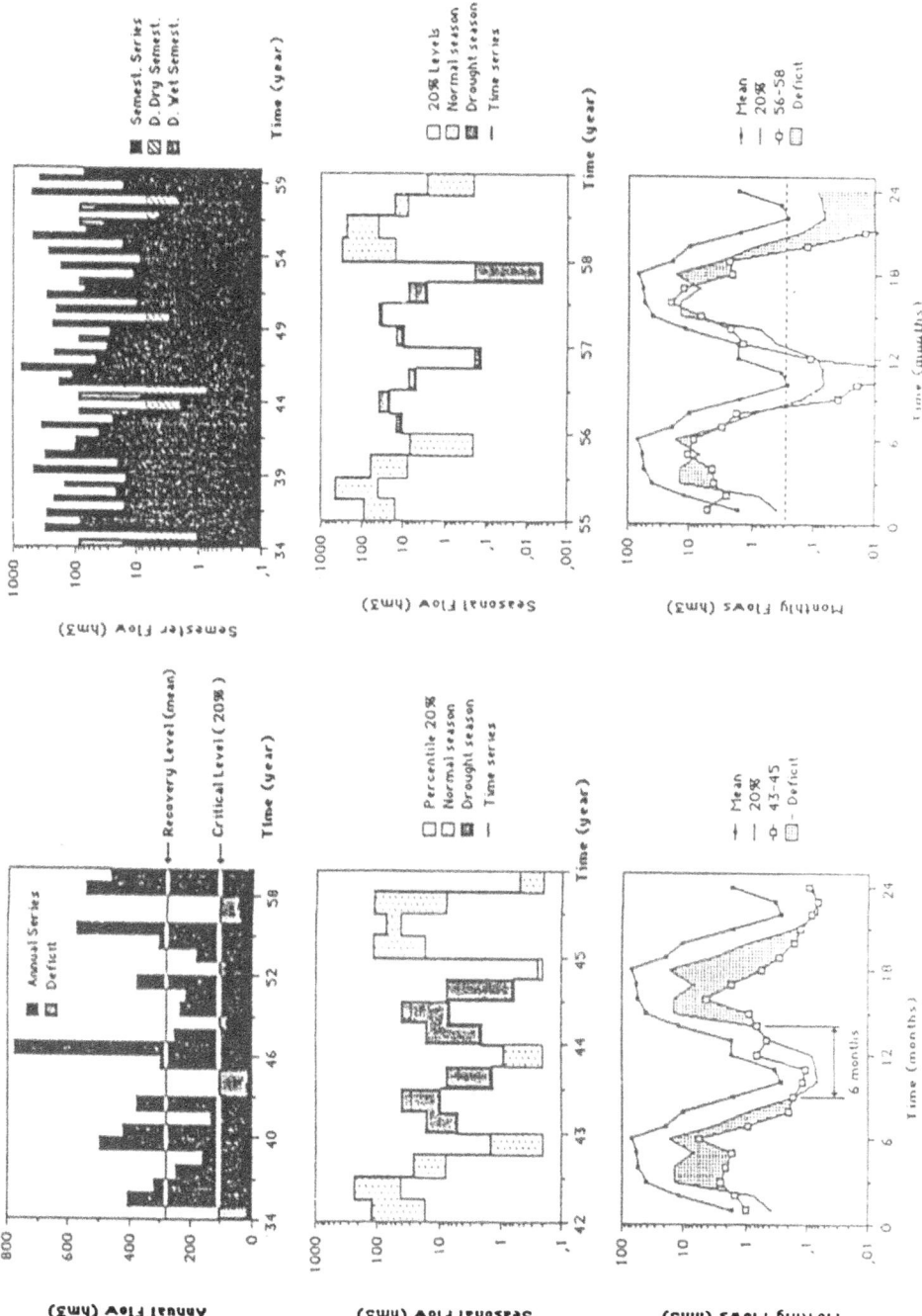

Figure 8. Historic drought events screening on different time units.

4.3. SENSITIVITY ANALYSIS

The second approach to the evaluation of the "ideal" discretization unit to be applied to the model input data obviously had to come from the analysis of the model outputs on each time resolution. These outputs do not present goodness-of-fit adjustments. Therefore, a very sharp-shaped configuration of the general features of the engineering risk functions occurs.

The sensitivity analysis was performed over the annual, six-monthly and seasonal time series validated by a sort of "human filter" described in the previous section. Three major blocks constitute the model output, namely the risk, resilience and drought vulnerability functions.

4.3.1. *Risk and Reliability Analysis.* Several runs of the model with different discretization time units provided data to study the influence of the truncation levels on the number of annual droughts thus enabling a risk analysis since the smaller the percentile used as the truncation level, the more severe the drought to be expected. Bearing in mind that the risk can be equated with the inverse of the return period T, it is possible to build a risk function (depending on the drought truncation level) for a fixed critical area and each discretization time unit.

Figure 9 resumes the information for the three time series used where the relevant drought periods are considered whenever they occur (a) within the year, or (b) only during a predefined mission period. The mission period used was the whole wet six-monthly period or, in seasonal terms, autumn and winter seasons. By the joint analysis of graphics (a) and (b) it can be inferred that the six-month's period is the most sensitive time unit in the mission risk approach. Since the mission risk function for wet six-monthly periods is very similar to the annual risk function, it can be inferred that the annual droughts are determined to a considerable extent by the wet six-monthly period performance. The insensitivity of the season time unit in the mission risk approach indicates that the seasonal discretization of flows weights, in a monotonous way, the four three-months periods, concentrating great amounts of moisture on the critical agricultural seasons for crop selection. This indicates that the seasonal time series of flows are most suitable for the assessment of agricultural droughts.

4.3.2. *Resilience Analysis.* Figure 10 summarizes the general shape of the resilience functions observed for drought definitions when the 20% percentile in the 50% critical area are used. From its analysis three general conclusions can be drawn:
   - the larger the discretization unit, the less detailed are the resilience functions,
   - the six-monthly period and the year are too crude as time units to give good estimates of the initial recovery times and the behaviour of the respective recovery functions is very much the same;
   - the seasonal recovery functions can reproduce both the shape and range of multiyear drought trends, making the season a very flexible discretization unit.

4.3.3. *Vulnerability Analysis.* Since all validation procedures of this study are based on the comparison of results from each time unit series,

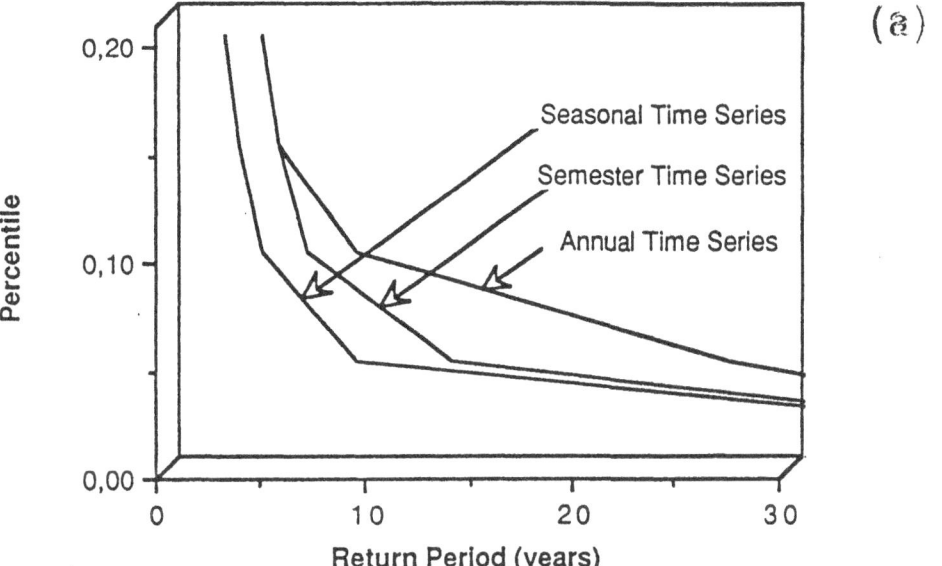

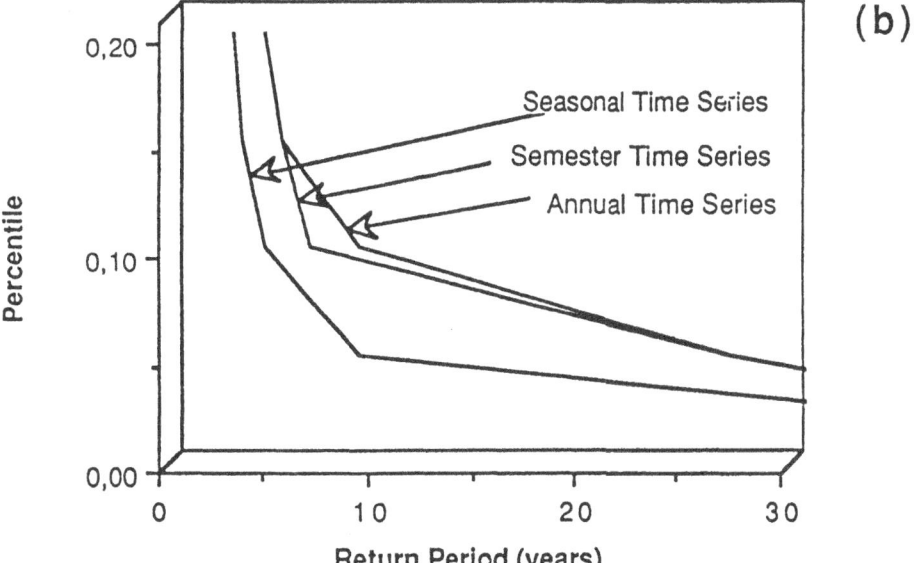

Figure 9. Risk (a) and mission-risk (b) drought functions for 50% critical areas.

262

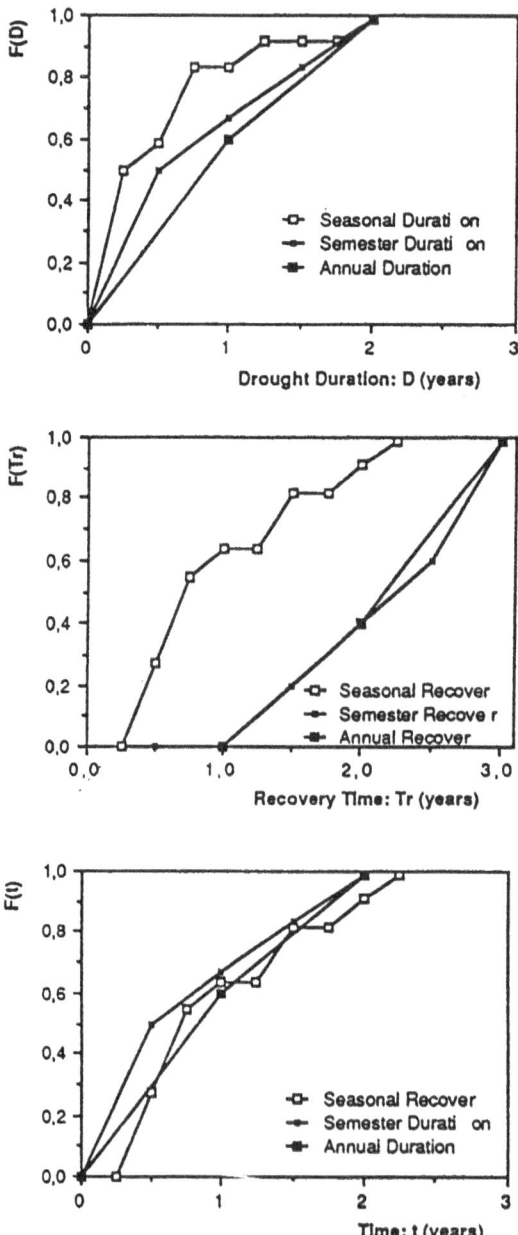

Figure 10. Probability distribution functions for drought duration, drought recovery time, and superimposition of duration and recovery times, for 50% critical areas and 0.20 critical levels.

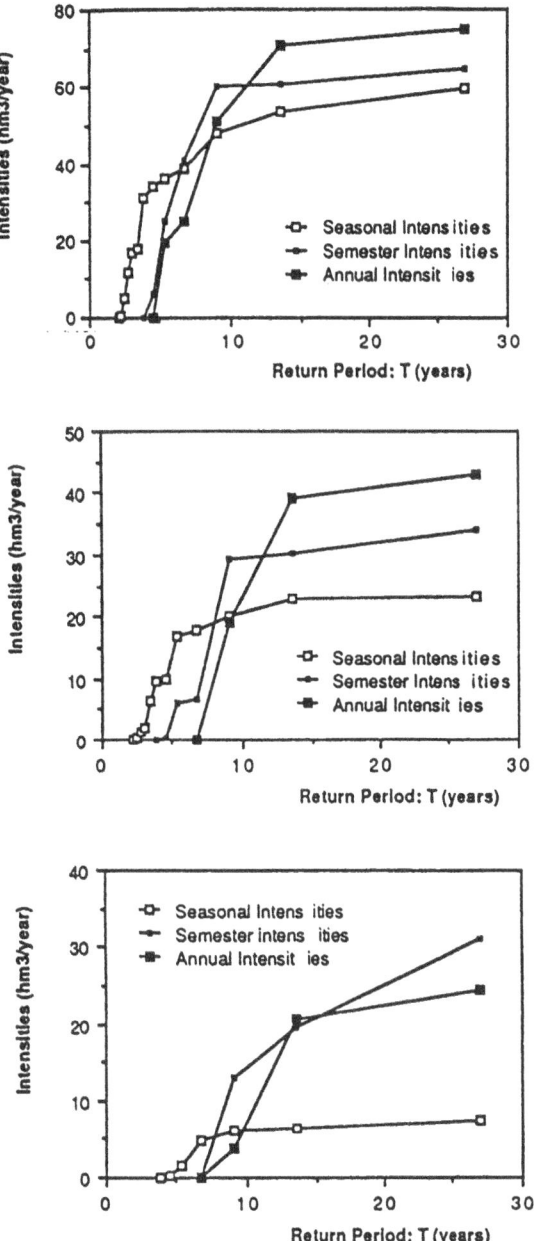

Figure 11. Drought vulnerability expressed in terms of maximum intensity for 0.20, 0.10, 0.05 critical levels and 50% critical areas.

drought intensity was chosen as the vulnerability factor once it measures
a standardized mean total deficit. Figure 11 presents the vulnerability
results for each discretization unit and for three truncation levels,
namely 20%, 10% and 5%. Although some close values and common behaviour
of all three vulnerability functions are observed when using the 20% and
10% percentiles as truncation levels, the seasonal vulnerability function
obtained from the 5% percentile yields smaller magnitudes than the
corresponding six-monthly and annual vulnerability functions. This might
be explained by the small number of droughts occurring in the sample when
such an extreme truncation level is used or even by the lack of detail in
the annual and six-monthly truncated series.

5. Concluding Remarks

A stochastic model for the risk assessment of regional droughts is applied
to the River Sado basin to illustrate the sensitivity of the different
discretizations performed over the same data series. Four discretization
time units, namely the year, the six-monthly period, the season and the
month, were used to develop some insight into the structure of drought
processes and its relation with area-time structures of differently
discretized time series. The final goal was to detect which time unit
could be used as optimal point from the intersection of two functions: one
where information is lost owing to larger discretization time units (Knisel
et al., 1967) and the other one where macroview trends are neglected for
small discretization units.

It was inferred that, in a run-theory environment, the monotony of the
series periodic component played an important role in the maintenance of
long-term trends since the basic reference parameters that define both the
truncation and recovery levels - derived from the statistical treatment of
data - are then coherent with each other. The seasonal discretization of
time series was therefore taken as the most appropriate approach to
exemplify this coherence. On the contrary, the truncation function for
monthly discretization shows a more unstable shape. These aspects need
nevertheless to be tested with percentile values derived from larger
samples. It is also important to run larger synthetic series through the
model (to derive detailed engineering risk functions) and then incorporate
information content concepts - as described by Dyhr-Nielsen (1972) - into
the risk functions in order to identify the cause of discrepancies between
different time units outputs. As to the large discretization units, it was
shown that the use of six-monthly discretized series can be a good
complementary approach for the annual assessment of regional droughts.

6. References

Correia, F.N., Santos, M.A., Rodrigues, R.R.,(1987) 'Engineering risk in
    regional drought studies', in Engineering Reliability and Risk in Water
    Resources', Martinus Nijhoff Pub. NATO ASI Series E, pp 61-86.
Dracup, J.A., Lee, K.S., Paulson, E.G., (1980) 'On the definition of
    droughts', Wat. Res. Res. 16(2), pp 297-302.
Dyhr-Nielsen, M. (1972) 'Loss of information by discretizing hydrologic

series', Colorado State University Hydrology Paper, No. 54.

Ferreira, J.F., (1973) 'Irrigation plan for the Alentejo region', Electricidade, No. 90, pp 184-188 (in Portuguese).

Knisel, W.G., Yevjevich, V. (1967) 'The statistical measure of hydrology time series', Proc. Intern. Hydrol. Symp., Fort Collins, pp 306-313.

Santos, M.A., (1983) 'Regional droughts: a stochastic characterization', Journal of Hydrology, 66, pp 183-211.

Yevjevich, V.M. (1967) 'An objective approach to definitions and investigations of continental hydrologic droughts', Colorado State University Hydrology Paper, No. 23.

PARAMETER ESTIMATION IN REAL-TIME FLOOD FORECASTING VIA CONCEPTUAL MODELS

ARMANDO BRATH
Institute of Hydraulics
University of Basilicata
Potenza
Italy

RENZO ROSSO
Institute of Hydraulics
Politecnico di Milano
Milan
Italy

ABSTRACT. The issue of the real-time adaptive use of conceptual models is analyzed by using a maximum likelihood framework for parameter estimation. The validity of the assumptions of the commonly-used ordinary least squares (OLS) estimation criterion is examined. The application in an on-line framework of maximum likelihood estimators for autocorrelated (AMLE) and heteroscedastic (HMLE) errors is also presented. Despite the presence of both a strong autocorrelation and a considerable heteroscedasticity in model residuals, the OLS estimator is found to be the most effective and suitable for real-time operation.

1. Introduction

In the last three decades a large number of conceptual models have been developed in order to describe the rainfall-runoff transformation at the basin scale. Whilst the goal of black-box modelling is the development of a relationship between the input and the output to the physical system, without any attempt to describe the processes occurring in the catchment, the conceptual approach is based on the introduction of some simple schematizations (conceptualizations) of the various hydrological sub-processes. Accordingly, hydrologists have extensively recognized the capability of a conceptual approach to improve the description of the hydrological phenomena with respect to black-box modelling (see e.g. Kitanidis and Bras, 1980; Bacchi et al., 1988).

Despite this, in the context of real time forecasting of floods, black-box modelling is usually preferred to the conceptual one. Many reasons for this can be found. A first reason is related to the unsatisfactory state of the art of parameter estimation for the latter, whereas highly efficient techniques are available for the former. Moreover, because of their simplified structure, black-box models can generally be easily formulated in a state-space framework, so allowing the application of a Kalman filter algorithm (O'Connell and Clarke, 1981). Although in recent years many hydrologists turned their attention to the application of the filtering theory to conceptual models (see e.g. Kitanidis and Bras, 1980), one must be aware, as pointed out by Sorooshian (1983), that such an approach

267

J.B. Marco et al. (eds.),
Stochastic Hydrology and its Use in Water Resources Systems Simulation and Optimization, 267–281.
© 1993 *Kluwer Academic Publishers.*

generally implies important modifications of the structure of the model, in order to cast it in a state-space framework. Thus, the behaviour of the model may be significantly modified.

A second reason for the rare application of conceptual rainfall-runoff models to flood forecasting is related to the difficulties which usually arise during the calibration step. Many researchers have extensively described the computational difficulties encountered in the off-line calibration of this kind of models (see e.g. Ibbit and O'Donnell, 1971 and 1974; Johnston and Pilgrim, 1973 and 1976; Pickup, 1977; Sorooshian and Gupta, 1983; Hendrickson et al., 1988). Calibration is usually achieved by looking into the hyperspace generated by parameters for the values which optimize (minimize or maximize, as appropriate) an objective function, the form of which is derived from the selected estimation criterion. The optimum set of parameters is sought by inspecting the response surface by means of numerical algorithms of optimization. The papers mentioned show the response surface of some complex conceptual models to be far from the ideal convex nature and to have an extremely irregular conformation, characterized by the presence of several local optima. The analysis of response surface maps, as performed by Sorooshian (1991), is a powerful tool for the detection of the presence of local optima and other irregularities of the response surface.

Furthermore, an on-going question in the real-time use of conceptual rainfall-runoff models is related to the choice between their on-line or off-line use. Off-line use is equivalent to the assumption of the time-invariance of model parameters, which may be inadequate even for the most complex models (Backie and Eeles, 1985). For instance, performing an off-line calibration of the Sacramento Soil Moisture Accounting model based on a one year data record, Sorooshian, Gupta and Fulton (1983) retrieved three different sets of parameters, using the data from three different years (one a wet, one an average and one a dry year) for the calibration.

Model adaptiveness is generally considered a very valuable feature for real-time flood forecasting procedures (see e.g. Nash and Sutcliffe, 1970; Wood and O'Connell, 1985). However, the difficulties experienced by the abovementioned hydrologists in the off-line calibration step have in some way discouraged a fully on-line use of conceptual models. Nevertheless, as pointed out by Brath (1989), it must be stressed that these difficulties seem to be ascribed especially to the cumbersome structure and redundant parametrization of the models under examination, involving the estimation of a large amount of parameters (often more than 20). Because of their overparametrization, high correlation between model parameters frequently exists. Then, the response surface is generally characterized by the presence of several line optima, extended valleys along which the objective function is relatively insensitive to the parameter values. This results in a poor structural identifiability (Sorooshian and Gupta, 1985) of the model and in the appearance of several problems in the calibration step. Due to the abovementioned reasons, large conceptual models (NWS, Stanford Watershed Model and so on) seem to be currently unsuitable for a fully on-line use. However, improvements in casting these models in a filtering context could be the solution for their implementation in an adaptive framework. These issues turn attention to parsimonious rainfall-runoff models, like the one presented here, which allow the adaptiveness requirement to be matched.

Finally, an open problem in the real-time use of conceptual rainfall-runoff models is the choice of the estimation criterion for model calibration. The lack of correct consideration of this problem can yield unsatisfactory results in model calibration. Although some papers exist in this field for off-line model calibration (see e.g. Sorooshian and Dracup, 1980; Troutman, 1985), little work has been done in the case of on-line model operation (Brath, 1989a). The aim of this paper is to assess the problems related to the estimation criterion choice in the context of real-time flood forecasting.

2. A Maximum Likelihood Framework for Conceptual Models Calibration

The discharge at the outlet section of the basin, computed by means of a conceptual rainfall-runoff model, for the i-th time step from the beginning of the storm event can be expressed as:

$$q_i = f (\underline{P}_i, \underline{\theta}) \tag{1}$$

where $\underline{P}_i$ is a matrix representative of the input data (rainfall data, evapotranspiration loss data, etc.) measured up to time step i and $\underline{\theta}$ is a vector representative of the set of the p model parameters: $\underline{\theta} = \{\theta_i, \theta_2, \ldots, \theta_p\}$.

Some sources of errors exist in rainfall-runoff modelling: (i) input errors, (ii) model errors, (iii) parameter errors, (iv) output errors. Input errors can be ascribed to errors in the point rainfall measurements, errors in the spatial characterization of the rainfall field, errors in the other input variables, and imperfect knowledge of the initial state of the system. Although some techniques are available to account for errors in input variables, it is customary to assume that input variables are measured without errors (see Troutman, 1985). It is also usual to assume that a correct set of parameters for the model exists. Thus, due to our lack of knowledge of this true set, an error of type (iii) arises. Besides these, errors due to model structural form and to noise in output data arise. Because f(.) is a simplified representation of the processes occurring in the catchment, a model error arises due to model inadequacy in representing the physical system under examination. Also the presence of errors in discharge measurements has been widely recognized. However, regardless of whether errors arise from source (ii) or (iv), the estimation problem is formally the same.

Accounting for the presence of errors, and assuming errors to have an additive nature, the observed discharge Q_i can be written as:

$$Q_i = q_i + \epsilon_i \tag{2}$$

Calibration of a rainfall-runoff model is generally achieved by optimizing a function (objective function) of the whole set of the errors $\underline{\epsilon} = \{\epsilon_1, \epsilon_2, \ldots, \epsilon_n\}$, where n denotes the number of available observations of the discharge. The choice of the objective function is a crucial point in rainfall-runoff modelling. As pointed out by Sorooshian and Dracup (1980), in almost all of the existing literature on conceptual models calibration the selection of the objective function has been rather subjective (see e.g. Ibbit and O'Donnell, 1971; Johnston and Pilgrim, 1976; Pickup, 1977, among others). The important fact has been overlooked that this selection implies assumptions concerning the probability distribution of the model residuals $\underline{\epsilon}$. Thus, if these assumptions are not in agreement with the actual probability distribution of the residuals, the calibration procedure generally leads to poor results and biased parameter estimates are obtained. Thus, unsatisfactory performances are to be expected in the forecasting step.

A quite general and suitable framework for parameter estimation is the Maximum Likelihood approach. Maximum likelihood estimators have some relevant characteristics of optimality: under some general conditions it can be proved that they are asymptotically efficient estimators (Chow, 1983). However, the maximum likelihood approach requires prior specification of the probability structure of the $\underline{\epsilon}$ errors. Assuming that additive errors in Equation 2 (i) have zero mean, $E[\epsilon_i]=0$, and (ii) come from a multinormal distribution with known statistics, the probability density function of the observed discharges $Q = \{Q_1, Q_2, \ldots, Q_n\}$ given $\underline{\theta} = \{\theta_1, \theta_2, \ldots, \theta_p\}$ is:

$$f(Q|\underline{\theta}) = (2\pi)^{-n/2} \; |\underline{\phi}|^{-\frac{1}{2}} \; \exp[-\underline{\epsilon}^T \; \underline{\phi}^{-1} \; \underline{\epsilon} \, /2] \tag{3}$$

where $\underline{\phi}$ denotes the variance-covariance matrix of residuals:

$$\underline{\phi} \; = \; \begin{bmatrix} \text{Var}[\epsilon_1] & \text{Cov}[\epsilon_1,\epsilon_2] & \cdots & \text{Cov}[\epsilon_1,\epsilon_n] \\ \cdot & \cdot & \cdots & \cdot \\ \cdot & \cdot & \cdots & \cdot \\ \text{Cov}[\epsilon_1,\epsilon_n] & \text{Cov}[\epsilon_2,\epsilon_n] & \cdots & \text{Var}[\epsilon_n] \end{bmatrix} \tag{4}$$

and $|\underline{\phi}|$ represents the determinant of matrix $\underline{\phi}$. The function $f(Q|\underline{\theta})$ defines the probability density of a set of measurements Q, given $\underline{\theta}$. The likelihood function $L(\underline{\theta}|Q)$ has the same form of $f(Q|\underline{\theta})$, but it is considered to be a function of parameters given Q (Mood et al., 1974). Then, from Equation 3 the natural logarithm $l(\underline{\theta}|Q)$ of the likelihood function is easily obtained as:

$$l(\underline{\theta}|Q) = \ln L(\underline{\theta}|Q) = - (1/2) \, [n \, . \, \ln(2\pi) + \ln|\underline{\phi}| + \underline{\epsilon}^T \, \underline{\phi}^{-1} \, \underline{\epsilon}] \tag{5}$$

In the maximum likelihood approach, the parameter estimates are the values that maximize the likelihood function $L(.)$ or its natural logarithm $\ell(.)$. From Equation 5, this is equivalent to minimizing the quantity:

$$S_{ML} = \underline{\epsilon}^T \, \underline{\phi}^{-1} \, \underline{\epsilon} + \ell n|\underline{\phi}| \tag{6}$$

Then, maximum likelihood estimation requires specification of the covariance matrix of the residuals $\underline{\epsilon}$. Assuming (iii) in correlation between errors $\underline{\epsilon} = \{\epsilon_1, \epsilon_2, \ldots, \epsilon_n\}$, that is $E[\epsilon_i \, . \, \epsilon_{i+k}] = 0$ if k is different from 0, and (iv) homoscedasticity that is $\text{Var}[\epsilon_i] = E[\epsilon_i^2] = \sigma^2$ for each $i=1,2,\ldots,n$, the covariance matrix defined by Equation 4 becomes:

$$\underline{\phi} = \underline{I} \, \sigma^2 \tag{7}$$

where $\underline{I}$ denotes the identity matrix. Then, if the assumptions (i) to (iv) hold, from Equation 6 one can obtain the fact that minimizing S_{ML} is equivalent to minimizing:

$$S_{OLS} = \underline{\epsilon}^{T} \underline{\epsilon} \tag{8}$$

that is, maximum likelihood estimation is equivalent to the ordinary least squares estimation. Then, if the assumptions (i) to (iv) hold, the ordinary least squares is an optimal procedure. Conversely, if these assumptions are violated, the calibration procedure may result in unsatisfactory estimates, as stressed by Clarke (1973) and Sorooshian and Dracup (1980). Many techniques for checking the validity of the assumptions (i)÷(iv) are discussed by Draper and Smith (1981).

However, two violations of the above assumptions very frequently arise: a) the residuals are heteroscedastic, violating assumption (iv) and b) the residuals are autocorrelated, violating assumption (iii). The problem of error heteroscedasticity is mainly due to the inhomogeneous nature of errors in streamflow measurements. Taking into account the concavity of a typical stage-discharge curve, it is clear that errors in stage measurements yield errors in discharge that are larger in the region of the high discharges than in the region of the low ones. Thus, discharge measurements via stage-discharge rating curves are usually affected by errors with inhomogeneous variance, increasing with the stage measured (see e.g. Potter and Walker, 1981), and so greater prediction errors are to be expected for larger discharges. Besides, the presence of correlation between model residuals in rainfall-runoff models has been recognized by many researchers. As suggested by modelling practice, a discharge underestimation is more likely to be followed by another underestimation for the subsequent time step than by an overestimation.

The lack of correct consideration of the stochastic nature of model residuals can yield to unsatisfactory results in model calibration. In order to overcome the problems due to the heteroscedasticity and correlation, Sorooshian and Dracup (1980) proposed two maximum likelihood estimators suited for: a) heteroscedastic incorrelated errors (HMLE) and b) homoscedastic first-lag autocorrelated errors (AMLE). Sorooshian, Gupta and Fulton (1983) used both HMLE and AMLE estimators in comparison with the OLS one for off-line calibration of the soil moisture accounting model of the US National Weather Service river forecasting system (SMA-NWSRFS). Calibration was performed using daily streamflow discharge measurements. It was found that, among the three estimators, HMLE was able to provide the most realistic parameter values and the most reliable performances in the forecasting period. Conversely, AMLE was found to perform no better than OLS. They concluded that the lack of residual homoscedasticity has more severe effects than the presence of residual autocorrelation. These results seem to be confirmed by the ones obtained by other researchers (see Sorooshian, 1991). However, it must be stressed that the degree of correlation is affected by the data time interval used: for a given catchment, the correlation between model residuals become stronger when the time step used become shorter. Accordingly, by decreasing the interval, the problems posed by the presence of residuals autocorrelation may tend to overcome those due to heteroscedasticity.

There follows an analysis of the performances of the on-line model calibration using ordinary least squares estimator and maximum likelihood criteria for autocorrelated lag-one and heteroscedastic errors.

3. Outline of the Conceptual Model Used

The conceptual rainfall-runoff model used basically comprises three components. The first one is designed to estimate absorption at the basin scale, so determining the separation between surface and subsurface runoff during the course of a storm. This first component provides the input to both surface and subsurface runoff models, which represent the remaining two components of the model, as sketched in Figure 1.

Because Hortonian flow was recognized as the major process of runoff production, infiltration at the basin scale is modelled by

$$f(t) = f_c + (f_0 - f_c) \cdot \exp(-t/Dcay) \tag{9}$$

where $f(t)$ is the rate of infiltration capacity, f_c is the saturated final infiltration capacity and f_0 its initial value respectively, while Dcay represents the time scale of the infiltration capacity rate. Denoting $i(t)$ as rainfall intensity during the time interval $(t, t+dt)$, the corresponding net rainfall intensity, $i_e(t)$, is given by:

$$i_e(t) = i(t) - f(t) \qquad \text{if } i(t) > f(t) \tag{10}$$

$$i_e(t) = 0 \qquad \text{if } i(t) \leq f(t) \tag{11}$$

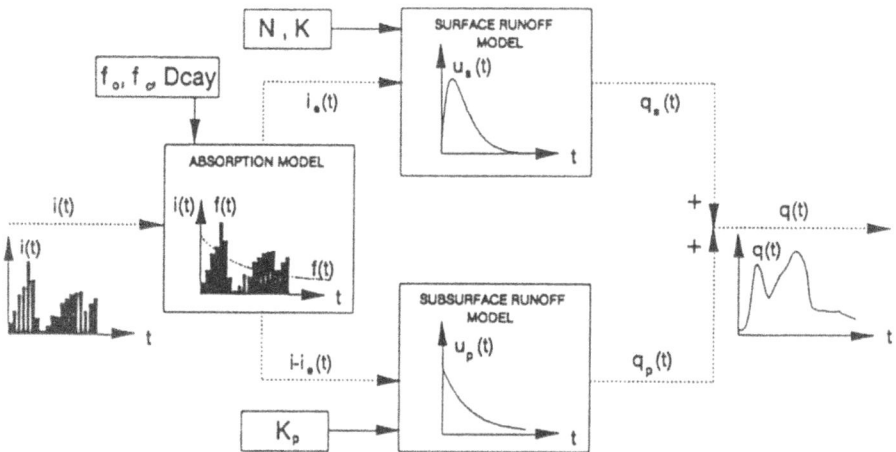

Figure 1. Sketch of the conceptual rainfall-runoff model structure.

The transformation of rainfall excess $i_e(t)$ into surface runoff at the outlet section of the basin, $q_s(t)$, is modelled by a linear approach, that is:

$$q_s(t) = A \int_0^t i_e(t-\tau) \cdot u_s(\tau) \cdot d\tau \qquad (12)$$

where A represents the area of the catchment. According to the analogy of the linear reservoir cascade, the impulse response function $u_s(t)$ is taken to be:

$$u_s(t) = (t/K)^{N-1} \cdot \exp(-t/K) / [K.\Gamma(N)] \qquad (13)$$

where K and N denote the time scale and the shape factor of surface runoff IUH respectively and $\Gamma(.)$ represents the complete gamma function. Rainfall losses are transformed in subsurface runoff at basin outlet, $q_p(t)$, again according to a linear approach

$$q_p(t) = A \int_0^t [i(t-\tau) - i_e(t-\tau)] \cdot u_p(\tau) \cdot d\tau \qquad (14)$$

with impulse response function $u_p(t)$ descending from the analogy with the linear reservoir model, that is:

$$u_p(t) = \exp(-t/K_p)/K_p \qquad (15)$$

where K_p denotes the time scale of subsurface runoff IUH.

4. On-line Model Calibration

The model presented above results in a parsimonious formulation, which is characterized by just six parameters; namely three parameters (f_0, f_c, Dcay) describing the absorption process, two (N, K) the transformation of rainfall excess into surface runoff, and one (K_p) the transformation of infiltration losses into subsurface runoff. Because of its simple structure and parsimony, the model was found to be very suitable for fully on-line use, so matching the adaptiveness required. As stressed above, this represents a very valuable feature in view of the real-time use of the model.

On-line model use during a flood is achieved by recalibrating all model parameters at each time step, whenever new measurements of rainfall and discharge become available. At time t, the measurements of the rainfall intensity occurred in the time interval (t-dt,t) and of the discharge at the basin outlet section occurred at time t, Q_t, becoming available, the implemented procedure searches for the optimal set of parameters, θ_t, assuming as initial set of parameters the one retrieved as optimal at the last time step, θ_{t-1}. The optimal set of parameters is defined as the one which minimizes (or maximizes, as appropriate) a function of model parameters named the objective function. The search for the optimal set of parameters, θ_t, is achieved by means of a constrained version of the Rosenbrock optimization algorithm (Kuester and Mize, 1973).

The characterization of the form of the objective function requires the specification of the estimation criterion. Some recent papers (Brath, 1989 and 1989a) have dealt with the problem of the presence of autocorrelation in model residuals, by comparing the OLS and AMLE performances in an adaptive framework. In this paper, the problem of heteroscedasticity is also accounted for. Accordingly, OLS, AMLE and HMLE criteria are used. Adaptive OLS calibration results in the minimization at each time step of the sum of the squared differences between the observed Q_i and computed discharge q_i, from the beginning of the flood event (i=1) to the actual time (i=n), according to Equation 8. In the case of AMLE and HMLE, the procedures described by Sorooshian and Dracup (1980) have been implemented in an adaptive framework, so retrieving the optimal values of the model parameters.

5. Forecasting Performances with OLS, AMLE and HMLE Adaptive Calibrations

At each time step, once the adaptive model calibration has been performed according to the estimation criterion selected, the forecasts for the requested lead-time D, q_{t+D}, are issued. In the forecasting step, the usual assumption of the availability of a perfect rainfall predictor has been taken: then, rainfall occurrences are supposed to be known up to the time t+D for which streamflow forecasts are issued. This allows a more correct analysis of forecasting procedure performances. Some applications are presented below, in order to analyze the performances of the forecasting procedure in the case of OLS, AMLE and HMLE adaptive calibrations.

The catchment considered is the Sieve drainage basin (830 kmq), a right-side tributary to the River Arno (Central Italy). The flood event under examination occurred in February, 1960. According to the data available, the model operates on an hourly basis. The discharge observed is compared in Figure 2 with the three hours ahead forecasts retrieved by OLS and AMLE adaptive calibrations; in Figure 3 with the three hours ahead forecasts retrieved by OLS and HMLE.

One can observe that the calibration according to the AMLE criterion is more effective than the OLS one in reproducing the recession limbs of the hydrograph. However, this does not represent an interesting feature for flood forecasting. In the remaining parts of the hydrograph, OLS and AMLE estimators do not lead to noticeably different results. Indeed, OLS estimation appears to yield more satisfactory performances in the region of the highest discharges. In fact in this region AMLE estimation results in a noticeable discharge overestimation and in a shifting of the time of peak.

With regard to the effects of residual heteroscedasticity, one can observe that the performances of the HMLE estimator are quite similar to those of the OLS one (Fig. 3). The latter appear to be a little more effective in forecasting the main rising limb of the hydrograph and the region around the peak. The superiority of both OLS and HMLE criteria with respect to the AMLE one is confirmed by the analysis of the root mean square error values of the discharge predictions reported in Table 1.

Increasing the forecasting lead time up to six hours (Figs 4 and 5), the performances of both OLS calibration and HMLE do not appear to deteriorate to any considerable extent (Fig. 5). These two estimation criteria still continue to perform quite similarly to each other, as confirmed by the root mean square values of the forecasted discharge (Table 1). Conversely, the AMLE calibration performances tend to become worse.

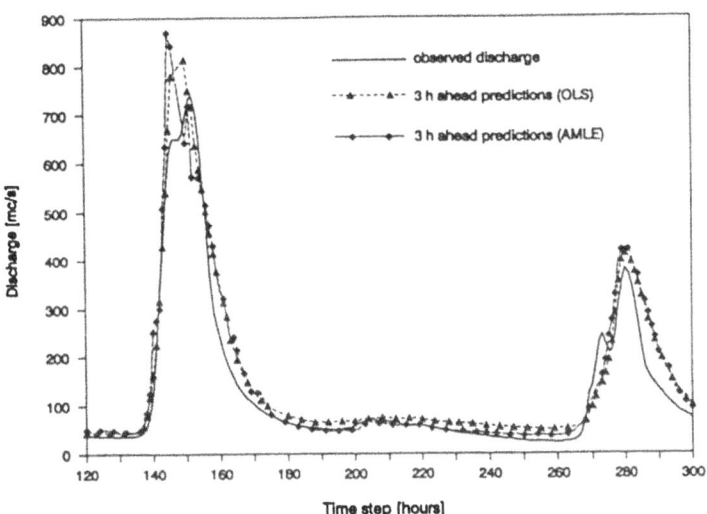

Figure 2. Observed discharge and three hours ahead streamflow predictions. Adaptive model calibration with both OLS and AMLE estimators. River Sieve (Italy).

In this regard, Figure 4 shows that AMLE is not capable of forecasting satisfactorily, in the region of the highest discharges. Indeed, its application leads to a shifting in time of the peak similar to the one obtained for a three hours lead-time but discharge overestimations in this region become very remarkable, thus leading to a notable increase in the root mean square error value.

TABLE 1. Root mean square error of the forecasted discharge [mc/s]

	OLS	HMLE	AMLE
3 h ahead	40.1	42.8	48.6
6 h ahead	41.5	43.6	60.5

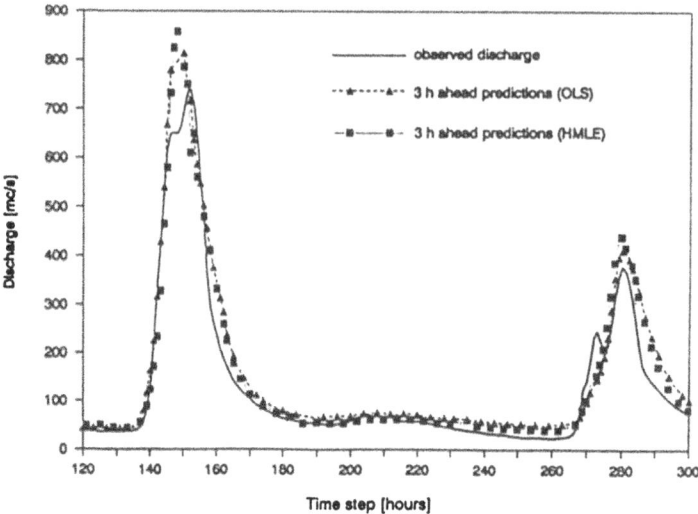

Figure 3. Observed discharge and three hours ahead streamflow predictions. Adaptive model calibration with both OLS and HMLE estimators. River Sieve (Italy).

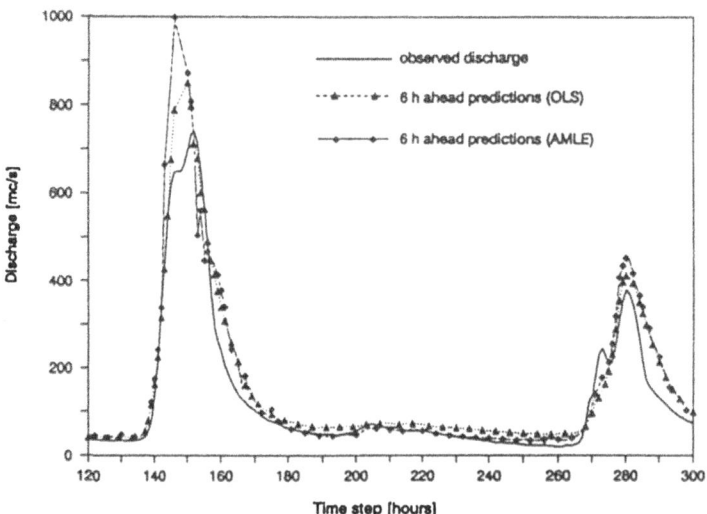

Figure 4. Observed discharge and six hours ahead streamflow predictions. Adaptive model calibration with both OLS and AMLE estimators. River Sieve (Italy).

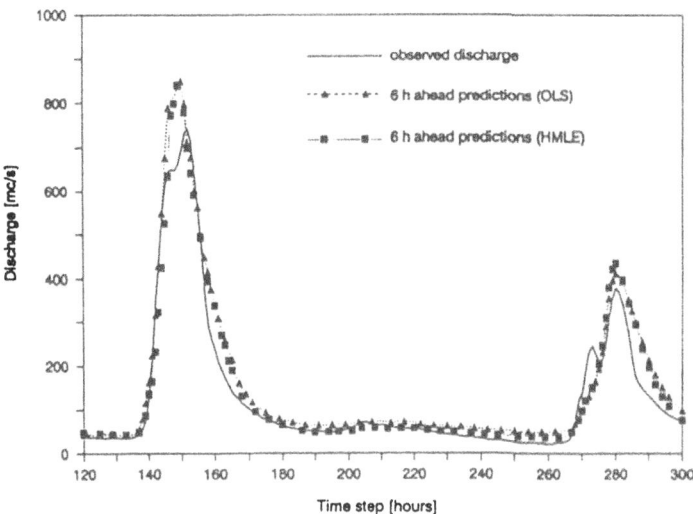

Figure 5. Observed discharge and six hours ahead streamflow predictions. Adaptive model calibration with both OLS and HMLE estimators. River Sieve (Italy).

Concerning the comparison of the performances of the three estimation criteria considered, some interesting issues arise from the analysis of the optimal sets of parameters retrieved from the automatic calibration. In this regard, Figure 6 shows the estimates of the time scale of surface runoff, K, according to OLS, AMLE and HMLE adaptive calibrations.

One can observe that the AMLE criterion results in a highly unstable pattern of the time scale parameter estimates, yielding a relevant variability even between two subsequent estimates. Probably, this high variability is due to the presence of several local minima in the response surface generated from the AMLE criterion. This would result in the convergence of the optimization algorithm onto different optimal sets of parameters, even in two subsequent time steps. The HMLE estimator shows similar behaviour, but the fluctuations of the retrieved parameter estimates, even if bigger than the AMLE ones at the beginning of the event, tend to damp down more quickly.

Conversely, the estimates of the time scale parameter K retrieved according to the OLS criterion fluctuate to a much less degree revealing the presence of a well conformed response surface. Similar considerations can be drawn from the analysis of the estimates of the shape parameter of the surface response, N, shown in Figure 7.

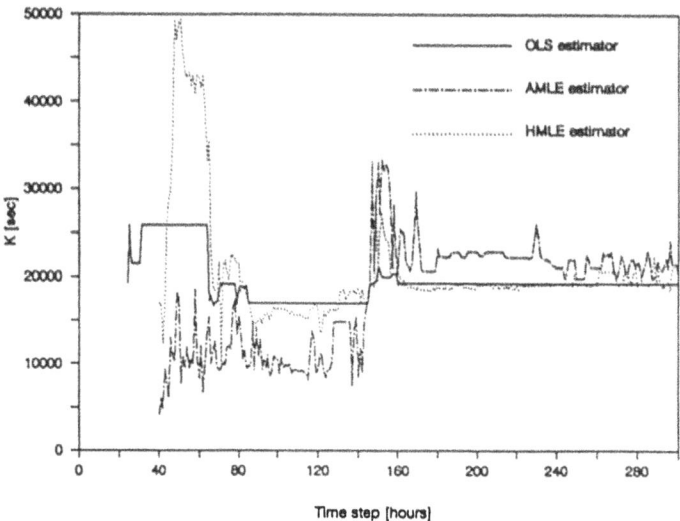

Figure 6. Estimates of the time scale parameter of the surface response, K, retrieved from adaptive calibration according to OLS, AMLE and HMLE criteria. River Sieve (Italy).

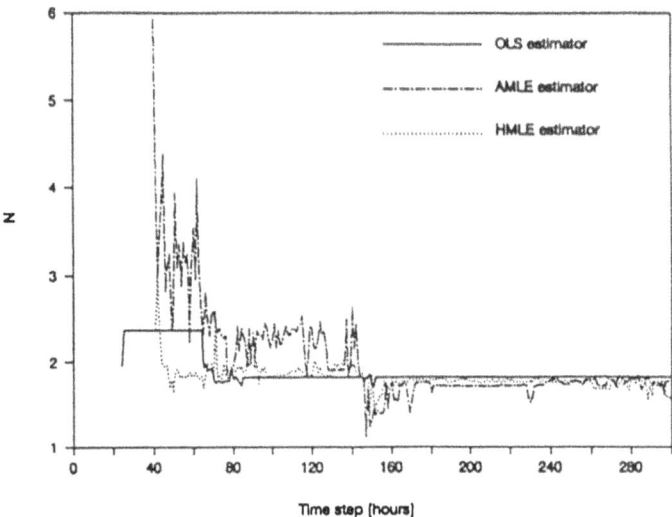

Figure 7. Estimates of the shape parameter of the surface response, N, retrieved from adaptive calibration according to OLS, AMLE and HMLE criteria. River Sieve (Italy).

Finally, the conclusion that can be drawn is that, although a strong autocorrelation between model residuals is present in the case examined (Brath, 1989), the AMLE calibration criterion shows the worst performances. Indeed, the performances of OLS and HMLE criteria are quite comparable. However, the former appears to be a little more effective with regard to a global index, as the mean square error, and it also seems to provide more reliable forecasts in the region of the main rising limb of the hydrograph. Moreover, OLS calibration leads to much more stable parameter values with respect to the HMLE procedure. With regard to the computational requirements involved, the model was used in an adaptive framework with an AT-80386 personal computer working at 25 MHz. The time taken for each time step by the adaptive calibration is very short in the case of OLS calibration (always less than two minutes for a generic time step), while both AMLE and HMLE estimation procedures generally result in a much more time-consuming procedure. This represents another valuable advantage for the OLS criterion over the HMLE one. The importance of the availability of a quick procedure for issuing streamflow forecasts should not indeed be overlooked in view of the supporting fast decision-making as required for flash-flood prone areas.

6. Conclusions

In the context of the real-time flood forecasting, a conceptual approach to rainfall-runoff transformation was shown to offer several advantages with respect to a black-box one. In spite of this, some major problems must be solved in order to cast conceptual models satisfactorily in a real-time framework. Adaptiveness, which represents a very valuable feature in real-time model use, can hardly be achieved in the case of the most used conceptual models, because of their cumbersome structure and redundant parametrization. In this regard, a parsimonious conceptual model has been presented, which was found to be particularly suitable for a fully on-line use.

The issues arising in conceptual model calibration have been outlined by using a maximum likelihood framework for parameter estimation. The validity of the assumptions of the commonly used ordinary least squares (OLS) estimation criterion has been analyzed. The presence of a strong autocorrelation between model residuals has been recognized to represent a serious violation of the abovementioned assumptions (see e.g. Brath, 1989). Then, the use in a real-time context of maximum likelihood estimators suited for the case of autocorrelated errors (AMLE) has been taken into account.

Moreover, the paper deals with the effect of lack of homoscedasticity in model residuals, which leads to another serious violation of the OLS assumptions. Indeed, the problem of heteroscedasticity is very frequent, because it is mainly due to the commonly inhomogeneous nature of errors in streamflow measurements. In order to take into account its effects, a maximum likelihood estimator suited to heteroscedastic errors (HMLE) has been considered in addition to the OLS and AMLE estimation criteria.

The applications referred to concern the use, in an adaptive framework, of OLS, AMLE and HMLE estimation criteria. Despite the presence of a strong autocorrelation structure in model residuals, the forecasting procedure based on the OLS criterion was found to be much more reliable than the AMLE one, especially in the case of high lead-times.

With regard to the effects of error heteroscedasticity, HMLE performances were found to be quite similar to the OLS ones; however, the adaptive OLS calibration performed more satisfactorily than the HMLE one in forecasting the rising limbs. Moreover, the OLS provided much more stable parameter estimates, whereas the HMLE criterion could result in a highly unstable behaviour of the model parameters. Finally, it must be stressed that both maximum likelihood criteria considered generally result in a much more time-consuming procedure with respect to the OLS one.

Indeed, because reliability and quickness are invaluable features for operational use, the choice of an OLS estimation criterion was recognized to be the most suitable for adaptive real-time flood forecasting.

7. References

Bacchi, B., Brath, A., Burlando, P. and Mancini, M. (1988) 'Application of a lumped conceptual model to real-time flashflood forecasting', Proc. 19th Conference on Modeling and Simulation, University of Pittsburgh, pp 1527-1534.

Backie, J.R. and Eeles, C.W. (1985) 'Lumped catchment models', in: M.G. Anderson and T.P. Burt (Eds), Hydrological Forecasting, J.Wiley and Sons, Chichester, UK.

Brath, A. (1989) 'Conceptual models for real-time flood forecasting. Adaptive calibration, uncertainty analysis, sensitivity to rainfall predictions', Ph.D. Thesis (in Italian), Politecnico di Milano, Milan, Italy.

Brath, A. (1989a) 'Real-time calibration of a conceptual rainfall-runoff model using least squares and maximum likelihood approaches', Proc. 20th Annual Conference on Modeling and Simulation, 4-5 May, Pittsburgh, USA, pp 1527-1534.

Chow, G.C. (1983) Econometrics, McGraw Hill, New York.

Clarke, R.T. (1973) 'A review of some mathematical models used in hydrology, with observations on their calibration and use', Journal of Hydrology, 19, pp 1-20.

Draper, N.R. and Smith, H. (1981) Applied Regression Analysis, 2nd Ed., J. Wiley and Sons, New York.

Hendrickson, J.D., Sorooshian, S. and Brazil, L.E. (1988) 'Comparison of Newton-type and direct search algorithms for calibration of conceptual rainfall-runoff models', Water Resources Research, 24(5), pp 691-700.

Ibbit, R.P. and O'Donnell, T.E. (1971) 'Fitting methods for conceptual catchment models', Journal of Hydrologic Division ASCE, HY9, pp 1331-1342.

Ibbit, R.P. and O'Donnell, T.E. (1974) 'Designing conceptual catchment models for automatic fitting methods', in Mathematical Models in Hydrology, Vol. 2, IAHS Publ. N.101, pp 461-475.

Johnston, P.R. and Pilgrim, D.H. (1973) 'Optimizing the parameters of a rainfall-runoff model', in Hydrology Papers 1973, The Institution of Engineers, Sydney, Australia, pp 173-180.

Johnson, P.R. and Pilgrim, D.H. (1976) 'Parameter optimization for watershed models', Water Resources Research, 12(3), pp 477-486.

Kitanidis, P.K. and Bras, R.L. (1980) 'Real-time forecasting with a conceptual hydrologic model, 1. Analysis of uncertainty, 2. Application and results', Water Resources Research, 16(6), pp 1025-1044.

Kuester, J.L. and Mize, J.H. (1973) Optimization Techniques with Fortran, McGraw Hill, New York.

Mood, A.M., Graybill, F.A. and Boes, D.C. (1974) Introduction to the Theory of Statistics, McGraw Hill, New York.

Nash, J.E. and Sutcliffe, J.V. (1970) 'River flow forecasting through conceptual models. 1: A discussion of principles', Journal of Hydrology, 10, pp 282-290.

O'Connell, P.E. and Clarke, R.T. (1981) 'Adaptive hydrological forecasting - a review', Hydrol. Sci. Bull., 26(2), pp 179-205.

Pickup, G. (1977) 'Testing the efficiencies of algorithms and strategies for automatic calibration of rainfall-runoff models', Hydrol. Sci. Bull., 22(2), pp 257-274.

Potter, K.W. and Walker, J.F. (1981) 'A model of discontinuous measurement error and its effects on the probability distribution of flood discharge measurements', Water Resources Research, 17(5), pp 1505-1509.

Sorooshian, S. and Dracup, J.A. (1980) 'Stochastic parameter estimation procedures for hydrologic rainfall-runoff models: correlated and heteroscedastic error cases', Water Resources Research, 16(2), pp 430-442.

Sorooshian, S. (1983) 'Surface water hydrology: on-line estimation', Rev. Geophys. Space Physics, 21(3), pp 706-721.

Sorooshian, S., Gupta, V.K. and Fulton, J.L. (1983) 'Evaluation of maximum likelihood parameter estimation techniques for conceptual rainfall-runoff models: influence of calibration data variability and length on model credibility', Water Resources Research, 19(1), pp 251-259.

Sorooshian, S. and Gupta, V.K. (1983) 'Automatic calibration of conceptual rainfall-runoff models: the question of parameter observability and uniqueness', Water Resources Research, 19(1), pp 260-268.

Sorooshian, S. and Gupta, V.K. (1985) 'The analysis of structural identifiability: theory and application to conceptual rainfall-runoff models', Water Resources Research, 21(4), pp 487-495.

Sorooshian, S. (1991) 'Parameter estimation, model identification, and model validation: conceptual-type models', in D.S. Bowles and P.E. O'Connell (Eds), Recent Advances in the Modeling of Hydrological Systems, pp 443-467, Kluwer.

Troutman, B.M. (1985) 'Errors and parameter estimation in precipitation-runoff modeling. 1: theory, 2: case study', Water Resources Research, 21(8), pp 1195-1222.

Wood, E.F. and O'Connell, P.E. (1985) 'Real-time forecasting', in: M.G. Anderson and T.P. Burt (Eds), Hydrological Forecasting, pp 505-558, J. Wiley and Sons, Chichester, UK.

LINEAR STOCHASTIC NON-STATIONARY RAINFALL-RUNOFF MODELLING AND FORECASTING

P.A. TROCH; F.P. DE TROCH & J. VAN HYFTE
Laboratory of Hydrology and Water Management
Ghent University
Coupure Links 653
B-9000 Ghent
Belgium

ABSTRACT. Many researchers in the field of real-time flood forecasting recognize the time-dependent nature of the rainfall-runoff relationship. However, for isolated storm events, the effects of this time-variant behaviour are very often neglected, leading to the definition of linear time-invariant systems or non-linear time-invariant systems. In the latter case, the non-linearity can be introduced using some form of rainfall separation (e.g. threshold parameters) or as an additional deterministic signal (e.g. Catchment Wetness Index). More advanced modelling techniques to deal with non-linearity are Volterra and Wiener series expansion (Napiorkowski, 1986).

This paper deals with linear time-variant (adaptive) modelling and forecasting. This modelling procedure is based on an on-line identification of the model parameters using the recursive instrumental variable estimator. The variation of the parameters is modelled using a random walk model. Some objective hydrological criteria for the evaluation of forecasting performance are introduced. This modelling technique is then compared, using real-world data from the catchment of the River Vesdre, with linear time-invariant modelling and forecasting. It can be concluded that, in general, adaptive modelling can improve the real-time forecasting performance within the linear framework.

1. Linear Modelling and Forecasting

The most general framework for linear forecasting is afforded by the non-stationary state-space discrete-time model:

$$x_{k+1} = A_k x_k + B_k u_k + D_k w_k \tag{1}$$

$$y_k = C_k x_k + v_k \tag{2}$$

where x_k represents the state vector (dimension n) of the system under study. The sequences $\{y_k\}$ and $\{u_k\}$ are the observed output and input vectors with dimension p and m. Usually the input and output series are scalar $(m = p = 1)$. A_k, B_k and C_k represent time-varying matrices of

283

J.B. Marco et al. (eds.),
Stochastic Hydrology and its Use in Water Resources Systems Simulation and Optimization, 283–292.
© 1993 *Kluwer Academic Publishers.*

appropriate dimension. The disturbance w_k and the measurement noise v_k are zero mean, statistically independent, white noise vectors with possible time-variable covariance matrices W_k and V_k. To prevent singularity, V_k is assumed to be positive definite for all k, so that no measurement is fully deterministic.

In his 1960 paper, Kalman solved the problem of state variable estimation since he assumed that the model parameters are known, but he acknowledged the problem of parameter estimation. Methods for joint recursive estimation of the states and parameters have since become available. The extended Kalman filter (EKF) is one relatively straightforward approach in which the state vector x_k is augmented to include an unknown parameter vector θ_k which includes all the unknown elements in the model matrices A_k, B_k and C_k. Estimation then proceeds using a Kalman filter-like algorithm obtained by linearizing the now non-linear relationship about the current estimates at each recursive step. We refer to Wood and O'Connell (1985) for a detailed description of the extended Kalman filter algorithms applied to hydrological time series. Young (1984) states that, although the method has been applied relatively successfully to the modelling of various real dynamic systems, it does not always work satisfactorily. It is known to have poor statistical efficiency, in the sense that its parameter estimates may have rather high error variance when compared with the theoretically possible minimum variance. Superior results in this sense can be obtained by considering the problem from a maximum-likelihood standpoint (Mehra and Tyler, 1973; Aström and Kallström, 1973).

The major difficulty with the estimation of parameters in the state-space model arises from its relative complexity as a description when viewed in parametric estimation terms since the problem is clearly non-linear (Young, 1984). What we would prefer is a representation in which the unknown parameters are associated with the variables measured in a linear model. This representation is far from easy to obtain in the time-variant case. The derivation of the observation space transfer function in the univariate time-invariant case is quite straightforward. We refer to Marco (1989) for a review of univariate modelling techniques.

The most general time series model is usually referred to as the transfer function noise (TFN) model or Box-Jenkins model (Box and Jenkins, 1970):

$$y_k = \frac{B(z^{-1})}{A(z^{-1})} u_{k-d} + \frac{D(z^{-1})}{C(z^{-1})} e_k \tag{3}$$

where $A(z^{-1})$, $B(z^{-1})$, $C(z^{-1})$, $D(z^{-1})$ are polynomials of appropriate order and d represents the dead time of the system. The sequence $\{e_k\}$ represents white noise with variance σ^2. Since it is clear from Equation 3 that neither the system model nor the noise model can be written in a regression relationship, the recursive least squares analysis is generally not applicable to the estimation of the model parameters. This problem can be overcome by the generation of an instrumental variable (IV) vector $\tilde{x}_k$ at each recursive step. The instrumental variable ξ_i is chosen to be independent of the noise inputs v_i and w_i. The statistical efficiency of the solution is highly dependent upon the degree of correlation between ξ_k and the deterministic output of the system (Durbin, 1954). The major problem with the IV method is the generation of suitable instrumental variables. Young (1965) suggests to use an auxiliary model of the process to generate ξ_i. The IV vector then takes the form:

$$\tilde{x}^T_k = [-\xi_{k-1} \ldots - \xi_{k-n}; \; u_k \ldots u_{k-n}] \tag{4}$$

where ξ_{k-i} can be calculated using the following auxiliary model:

$$\tilde{A}(z^{-1})\xi_k = B(z^{-1})u_k \tag{5}$$

with $\tilde{A}(z^{-1})$ and $B(z^{-1})$ polynomials with parameters chosen in some reasonable manner. An on-line IV procedure can be obtained by utilizing the recursive solution to the IV equations and then updating the auxiliary model continuously on the basis of these recursive estimates:

$$\theta_k = \theta_{k-1} - \hat{K}_k[z^T_k \; \theta_{k-1} - y_k] \tag{6}$$

$$\hat{K}_k = P_{k-1} \; \hat{x}_k \; [\delta_k + z^T_k \; P_{k-1}\hat{x}_k]^{-1} \tag{7}$$

$$P_k = \frac{1}{\delta_k} \{P_{k-1} - P_{k-1} \; \hat{x}_k \; [\delta_k + z^T_k \; P_{k-1}\hat{x}_k]^{-1} \; z^T_k \; P_{k-1}\} \tag{8}$$

with the scalar δ_k defined as:

$$\delta_k = \lambda_0 \; \delta_{k-1} + (1-\lambda_0)\delta \tag{9}$$

and $\delta = 1.0$. λ_0 and δ_0 are the additional parameters to be chosen by the analyst. Equation 9 is introduced into the IV algorithms in order to improve convergence characteristics. A modified recursive approximate maximum likelihood (AML) procedure, suggested by Panuska (1968), can be used to estimate the parameters in the ARMA structure noise model, based on the estimation of the noise sequence.

2.   Some Comments on Structure Characterization

One of the steps in the model building process is the specification of the structure, which in this case is the order of the polynomials of the deterministic and stochastic transfer function as well as the size of the dead time. In this case the structure has to be inferred from the data. An inductive structure characterization method is therefore needed. The classical approach was presented by Box and Jenkins (1970) and consists of partly systematic, partly heuristic techniques. This structure character- ization method splits the problem into two parts. First, the structure of the deterministic part of the model is determined. Then, the details of the structure of the noise model are inferred. Proof and description of this procedure can be found in Box and Jenkins (1970). In the last two decades, an attempt has been made to derive objective structure characterization methods based on statistical or information theoretic grounds. The optimum structure is based on a certain criterion. At the present, the class of methods which is most often advocated is based on a criterion which trades off fit against complexity. For each candidate model structure the following criterion is computed:

$$\Lambda_i = N \log 1/N \sum_{k=1}^{N} \hat{e}_k^2 + f(n_i, N) \tag{10}$$

where N is the number of data points in the calibration set, n_i is the total number of unknown parameters and $\hat{e}_k$ is the one-step-ahead prediction error at time k. The Bayesian information criterion (BIC) is obtained by choosing:

$$f(n_i, N) = n_i \log N \tag{11}$$

This criterion was selected to determine the model structure for the rainfall-runoff process in the catchment of the River Vesdre (see below). In Young (1986) another method is presented based on two statistics:

$$R_d^2 = 1 - \frac{\sum_{k}^{N} \hat{e}_k^2}{\sum_{k}^{N} (y_k - \bar{y})^2} \tag{12}$$

$$EVN = \frac{1}{n_i} \sum_{k}^{n_i} \hat{p}^*_{kk} \tag{13}$$

where $\bar{y}$ is the mean value of y_k and $\hat{p}^*_{kk}$ is the kth diagonal element of the $\hat{P}^*_k = \hat{\sigma}^2 P_k$ matrix with $\hat{\sigma}^2$ the estimated noise variance. R_d^2, the coefficient of determination, measures the model fit. EVN, the error variance norm, is an indication of the overall variance of the parameter estimates in the n_i-th order model. It is a sensitive indicator of over-parameterization. This is clearly demonstrated in Figure 1.

This figure is obtained as follows: using a known TF(2,2) model, a random input sequence was selected and an output sequence was generated. The input and output time series were then used to identify different TF models, using the recursive IV algorithm described above. For each candidate model the EVN was then calculated. In Figure 1 the x-axis indicates the model structure (1.2 means 1st order A polynomial and 2nd order B polynomial). In practice, the analyst monitors both R_d^2 and EVN and chooses the model which has the best combination of the two statistics: usually R_d^2 will reach a plateau level, with little further increase when increasing the model order, while EVN should have a low value compared to that obtained for higher order models.

3. Parametric Variations

Suppose that, instead of assuming constant parameters we assume that the parameters vary in a manner than can be described by a Gauss-Markov stochastic difference equation:

$$\theta_{k+1} = \Phi\theta_k + \Gamma q_k \tag{14}$$

where Φ is an nxn transition matrix, and Γ is an nxm input matrix, both of which may be time variable; and q_k is an mxl white noise vector of serially-independent random variables with zero mean and covariance matrix Q_p. The simplest example of this model is the random walk model:

$$\theta_{k+1} = \theta_k + q_k \tag{15}$$

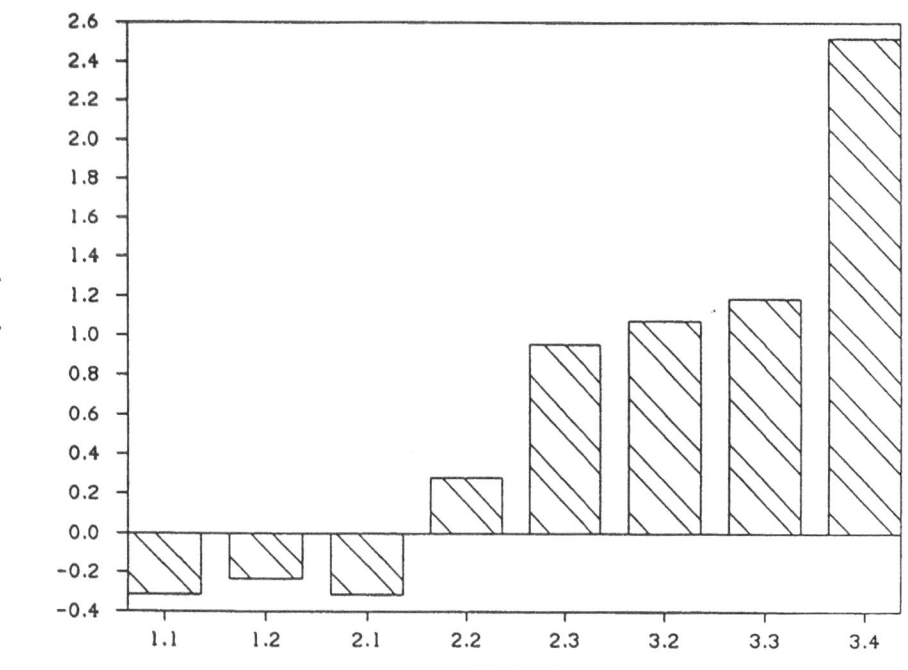

Evolution of ln(EVN)

Figure 1. Evolution of the error variance norm within the range of plausible model structures.

In the case of the random walk model of the parametric variations the IV algorithm can be modified to become (Young, 1986):

$$\theta_k = \theta_{k-1} - \hat{k}_k [z^T_k \theta_{k-1} - y_k] \tag{16}$$

$$\hat{k}_k = P_{k/k-1} \hat{x}_k [1 + z^T_k P_{k/k-1} \hat{x}_k]^{-1} \tag{17}$$

$$P_{k/k-1} = P_{k-1} + Q \tag{18}$$

$$P_k = P_{k/k-1} - P_{k/k-1} \hat{x}_k [1 + z^T_k P_{k/k-1} \hat{x}_k]^{-1} z^T_k P_{k/k-1} \tag{19}$$

Here Q is a $(2n+1)$ x $(2n+1)$ matrix which allows for possible parametric variations. Although the derivation of these algorithms is quite heuristic, Young (1984) finds that they make good sense and can be further justified in statistical terms.

4. Measures of Forecasting Performance

The most commonly used statistic for comparing models with respect to forecasting performance is the average maximum-step-ahead forecasting precision $\hat{\alpha}_d{}^2$:

$$\hat{\sigma}_d{}^2 = \frac{1}{N-n_0} \sum_{k=n_0+1}^{N} (y_k - \hat{y}_{k/k-d})^2 \tag{20}$$

$$n_0 = \max[(n_a + n_c), (n_b + n_c) + d]$$

where n_a, n_b, n_c represent the order of the polynomials $A(z^{-1})$, $B(z^{-1})$, $C(z^{-1})$, respectively. Parameter d represents the dead time of the transfer function model while N is the number of data points during the flood event. Parameter $\hat{y}_{k/k-d}$ represents the d step-ahead forecast of y_k, the observed discharge at time k. The maximum-step-ahead forecasting precision gives an indication of the global performance of the model during a flood event. This can be misleading for evaluating flood forecasting models since the main purpose of these models is to forecast accurately the time-to-peak, the maximum value of the observed discharge/water level and the prediction of the rising limb of the hydrograph. These hydrologically important characteristics are not explicitly taken into account in the statistic represented in Equation 20.

In order to be able to evaluate objectively the forecasting performance of the time-invariant and the time-variant rainfall-runoff model, other statistics are needed:

a. Rising limb maximum-step-ahead forecasting precision:

$$\hat{\sigma}^2{}_{d.[N1,N2]} = \frac{1}{N_2-N_1} \sum_{k=N_1+1}^{N_2} (y_k - \hat{y}_{k|k-d})^2 \tag{21}$$

where $[N_1,N_2]$ represents the rising limb interval with the following characteristics:

$$N_1 \in [1,N]; \quad N_2 \in [1,N]$$

$$N_2 > N_1$$

with $N_2 \Delta t = t_p$ (time-to-peak) and $N_1 \Delta t = t_0$ (time at which hydrograph starts to rise). Δt represents the time interval between successive sampling.

b. Normalized peak error

$$\Psi = \frac{|y_{max}-\hat{y}_{max}|}{y_{max}} \tag{22}$$

with y_{max} the maximum observed discharge and $\hat{y}_{max}$ the maximum forecasted discharge.

c. Time-to-peak error

$$\alpha = |t_p - \hat{t}_p| \tag{23}$$

with t_p the time-to-peak and $\hat{t}_p$ the forecasted time-to-peak.

These three additional statistics allow us to compare modelling techniques with respect to important characteristics of the hydrograph.

5. The Rainfall-runoff Process for the River Vesdre

The real-time flood forecasting model for the River Meuse (Belgium), developed by the Laboratory of Hydrology and Water Management of Ghent University, consists of two major parts: the hydrological module and the hydraulic module. The hydrological module is based on a linear time-invariant representation of the system under study (Equation 3). For the most important subcatchments of the River Meuse up to Liege these linear time-invariant models are identified using time-series analysis. The forecasting performance is in most cases satisfactory from the standpoint of operational use (Troch, Spriet and De Troch, 1988).

It is, however, recognized by the authors that the basic assumption made throughout the modelling procedure, namely that the rainfall-runoff process is linear and time-invariant, is not always valid. A more conceptual justified assumption would be that the process is inherent time-variant (see introduction). Therefore, the non-stationary time-series analysis presented here can be used to identify this time-varying behaviour.

The technique of an additive Q matrix is applied to the catchment of the River Vesdre, one of the major tributaries of the Meuse. The structure of the TFN model was deduced from the estimated cross-correlation function between pre-whitened rainfall data from selected storms and the coresponding filtered runoff data (Box and Jenkins, 1970). The structure of the noise model was obtained using an objective BIC search.

The time-invariant TFN model for the Vesdre subcatchment (drainage basin: 677 km^2 at Chaudfontaine) has the following structure: first order A polynomial, fifth order B polynomial with dead time d=4, second order C polynomial and zero order D polynomial. The parameters were estimated using the recursive-iterative IV-AML procedure described in Young (1984), based on a calibration set of flood events.

Table 1 summarizes a forecasting performance study for the time-invariant and the time-variant model, based on eight storm events not included in the calibration set. Forecasting performance is measured using the statistics introduced in the previous section (1: $\hat{\sigma}_d[m^3/s]$; 2: $\hat{\sigma}_{d[N1,N2]}[m^3/s]$; 3: $\psi[\%]$; 4: $\alpha[hrs]$). Each of the eight storm events that are selected for this study (except event No. 8) contain more than one peak. Based on the criteria set up for N_1 and N_2, different rising limbs within each storm event can be selected. These different periods are indicated in the table using the number of the storm event followed by a capital letter starting from A.

The following conclusions can be drawn from the table: the average maximum-step-ahead forecasting performance (1) for the time-variant model is significantly better than for the time-invariant model; the rising limb of the hydrograph (2) is also more efficiently forecasted (this is further illustrated in Figure 2); the normalized peak error (3) is much greater for the time-variant case (in general an overestimation is observed) whereas the forecasted time-to-peak (4) is, in most cases, significantly improved using an adaptive modelling approach.

The Q matrix has not been taken constant for the eight selected storm events. The value of the diagonal elements is chosen in order to optimize forecasting performance. For events 1, 2, 5 and 6, Q is chosen to be equal to 0.01*I (I: unity matrix); for events 4 and 7, Q = 0.001*I and for events 3 and 8, Q = 0.0001*I. Further research is needed in order to develop an objective selection criterion which is useful in real-time.

TABLE 1.   Comparison between the forecasting performance of the time-invariant model and the time-variant model for the River Vesdre.

Event	Time-invariant				Time-variant			
	1	2	3	4	1	2	3	4
1	3.5				1.0			
1A		3.1	13.6	1		1.2	12.0	2
1B		3.7	8.5	3		2.1	4.3	1
2	4.6				3.2			
2A		2.5	0.7	3		1.1	13.6	1
2B		3.7	17.7	4		3.2	35.4	1
2C		13.2	2.3	3		5.3	16.1	1
3	7.3				5.4			
3A		17.8	3.5	1		8.4	11.4	2
3B		15.4	39.2	2		9.8	48.1	0
4	12.9				7.9			
4A		30.4	3.7	2		11.1	21.3	0
4B		24.9	25.7	3		14.4	59.7	1
4C		10.8	0.8	3		4.5	8.5	1
5	2.4				1.4			
5A		4.8	2.7	3		2.6	8.6	1
5B		2.8	0.7	4		1.1	9.7	2
6	5.6				3.8			
6A		7.1	9.8	3		6.9	12.7	0
6B		9.2	5.1	3		4.5	3.6	1
7	2.4				1.5			
7A		3.5	14.0	2		2.2	11.3	0
7B		5.3	3.6	2		2.7	2.7	1
8	6.1				3.7			
8A		15.2	10.6	3		6.6	20.3	1

Figure 2 illustrates the benefits that can be obtained in real-time flood forecasting from adaptive modelling of the rainfall-runoff process. During flood forecasting it is very important that the rising limb of the hydrograph is forecasted efficiently. Using linear time-invariant models it is very often observed in practice that the predicted rising limb is shifted in time over a distance equal to the forecasting horizon or dead time. As demonstrated in Figure 2, this error can be reduced using on-line identification. However, the price that has to be paid for this improvement is overestimation of the maximum value of the hydrograph. One can argue that, from the standpoint of flood forecasting, overestimation of the peak is less harmful than underestimation.

Rising Limb Forecasting Performance

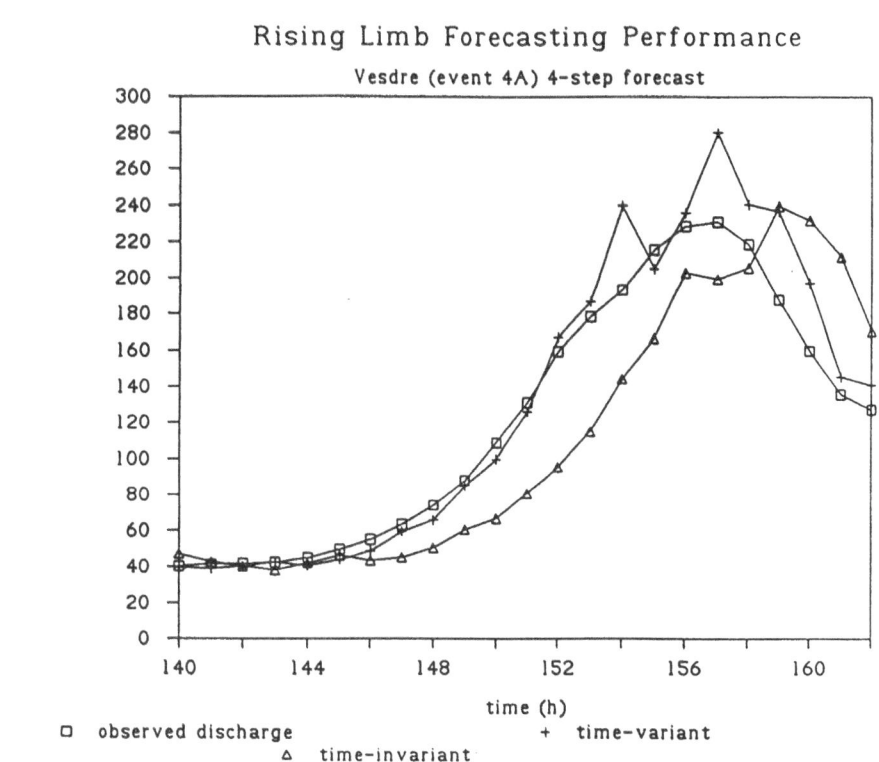

Figure 2. Rising limb forecasting performance for the time-invariant TFN model and for the time-variant TFN model.

6. Conclusions

It is shown that the on-line identification procedure presented in this paper can improve the real-time forecasting performance of linear stochastic rainfall-runoff models. Special attention is drawn to the ability of efficiently forecasting specific characteristics of the hydrograph, such as the rising limb and time-to-peak. It is observed that, in general, these hydrologically important characteristics are better forecasted in real-time using a time-variant model. The value of the peak of the hydrograph, however, is usually overestimated.

7. References

Aström, K.J. and Kallström, C.G. (1973) 'Application of system identification techniques to the determination of ship dynamics', in P. Eykhoff (Ed), Identification and System Parameter Estimation, Proc. of the 3rd IFAC Symp., The Hague, pp 415-424.

Box, G. and Jenkins, G. (1970) Time Series Analysis, Forecasting and Control, Holden-Day, San Francisco.

Durbin, J. (1954) 'Errors in variables', Ev. Int. Statist. Inst., 22, pp 23-32.

Kalman, R.E. (1960) 'A new approach to linear filtering and prediction problems', ASME trans., Journal Basic. Eng., 82-D, 35.

Kalman, R.E. and Bucy, R.S. (1961) 'New results in linear filtering and prediction theory', ASME trans., Journal Basic. Eng., 83-D, 95.

Marco, J.B. (1989) 'ARMAX and transfer function modeling in hydrology', in: Stochastic Hydrology in Water Resources Systems: Simulation and Optimization, NATO ASI, September 1989, Peñiscola, Spain.

Mehra, R.K. and Tyler, J.S. (1973) 'Case studies in aircraft parameter identification', in: P. Eykhoff (Ed), Identification and System Parameter Estimation, Proc. of the 3rd IFAC Symp., The Hague, 117.

Napiorkowski, J.J. (1986) 'Application of Volterra series to modelling of rainfall-runoff systems and flow in open channels', Hydrological Sciences Journal, 31(2), pp 187-203.

Panuska, V. (1968) 'A stochastic approximation method for the identification of linear systems using adaptive filtering', Proc. Joint Automatic Control Conf., Ann Arbor, Michigan.

Troch, P.A., Spriet, J.A. and De Troch, F.P. (1988) 'A methodology for real-time flood forecasting using stochastic rainfall-runoff modelling', in: Computer Methods and Water Resources: Computational Hydrology, D. Ouazar et al. (Eds), pp 243-255.

Young, P.C. (1965) 'Process parameter estimation and self-adaptive control', Proc. IFAC Symp., Teddington, P.H. Hammond (Ed), Theory of Self-adaptive Control Systems, Plenum Press, New York.

Young, P.C. (1984) Recursive Estimation and Time-series Analysis: An Introduction, Springer-Verlag, Berlin.

Young, P.C. (1986) 'Time-series methods and recursive estimation in hydrological systems analysis', in: D.A. Kraijenhof and J.R. Moll (Eds), River Flow Modelling and Forecasting.

Wood, E.F. and O'Connell, P.E. (1985) 'Real-time forecasting' (chapter 15), in: Anderson and Burt (Eds), Hydrological Forecasting, Wiley and Sons Ltd.

PART III

INVITED PAPERS ON WATER RESOURCES SYSTEMS
OPTIMIZATION AND SIMULATION

EXPLICIT STOCHASTIC OPTIMIZATION

R. HARBOE
Division of Water Resources Engineering
Asian Institute of Technology (AIT)
G.P.O. Box 2754
Bangkok 10501
Thailand

ABSTRACT. The paper includes the theory behind mathematical models (linear and dynamic programming) with explicit consideration of the stochasticity of river flows into a reservoir system. Three examples of application are presented in increasing order of complexity.

1. Introduction

In linear programming, the so-called linear decision rule was introduced in order to transform probabilistic constraints into deterministic ones, based on the probability distribution function of the inflows to a reservoir. The rule constitutes in itself an additional constraint to the system and therefore results show overdimensioning of reservoirs, a local optimum. Dynamic Programming models can use the complete probability distribution function of inflows and there are several ways to do it: probabilistic or stochastic approaches. Three examples will show the applicability of the model. Nevertheless, computational time is very high even for single reservoir systems with multiple purposes. The examples are taken from previous experiences by the author in Chile and the former F.R. of Germany.

2. The Linear-decision Rule Approach

When considering a multipurpose single reservoir, several constraints can be formulated in probabilistic form such as (Revelle et al., 1969):

$$P[C-S_t \geq v_i] \geq \alpha \tag{1}$$

where: P = probability
  C = capacity of reservoir
  S_t = amount of water in storage at the beginning of time period t $(t=1,\ldots,T)$
  v_i = flood control reservation for month type i $(i=1,\ldots,12)$

295

J.B. Marco et al. (eds.),
Stochastic Hydrology and its Use in Water Resources Systems Simulation and Optimization, 295–306.
© 1993 *Kluwer Academic Publishers.*

α = probability of satisfying the constraint on the left-hand side

This probabilistic constraint can be transformed into a deterministic one if we consider the following linear decision rule:

$$x_t = S_{t-1} + b_i \tag{2}$$

where: x_t = release from the reservoir in time period t
b_i = a new decision variable that can be positive or negative (parameter of the linear-decision rule)

The operation according to the rule is presented in Figure 1. If we consider r_t as the inflow during time period t, Equation 1 becomes (constraint of a normal LP model):

$$C + b_i \geq r_i(\alpha) + v_i \tag{3}$$

where: C and b_i are decision variables of an LP model
$r_i(\alpha) = 100.\alpha$ percentile of the probability distribution of inflow r_i in month type i

3. The Dynamic Programming Approach

Basically DP can consider the complete probability distribution of inflows in two ways:
- probabilistic model
- stochastic model (with or without forecasting)
The probabilistic problem form is formulated in the following way:

$$f_i(S_i) = \max_{x_i} E[b_i(S_i, X_i) + f_{i-1}(S_{i-1})] \tag{4}$$

where this recursive equation considers the expected value of returns as an objective function and:

$F_i(S_i)$ = state evaluation function
E = expected value over the inflows (random variable)
S_i = amount of water in the reservoir at the beginning of period i (state variable)
X_i = release from reservoir (decision variable)
b_i = return in time period i; a function of state and decision

The stochastic model has two alternatives:

$$f_i(S_i, Q_i) = \max_{x_i} E[b_i(S_i, X_i, Q_i) + f_{i-1}(S_{i-1}, Q_{i-1})] \tag{5}$$

where: Q_i = forecasted inflow for time period i (this inflow is assumed to be available for release in time period i)

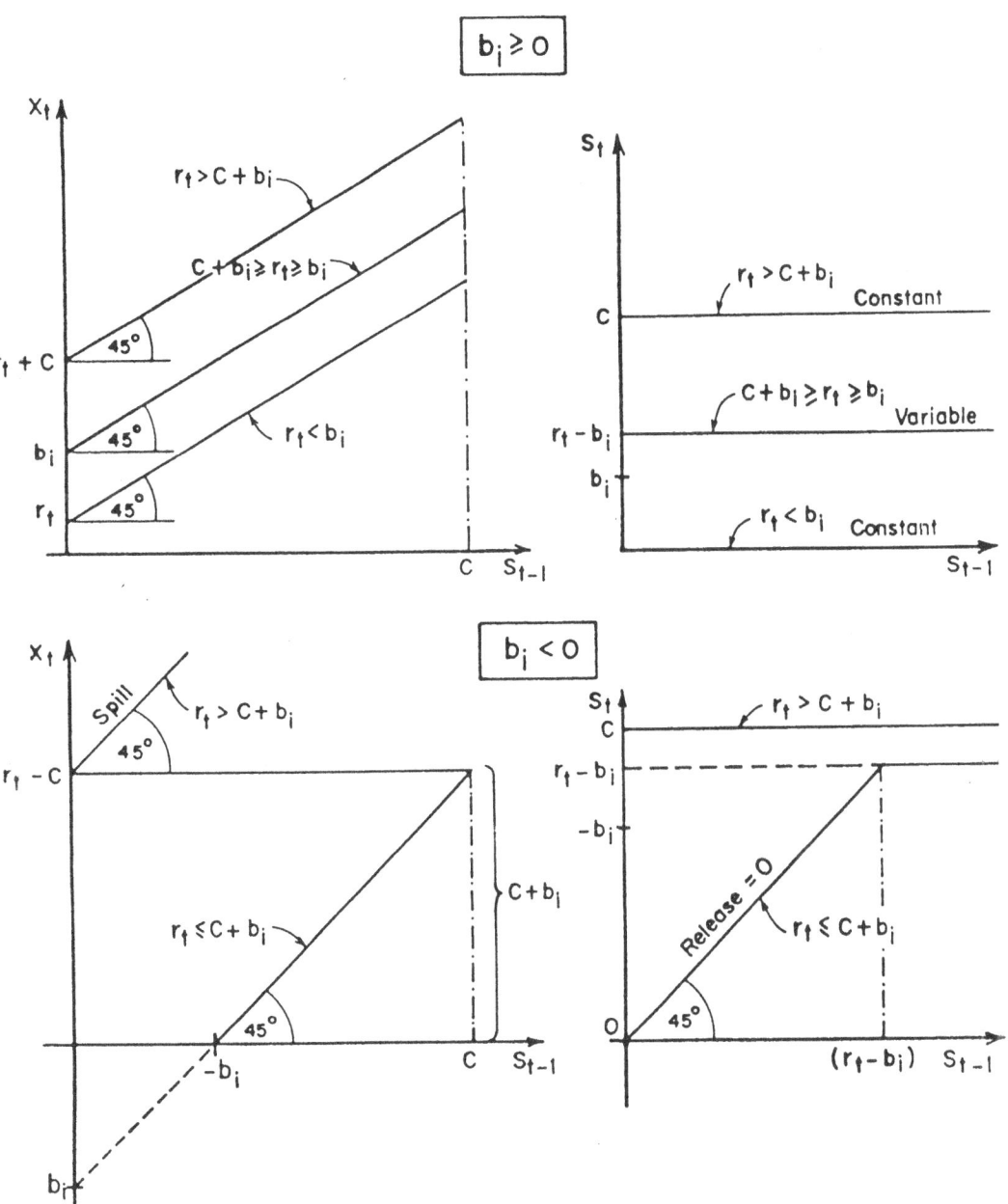

Figure 1. The Linear Decision Rule for $b_i \geq 0$ and $b_i < 0$.

or:

$$f_i(S_i, Q_{i+1}) = \max_{x_i} E[b_i(S_i, X_i, Q_{i+1}) + f_{i-1}(S_{i-1}, Q_i)] \qquad (6)$$

where: Q_{i+1} = inflow recorded in previous time period (known at the moment of making a decision for operation)

4. Example Applications

Three stochastic DP applications to the operation of reservoir systems will be included: the Maule-River Model (Chile), the Lech-River Model (former F.R. Germany), and the Wupper-River Model (former F.R. Germany). These models start with a simple application (probabilistic) on a half-yearly basis of a reservoir for irrigation and power production (Maule). Then a complex model mainly for hydropower considering serial correlation among inflows is presented (Lech, stochastic model). Finally a complex, multiobjective model is used for the operation of a multipurpose reservoir (Wupper). In all three cases, a reservoir operating rule which can be used in real-time operation of the system is obtained.

4.1. THE MAULE-RIVER MODEL

The system located upstream on the Maule-River has a reservoir which consists of a dam at the outlet of a large lake. The purposes are irrigation and hydropower production (Fig. 2). The DP Model (Harboe, 1967) has only two six-month time periods per year: a summer period (October-March) and a winter-period (April-September). Irrigation takes place only in summer and hydropower over the whole year. The probability distribution of inflows are shown in Table 1 based on the discretization of the historical probability distribution.

The objective function of this example was expressed in monetary terms for the benefits from irrigation, benefits from selling hydropower and costs (construction + OMR costs) of the reservoir. With the optimization according to Equation 4 (maximization of the expected value of the objective function), the following operating rule was obtained (Fig. 3). This operating rule can be recommended for the real-time operation of the system, because the release from the reservoir in each season is only a function of the reservoir storage at the beginning of that season.

4.2. THE LECH-RIVER MODEL

The Lech-System consisted of one head-reservoir (Forggensee) and fifteen hydropower plant (Fig. 4). The purposes of the system are: energy-production, flood control and low-flow augmentation (for water quality control, navigation and flora and fauna enhancement).

A stochastic model based on Equation 6 was developed, considering a physical objective function (maximization of the expected value of the on-peak energy production) (Harboe, 1976b). This model considers the serial correlation of monthly inflows in a simple way. For each month, three probability distributions are considered: one for low flow in the previous month, i.e.:

$$Q < \bar{Q} - \tfrac{1}{2}\sigma \tag{7}$$

where: Q = flow in previous month
          $\bar{Q}$ = average flow in previous month
          σ = standard deviation of flow in previous month:

one for medium flows in the previous month:

$$\bar{Q} - \tfrac{1}{2}\sigma \le Q \le \bar{Q} + \tfrac{1}{2}\sigma \tag{8}$$

and one for high flows in the previous month:

$$Q > \bar{Q} + \tfrac{1}{2}\sigma \tag{9}$$

TABLE 1. Discrete probability distribution of inflows

Winter Season		Summer Season	
Inflow Q in 10^6 m^3	Probability P	Inflow Q in 10^6 m^3	Probability P
150	0.419	150	0.233
200	0.488	200	0.186
250	0.093	250	0.139
		300	0.279
		350	0.163

An example of one of these conditional probability distributions is included in Figure 5. The three discretized probability distributions for February are presented in Table 2.

TABLE 2. Conditional discrete probability distribution of inflows in February

January Inflow Low ($Q_{JAN} < \bar{Q}_{JAN} - \tfrac{1}{2}\sigma$)		January Inflow Medium (Figure 5)		January Inflow High ($Q_{JAN} > Q_{JAN} + \tfrac{1}{2}\sigma$)	
Q_{PEB}	P(%)	Q_{PEB}	P(%)	Q_{PEB}	P(%)
16.3	24	17.9	27	18.9	22
20.1	41	24.8	46	30.4	45
25.8	35	38.9	27	44.2	33

The result of this model consists of three types of operating rules, one for each kind of flow in the previous month. As an example, Figure 6 presents all the operating rules (for all months) in the case of a medium flow occurring in the previous month. The small releases in the winter season in comparison to those of the summer can clearly be seen.

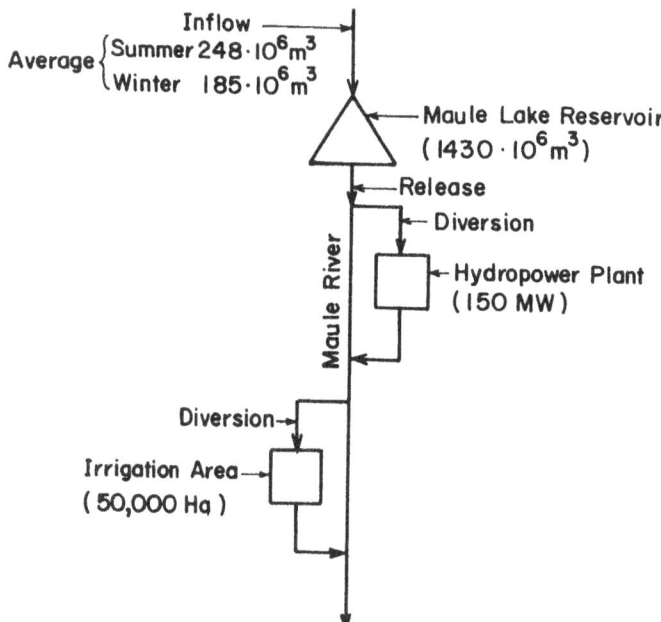

Figure 2. The Maule System

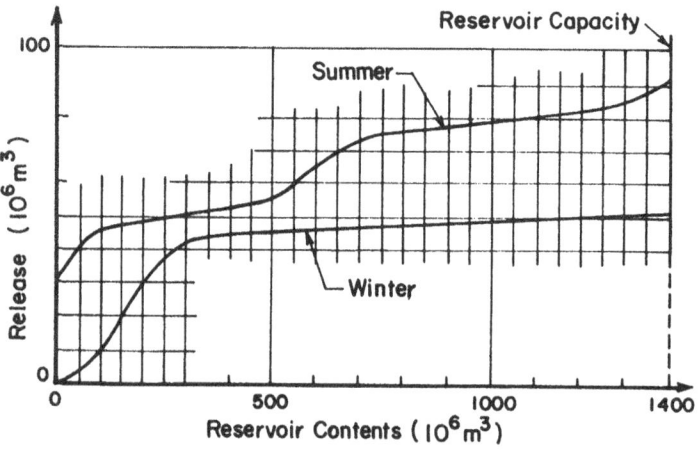

Figure 3. Reservoir release in one half-year period as a function of the reservoir contents at the beginning of the corresponding six months.

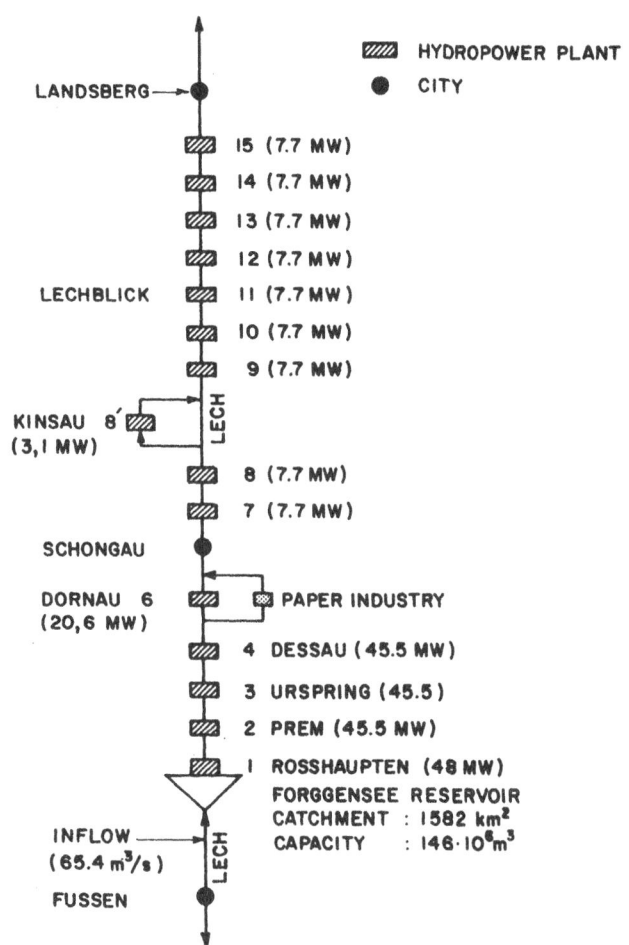

Figure 4. The Lech-River System

A test of these operating rules was made by comparing the energy generation with the historical generation. In the monthly simulation with optimal operating rules, the historical inflow record was used. Figure 7 shows this comparison. If we consider the development of the system over time (as new power units were added) the simulation yields 5% more energy over the 20 years. This positive value can be explained in part due to the high releases to leave space for flood emptying, and the low probability of occurrence of these floods. In all these years, the first power plant was operated for long terms with a low head. Furthermore, the quality of the historical operation is deteriorating over time as more power plant are added to the system.

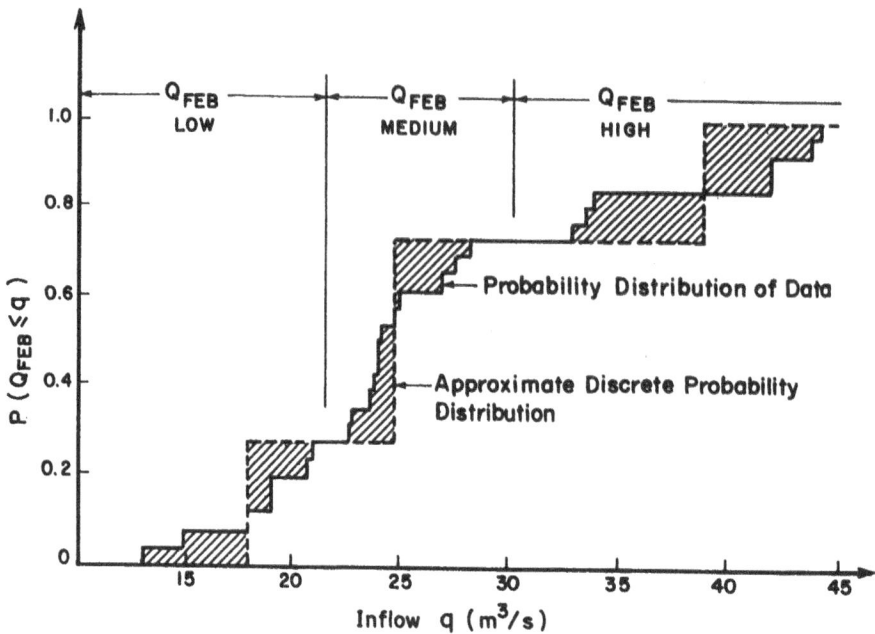

Figure 5. Conditional probability distribution of inflows in February (Q_{FEB}) for a medium flow in January.

$$(\bar{Q}_{JAN} - \tfrac{1}{2}\sigma \leq Q_{JAN} \leq \bar{Q}_{JAN} + \tfrac{1}{2}\sigma)$$

4.3. THE WUPPER-RIVER MODEL

The aim of the example is to show how to find a Pareto-optimal solution in a stochastic framework. A simple reservoir with multiple purposes (Wupper Reservoir) such as:
  - low flow augmentation
  - recreation
is considered to have multiple targets. The dynamic programming models (probabilistic and stochastic) are considered.

The probabilistic model has the following recursive equations (in discrete terms):

$$f_i(S_i) = \max_{x_i} \sum_{k=1}^{K} [b_i(S_i, I_{ik}, X_i) + f_{i-1}(S_{i-1,k})] \cdot P(I_{i,k}) \qquad (10)$$

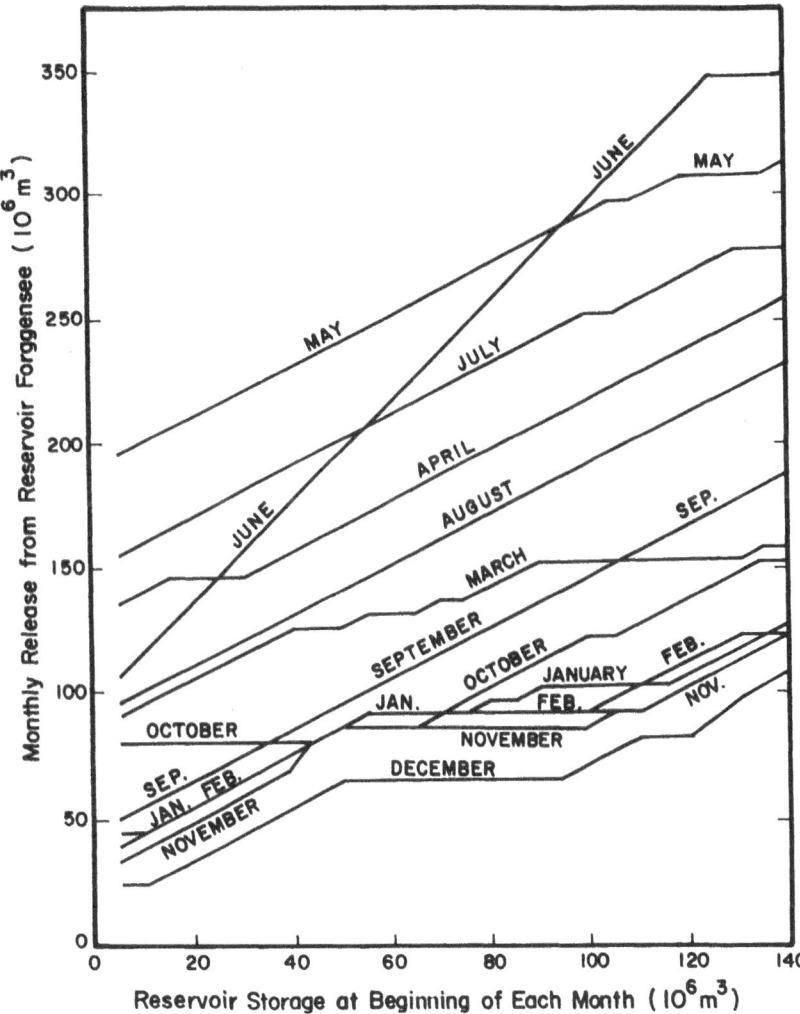

Figure 6. Monthly operating rules for Forggensee Reservoir if inflow in previous month was medium

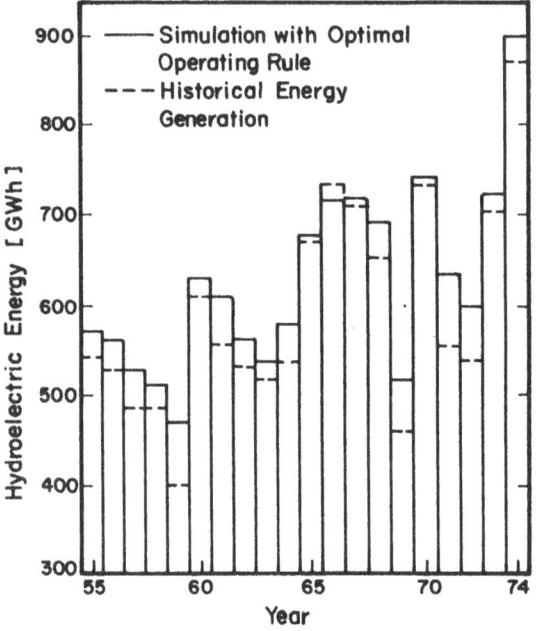

1955 - 1959 : HYDROPOWER PLANT   I   and 7-15
1960 - 1970 : HYDROPOWER PLANT   1,6,8´and 7-15
1971 - 1974 : HYDROPOWER PLANT 1-4,6,8´and 7-15

Figure 7. Comparison between historical and simulated energy generated with optimal operating rules

where: $I_{i,k}$ = one of K discrete inflows in month i

$$S_{i-1,k} = S_i + I_{i,k} - X_i \tag{11}$$

$P(I_{i,k})$ = probability of occurrence of flow $I_{i,k}$

b_i = net benefits function

The stochastic model has the following recursive equation:

$$f_i(S_i, I_{i+1}) = \max_{x_i} \sum_{k=1}^{K} [b_i(S_i, I_{i,k}, X_i) + f_{i-1}(S_{i-1,k} I_{i,k})] \cdot P[I_{i,k}/I_{i+1}] \tag{10}$$

where: $P[I_{i,k}/I_{i+1}]$ =   conditional probability of flow $I_{i,k}$ if flow I_{i+1} occurred in the previous time interval

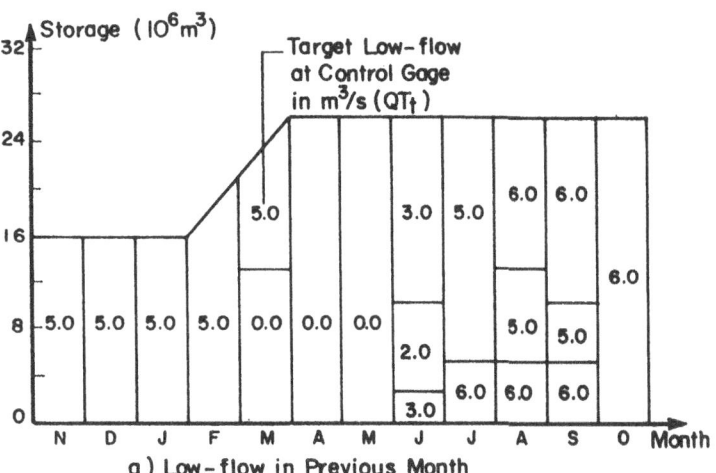

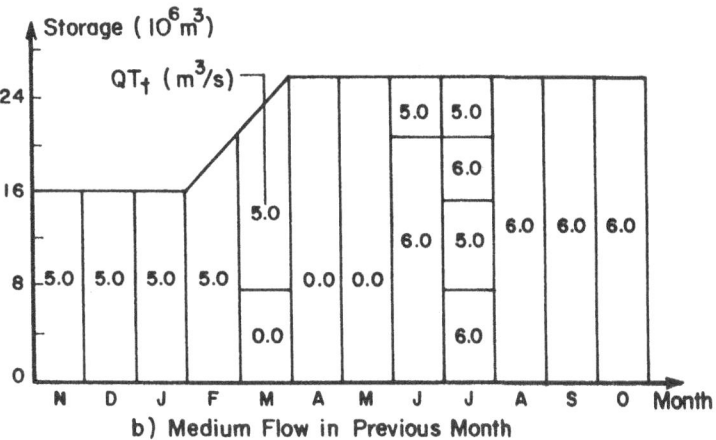

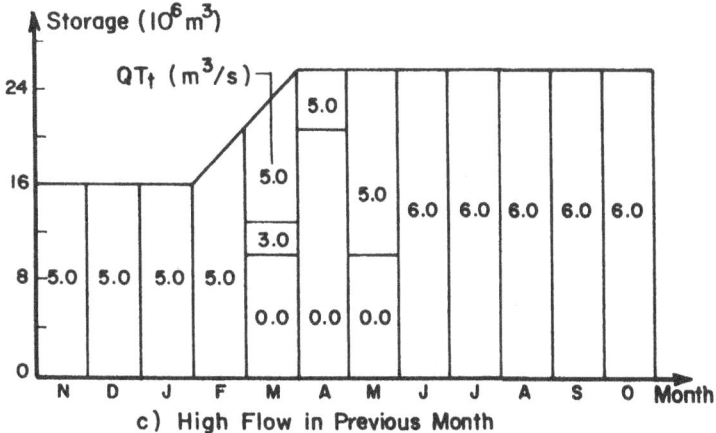

Figure 8. Operating rules for low, medium and high flow in the previous month

The equations were solved for the following five objective functions:

(a) minimizing the squared deviation from low-flow target
(b) maximizing the probability of attaining low-flow target
(c) maximizing the recreation target
(d) maximizing the water quality
(e) compromise solutions

Example results are shown in Figure 8 where the objective was a compromise between all main targets. In order to select a satisfying solution among all the Pareto optimals obtained, a second step using the same or another multiobjective approach should take place (see Laabs et al., 1988).

5. Conclusions and Recommendations

Several models were presented in which the stochastic nature of the inflows to a reservoir is considered explicit. The linear decision rule in reality uses only one point of the probability distribution to transform probabilistic constraints into equivalent deterministic ones. The probability constraint refers though to each time interval which, together with the linear rule, constitute an additional constraint for the model. Most models require reliability constraints over the whole planning horizon. The dynamic programming model shows large flexibility for finding any optimal release policy. The model is constrainted to only one or two reservoirs in the system (dimensionality curse), where the capacity of the reservoir has to be given. The stochastic DP model considers, for the state transition, the assumptions of an autoregressive Markow chain (lag 1). More complex models underlying the stochastic transition probability matrix could be included with little additional computational effort.

6. References

Harboe, R. (1976) 'A stochastic optimization and simulation model for the operation of the Lech River System', Lehrstuhl für Hydraulik and Gewässerkunde, TU München Heft 21.

Harboe, R. (1983) 'Reservoir operation in the Wupper-River System', Chapter 23, Operation of Complex Water Systems. E. Guggino, G. Rossi & D. Hendricks (Eds), Martinus Nijhoff, Den Hagg.

Laabs, H. and Harboe, R. (1988) 'Generation of operating rules with stochastic dynamic programming and multiple objective', Water Resources Management, Vol 2, pp 221-227.

Revelle, C.S., Joeres, E. and Kirby, W. (1969) 'The linear decision rule in reservoir management and design. 1: development of stochastic model', Water Resources Research, Vol. 5, No. 4.

Yakowitz, S. (1982) 'Dynamic programming applications in water resources', Water Resources Research, Vol. 18, No. 4, pp 673-696.

Yeh, W.W.G. (1985) 'Reservoir management and operation models: a state-of-the-art review', Water Resources Research, Vol. 21, No. 12.

IMPLICIT STOCHASTIC OPTIMIZATION AND SIMULATION

D.P. LOUCKS
School of Civil and Environmental Engineering
Cornell University
Hollister Hall
Ithaca, New York 14853-3501
USA

ABSTRACT.   This   paper   will   review   how   historical   and   stochastic
hydrological time series data have been incorporated into optimization and
simulation models designed to aid water resources development planning and
management.   Optimization modelling, used most often for preliminary
design, allocation, and sequencing or scheduling analyses, can be either
deterministic   or   stochastic   with   respect   to   the   incorporation   of
hydrological time series. Deterministic models can be solved for numerous
alternative and equally likely hydrological sequences to obtain at least
some idea of the probability distributions of various design or operating
variable values required to achieve specific objectives.   Alternatively,
hydrological   uncertainty   can   be   incorporated   directly   into   some
formulations of stochastic optimization models.   Both approaches using
optimization have their advantages and limitations. Regardless of the
sophistication of the optimization modelling, the output from these
preliminary analyses should be studied further and evaluated using
simulation. Recent work at a number of research institutes in Europe and
North America over the past three years has resulted in an approach to
dynamic   river   system   simulation   that   can   incorporate   multiple
hydrologicalal sequences in each time step of the simulation. This permits
some estimation of the reliability of various system performance indicators
in each time period of the simulation.

1.   Introduction

There is ample evidence that optimization of simulation models can be
useful tools for identifying and evaluating the impacts of various
alternative systems that are designed and operated to solve a variety of
water resources problems.   Yet, these models are not likely to be very
useful unless they consider the uncertain conditions affecting the future
performance of those systems. This includes the uncertain future demands
imposed on those systems, the uncertain future costs of those systems, and
the uncertain quantities and qualities of the flow within those systems.
Assumptions made regarding each of these sources of uncertainty can have
a major impact on system design, operation and performance.

J.B. Marco et al. (eds.),
Stochastic Hydrology and its Use in Water Resources Systems Simulation and Optimization, 307–318.
© 1993 *Kluwer Academic Publishers.*

These facts have served to motivate the development of stochastic models; models that take into consideration at least some of the important sources of uncertainty and its impact on system design and operation. All existing stochastic models can be classified as one of two types: explicit or implicit. For the purpose of these notes, explicit stochastic models will be defined as those that incorporate, within the model, at least some of the uncertainty associated with at least some of the inputs or outputs. All models that consider, for example, multiple possible flows with their associated probabilities, or multiple possible demands or costs with their estimated probabilities, would be examples of explicit stochastic models. Models whose variables are joint probabilities, or whose constraints include chance constraints with specified probabilities less than one, are also explicit stochastic models.

Implicit stochastic models are deterministic models designed and used for identifying and evaluating system performance in its uncertain or stochastic environment. Most, if not all, simulation models can be placed in this category if they are used, for example, with multiple flow sequences to obtain multiple outputs that can be converted to probability distributions of future system performance.

Take, for example, the sequent peak algorithm for sizing a reservoir to produce specified releases in each period t. The algorithm computes accumulated storage volume deficits in each period t, given known Inflows and Releases.

$$\text{Initial Deficit}_t - \text{Inflow}_t + \text{Release}_t = \text{Final Deficit}_t$$

which is equal to the Initial Deficit$_{t+1}$. If the resulting deficits are less than 0, then

$$\text{Final Deficit}_t = \text{Initial Deficit}_{t+1} = 0$$

Assuming an initial deficit of 0 and solving these series of equations for two cycles of periods t, where in each cycle the Inflow and Release vectors are identical, will result in a finite Required Capacity if the mean Release is no more than the mean Inflow

$$\text{Required Capacity} = \max_t (\text{Final Deficit}_t)$$

Clearly, different inflow sequences will result in different Required Capacities. Even more clearly, if we consider different Release (or demand) sequences as well as Inflow sequences, we will likely obtain an even wider range of Required Capacities. Using the sequent peak algorithm in multiple simulations using multiple, but equally likely, Inflow and Release sequences will provide a basis for defining a distribution function of Required Capacities.

Now, this analysis can be extended to find the capacity that should be selected based on some economic or other criterion and, very importantly, based on a more realistic simulation of the reservoir system (see Figure 1). A realistic simulation of this problem would likely indicate that the reservoir will have to be a little bigger to achieve the same probabilistic performance as predicted by the sequent peak algorithm simulations. This is because in the sequent peak simulations discussed above, no consideration is given to evaporation or to a realistic operating policy. It is this latter issue, how to define an effective reservoir operating policy, possibly along with an appropriate active storage capacity, that has attracted, and still attracts, the attention of numerous stochastic modellers.

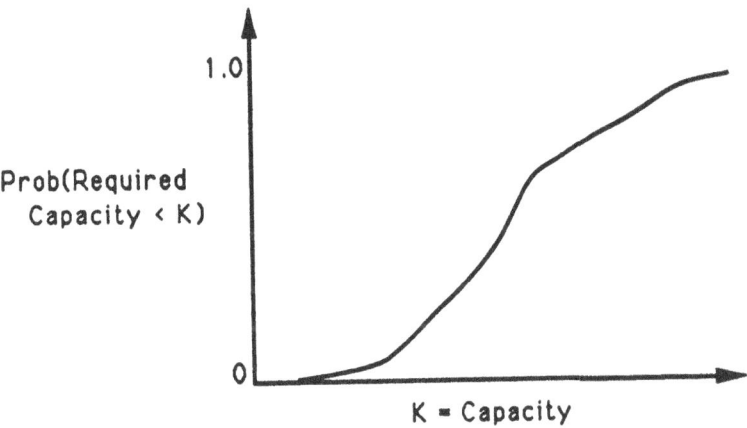

Figure 1. Outcome of multiple simulations of the sequent-peak algorithm.

Implicit stochastic optimization models are deterministic models used to identify and evaluate reservoir release (or groundwater pumping) rules and water use allocation rules, perhaps together with design parameters for a hydrologicalally stochastic environment. Reservoir release rules typically define the releases as a function of storage volume and time of year. Similarly, groundwater pumping rates may be functions of aquifer head and time of year. Allocation or diversion functions are typically functions of streamflow at the withdrawal site, or of storage in upstream reservoirs. All of these implicit models used to identify these release and allocation functions assume that the probability of all unknown variables is one. This is clearly not the case. So why use deterministic models? It seems to me there is only one reason: they are simpler and cheaper to develop and solve. Hence if they can be used in a way that will provide some good operating policy and design variable values for more thorough evaluation using a stochastic simulation model of that system, why not?

2. Optimization of Reservoir Operation

One of the first applications of implicit stochastic optimization was that of Young (1967) who used deterministic dynamic programming with streamflow generation and regression techniques to derive stochastic operating policies for an existing reservoir of known capacity. Without attempting to review all the details of Young's excellent work, which others have extended over the past twenty years, consider the following introductory example.

If one can assume that the inflows and the objective function OBJ_t (S_t, Q_t, R_t) of initial storage S_t, inflow Q_t and Release R_t in each within-year period t ($t = 1,2,...,T$), are constant from year to year, then deterministic dynamic programming can be used to find a steady-state release rule that is a function of reservoir storage S_t and the within-year period t ($t = 1,2,...,T$). For example, a backward-moving approach would involve solving for the cumulative optimal (maximum or minimum) sum of objective function values $F_t^n(S_t)$ for an increasing number n of remaining periods given each initial storage S_t. The sequence of recursion equations, beginning with any $F_t^0(S_t) = 0$ for all S_t,

$$F_t^n(S_t) = \max_{R_t} \text{ or } \min \{OBJ_t(S_t, Q_t R_t) + F_{t+1}^{n-1}(S_{t+1})\}$$
$$(R_t \text{ feasible})$$

must be solved for each value of S_t and for increasing values of n as is necessary to reach a steady state. Steady state is achieved when

$$F_t^n(S_t) - F_t^{n-T}(S_t) = \text{constant}$$

and the particular release R_t that defines $F_t^n(S_t)$ are constant from one year, n, to the next, n-T, for all within-year periods t and S_t.

Now, repeating this procedure for m different, but equally likely, hydrologicalal yearly sequences would produce m different reservoir release policies for each within-year period t. One can then compute the mean release for each initial storage level S_t in each within-year period t, or find a least-squares fit to a piecewise linear, quadratic, or other type of continuous release rule function $R_t = f_t(S_t)$ if desired, to obtain an improved steady-state operating policy for a more thorough evaluation using simulation.

If it is appropriate to define the operating policy as a function of the inflow Q_t as well as initial storage S_t, the deterministic recursive equations can be expressed as functions $F_n^h(S_t,Q_t)$ of storage S_t and inflow Q_t just as they would be if explicit stochastic dynamic programming were used. In the explicit case, the sequence of recursion equations needs to be solved only once to define a policy. In this implicit stochastic dynamic programming case, numerous policies would have to be derived for numerous annual hydrologicalal sequences, as before. Then some averaging or regression technique could be used to define the policy. This policy would require a forecasting of the flow for the current period. In real time applications, the period of forecasting could be daily even though the model time step might be monthly, since the model results can be converted to a daily release rate.

3. Multi-purpose, Multi-reservoir Design and Operation

The so-called yield model, introduced in Biswas (1976) and Loucks et al.

(1981), and extended by Lall and Miller (1988), is an example of a deterministic optimization model that can identify both design and operating policy variable values that influence the reliability of water allocations, or yields, as well as their amounts. Without going into all the extensions and variations of this class of models, a simple example can illustrate the approach.

Consider a three-reservoir system as shown below in Figure 2. Reservoirs A, B and C are to be operated jointly to provide a reliable water supply or yield at Site D. Part of this yield is diverted to a water use. The remainder flows downstream along with any excess flows. The allocation or diversion to the water use will be a function of the total yield available at Site D. The combined reservoir release rule, at Site D, will specify how much water is to be released from the upstream reservoirs as a function of their total storage. The reliability of the flow to be achieved at Site D is specified. Assume the quantity desired with that reliability is also known.

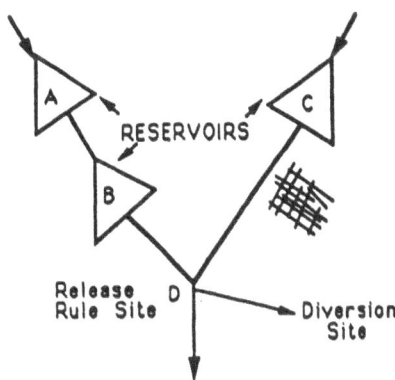

Figure 2.   A multi-reservoir water resource system.

To be determined are the reservoir capacities and operating policies that will provide those reliable flows, or "yields", at Site D. To reduce the number of constraints as well as to identify (for those who are used to distinguishing between) over-year and within-year storage requirements, two sets of storage continuity and reservoir capacity constraints are used. The over-year continuity constraints at Sites s = A and C equate initial storage S_y^s, in year y, plus the unregulated inflow, Q_y^s, less the total release R_y^s to the final storage in Year y, or the initial storage S_{y+1}^s in the following year. The total release is equated to a yield component Y_s which is the same in all years the yield is to be achieved, plus an excess release E_y^s. Hence:

$$S_y^s + Q_y^s - Y^s + E_y^s = S_{y+1}^s$$

for those Years y in which the yield is to be made available, and

$$S_y^s + Q_y^s - E_y^s = S_{y+1}^s$$

for those Years y in which the yield is not to be made available. The reliability of the yield is approximately equal to the number of years the yield is available divided by one plus the number of years of record, providing, and this is important, that the excess release E_y^s in those failure years is less than the yield Y^s.

At Reservoir Site B, the incremental inflow $Q_y^B - Q_y^A$ and excess release from Reservoir A must be included in the inflow.

$$S_y^B + Q_y^B + Q_y^A + E_y^A - Y^B - E_y^B = S_{y+1}^B \qquad \text{for all successful y}$$

and

$$S_y^B + Q_y^B + Q_y^A + E_y^A - E_y^B = S_{y+1}^B \qquad \text{for all unsuccessful y}$$

By the selection of the number of critical years in which yield failure is permitted, the reliability (probability of success) is determined. Clearly, each sequence of years in which for all Sites s the Q_y is greater than the mean annual flow can be grouped into a single constraint if desired for model size reduction.

These yield variables Y are annual values having annual reliabilities. Their distribution within the year may require additional within-year storage. For each within-year period t, the within-year continuity constraints are:

$$s_t^s + \beta_t Y^s - y_t^s = s_{t+1}^s \qquad \text{for s = A, C}$$

$$s_t^s + \beta_t Y^B + y_t^A - y_t^B = s_{t+1}^B$$

where β_t are the distribution coefficients for within-year yield inflows $(\Sigma_t^T \beta_t = 1.0)$. Since excess releases are not to be allocated at Site D, there is no value building capacity to regulate them.

The over-year capacity K_0^s is added to the within-year capacity K_w^s to provide an estimate of required total capacity K^s at each reservoir.

$$S_y^s \leq K_0^s \qquad \text{for all s}$$

and

$$K_0^s + S_t^s \leq K^s \qquad \text{for all s, t}$$

or

$$s_t^s \leq K_w^s \qquad \text{for all t, s; and } K_0^s + K_w^s = K^s \text{ for all s}$$

At Site D the total yield y_t^D available in each within-year period t is

$$y_t^D = \min_{y} \{q_{yt}^D - q_{yt}^B - q_{yt}^C\} + y_t^B + y_t^C$$
$$y \in \text{success years}$$

where q_{yt}^s is the unregulated flow at Site s in Year y, Period t. Each within-year period yeild y_t^D at Site D can be allocated among the diversion use and the downstream uses, as required. Now, if all y_t^D are predefined, i.e. are known variables, then one could find the least-cost combination of reservoir capacities required. Other objectives might include the yields y_t^D as unknowns. Nevertheless, the resulting model is deterministic, yet it provides probabilistic information. It recognizes the stochastic nature of the inflows and hence of the releases too.

To define the operating policy that will produce these series of yields y_t^D and maintain their reliabilities, a storage-release rule can be constructed at each reservoir site using the required over-year storage K_0 and within-year storage volumes, s_t. For this single-capacity yield problem, there will be only two zones, at most, of storage for each time period t. Figure 3 shows what this reservoir release rule policy may look like.

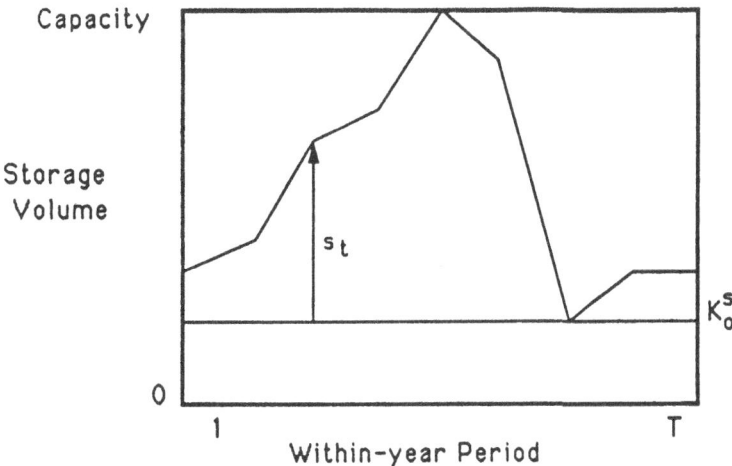

Figure 3. A reservoir release rule derived from the yield model.

If the actual storage volume in the reservoir in a period t is within the lower zone, only the yield y_t should be released. One should be able to do this successfully with the prespecified reliability. If the storage volume is in the upper zone, additional water could be released if desired. This policy can be simulated, and adjustments to capacity and to the release rule can be made as required to improve the design and operation of the system. If additional incremental yields at lower reliabilities are included in the model, additional zones of storage with their increased releases would be defined in the release rule shown in Figure 3.

For a more coordinated operation, it would be desirable to define a single-release rule at Site D as a function of total storage upstream. This can be done in the same manner as the individual reservoir release rules were constructed, except that K_0^s is replaced with $\Sigma s\ K_0^s$ and each s_t^s is replaced with $\Sigma s\ s_t^s$. Next, balancing functions are needed to define the desired distribution of upstream storage given all possible values of total upstream storage. Regretfully, very little information can be obtained from the model solution to help one derive these functions. Hence for reservoirs in parallel, a combination of space rules that equalize the probability of filling during filling conditions, storage rules that equalize the probability of emptying during drawdowns and common sense need to be used to derive some balancing functions before a simulation can be made of the coordinated joint operation of multiple reservoir systems using a single release rule. In any event, it is unlikely the system as defined by the optimization model will perform as indicated without some modification based on a more realistic simulation.

4. Integrated River System Operation

A number of deterministic models combining optimization and simulation have been developed to analyze the operation of multiple river basin projects subjected to stochastic flows. Kuczera (1989) describes one of them that is typical of many. The river system is represented as a network linear program, and as such specialized linear programming algorithms designed to solve network problems can be used to reduce the model running time. This is useful as the number of components and/or periods being simulated increases, especially if these models are to be used in an interactive environment. On the network, each link has a cost (say of pumping) or penalty (say for a deficit or shortfall in meeting a demand). Optimization is used to direct flows throughout the network in a way that minimizes the total cost plus total penalty over a specified planning period.

To illustrate the basic network formulation, consider the single reservoir and demand situation shown in Figure 4. The reservoir capacity and the demand for the use must be specified. The reservoir supplies the user through the conduit, where pumping could be required. Releases downstream are also possible. The inflow to the reservoir is also assumed known.

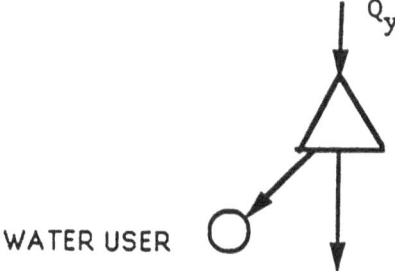

WATER USER

Figure 4. A portion of a river system showing a reservoir used for water supply.

Figure 5 shows the network representation of the system in Figure 4. The network shown is for two contiguous periods, t and t+1, within the total number of periods of simulation.

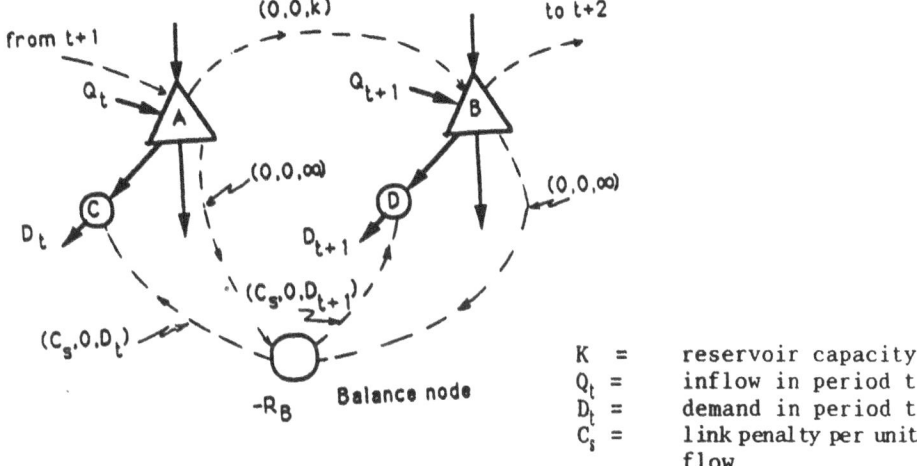

$$K = \text{reservoir capacity}$$
$$Q_t = \text{inflow in period t}$$
$$D_t = \text{demand in period t}$$
$$C_s = \text{link penalty per unit flow}$$

Figure 5. Network representation of Figure 4 for periods t and t+1. (Link triplets are cost or penalty per unit flow and minimum and maximum flows.)

For each period a shortfall network (links to Nodes C and D from the balance node) is added to permit demand shortages at the demand site. The balance node requirement (R_B) is the difference between the sum of all known nodal inflows less the sum of all known nodal outflows or demands. If the sum of the known model outflows or demands is less than the sum of known nodal inflows, R_B is negative and unavoidable spills into the stream channels (the links from nodes A and B to the balance node) will occur. If the sum of known demand outflows is greater than the sum of known inflows, all demands will not be satisfied, and R_B will be positive.

Spills from the system for every period t are sent (e.g. for periods t and t+1 on the links from Nodes A and B) to the balance node. If no demand shortages occur, the spills arriving at the balance node will satisfy exactly the balancing node requirement R_B. If demand shortages occur, the need to maintain a water balance over all simulation periods results in an excess of water at the balance node exactly equal to the demand shortages. The shortfall links, such as those from the balance node to the demand zone Nodes C and D, convey this excess water to the demand sites experiencing shortages, thereby establishing a mass balance. Their shortfall links are used only when it is not possible to meet the demands. Hence a very large penalty, C_s, is assigned to all shortfall links.

Carryover storage between successive periods is modelled using carryover links, for example the link from Node A to Node B, where A and B represent the same reservoir site, but in two succeeding periods. The flow in the link represents the storage volume in the reservoir at the end of period t or the initial storage at the beginning of period t+1. This flow is limited to the storage capacity of the reservoir. Avoidable spills from a reservoir are implicitly penalized through the need to use the shortfall network. These unavoidable spills will induce a demand shortage which must then be satisfied by flow in the high penalty shortfall arcs.

Nonlinear costs and penalties are incorporated into models like these by adding multiple links between any two nodes.

This deterministic network linear optimization model is used to simulate the distribution of a set of known inflows through multiple stream reaches and reservoirs to multiple demand sites over multiple time periods. Sensitivity analyses permit the identification of tradeoffs between capacities and system performance, for example between capacities and demand satisfaction reliabilities. The system operation, however, is not necessarily how it would happen in practice, unless of course this type of model, with continuously updated data, is run periodically in real time to determine all reservoir releases and demand allocations. It would be difficult, even with multiple solutions of the model for multiple streamflow sequences, to infer reservoir operating policies for stochastic inflows, if indeed such policies were needed instead of using the model itself for real time operation.

Another approach to implicit stochastic modelling for long-term planning of multiple reservoir operation is presented by Simonovic (1987). This approach combines deterministic optimization and simulation in an iterative three-level algorithm. At the first level, a simulation approach is used for computing non-economic loss functions associated with water demand deficits. These loss functions are used in an optimization model to derive reservoir operating rules in the second level. The third level uses optimization to determine the reservoir yield functions that can be used for long-term planning.

Each of these approaches just discussed for deriving operating policies for complex integrated river systems has combined deterministic optimization and simulation modelling techniques. Through their appropriate use over multiple-flow sequences, they provide information that can help define multiple reservoir release and demand allocation rules. In an attempt to develop a method of realistically simulating and evaluating such policies subject to stochastic flows and a possibly changing demand environment, a simulation model called IRIS has been developed (Pegram et al., 1989). This interactive river system simulation model is deterministic, uses no optimization, and requires as input data all reservoir capacities and release rules for single or groups of reservoirs operated together, allocation functions at each site where diversions are made, and reservoir storage balancing functions for all reservoirs operated as a group using a single combined reservoir release rule. IRIS is only able to compare alternative operating policies and decisions, not define them (similar to HEC models).

There are a few features that distinguish IRIS from most other deterministic simulation models of river systems, but perhaps the significant one to be mentioned here is that it is designed to be run over real time, say from now to 2020 or some other year in the future. This simulation into the future can include a scenario of changing demands, operating policies, and even system capacities. Since the model is indicating what might be happening in specific simulation periods in the future, say in January of 1997 or in July of 2003, given an operating policy scenario, and since we clearly do not know what the future streamflows will be, say from now until 1997 or 2003, the model permits the simulation in each time step of multiple streamflow sequences. Hence each simulation flow, storage or energy variable in each period of simulation takes on as many values as there are flow sequences incorporated within the simulation. This permits some estimates of the reliability and vulnerability of system performance associated with a given stochastic operating policy and design scenario. Of course the end result after a full simulation of n time steps is the same as if a single but different flow sequence were used in n separate simulations. The latter approach, however, does not allow an examination of the variability of any simulated variable at any intermediate time step until the end of all simulations.

5. Conclusions

To be useful to water resource system managers and operators, any system design or operating policy must be defined for, and adaptive to, changing and uncertain future demands. Various modelling approaches have been proposed for deriving such operating and design policies using both explicit and implicit stochastic optimization and simulation.

This paper has been a brief introductory review of some implicit, i.e. deterministic, modelling approaches developed and used to derive stochastic design and operating policies. These can be compared with models that include the explicit consideration of multiple flows and storage volumes, and their probabilities, discussed elsewhere in this Institute's proceedings.

One basis for the comparison of all explicit and implicit stochastic models is how well each of them are able to define and/or evaluate policies of the type actually used by the planners, managers and operators of the actual system under analysis. Another basis for comparison is how easy each of the modelling approaches can be implemented and the results evaluated, possibly using a realistic static or dynamic simulation model. A third basis for comparison is how each of the modelling approaches fits the problem needing analyzing, and the time and money available to carry out the analysis.

Just as multiple system performance criteria can lead to multiple policies and designs, multiple modelling selection criteria will likely lead to multiple modelling approaches. In this case, the analysts just might have to be the decision makers.

6. References

Bhaskar, N.R. and Whitlatch, E.E. (1980) 'Derivation of monthly reservoir release policies', Water Resources Research 16, No. 6, pp 987-993.

Biswas, A.K. (1976) 'Systems approach to water management', McGraw-Hill Book Co., New York, pp 164-215.

Kuczera, G. (1989) 'Fast multireservoir multiperiod linear programming models', Water Resources Research 25, No. 2, pp 169-176.

Lall, U. and Miller, C.W. (1988) 'An optimization model for screening multipurpose reservoir systems', Water Resources Research 24, No. 7, pp 953-968.

Loucks, D.P., Stedinger, J.R. and Haith, D.A. (1981) 'Water resources systems planning and analysis', Prentice-Hall, Englewood Cliffs, New Jersey, pp 339-355.

Pegram, G.C. et al. (1989) 'IRIS - An interactive river system simulation program: general introduction and description and user's manual', Cornell University, Ithaca, NY.

Simonovic, S. (1987) 'The implicit stochastic model for reservoir yield optimization', Water Resources Research 23, No. 12, pp 2159-2165.

Young, G.K. (1967) 'Finding reservoir operating rules', ASCE Journal of the Hydraulics Division 93, No. HY5, pp 297-321.

MULTIOBJECTIVE ANALYSIS IN WATER RESOURCES
PART I NUMERICAL VERSUS NON-NUMERICAL CRITERIA AND DISCRETE
VERSUS CONTINUOUS ALTERNATIVES

L. DUCKSTEIN
Professor
Dept of Systems Engineering
University of Arizona
Tucson, Arizona 87521
USA

A. TECLE
School of Forestry
Northern Arizona University
Flagstaff, Arizona 86011
USA

1. Introduction

In the design and operation of a water resources system, such as a
reservoir complex or a wastewater treatment facility, it often happens that
more than one objective has to be satisfied (Hipel, 1992). For example,
several of the following objectives may be appropriate:
   - to satisfy water supply requirements
   - to produce hydropower
   - to augment low flows
   - to ensure recreation
   - to protect groundwater resources
   - to maintain the ecological balance
   - to be modular
   - to be reliable
   - to be resilient.
Assume, first, that all the criterion functions depend on a continuous
decision vector $\underline{x}$. For example, $\underline{x}$ is a vector of embankment heights or
slopes, or else a set of reservoir releases; the criterion functions $f_i(x)$,
$i=I$, may be deviations from flow targets t_i, or costs. Furthermore, let
the functions $f_i(x)$ and the constraints be linear in x. Then if $I=1$ (one
criterion function), the maximization of $f_1(x)$ subject to the constraints
constitutes a familiar linear programming problem. If $I>1$, then maximizing
(or attempting to maximize) the I criterion functions subject to the common
constraint set is a multiobjective linear programming problem, the solution
of which should be studied in the so-called payoff space $(f_1, ..., f_I)$.
The dimension of this payoff space is usually much smaller than that of the
decision space (Cohon, 1978; Goicoechea et al., 1982; Steuer, 1986); also,
the visualization of a compromise solution is much easier. If either an
objective or a constraint is non-linear, then a non-linear multiobjective
programming problem is obtained. If $\underline{x}$ is integer, then the multiobjective
problem is also integer.
  On the other hand, objectives such as ecological balance, modularity and
resiliency usually cannot be defined on a continuous numerical scale.
Consider such a case of the presence of non-numerical criteria defined on
an ordinal scale (say A, B, C, D, A = best, D = worst); then only consider-

319

J.B. Marco et al. (eds.),
Stochastic Hydrology and its Use in Water Resources Systems Simulation and Optimization, 319–332.
© 1993 Kluwer Academic Publishers.

ations of discrete alternatives make sense, regardless of whether or not the decision variables are continuous. For example, the release x from a reservoir may be continuous in m^3/sec, but only a release between 10 and 15 may qualify for an ecological criterion rating of A. The remainder of this section is devoted to presenting the multiobjective case study of the design of a wastewater treatment facility to illustrate a modelling approach for discrete design problems with non-numerical criteria (Tecle et al., 1988). The references cited in that paper and in this chapter contain other examples of the application of this approach. Two multi-criterion decision-making techniques are then applied to the example: distance-based compromise programming, as in Duckstein and Opricovic (1980) and the outranking technique ELECTRE I, as in Gershon et al. (1982). The stochastic elements of the analysis are identified and their possible effects on the decision are recognized.

In 1951, a wastewater treatment plant of the activated sludge type was constructed to provide primary and secondary treatment. The plant with a 6,100 m^3/day capacity was to serve a population of 20,000 straddling the border. In the early 1960's rapid growth of the twin cities resulted in the plant being overloaded. In 1972, a new plant, the Nogales International Wastewater Treatment Plant, was designed and constructed 14.2 km north of the International Boundary line, with a capacity of 31,000 m^3/day for a combined population of 102,000. Again, because of rapid population growth in the area, the plant capacity became insufficient in 1985: for a twin cities population of about 250,000, the average discharge of waterwaste inflow of 34,000 m^3/day (directly into the treatment plant) was 9% above design capacity (Arthur Beard Engrs, 1982). Furthermore, because of the accumulation of sludge deposits in about one-third of the available pond space, the plant had to operate at less than 70% of its design capacity. As a result, up to 40% of untreated wastewater had to be discharged to the Santa Cruz Rivel channel.

In parallel, as explained in Tecle and Fogel (1986), several facility plans were then prepared for the City of Nogales, Arizona, to cope with the deficiencies created by new regulations pertaining to the Santa Cruz River. Each study plan proposed a number of feasible alternative schemes but gave no ranking of the alternatives. Furthermore, the plans mentioned the presence of non-numerical criteria. Thus, an MCDM methodology for selecting a wastewater management option from a set of feasible alternatives that satisfy local, state and federal requirements, needed to be developed so as to be able to accommodate non-numerical criteria. Here, the multiobjective optimization approaches presented in Duckstein and Opricovic (1980) and Gershon et al. (1982) are applied. The techniques of compromise programming (CP) and ELECTRE I are briefly discussed and applied to the example problem.

2. Multicriterion Problem Formulation

The multicriterion procedure consists of the following steps:
1. Define the four-tuple objectives - specifications - criteria - scales.
    a. The desired objectives are identified as the five broad categories in the first column of Table 1.
    b. The mission requirements, desired or mandated engineering specif-

ications and constraints for the objectives are explicitly stated.
c. Evaluation criteria that relate system capabilities to specifications and hence to objectives are defined.
d. Measurement scales are defined to describe the range of possible values (quantitative) or relative position (qualitative ordinal) which an alternative system can attain in terms of a given criterion. For the case study, the results of applying steps 1a through 1d are summarized in Table 1.

TABLE 1. Objectives, specifications, criteria and criterion scales

Objectives	Specifications	Criteria	Criterion Scales
Groundwater	Pollutant transport Ordinal (A to G)		Pollution level
Protection	Quality requirement Ordinal (A to G)		Water quality
Effluent quality	Required effluent quality level	Level of treatment achieved	Ratio [0,1]
Wastewater reuse	Effluent volume reused	Amount of effluent used	Ordinal (A to G)
	Reliability	Ratio [0,1] Compatibility	Ordinal (A to G) Ordinal (A to G)
System reliability	Resiliency	Resiliency Flexibility	
Resources	Monetary cost	Capital cost O & M cost	$/1000 Gal/Day $/1000 Gal/Day
	Resource need	Land Manpower	Ordinal (A to G) Ordinal (A to G)

Note: Ordinal implies a 7-point qualitative scale: A= best and G = worst

2. Alternative schemes for attaining the desired objectives are now generated. Note that this step takes place only after the criteria have been defined. Here 15 different alternative schemes consisting of different mixes of seven different activities are generated (Table 2). The first three consist mostly of secondary wastewater treatment. The four other activities are added to each of the first three, either individually or pairwise, in order to reach a combined tertiary level of treatment capability.
3. An evaluation matrix is constructed, the elements of which represent particular values or relative ratings of an alternative in terms of the criteria (Table 3).
4. Multiobjective-multicriterion decision-making (MCDM) and sensitivity analysis are performed, and the study is documented.
5. A solution is accepted as the most "satisficing" one. "Satisficing" implies a preferred but usually not an "optimal" solution which is not a

relevant concept in MCDM.

6. In the event of the final solution not being acceptable to the DM, iteration is performed by gathering more information and starting the procedures at Step 1.

This step gives the process a dynamic behavior which is thus responsive to changing inputs.

Steps 1a, 5 and 6 are performed at the highest policy level in which decision makers have a central role. All other steps are analytical tasks. It may also be noted that, in proceeding from Step 1a to 1b and from 1b to 1c, a one-to-one correspondence is usually not the rule.

TABLE 2.   Generated alternatives

Code	Description
A1	facultative lagoons
A2	aerobic lagoons
A3	oxidation ditches
A4	A1 + chemical algae removal
A5	A2 + chemical algae removal
A6	A1 + filtration algae removal
A7	A2 + filtration algae removal
A8	A4 + nutrient removal
A9	A5 + nutrient removal
A10	A6 + nutrient removal
A11	A7 + nutrient removal
A12	A3 + nutrient removal
A13	A1 + land application
A14	A2 + land application
A15	A3 + land application

3. Methodology

The concept of nondominated solution best visualized in the payoff space is used to obtain a solution for the problem under consideration.   A feasible solution of a multiobjective problem is said to be nondominated if there exists no other feasible solution that will cause an improvement in any one of the objectives without making at least one other objective worse (Cohon, 1978; Goicoechea, et al., 1982; Szidarovszky et al., 1986; Yakowitz, et al., 1992). Compromise programming, a distance based technique, and ELECTRE I, an outranking type technique, essentially use this concept of nondominated solution.   Each one of these techniques is briefly presented below with an emphasis on the underlying mathematical algorithms used.

3.1. COMPROMISE PROGRAMMING

Compromise programming is a distance-based technique designed to identify nondominated solutions which are closest to an ideal solution by some distance measure (Duckstein and Orpicovic, 1980; Starr and Zeleny, 1977; Tecle, 1992; Tecle and Fogel, 1986; Zeleny, 1973; 1974; 1982).   An ideal solution, in general, can be defined as the vector $f^* = (f_1^*, f_2^* \ldots f_I^*)$ where the f_i^* are the solutions to the so-called "marginal" problem stated as Max $f_i(x)$, $i=1,2, \ldots, I$. In a discrete setting such as the case problem under consideration, however, the ideal solution is defined as the vector of best values selected from the payoff matrix. Such a payoff matrix is shown in Table 4. The vector of worst values represents the minimum objective function values denoted as f_i^{**} as shown in Table 5. Both the f_i^* and f_i^{**} values are used in determining the degree of closeness of an alternative to the ideal solution. We consider f_i^* to be the best objective function value (or best alternative), which is sometimes the minimum value of f_i, and f_i^{**} to be the worst objective function value, sometimes the minimum value of f_i.

One of the most commonly used measures of closeness is a family of ℓ_p metrics (Duckstein and Oprovic, 1980; Goicoechea et al., 1982; Zeleny, 1982), that are expressed here as

$$
\ell_p = \left[\sum_{i=1}^{I} \lambda_j^p \left[\frac{f_1^* - f_1(x)}{f_i^* - f_i^{**}} \right] \right]^{1/p} \tag{1}
$$

where $f_i^{**} = \min_i f_i(x)$, $i = 1,2,\ldots,I$ is the minimum objective function in terms of criterion i as illustrated in Table 5 and λ_i's are the criterion weights. As a last step in this technique, upon determination of the ℓ_p distance of each alternative x from the ideal solution, the alternative with the minimum distance ℓ_p is selected as the compromise solution:

$$
\min_x \ell_p(x) = \ell_p(x^*) \tag{2}
$$

The set of all compromise solutions for the particular set of weights $(\lambda_1, \lambda_2, \ldots, \lambda_I)$ and for all $1 \leq p \leq \infty$ constitutes a compromise set. As it is usually done, only three points of the compromise set, that is, those corresponding to $p = 1$, 2 and ∞ are calculated. These p values represent the attitude of the decision maker regarding the compensation between objectives. A small value of p represents strong compensation ($p = 1$ means perfect compensation) and a large value of p means little or no compensation is present.

3.2. ELECTRE I

The outranking type technique ELECTRE I which stands for "Elimination and (Et) Choice Translating Reality" was initially developed by Benayoun, et al. (1966) and improved by Roy (1971). A fundamental feature of ELECTRE I is the use of pairwise comparisons among members of a set of alternative systems in order to eliminate a subset of less desirable (outranked) alternatives while choosing those systems which are preferred for most of the criteria without causing an unacceptable level of discontent or discordance for any one criterion (Gershon, et al., 1982; Nijkamp and van Delft, 1977). Both elimination and choice are therefore essential ingredients of ELECTRE I. The methodology involves the following three concepts: concordance, discordance and threshold values.

The concordance between any two alternative actions m and n is a weighted measure of the number of criteria for which action m is weakly preferred to action n: m.O.n or m.E.n, that is, action m is preferred to or equivalent to action n. Now, letting λi, $i=1,2,\ldots,I$ represent the weight given a priori to criterion i by the decision maker for use in a specific algorithm, then the concordance index between actions m and n can be determined as follows:

$$c(m,n) = \sum_{i \in s(m,n)} \lambda_i / (\sum_{i=1}^{I} \lambda_i) \qquad (3)$$

where $s(m,n) = \{i \mid (m.P.N).U.(m.E.n)\}$, that is, the set of all criteria for which m is preferred to n or equivalent to n. The weights λ_i which are elicited from the decision maker reflect his/her preference structure. The concordance matrix can be thought of as representing the weight percentage of all criteria for which one action would be outranking another one. By definition $0 \leq c(m.n) \leq 1$.

Discordance and concordance concepts are complementary. Discordance represents the maximum discomfort one experiences when confronted with a ranking of a pair of alternatives. To compute the discordance matrix, each criterion is assigned a scale range with an upper value, here, between 70 and 210 (Tables 5 and 6). The interval scale is subjectively determined to represent the degree of dissatisfaction the DM may experience in moving from one point scale to the next less desirable point scale for one criterion as compared to a similar move for another criterion. A seven-point scale has been selected for each criterion. The discordance index, $d(m,n)$, can thus be defined as

$$d(m.n) = 1/R^* \max_{i \in \{1,I\}} (f(n,i)-f(m,i)) \qquad (4)$$

where $f(n.i)$ is the evaluation of alternative n with respect to criterion i, and R^* is the largest of the I criterion scale ranges. Thus the normalized discordance interval is calculated for each criterion where alternative n is preferred to alternative m, and the largest normalized discordance interval of these criteria is defined as the discordance coefficient for alternatives m and n (Goicoechea, et al., 1982). Again, $0 \leq d(m,n) \leq 1$.

To synthesize both the concordance and discordance matrices and determine an outranking relationship among the nondominated alternatives, threshold values (p,q), both between 0 and 1, are selected. p specifies the minimum limit of concordance level whereas q defines the maximum level of discordance the decision maker is willing to accept. A value of $p = 1$ represents full concordance while a value of $q = 0$ indicates no discordance at all. It is possible that some choices of p and 1 may yield no feasible alternatives. If this is the case, the values of p and/or q must be relaxed (lower p, higher q). It is also possible for intransit-ivities among three alternatives to occur, that is, m.P.n, n.P.r, and r.P.m. This is said to form a cycle, in which case the cycle consisting of the three alternatives is collapsed into a single node and receives the same rank (that is, none of the three alternatives is preferred to the others). The final result of ELECTRE I is an outranking relationship which yields a partial ordering of the alternatives.

Alternative Systems

Criteria	A1	A2	A3	A4	A5	A6	A7	A8	A9	A10	A11	A12	A13	A14	A15
Vulnerability to pollution	D	D	C	C	C	C	C	B	B	B	B	B	D	D	C
Water Quality	D	D	C	C	C	C	C	B	B	A	A	A	C	C	B
Level of Treatment	.40	.52	.65	.61	.60	.60	.65	.86	.86	.91	.91	.91	.86	.86	.86
Effluent use	F	F	G	F	F	E	E	F	F	F	E	F	B	B	B
Reliability	.71	.66	.46	.66	.63	.71	.63	.63	.57	.71	.60	.50	.71	.69	.50
Compatibility	E	E	D	D	D	C	C	B	F	A	A	B	A	A	B
Resiliency	C	C	F	C	C	C	C	B	B	B	R	A	B	B	C
Flexibility	D	D	B	C	C	C	C	B	B	B	B	A	B	B	B
Capital cost	.54	.61	1.85	.80	.87	.69	.77	1.51	1.58	1.40	1.47	1.95	.94	1.01	2.25
O & M cost	.21	.37	.38	.28	.44	.24	.39	.45	.61	.40	.55	.48	.26	.42	.43
Land	E	D	B	D	D	D	C	D	D	D	D	C	F	F	E
Manpower	B	C	E	C	C	C	D	C	D	C	E	F	B	C	E

Note: 1. In this table ordinal scale ranges from A to G with A = best and G = worst (in a 7-point scale).
2. Cost is in dollars per 1000 gallons per day.

TABLE 4. Payoff matrix

Alternative Systems

Criteria	A1	A2	A3	A4	A5	A6	A7	A8	A9	A10	A11	A12	A13	A14	A15
1	60.00	60.00	75.00	75.00	75.00	75.00	75.00	90.00	90.00	90.00	90.00	90.00	60.00	60.00	75.00
2	80.00	80.00	100.00	100.00	100.00	100.00	100.00	120.00	120.00	140.00	140.00	140.00	100.00	100.00	120.00
3	60.00	90.00	120.00	120.00	120.00	120.00	120.00	180.00	180.00	210.00	210.00	210.00	180.00	180.00	180.00
4	30.00	30.00	15.00	30.00	30.00	45.00	45.00	30.00	30.00	45.00	45.00	30.00	90.00	90.00	90.00
5	125.00	125.00	75.00	125.00	125.00	150.00	125.00	125.00	100.00	150.00	150.00	100.00	90.00	125.00	100.00
6	30.00	30.00	75.00	40.00	40.00	50.00	50.00	60.00	60.00	70.00	70.00	60.00	70.00	70.00	60.00
7	100.00	100.00	40.00	40.00	100.00	100.00	100.00	120.00	120.00	120.00	120.00	120.00	120.00	120.00	100.00
8	60.00	100.00	40.00	100.00	75.00	75.00	75.00	90.00	90.00	90.00	90.00	105.00	90.00	90.00	90.00
9	60.01	60.00	90.00	75.00	150.00	75.00	150.00	75.00	75.00	100.00	100.00	50.00	90.00	90.00	90.00
10	175.00	175.00	50.00	150.00	75.00	175.00	150.00	75.00	75.00	100.00	100.00	50.00	150.00	125.00	125.00
11	100.00	80.00	80.00	100.00	60.00	100.00	80.00	60.00	40.00	40.00	40.00	60.00	100.00	60.00	60.00
12	45.00	50.00	30.00	50.00	40.00	50.00	40.00	50.00	60.00	50.00	30.00	75.00	60.00	50.00	30.00

TABLE 5.   Criterion bounds and weights

Criteria	Best Values	Worst Values	Weights
1	90.00	60.00	5.00
2	140.00	80.00	6.00
3	210.00	60.00	7.00
4	90.00	15.00	6.00
5	150.00	75.00	7.00
6	70.00	30.00	4.00
7	120.00	40.00	5.00
8	105.00	30.00	3.00
9	175.00	25.00	5.00
10	100.00	40.00	4.00
11	105.00	30.00	3.00
12	60.00	20.00	2.00

TABLE 6. Point values used in discordance matrix determination

Criteria	Levels	Value	Criteria	Levels	Value
1,4,8,11	G	15	6,12	G	10
	F	30		F	20
	E	45		E	30
	D	60		D	40
	C	75		C	50
	B	90		B	60
	A	180		A	70
2,7	G	20	9	0.50-0.75	175
	F	40		0.75-1.00	150
	E	60		1.00-1.25	125
	D	80		1.25-1.50	100
	C	100		1.50-1.75	75
	B	120		1.75-2.00	50
	A	140		2.00-2.25	25
3	0.3-0.4	30	10	0.00-0.10	140.
	0.4-0.5	60		0.10-0.20	120
	0.5-0.6	90		0.20-0.30	100
	0.6-0.7	120		0.30-0.40	80
	0.7-0.8	150		0.40-0.50	60
	0.8-0.9	180		0.50-0.60	40
	0.9-1.0	210		0.60-0.70	20
5	0.2-0.3	25			
	0.3-0.4	50			
	0.4-0.5	75			
	0.5-0.6	100			
	0.6-0.7	125			
	0.7-0.8	150			
	0.8-0.9	175			

4. Application to the Case Study

Two separate computer algorithms, one for each technique, have been used to analyze the problem. The payoff matrix in Table 4 has been used as a standard dimensionless input into the programs. This payoff matrix is a quantified version of the evaluation matrix in Table 3. The quantification is based upon the range of the scale assigned to each criterion, using, here, a linear transformation. Note that a value function transformation can (and often should) be used.

4.1. COMPROMISE PROGRAMMING

In the application of this technique, the vector of best values for all criteria in Table 4 is assumed to be the "ideal point" and the worst values for each criterion define the other limit of the range of the criterion (Table 5). Two different sets of weights (Table 5) are used to test the CP technique's sensitivity to the criteria weights used. One of these sets represents uniform weighting while the other consists of varying weights to represent the DM's preference structure. Then the CP algorithm computer program is used to calculate the distance from the "ideal point" for each set of weights, using Equation 1. The computation is repeated for p = 1,2, and ∞. Various compromise solutions are presented in Table 7. It is observed that Alternative 10, that is, facultative lagoons supplemented by filtration algae removal and nutrient removal facilities, is the most preferred management plan for all three p values and using both sets of weights (see the asterisk-marked values in Table 7). It is interesting to note that when the weights are equal (i.e., $\lambda i = 1$), Alternatives 6 and 7, which have lower treatment capabilities than Alternative 10, are equally preferred for p = ∞. This may be due to the fact that for p=∞ and λi=1 for every i, CP becomes a minimax problem (Duckstein and Opricovic, 1980), also, when no preferential differences are given, economic consequences and quality criteria are balancing each other. Alternative 10 is more costly but has better qualitative ratings whereas 6 and 7 are less costly and also have lower qualitative ratings.

4.2. ELECTRE I

The second technique applied to the problem is the outranking technique ELECTRE I. Equations 3 and 4 are used to generate the concordance matrix (Table 8) and discordance matrix (Table 9). Both concordance and discordance indices are calculated for every pair of alternatives using all the criteria. The result in both cases are square concordance and discordance matrices shown in Tables 8 and 9. Threshold values (p,q) are then introduced to represent the acceptable levels of concordance (p) and of discordance (q). Twenty-one pairs of values of (p,q) have been tried to determine the effect of varying threshold values upon alternatives that pass screening for acceptability. The results are shown in Table 10. The most often preferred alternatives are Alternative 13 (19 times), Alternative 10 (17 times), and Alternative 11 (13 times). As can usually be expected, narrowing the interval between p and q tends to limit the choice to the alternatives with high qualitative ratings, here 10, 11, 13, and 14, while spreading the interval tends to include alternatives with lower treatment capacities, such as Alternatives 3 and 6 (Tables 2 and 10). Furthermore, even though Alternatives 13 and 10 are the most often preferred alternatives, overall, the former tends to have first preference when the interval between p and q is narrow at the low end of their range and the latter at higher values of p and q (Table 10).

TABLE 7. Values of ℓ_p metrics

Alternative	Equal Weights			Varying Weights		
	p = 1	p = 2	p = ∞	p = 1	p = 2	p = ∞
1	7.133	6.376	1.000	5.233	3.614	1.000
2	7.267	5.877	1.000	5.188	3.238	0.857
3	7.767	6.096	1.000	5.719	3.715	1.000
4	5.483	3.215	0.800	3.940	1.821	0.686
5	6.150	3.660	0.800	4.286	1.947	0.686
6	4.533	2.484	0.667*	3.174	1.388	0.600
7	5.367	2.734	0.667*	3.781	1.531	0.600
8	4.333	2.277	0.800	2.981	1.183	0.686
9	5.250	3.354	1.000	3.576	1.713	0.686
10	2.850*	1.478*	0.667*	1.729*	0.6243*	0.514*
11	3.683	2.534	1.000	2.062	0.847	0.571
12	5.217	3.911	1.000	3.400	1.815	0.688
13	3.367	2.623	1.100	2.224	1.111	0.714
14	4.783	3.325	1.000	3.129	1.415	0.714
15	5.700	3.651	1.000	3.652	1.587	0.714

5. Discussion and Conclusions

A multiobjective design case which originally possessed some non-numerical criteria has been illustrated by means of a wastewater disposal example. Wastewater is a potential nuisance or even hazard that must be dealt with in the best possible way, in the sense of technical feasibility, economic viability, environmental quality, socio-political acceptability, resources availability and post-treatment usability. Thus, wastewater management is a complex activity that must satisfy many constraints and satisfy a number of conflicting objectives. This procedure has motivated use of an MCDM approach. In addition, the use of more than one technique may enhance the selection process, especially when uncertainties and imprecision are present. How can such uncertainties be accounted for? First, one should recognize the criteria subject to high uncertainty or imprecision. Practically, the only criteria that may be deterministic (or crisp) are: level of treatment, capital cost and land used. A sensitivity analysis should then be performed to find out changes in ranking due to uncertainty/ imprecision. Alternatively, techniques such as fuzzy compromise programming can be used (Bardossy and Duckstein, 1992).

Comparison of the solutions shows that Alternative 10 is the system most often preferred using either technique (Table 11). Compromise programming shows only Alternative 10 to be the most preferred one while ELECTRE I

Indices of Concordance Matrix

	1	2	3	4	5	6	7	8	9	10	11	12	13	14	15
1		.828	.500	.500	.500	.276	.397	.414	.414	.190	.190	.500	.328	.448	.448
2	.897		.500	.483	.500	.259	.397	.466	.466	.241	.241	.500	.224	.448	.448
3	.500	.500		.500	.569	.431	.500	.190	.190	.190	.224	.241	.310	.379	.397
4	.879	.879	.879		.948	.621	.776	.466	.466	.241	.241	.500	.397	.552	.534
5	.810	.810	.810	.897		.517	.759	.431	.466	.207	.241	.552	.328	.517	.534
6	.966	1.000	.8791	.000	.948		.948	.466	.466	.466	.166	.500	.517	.552	.534
7	.810	.879	.879	.897	1.000	.690		.431	.466	.310	.345	.552	.328	.517	.534
8	.810	.845	.879	.845	.862	.621	.569		1.000	.397	.397	.655	.517	.741	.897
9	.690	.690	.879	.690	1.000	.586	.569	.776		.293	.397	.586	.517	.517	.828
10	.810	.845	.879	.845	.672	.845	.793	1.000	1.000		1.000	.579	.707	.810	.897
11	.810	.810	.879	.810	.759	.810	.759	.897	.966	.897		.810	.707	.707	.828
12	.603	.603	.845	.603	.672	.500	.500	.672	.793	.500	.500		.431	.500	.776
13	.862	.862	.862	.862	.862	.776	.862	.759	.759	.638	.638	.569		1.000	.759
14	.759	.793	.793	.707	.776	.586	.707	.759	.759	.517	.517	.569	.690		.759
15	.690	.638	.793	.638	.707	.638	.638	.534	.655	.241	.276	.483	.534	.603	

TABLE 9. Discordance Matrix

	1	2	3	4	5	6	7	8	9	10	11	12	13	14	15
1		.143	.286	.286	.286	.286	.286	.571	.571	.714	.714	.714	.571	.571	.571
2	.095		.143	.143	.143	.143	.143	.429	.429	.571	.571	.571	.429	.429	.429
3	.595	.595		.476	.476	.595	.476	.381	.381	.429	.429	.429	.476	.381	.357
4	.119	.119	.143		.071	.119	.071	.286	.286	.429	.429	.429	.286	.286	.286
5	.190	.119	.095	.190		.190	.095	.286	.286	.429	.429	.429	.286	.286	.286
6	.0480	.000	.1430	.000	.071		.071	.286	.286	.429	.429	.429	.286	.286	.286
7	.119	.119	.071	.0950	.000	.119		.286	.000	.429	.429	.429	.357	.286	.286
8	.476	.476	.143	.357	.357	.476	.357		.119	.143	.143	.143	.357	.286	.286
9	.476	.476	.190	.357	.357	.476	.357	.119		.238	.238	.143	.357	.286	.286
10	.357	.357	.143	.238	.238	.357	.2380	.000	.000		.000	.071	.238	.214	.214
11	.357	.357	.190	.286	.238	.357	.238	.095	.048	.095		.095	.286	.214	.214
12	.595	.595	.095	.476	.476	.595	.476	.286	.286	.286	.286		.476	.357	.286
13	.119	.143	.286	.143	.214	.143	.214	.143	.143	.190	.190	.214		.000	.095
14	.238	.238	.286	.190	.214	.238	.214	.143	.143	.190	.190	.214	.190		.095
15	.714	.714	.214	.595	.595	.714	.595	.238	.238	.357	.357	.143	.595	.476	

indicated that 10 and 13 are the alternatives most often preferred. For a complete ordering of the preferences in the latter case the procedure may be supplemented by ELECTRE II (Gershon, et al., 1982). Note that another distance-based technique, based on cooperative game theory, has been applied in Tecle et al. (1988) yielding results similar to compromise programming.

TABLE 10. ELECTRE I results

Threshold Values		Selected Alternatives	Threshold Values		Selected Alternatives
P	Q		P	Q	
.3	.1	6,10,11,13	.7	.3	10,11
.3	.2	13,15	.7	.4	10,11
.4	.1	6,10,11,13	.8	.1	6,10,11,13,15
.4	.2	13,14	.8	.2	6,10,11,13,15
.5	.1	6,10,11,13	.8	.3	6,10,11,13
.5	.2	13,14	.8	.4	10,11,13
.6	.1	6,10,11,13	.9	.1	3,6,10,12,13,15
.6	.2	13,14	.9	.2	3,6,10,12,13,15
.6	.3	10,11,12,13,14,15	.9	.3	3,6,10,12,13,15
.7	.1	6,10,11,13	.9	.4	3,6,10,12,13,15
.7	.2	10,11,13			

Concerning the case study, the most preferred Alternative 10 can be constructed at an initial cost of about \$5,000,000 (1984 dollars) (Arthur Beard Engineers, 1984). Upon construction, it will have a design capacity of 43,500 m^3/day of wastewater inflow and a performance capacity equivalent to a tertiary treatment plant. Alternative 13 has the same capacity as 10 with an initial cost of \$4.6 million dollars but higher future polluting effect.

The following concluding remarks can be made about the comparative performances of the two techniques. It is observed that both of them can be conveniently used to determine the preference structure of problems having a finite number of discrete alternative systems, noting, however, that CP can also be used for problems with continuous alternatives. It is also found that the two techniques follow different paths to accomplish the same task. Finally, accounting for uncertainty and imprecision may be easier with compromise programming, where each compromise point in the non-dominated set may be defined as a fuzzy number.

1. Compromise programming uses weights and determines solutions with respect to ℓ_p-norm, and ELECTRE I uses pairwise comparisons among alternatives to determine their partial ranking.

2. In CP, the DM can choose the weights and specify the value of the metric parameter p; in ELECTRE I the DM can do the same with weights, scales, threshold value p and, after an initial calibration, threshold value q.

3. CP requires a cardinal scale for criteria whereas ELECTRE I can use ordinal ranking with an interval scale to rank, at least partially, the alternatives.

TABLE 11. Comparison of results

Technique		Alternative Selected
Compromise programming	p = 1	10
	p = 2	10
	p = ∞	10
ELECTRE I		10,13

6. References

Arthur Beard Engineers, Inc. (1982) 'Effluent land disposal systems', section VI, Addendum NØ 2 to facility plan, 1979, Nogales Wastewater Project EPA Project NØ C-04-0181-01, City of Nogales, Arizona, 33 pp.

Arthur Beard Engineers, Inc. (1984) 'Facility plan improvements and expansion of international wastewater treatment plant', Nogales, Arizona, report prepared for City of Nogales.

Bardossy, A. and Duckstein, L. (1992) 'Analysis of a karstic aquifer management problem by fuzzy composite programming', Water Resources Bulletin, 28(1), pp 63-74.

Benayoun, R., Roy, B. and Sussman, B. (1966) 'ELECTRE: une méthode pour guider le choix en présence de points de vue multiples', SEMA (Metra International), Direction Scientifique, Note de Travail NØ 49, Paris.

Cohon, J.L. (1978) Multiobjective Programming and Planning, Academic Press, New York.

Duckstein, L., and Opricovic, S. (1980) 'Multiobjective optimization in river basin development', Water Resources Research, 16(1), pp 14-20.

Gershom, M., Duckstein, L., and Mc Aniff, R. (1982) 'Multiobjective river basin planning with qualitative criteria', Water Resources Research, 18(2), pp 193-202.

Goicoechea, A., Hansen, D.H. and Duckstein, L. (1982) 'Multiobjective decision analysis with engineering and business applications', John Wiley and Sons, Inc., New York.

Hipel, K.W. (1992) 'Multiple objective decision making in water resources', Water Resources Bulletin, 28(1), pp 3-12.

John Corollo Engineers (1979) 'Nogales wastewater facility plan', EPA Project NØ C-04-0181.

Nijkamp, P., and van Delft, A. (1977) 'Multicriteria analysis and regional decision-making', Martinus Nijhoff Social Sciences Division, Leiden, 135 pp.

Roy, B. (1971) 'Problems and methods with multiple objective functions', Mathematical Programming, 1(2), pp 239-266.

Starr, M.K. and Zeleny, J. (1977) 'MCDM - State and future of the arts' in: Multiple Criteria Decision-making, edited by M.K. Starr and M. Zeleny, North-Holland, New York, pp 5-29.

Steuer, R.E. (1986) Multiple Criteria Optimization: Theory, Computation and Application, John Wiley and sons, NY.

Szidarovszky, F., Gershon, M.E. and Duckstein, L. (1986) Techniques for Multiobjective Decision Making in Systems Management', Elsevier Publ., Amsterdam.

332

Tecle, A. (1992) 'Selecting a Multicriterion decision making technique for watershed resource management', Water Resources Bulletin, 28(1), pp 129-140.

Tecle, A., and Fogel, M. (1986) 'Multiobjective wastewater management planning in a semiarid region', Hydrology and Water Resources in Arizona and the Southwest, vol. 16, pp 43-61, April, Glendale, Arizona.

Tecle, A., Fogel, M. and Duckstein, L. (1988) 'Multicriterion selection of wastewater alternatives', ASCE, Journal of Water Resource Planning and Management, 114(4), 383-398.

Yakowitz, D.S., Lane, L.J. and Szidarovszky, F. (1992) 'Multiattribute decision making: dominance with respect to an important order of the attributes', to appear, special MCDM issue of Applied Mathematics and Computation.

Zeleny, M. (1973) 'Compromise programming' in: Multiple Criteria Decision-Making, edited by J.L. Cochrane and M. Zeleny, University of South Carolina Press, Columbia, pp 263-301.

Zeleny, M. (1974) 'A concept of compromise solution and the method of displaced ideal', Computers and Operations Research, 1(4), pp 479-496.

Zeleny, M. (1982) Multiple Criteria Decision-making, McGraw-Hill, New York, 563 pp.

List of symbols

A to G	= ordinal scale for non-numeric criteria where A=best, G=worst and B, C, D, E, F are intermediate scale values between A and G.
c	= left hand side of concordance index equation
CP	= acronym for compromise programming
d	= left hand side of discordance index equation
DM	= acronym for decision-maker
E	= "equivalent to" in preference situation
ELECTRE	= acronym for Elimination and (Et) Choice TRanslating REality
$\in$	= belongs to (or is a member of)
f	= an element of a payoff evaluation matrix
I	= number of criteria
ℓ_p	= distance norm in compromise programming equation
m.n	= adjacent matrix elements in a payoff or evaluation matrix
MCDM	= acronym for multicriterion decision-making
p	= concordance threshold
.P.	= preferred to
q	= discordance threshold
R*	= largest of given criterion scale ranges
s	= preferential set notation
t_j	= flow target
.U.	= union of two sets
x	= vector of decision variables
λ	= relative criterion weight

Subscripts

i	= an element of the I criteria
p	= dimension of the distance norm

Superscripts

*	= vector of best values in a payoff matrix
**	= worst values in a payoff matrix

MULTIOBJECTIVE ANALYSIS IN WATER RESOURCES
PART II - A NEW TYPOLOGY OF MCDM TECHNIQUES

L. DUCKSTEIN
Professor
Dept of Systems Engineering
University of Arizona
Tucson, Arizona 87521
USA

A. TECLE
School of Forestry
Northern Arizona University
Flagstaff, Arizona 86011
USA

1. Introduction

In this section, the numerous MCDM techniques available to analyze water resources problems are clustered into five types, corresponding to the philosophy of the decision-making process. As pointed out in Duckstein (1978) and Tecle (1988, 1992), there are many possible classification schemes for MCDM problems, possibly leading to a combinatorial number of categories. In water resources, the grouping used in Goicoechea et al. (1982) seems to be useful, as it is based on the timing of the decision-maker's articulation of preferences, which can be prior, progressive or posterior. Other criteria for classification may include such classification characteristics as single decision maker versus multiple decision makers, fuzzy versus crisp (i.e. non-fuzzy criteria) or alternatives, integer versus noninteger programming algorithms, assumptions of the existence of a multiattribute value or utility function, implicit versus explicit constraints, single or multiperiod problems, and continuous versus discrete procedures (Korhonen et al., 1984). The problem with all these kinds of classification schemes is that each one is mostly based on a pair of characteristics and the scheme, as a result, can lead to as many classifications categories as there are different pairs of characteristics of MCDM techniques. Besides, many of the above classification schemes can be biased, usually reflecting the particular author's interpretation of the world of MCDM (Zionts, 1982, 1985; Haimes et al., 1975). As an improvement to this kind of classification, we propose in this study a classification scheme of MCDM techniques that may potentially have fairly general acceptability and be helpful in narrowing down the identification of techniques for application to hydrological design and operation problems. This new classification scheme is a modified and augmented version of the one presented in Tecle (1988) and Bogardi et al. (1988).

2. A New Typology of MCDM Techniques

The characteristics of the decision making process used to classify MCDM techniques take into account the basic structural formulation of the algo-

J.B. Marco et al. (eds.),
Stochastic Hydrology and its Use in Water Resources Systems Simulation and Optimization, 333-343.
© 1993 Kluwer Academic Publishers.

rithms. In addition, the classification characteristics are based on the final solutions desired by the DM. Five different types of MCDM techniques are identified, namely:
(1) Outranking
(2) Distance-based
(3) Value or utility
(4) Direction-based
(5) Mixed
Table 1 provides a classification of some of the available MCDM techniques into the above four categories. Moreover, each one of these categories is described below as a class to demonstrate the basic differences inherent among the techniques that belong to each of the five categories.

2.1. OUTRANKING TYPE OF MCDM TECHNIQUES

These techniques use outranking relationships among alternatives to select the most satisfying alternative. An outranking relation is conceived to represent the preference ordering of a finite set of alternatives. Four different preference relations between alternatives can be recognized: a strict preference, indifference, weak preference and incomparability (Roy, 1973; Roy and Vincke, 1984; Haimes et al., 1975, 1984).

Given two alternatives a and b belonging to a finite set of available alternatives A, a strict preference between a and b implies there exist clear and positive reasons to justify that one of them is significantly preferred to the other, that is a.P.b or b.P.a but not both. Here .P. stands for the strict preference relation and is considered to be asymmetric and irreflexive. In contrast, we have the indifference relation when the difference between alternatives a and b becomes too small to be recognizable. In this case, the two actions are indifferent in the sense that there exist clear and positive reasons to choose equivalence, that is, a.I.b and b.I.a, where .I. represents the equivalence relation and is both symmetric and reflexive (Roy and Vincke, 1984). When the difference between alternatives is neither sufficiently small so as to be indifferent nor sufficiently large as to constitute a strict preference, a concept known as weak preference introduced in Roy (1973) may be used to describe the situation. The symbolism and properties of strict preference are supposed to hold for this relation with the exception of replacing the letter P by Q to represent the weak preference relation. On the other hand, when the attributes between alternatives are significantly different from each other (Haimes et al., 1975, 1984) or the DM <u>does not have</u> adequate information for comparing them, then the alternatives are said to be incomparable with each other. The relation of incomparability represented by R is both symmetric and irreflexive, that is, a.R.b⟹b.R.a but not a.R.a.

Now, using the above four distinct preference relations, the outranking relation between any two alternatives a and b in A can be formally defined. Specifically, two types of outranking relations can be defined (Roy, 1974, 1977). A deterministic outranking relation asserts that, given alternatives (a, b).E.A., a outranks b (a.S.b.) if there exists sufficient evidence to support that "alternative a is at least as good as alternative b" and there is no good reason to reject it. A crisp outranking relation may thus represent either strict preference, or weak preference or indifference (Roy and Vincke, 1984). On the other hand, a fuzzy outranking relation is finer than the deterministic one, since a "credibility level" of outranking of one alternative on another is given. In other words, a fuzzy outranking relation is a fuzzy subset B (Zadeh, 1965; Nakamura, 1984) of the set AxA of pairs of alternatives, characterized by a membership function u_B (A, A), called by Roy (1977), the "credibility degree" of outranking.

Several examples of outranking types of techniques are provided in the first column of Table 1. These techniques exhibit varying complexities during the solution process as indicated in Jacquet-Lagreze (1974), Saaty (1977, 1980), Goicoechea et al. (1982), Chankong and Haimes (1982), Fraser and Hipel (1984), Tecle et al. (1978a, b), Brans et al. (1985, 1986) and Vincke (1992) among others.

Outranking techniques constitute excellent tools for hydrological and water resources design under conflicting objectives. As stated earlier, the key element of such schemes is a table of alternatives versus criteria; the criteria may include non-numerical measures, in which case techniques that can handle such non-numerical criteria should be selected for use.

2.2. DISTANCE-BASED TYPE OF MCDM TECHNIQUES

Some MCDM techniques are based on the concept of distance to arrive at the most satisfying solution. This distance is not limited to the geometrical sense of distance between two points, but it is used as a proxy measure for human preference (Zeleny, 1974b, 1982). This latter interpretation of distance is used as a measure of resemblance, similarity or proximity with respect to individual coordinates, dimensions and attributes. This concept of distance is, therefore, used for the purpose of determining solutions in reference to some point in the payoff space.

As shown in Column 2 of Table 1, several distance-based techniques have been developed to date. Generally, the solution procedure in these techniques proceeds by first defining some reference point which is usually an unfeasible alternative, that one relates to the solution one desires to achieve. One major difference among the techniques that belong to this group is the way they relate to the reference point. Compromise programming (CP) uses optimization procedures to find the feasible solution that is closest to the reference point which is known in this case as the ideal solution (Zeleny, 1977, 1974a, 1982; Yu, 1973; Starr and Zeleny, 1977; Duckstein and Opricovic, 1980; Szidarovszky et al., 1986; Tecle and Duckstein, 1992; Tecle et al., 1987b, c). Other techniques such as cooperative game theory (CGT) (Szidarovszky, et al., 1984) use quite a different concept of distance measure to determine an acceptable solution. In this case, the geometrical distance from an undesirable point, called "status quo" point (in the payoff space), is maximized. Another group of techniques which also use the concept of minimum-distance from the target or "goal" is goal programming and its variants. But, the distance here is usually measured as a deviation from an ideal solution. Distance-based techniques are intuitively satisfactory and can readily be used in water problems.

TABLE 1. A new typology of MCDM techniques (after Tecle, 1988)

Outranking Type	Distance-based Type	Value or Utility Type	Direction-based Type	Mixed Type
ELECTRE I, II,III & IV	Compromise Programming(CP)	Method of Stewart	Dynamic MOP	Ellipsoid Algorithm
PROMETHEE I,II,&III	Composite Programming	Weighted Average Method	Visual Interactive Method	Lagrangian Method
Conjunctive Ranking	∈-Constraint Method	Delphi Goal Programming	Zionts' Interactive Method	Stochastic MOP
Exclusionary Screening	Game Theory (Coop & Noncoop)	MO Statistical Method	MO Simplex Method (Philips/Zeleny)	MCQA I & II
Lexicographic Method	Interactive Goal Programming	Method of Geoffrion	Adaptive Search Method	MISE
Concordance Method	Sequential SEMOPS Method	Method of Zionts Wallenius	Reference Objective	
Copeland's Reasonable SWF	Nonlinear Goal Programming	Measurable Value Method	Method of Yu	
Metagame	Goal Attainment Method	Method of White	PARETO RACE	
Analytical Hierarchical Process(AHP)			Visual Interactive Method	
Hypergame	Weighting Method	Method of GDF	PROTRADE	
Surrogate Worth Trade-off (SWT)	Goal Programming (GP)	Local MAUT	STEP Method	
	Linear Integer (GP)	Global MAUT	PASEB	
	Displaced Ideal	STRANGE	VODCA	
	Filtering Approach	SMOP	TRADE	
	Dynamic CP	ESAP	MOLPF	
	SIGMOP	HOPE		
	GPSTEM			
	LINMAP			

2.3. VALUE OR UTILITY TYPE OF MCDM TECHNIQUES

Many MCDM procedures are based on the preference order of the decision maker, which is assumed to be known (Szidarovszky et al., 1986), and on the hypothesis that the DM's reference structure can be formally and mathematically represented by a value function if the problem is deterministic, or utility function if there is any risk involved in the problem. However, an individual's preference structure must satisfy certain axiomatic conditions before the existence of a value or utility function can be hypothesized.

Different sets of axioms were developed and are available for this purpose (see, for example, von Neumann and Morgenstern, 1944; Markowitz, 1959; Keeney, 1969; Fishburn, 1970, 1976; Szidarovszky et al., 1984). Four axioms that relate both the choices among certain (riskless) and uncertain (risky) outcomes are provided below (Markowitz, 1959; Goicoechea et al., 1982); and if an individual behaves according to these axioms, a utility function can be derived to indicate the individual's preferences for both the certain outcomes and for choices in a risky situation. Given any three elements a, b and c as members of a set of alternatives, A, the axioms can be stated as follows:

(1) <u>Completeness:</u> For any two alternatives (a, b).E.A, one of the following must be true: either a is preferred to b, or be is preferred to a, or there is indifference between them.

(2) <u>Transitivity:</u> For any alternatives (a, b, a).E.A, one's preference among them is transitive, that is, if a is preferred to b and b to c, then a is preferred to c.

(3) <u>Continuity:</u> Given that a is preferred to b and b to c, then there exists some probability p, where $0 < p < 1$, such that there is indifference between having outcome b with certainty or getting a with probability p and c with probability (1-p). In other words, there is a certainty equivalent to any gamble or lottery.

(4) <u>Independence:</u> Given that there is indifference between a and b, and c is any third alternative, then there will be indifference between the following strategies: Strategy 1 offers a probability p of receiving a and a probability (1-p) of receiving c, and Strategy 2 offers a probability p of receiving b and a probability (1-p) of receiving c.

If one conforms to these four axioms, a utility function can be constructed (Goicoechea et al., 1982, pp 26-27). For multicriterion problems, utility functions can, in principle, provide a complete ordering of the set of nondominated alternatives. The nondominated alternative which yields the highest utility is then taken to be the preferred solution.

The multiattribute utility theory (MAUT), as described in Keeney and Raiffa (1976) is one example of a utility-type MCDM technique which is based on the above axioms of the utility theory. Note the above four axioms are augmented by two additional axioms to ensure a functional form for the utility function (additive or multiplicative form). The multiattribute utility is generally of the form

$$u(c_1(x,a),\ldots,c_I(x,a)) = f(u_1(c_1(x,a)),\ldots,u_I(c_I(x,a)))$$

where $c_i(x,a)$ is the ith attribute, $i=1,\ldots,I$.

Thus, u is a function of random variable x and action or decision variable a. The action a* is sought, which minimizes the expected value of the utility functions $u(x,a)$. Other MCDM techniques also belonging to this group are shown in Column 2 of Table 1.

The value or utility-type techniques are certainly well adapted to the modelling and resolving of conflicts in water resources management (Krzysztofowicz, 1978; Krzysztofowicz and Duckstein, 1979; Yakowitz, et al., 1992). Yet the techniques may require long and involved assessment procedures which include verification of axioms and consistency checks. In any case, the concepts of value or utility function to represent preferences (and risk attitude in the case of utilities) is recommended especially in those cases where only two or at most three attributes are present. An example of the use of value functions in a multiobjective analysis of small hydropower plant operation can be found in Duckstein et al., (1989).

2.4. DIRECTION-BASED TECHNIQUES

Most interactive schemes include a step during which the decision-maker is asked to state a preferred direction for the search for a compromise solution. For example, the decision-maker may prefer that high emphasis be given to the flood control objective and next, to the water supply objective: this defines a direction of search in the payoff space. Techniques such as evolutionary SEMOPS (Bogardi et al., 1988a, 1988b), PARETO RACE (Korhonen, et al., 1984), STEM, Zionts and Wallenius' approach (Zionts, 1985), the method of Malakooti (Al-Alwani, et al., 1992) and even SWT (Haimes, et al., 1975) certainly belong to this category as indicated in Column 4 of Table 1. The direction-based techniques are recommended whenever a decision-maker is willing to provide information for a directional search of a "satisfactum" and feels comfortable to communicate with a computer screen.

2.5. MIXED TYPE OF MCDM TECHNIQUES

Apart from the above four types of MCDM techniques, there is a certain number of miscellaneous techniques listed in part in Column 5 of Table 1 that cannot at this moment be placed under any one category. Many of them, however, can be considered as generating techniques. According to Goicoechea et al. (1982), a generating technique uses the vector of objective functions in an MODP to identify and generate the subset of nondominated solutions in the initial feasible region. This is done by dealing strictly with the physical realities of the problem without making any attempts to consider the DM's preference structure. In the process, these techniques identify a set of nondominated solutions to help the DM gain insight into the physical realities of the problem under consideration. Examples of the generating type of techniques include the Multiobjective Simplex Method (MOSM) of Zeleny (1974), the Vector Maximum Approach (VMA) of Steuer (1977, 1986) and PASEB of Thanassoulis (1983, 1985) to mention a few.

Other techniques that are included in this category have characteristics that cannot be clearly identified with any of the above four categories, for example, the Lagrangian Method (Neuman and Krzysztofowicz, 1977) the ellipsoid algorithm (Zeleny, 1982), and the adaptive search method (Beeson and Meisel, 1971). Moreover, techniques that include a combination of one or more characteristics of any of the above four categories and other additional characteristics are also placed in this category. Examples of the latter type are PROTRADE (Goicoechea et al., 1979) which combines both utility assessment and the distance concept to reach a compromise solution and Multicriterion Q-Analysis (MCQA) which combines outranking and value concepts (Hiessl et al., 1985; Chin et al., 1991).

3. A Prescription for the Use of MCDM Techniques

As can be expected and as shown in many of the references cited, in particular Gershon and Duckstein (1983, 1984) and Tecle et al. (1988), the application of different techniques leads to different rankings of the alternatives. Sensitivity analyses show that weights and scales influence the ranking for any given MCDM technique (Gershon et al., 1982). To use only one technique and ignore the information that would be provided by other techniques is not advisable. In fact, one should take advantage of the change in ranking from one technique to the other one, to test the robustness of the ranking for most preferred alternatives, as well as the few ranked list. Thus, the procedure would be as follows:

1. Define and model the problem on hand, as prescribed in Part I.
2. Select appropriate MCDM techniques, preferably of different types following for example the method developed in Tecle and Duckstein (1992).
3. Apply the techniques compare the rankings, and perform a sensitivity analysis.
4. Alternatives ranked high by several techniques should be selected for further study, they form the "short list".

Sensitivity analyses should be included in Phase 3 of the analysis, and can be repeated once the few preferred alternatives have been selected. In particular, the sensitivity to errors, uncertainties and imprecision.

Finally, casting the hydrological problem into a larger water resources development framework may be instructive, in terms of clarifying the origin of the conflicts between objectives. Thus, flood control, water supply, power production can be provided by means other than building new reservoirs that may be subject to highly conflicting interests. First, a better operation of existing systems can be developed. Next, alternative schemes for reaching the desired goals or at least approaching them exist, as illustrated throughout this Institute and described in the cited references.

4. References

Al-Alwani, J.E., Hobbs, B.F. and Malakooti, B. (1992) 'An interactive integrated multiobjective optimization approach or quasiconcave quasiconvex utility functions', to appear, special MCDM issue of Applied Mathematics and Computations.

Beeson, R.M. and Meisel, W.S. (1971) 'The optimization of complex systems with respect to multiple criteria', in: Proceedings, Systems, Man and

Cybernetics Conference, IEEE, Anaheim, California.

Bogardi, J., Budhakooncharoen, S., Duckstein, L., and Sutanto, A. (1988a) 'Evolutionary interactive MCDM for discrete water resources systems planning', Proceedings, XI Int. Symposium MCDM, Manchester, UK, 21-26.

Bogardi, J. et al. (1988b) 'Improved large scale water resources development planning', Progress Report, Water Resources Engineering Division, AIR, Bangkok.

Brans, J.P., Mareschal, B. and Vincke, P.H. (1985) 'PROMETHEE: A new family of outranking methods in multicriteria analysis, in: Brans, J.P. (ed.), Operational Research, 84. North-Holland, Amsterdam.

Brans, J.P., Vincke, P.H. and Mareschal, B. (1986) 'How to select and how to rank projects: the PROMETHEE method', European Journal of Operational Research, Vol. 24(2), pp 228-238.

Chankong, V. and Haimes, Y.Y. (1982) 'Multiobjective decision making theory and methodology', Elsevier, North Holland, N.Y., p 472.

Chin, C., Duckstein, L. and Wymore, M.L. (1991) 'Factory automation project selection using multicriterion Q-analysis', Applied Mathematics and Computation, 46(2), pp 107-128.

Duckstein, L., Tecle, A., Nachtnebel, H.P. and Hobbs, B.F. (1989) 'Multicriterion analysis of hydropower operation', Working Paper, Systems Engineering Dept., Case Western Reserve University, Cleveland, Oh.

Duckstein, L. (1978) 'Imbedding uncertainties into multiobjective decision models in water resources', Proceedings, International Symposium on Risk & Reliability in Water Resources, June, Waterloo, Ontario, Canada.

Duckstein, L. and Opricovic, S. (1980) 'Multiobjective optimization in river basin development, Water Resources Reserach, 16(1), pp 14-20.

Fischburn, P.C. (1970) 'Utility theory for decision making', John Wiley and Sons, New York, p. 234.

Fischburn, P.C. (1976) 'Utility independence on subsets of products sets', Operations Reserach, Vol. 124, pp 245-255.

Fraser, N.M. and Hipel, E.W. (1984) 'Conflict analysis in models and resolutions', North Holland, NY.

Gershon, M.E. and Duckstein, L. (1984) 'A procedure for selection of a multiobjective technique with application to water and mineral resources, Applied Mathematics and Computation, Vol 14, No. 3, pp 245-271.

Gershon, M.E. and Duckstein, L. (1983) 'Multiobjective approaches to river basin planning, ASCE, Journal of Water Resources Planning and Managing, Vol. 198(1), pp 13-28.

Gershon, M., Duckstein, L. and McAniff, R. (1982) 'Multiobjective river basin planning with qualitative criteria, Water Resources Research, 18(2), pp 193-202.

Goicoechea, A., Duckstein, L. and Fogel, M. (1979) 'Multiple programming in watershed management: an illustrative application of PROTRADE, Water Resources Research, 15(2), pp 203-210.

Goicoechea, A., Hansen, D.H. and Duckstein, L. (1982) 'Multiobjective decision analysis with engineering and business applications', John Wiley and Sons, Inc., New York, NY.

Haimes, Y.Y., Hall, W.A. and Freedman, H.T. (1975) 'Multiobjective optimization in water resources systems: the surrogate worth trade-off method', Elsevier Scientific Pub. Co., The Netherlands.

Haimes, Y.Y. and Alle, D.S. (Eds) (1984) 'Multiobjective analysis in water Resources, American Society of Civil Engineering, N.Y., p 237.

Hiessel, H., Duckstein, L. and Plate, E.J. (1985) 'Multiobjective analysis with concordance and discordance', Applied Mathematics and Computations, Vol. 17, pp 107-122.

Jacquet-Lagreze, E. (1974) 'How we can use the notion of semi-orders to build outranking relations in multi-criteria decision making', Revue Metra, 13(1).

Keeney, R.L. (1969) 'Multidimensional utility functions: theory, assessment and applications', Technical Report No. 43, Operations Reserach Center, MIT, Cambridge, Massachusetts.

Keeney, R.L. and Raiffa, H. (1976) 'Decisions with multiple objectives: preferences and value trade-offs', John Wiley, New York.

Korhonen, P., Moskowitz, M. and Wallenius, J. (1984) 'A sequential approach to modeling and solving multiple criteria decision problems', Working Paper No. F-70, Helsinki School of Economics, Finland.

Krzysztofowicz, R. (1978) 'Preference criterion and group utility model for reservoir control under uncertainty', Department of Hydrology and Water Resources, Technical Report No. 30, University of Arizona, Tucson, Arizona 85721.

Krzysztofowicz, R. and Duckstein, L. (1979) 'Preference criterion for flood control under uncertainty', Water Resources Research, Vol. 15(3), pp 513-520.

Markowitz, H. (1959) 'Portfolio selection', John Wiley, New York.

Nakamura, K. (1984) 'Preference relations on a set of fuzzy utilities as a basis for decision making', Fuzzy Sets and Systems, Vol 20, pp 147-162.

Neuman, S. and Krzysztofowicz, R. (1977) 'An interactive algorithm for interactive multiobjective programming, Advances in Water Resources, Vol. 1(1). pp 1-14.

Roy, B. (1973) 'How outranking relation helps multiple criteria decision making', in: Cochrane, J.L. and Zeleny, M. (Eds), Multiple Criteria Decision Making, University of South Carolina Press, Columbia, SC.

Roy, B. (1974) 'Critères multiples et modelisation des préférences - l'apport des relations de surclassement', Revue d'Economie Politique, No. 1.

Roy, B. (1977) 'Partial preference analysis and decision aid: the fuzzy outranking relation concept', in Bell, Keeney and Raiffa, Advances in Multiple Criteria Decision Making.

Roy, B. and Vincke, P.H. (1984) 'Relational systems of preference with one or more pseudo-criteria: some new concepts and results', Management Science, Vol. 30(11), pp 1323-1335.

Saaty, T.L. (1977) 'A scaling method for priorities in hierarchical structures, Journal of Mathematical Psychology, Vol. 15(3), pp 234-281.

Saaty, T.L. (1980) 'The analytic hierarchy process: planning, priority setting, resources allocation', McGraw-Hill International Book Co., New York, p 287.

Starr, M.K. and Zeleny, M. (1977) 'MCDM - state and future of the arts' in: Starr, M.K. and Zeleny, M. (Eds), Multiple Criteria Decision Making, North-Holland Publ., Amsterdam, pp 5-29.

Steuer, R.E. (1977) 'An interactive multiple objective linear programming procedure' in: Starr, M.K. and Zeleny, M. (Eds) 'Multiple Criteria

Decision Making, North-Holland, Publ., Amsterdam, pp 225-239.

Steuer, R.E. (1986) 'Multiple criteria optimization: theory, computation and application', John Wiley and Sons, N.Y.

Szidarovszky, F., Duckstein, L. and Bogardi, I. (1984) 'Multiobjective management of mining under water hazard by games theory', Eur. J. Oper. Res., 15(2), pp 251-258.

Szidarovszky, F., Gershon, M.E. and Duckstein, L. (1986) 'Techniques for multiobjective decision making in systems management', Elsevier Publ., Amsterdam, The Netherlands.

Tecle, A. (1992) 'Selecting a multicriterion decision making technique for watershed resource management', Water Resources Bulletin, 28(1), pp 129-140.

Tecle, A. and Duckstein, L. (1992) 'A procedure for selecting MCDM techniques for forest resources management', in: Goicoechea, A.L., Duckstein, L. and Zionts, S. (eds), Multicriteria Decision Making: Interface of Industry, Business and Finance, Springer Verlag, New York, pp 19-32.

Tecle, A., Duckstein, L. and Fogel, M. (1987a) 'Multicriterion decision making in wastewater management: problem formulation', Working Paper No. 87-007, April, Systems and Industrial Engineering Department, University of Arizona, Tucson, Arizona, p 89.

Tecle, A., Duckstein, L. and Fogel, M. (1987b) 'Multicriterion decision making in wastewater management: application and comparison of techniques, Working Paper No. 87-008, April, Systems and Industrial Engineering Department, University of Arizona, Tucson, Arizona, p. 57.

Tecle, A., Duckstein, L. and Fogel, M. (1987c) 'Multicriterion forest watershed resources management', Forest Hydrology and Watershed Management, Proceedings of the Vancouver Symposium, August 1987: IAHS-AISH Publ. No. 167, pp 617-625.

Tecle, A., Fogel, M. and Duckstein, L. (1988) 'Multicriterion selection of wastewater alternatives, ASCE, Journal of Water Resources Planning and Management, Vol. 114(4), pp 383-398.

Tecle, A. (1988) 'Choice of multicriterion decision making technique for watershed management, unpublished Ph.D. Dissertation, the University of Arizona, Tucson.

Thanassoulis, E. (1983) 'The solution procedure "PASEB" for multiobjective linear programming problems', in: French, S., Thomas, L.C., Hartley, R. and White D.J. (Eds), Multiobjective Decision Making, Academic Press, London, Great Britain, pp 145-177.

Thanassoulis, E. (1985) 'An adaptation of PASEB for the solution of multiobjective linear fractional programming problems', J. Opr. Res. Soc., Vol. 36(2), pp 155-161.

Vincke, P. (1992) 'Multicriteria decision-aid', Wiley, Chichester, England, 174 p.

Von Neumann, J. and Morgenstern (1944) 'Theory of games and economic behavior', 1st Edition, Princeton University Press, New Jersey (2nd Edition, 1947; 3rd Edition, 1967).

Yakowitz, D.S., Lane, L.J. and Szidarovszky, F. (1992) 'Multiattribute decision making: dominance with respect to an importance order of the attributes', to appear, Special MCDM Issue of Applied Mathematics and Computation.

Yu, P.L. (1973) 'Introduction to domination structure in multi-criteria decision problems', in: Cochrane, J.L. and Zeleny, M. (Eds), Multiple Criteria Decision Making, University of South Carolina Press.

Zadeh, L. (1965) 'Fuzzy sets', J. Information and Control, Vol. 8, pp 338-353.

Zeleny, M. (1977) 'Adaptive displacement of preferences in decision making', Multiple Criteria Decision Making, Starr, M.K. and Zeleny, M. (Eds), North-Holland Pub. Co., Amsterdam-New York, pp 147-157.

Zeleny, M. (1982) 'Multiple criteria decision making', McGraw-Hill, New York, N.Y.

Zeleny, M. (1974a) 'Linear multiobjective programming', Springer-Verlag, Berlin, West Germany.

Zeleny, M. (1974b) 'A concept of compromise solutions and the method of the displaced ideal, Computers and Operations Research,Vol.1(14),pp 479-496.

Zionts, S. (1982) 'Multiple criteria decision making: an overview and several approaches', Working Paper No. 545, School of Management, SUNY, Buffalo.

Zionts, S. (1985) 'Multiple criteria mathematical programming: an overview and several approaches', in: Fandel, G. and Spronk, J. (Eds), Multiple Criteria Decision Methods and Applications, Springer-Verlag, Berlin, pp 85-128.

COMBINING SIMULATION AND OPTIMIZATION IN RIVER BASIN MANAGEMENT

J.W. LABADIE
Department of Civil Engineering
Colorado State University
Fort Collins, Colorado 80523
USA

ABSTRACT. Historically, there has existed a dichotomy in relation to the use of descriptive models (i.e. simulation) and prescriptive models (i.e. optimization) in river basin management. This is unfortunate, since there are unique advantages with each modelling approach which can be greatly enhanced through their joint usage. Descriptive models allow accurate modelling of complex flow and storage mechanisms in a river basin, and are conducive to Monte Carlo analysis and assessment of risks. Descriptive models, on the other hand, cannot directly find the "best" solutions, which represents the primary advantage of prescriptive models. The latter, however, often require simplifying assumptions on model structure. A combined simulation-optimization strategy for river system management is presented. The procedure is evaluated and tested using the Nizao River system in the Dominican Republic as a case study.

1. Introduction

Many significant river basin management projects worldwide are in need of extensive reevaluation of operational goals and strategies. For example, many projects were originally constructed under a heavy debt service for providing primarily agricultural and flood control benefits. In ensuing years, the energy crisis of the 70's and rapid growth in population centres have placed new pressures on these systems to provide increased hydropower production and water supply for municipal and industrial use. In addition, growing public awareness of environmental and ecological issues, as well as recreational needs, have placed new constraints on water managers that were not considered when the project was originally planned and implemented.

Comprehensive reevaluation of the management of many water resources projects is needed in order to operate these systems in a truly multipurpose fashion. The need for integrated operational strategies confronts system managers with a difficult task. Expanding the scope of the working system for more integrated analysis greatly multiplies the potential number of alternative operational policies. This is further complicated by conflicting objectives, stochastic hydrology and uncertain

345

J.B. Marco et al. (eds.),
Stochastic Hydrology and its Use in Water Resources Systems Simulation and Optimization, 345–371.
© 1993 *Kluwer Academic Publishers.*

consumptive water use. Optimal coordination of the many facets of such a system requires the assistance of computer modelling tools to provide information on which to base rational operational decisions. These tools should be adaptable enough for both operational planning uses, as well as actual real-time decision support.

Computer simulation models are currently being applied for management and operations within a number of river basin systems in the US and around the world. Many are customized for the particular system, such as the water quantity/quality Colorado River Simulation System (CRSS) used by the US Bureau of Reclamation (BOR, 1985), but there is also substantial usage of public domain, general purpose models such as HEC 5 (Hydrologic Engineering Centre, 1988) and the Streamflow Synthesis and River Regulation Model (SSARR) developed by the US Army Corps of Engineers (COE, 1987). Many of these models have recently been adapted to MS-DOS based microcomputers with interactive, user-friendly interfaces that facilitate model usage (Eichert, 1989).

Simulation or "descriptive" models are particularly attractive for answering "what if" questions regarding the performance of alternative regulation strategies. Computer simulation models are usually better able to represent the operations of a system with a higher degree of accuracy and are useful for risk analysis in examining the long term reliability of proposed operating strategies. A major disadvantage of most existing river basin simulation models is that although the physical features of the system are reasonably accounted for, subject to the availability of an adequate data base for model calibration and verification, the administrative and institutional aspects of water allocation are often not properly accounted for. For example, it is difficult with a model such as HEC 5 to specify water allocation priorities based on water rights or policies such that flows are spatially and temporally allocated in accordance with those priorities. In many cases, it is these legal and institutional factors that play a dominant role in how water is distributed and used within a complex river basin system.

Unfortunately, "descriptive" models are not well suited to finding or "prescribing" the best or optimum strategies when flexibility exists in coordinated system operations. Prescriptive optimizing models, on the other hand, offer an expanded capability to systematically select optimum solutions, or families of solutions, under agreed upon objectives and constraints. Specification of allocation priorities is easily accomplished within the structure of prescriptive models. There are several generalized optimization packages available for linear, integer, and nonlinear programming which are available for a wide range of computing systems, such as MINOS (Murtaugh and Sanders, 1984). Use of these packages requires water managers to develop their own interfaces which can manage data, define objective functions and specify system constraints in a predefined structure. In addition, these are primarily deterministic tools which are incapable of directly incorporating probabilistic information.

Considering the respective strengths and weaknesses of simulation and optimization, it would seem that the best approach is to combine them together so as to accentuate their unique advantages. Optimizing models can be used for generating operational policies which can then be tested and refined with a more detailed simulation model. The purpose of this paper is to present a suggested procedure for combining simulation and optimization models together for analysis of multipurpose river basin systems which include hydropower, water supply, flood control, water quality management, and other beneficial uses.

A generalized river basin network flow model called MODSIM is described for satisfying simulation needs in river basin management, with the added advantage of the ability to consider administrative and legal priorities associated with water allocation. A generalized dynamic programming package called CSUDP is proposed for performing the necessary stochastic optimization analyses. The suite of models developed by Obeysekara et al. (1986a) provide stochastic analysis and generation of multivariate hydrological time series for input to the simulation and optimization models. A case study is presented for the Nizao River System in the Dominican Republic which includes risk analysis and assessment of the economic viability of the proposed simulation-optimization scheme in improving system performance levels and efficiencies.

2. Strategy for Combined Use of Simulation and Optimization

Computer models are needed for operational planning in a river basin under normal conditions (i.e. excluding short-term flood operations and emergencies) as well as actual real-time operation of the system. A methodology is presented herein for normal system operations in a multipurpose environment. The reader is referred to Obeysekara et al. (1986b) for the development of modelling tools for short-term flood control operations in reservoir systems. The basic steps in the methodology are listed as follows:

1. Develop optimal end-of-month storage guidecurves which maximize the expected value of a selected primary objective function for the system, subject to satisfying other system objectives based on specification of target performance levels. For example, a primary objective of maximizing expected energy production could be selected, subject to meeting irrigation and other water supply demands, as well as flood control targets and downstream minimum streamflow standards for water quality management and other instream uses. Explicit stochastic optimization is performed with full consideration of stochastic hydrology in relation to all possible system states attainable during real-time operations.

2. Apply a generalized river basin simulation model which can serve as a weekly real-time operational tool for synchronizing hydropower, agricultural and other water uses in the system, including allocation between the various water use sectors according to administrative and institutional priorities. This simulation model can serve to evaluate reservoir storage guidecurves from the monthly optimization analysis. The model can be extended to developing daily operational strategies, or simply provide weekly guidelines under which experienced operators can perform necessary daily or even hourly management.

3. Employ the simulation model for testing and evaluating the optimal guidecurves and perform a stochastic analysis to estimate the reliability of these optimal rules in meeting project requirements, including assessment of associated risks. The simulation may be performed on a weekly, or possibly daily basis, but is focused on attaining the targeted monthly guidecurves. A feedback process may be required whereby adjust-

ments and further interrogation of the stochastic optimization model are required due to the danger of unacceptable risks in system performance as ascertained through use of the simulation model.

4. Provide a realistic assessment of the expected economic benefits of applying the optimized operational guidelines to the system and compare the historical performance of the system with what would likely have occurred had the optimal operating rules been applied over the historical period. This analysis is important for justifying and documenting the value of utilizing the simulation-optimization tools in order to eventually implement them for actual real-time use by the water system managers and operators.

A hierarchical scheme for meeting these objectives is shown in Figure 1. A temporal decomposition approach is selected which begins with development of optimal feedback decision policies for each calendar month for the reservoir system that maximize a selected primary objective, subject to satisfying other system objectives, targets and constraints according to ranked priorities. For illustration purposes, Figure 1 shows energy output as the primary objective, subject to satisfying irrigation and other project requirements as closely as possible. Other objective priorities can be selected, depending on the nature and scope of the project being analyzed. A fully dynamic optimization approach employing stochastic dynamic programming is utilized at the monthly level, requiring inputs of:

1. Month-to-month transition probabilities describing the stochastic characteristics of natural river basin inflows; since the flows are random, an explicit stochastic approach is necessary for developing operational policies that will maximize long-term beneficial use of the system.

2. Hours of hydropower production, which may also have been random in the past due to variable daily peak load periods and use of the system for base load as necessitated by deficiencies or failure in various portions of the power network to which it is linked.

3. Physical system features including power plant characteristics, reservoir capacities, surface area-head-capacity tables, etc.

4. Irrigation demands estimated by methods such as the modified Penman model, as well as other demands for municipal and industrial water supply.

It is expected that the dynamic programming optimization need only be rerun if it becomes evident that current values of one or more of the above inputs are no longer valid or need adjustment. The operating policies generated by the optimization specify optimal end-of-month storage levels, conditioned on initial storage levels and previous period inflows to consider the persistence of successive monthly inflows. These kind of "feedback" policies are essential in that they represent optimal guidelines for a wide variety of storage and inflow conditions that may exist at any time, rather than just one inflexible optimal open loop policy. Most other mathematical optimization methods are only able to produce the latter type of solution.

Prior to actual real-time utilization of the optimal operating policies, they are tested through Monte Carlo analyses. These are performed using several hundred or thousand years of synthetically generated inflows to assess the probabilities of maintaining various acceptable levels of release for irrigation, power production and other uses, as well as determining the possibility of increasing target levels to improve beneficial use of the system. The same network simulation model is used for both weekly or daily operations.

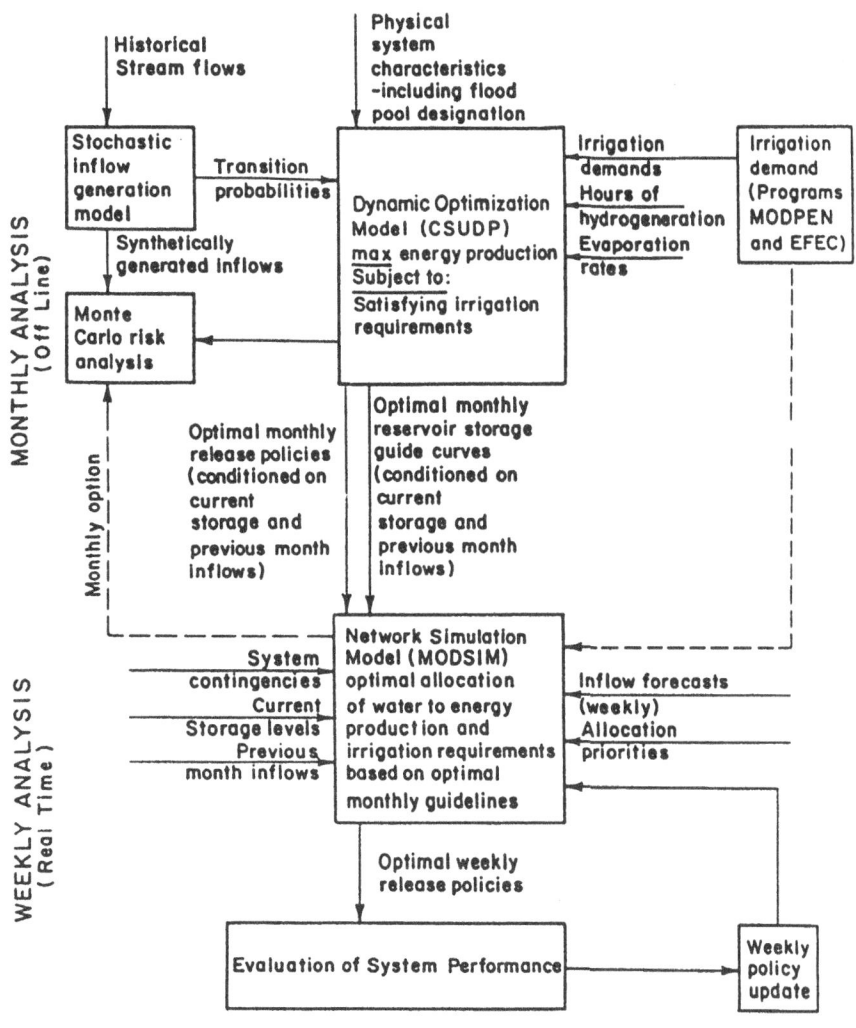

Figure 1. Hierarchical approach for developing optimal normal operation policies.

Optimal guidecurves from the monthly analysis can now be utilized for weekly or daily operation of the system. A more detailed network simulation model is employed for determining the optimal allocation of water according to specified priorities for hydropower, the various irrigation sectors in the system, and other uses, while attempting to conform to the monthly guidelines. Opportunities are available at the weekly or daily level to include forecast information on inflows. These forecasts can be updated on a regular basis and the model rerun to accommodate changing conditions as the system performance is monitored.

The goal is for the network simulation model to become a real-time operational tool that is tractible and easy to use by system operators. The guidecurves and computer models are intended as support mechanisms only for the system operators. Use of weekly and monthly targets allow flexibility in daily operations. They are regarded as targets only and operators are free to deviate from them as the need arises and their experience dictates.

In order to confirm the value of employing the guidecurves, policies, and computer models developed for normal operation of the reservoir system, attempts should be made to assess improvement in system performance that would have occurred historically had they been employed, versus the actual system performance. In addition, attempts should be made to assign an estimated economic value to these improvements.

3. Computer Models for Operational Decision Support

Two types of computer models are utilized for implementation of this proposed water project reevaluation scheme: (i) a generalized river basin network simulation model called MODSIM, and (ii) a general purpose dynamic programming code called CSUDP for developing optimal stochastic operational guidelines to be tested and employed in MODSIM. These two programs work together in that CSUDP gives optimal operation rules which can be input to MODSIM for basic operational decisions and risk assessment. Again, the idea is that simulation and optimization should be used together. The former can more accurately depict the operation of the entire system, whereas the latter can be useful in helping to find optimal operating guidelines for use in the simulation.

3.1. OPTIMIZATION MODEL FOR MONTHLY OPERATIONAL TARGETS: CSUDP

There are a wide variety of optimization techniques available, but for this proposed scheme, the optimization technique selected is dynamic programming. The reader is referred to Labadie (1980) and Yakowitz (1982) for applications of dynamic programming to water resources. The advantages of dynamic programming (DP) include:

a. Particular suitability in solving sequential decision problems over many time periods or stages. As seen in Figure 2, reservoir operations can be naturally viewed as a sequential decision process with release decisions at each stage resulting in beneficial returns. Monthly storage volumes represent the "state" transitions between stages. With DP, computer time requirements increase linearly with the number of stages. All other methods, except for optimal control theory, are characterized by a geometric computer time requirement with the number of stages.

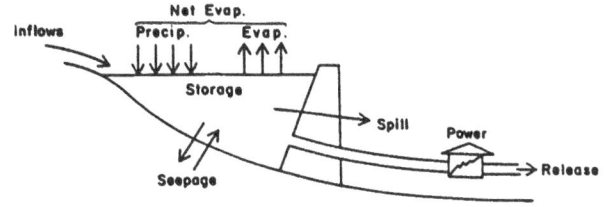

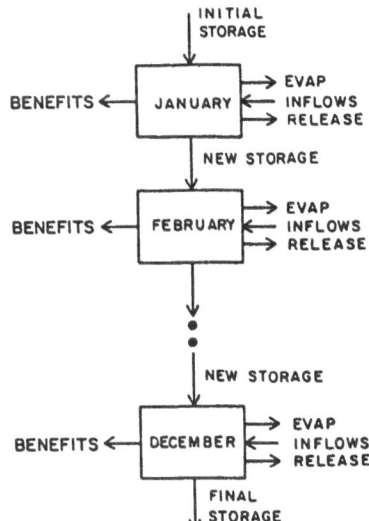

Figure 2. Illustration of reservoir operations as a sequential decision process.

b. Ease of considering nonlinear aspects of reservoir modelling including hydropower production functions, evaporation calculations, flow routing and pumping costs. Power calculations in particular introduce a high degree of nonconvexity into optimization problems which can cause severe difficulties with other optimization methods.

c. Determination of "feedback" operating rules for a wide range of conditions or states of the system, rather than just one "open loop" optimal solution that would be obtained with other methods.

d. Solution efficiency actually increases when a large number of operational constraints are included, whereas the opposite occurs with other methods.

e. Particular attractiveness for solving stochastic optimization problems with inclusion of conditional risk constraints defining the limits of probability of failure to meet certain operational criteria (Sniedovich, 1979).

The primary disadvantages are:

a. The so-called "curse of dimensionality", which asserts itself with a vengeance as the number of reservoirs considered surpasses three. Computer costs and rapid access storage requirements increase dramatically with each added reservoir.

b. Lack of readily available, generalized dynamic programming computer software; the usual approach is to write a specialized DP code for each new application. This requires both good programming skills and a thorough understanding of the theory and numerical aspects of DP.

The second disadvantage is overcome by the availability of Program CSUDP developed at Colorado State University (Labadie, 1990). Program CSUDP is used primarily for developing monthly optimal stochastic operating rules for a reservoir. The code allows inclusion of objective function and constraints for any system through user-supplied FORTRAN subroutines linked to the main program. A microcomputer version of the program has been developed for IBM PC or compatible microcomputers which includes user-friendly, menu-driven data entry and editing capabilities (Labadie, 1990).

The first disadvantage is a continuing problem with no completely satisfactory solution as yet developed. Trezos and Yeh (1989) and Georgakakos and Meeks (1989) have proposed interesting solutions to this problem which require further verification of their effectiveness. For the case study presented herein, there is a limited number of reservoirs, so the dimensionality problem does not apply.

The CSUDP dynamic programming model can be used in a deterministic sense to analyze optimal reservoir operations for any given period of input data. Since results of the analysis are based upon that specific period of input data, then generalized operational policies can be obtained only by analyzing a lengthy period covering a variety of hydrological conditions. Though this may be adequate for calibration purposes, it can be computationally expensive for developing general operating guidelines under a wide variety of hydrological conditions. In addition, it may be difficult to obtain unique operating rules in this case.

Since inflows to a reservoir are a stochastic process, operational policies that consider this stochasticity directly can be obtained by using explicit stochastic optimization procedures which incorporate inflow probability distributions rather than deterministic inflow levels. Stochastic dynamic programming, which was selected for this analysis, can find feedback policies that optimize the long-term expected value of the operational objective. Transition probability matrices are used to describe the discrete probability of a certain inflow conditioned on the previous period inflow.

Starting at stage N, the dynamic programming optimal return function is calculated backwards recursively, starting with $F_{N+1}(\cdot\cdot) = 0$, over stages N, N-1,...,1, for all discrete V_i and I_{i-1}:

$$F_i(V_i, I_{i-1}) = \max_{V_{i+1}(\text{or } R_i)} \sum_{k=1}^{K} p(I_{ik} | I_{i-1})[E_i(V_i, V_{i+1}, R_i) +$$

$$+ F_{i+1}(V_{i+1}, I_{ik})] \tag{1}$$

(for $i = N, N-1, \ldots, 1$)

where $E_i(\cdot\cdot) =$ energy generated during months $i=1,\ldots,N$, as function of releases and average head over the period; $V_i =$ initial storage for month i; $R_i =$ reservoir release during month i; $I_{i-1} =$ value of a specific inflow class for month $i-1$; $I_{ik} =$ discrete random inflow during month i at discrete probability level k; and $p(I_{ik} | I_{i-1}) =$ probability of occurrence of I_{ik} conditioned on the previous I_{i-1}, where

$$\sum_{k=1}^{K} p(I_{ik} | I_{i-1}) = 1 \tag{2}$$

There are two ways the reservoir mass balance can be considered:

- noninverted form:

$$V_{i+1,k} = V_i - R_i - EVAP_i(V_i, V_{i+1,k}) + I_{ik} \tag{3}$$

- inverted form:

$$R_{ik} = V_i - V_{i+1} - EVAP(V_i, V_{i+1}) + I_{ik} \tag{4}$$

Researchers such as Roefs (1968), Harboe (1977), and Stedinger et al. (1984) have applied stochastic dynamic programming to finding optimal reservoir operating rules, but always using the noninverted form.

Notice that in the noninverted form, the end-of-period volume is treated as a random variable, and releases R_i are directly optimized. This results in optimal release policies $R^*_i(V_i, I_{i-1})$. A disadvantage is that Equation 3 is an implicit function of $V_{i+1,k}$, thereby requiring several iterations to find an explicit solution. Strict bounds on release can be maintained with this approach, including constraints for satisfying downstream water demands and instream flow requirements. Since end-of-period storage is a random variable, however, there may exist a certain risk of failure to satisfy required end-of-period bounds on reservoir storage.

In the inverted form, optimization is performed directly over the end-of-period volume V_{i+1} and releases are now regarded as random variables. Release R_{ik} can be explicitly calculated in Equation 4, so there is no need for iterative procedures. This approach results in optimal storage guidecurves $V^*_{i+1}(V_i, I_{i-1})$. The latter are considered to be more flexible for reservoir operations, and also allow a direct "traceback" of optimal storage levels that is not possible using the noninverted form. Strict bounds can now be maintained on reservoir storage in the optimization, but there now may be a risk of failure to satisfy constraints on release. In this case, a penalty term may be substracted from the objective function to discourage violation of release constraints. Another alternative provided by Program CSUDP is to specify a maximum conditional risk of failure to satisfy downstream requirements, and allow violations up to but not exceeding these maximum risk levels.

Since the transition probabilities repeat every 12 months, the undiscounted stochastic dynamic programming calculations are repeated via successive approximations or "value iteration" (Dreyfus and Law, 1977) to determine if optimal guidecurves for each month are converging to stationary policies. Usually, three or four sets of yearly cycles are sufficient for reasonable convergence. The optimal seasonally stationary policies can then be applied each year over the entire operational horizon with any sequence of inflows. Ross (1983) proves that for the undiscounted case, sufficient conditions for optimality are defined such that if this procedure converges, then the solution must be optimal.

3.2. SIMULATION MODEL FOR WEEKLY REAL-TIME OPERATIONS

For development of normal weekly operational guidelines under nonemergency conditions, it is proposed to use Program MODSIM developed at Colorado State University (Labadie, 1987). MODSIM has been successfully applied to a number of river basin systems, such as the Rio Grande River Basin, Colorado (Graham et al., 1986); the Cumberland River Basin, Tennessee (Labadie, 1983); and the Upper Pampanga River Basin, the Philippines (Faux et al., 1983). A version of this program for IBM PC or compatible microcomputers has been developed which provides a menu-driven format for data entry, editing and error checking. A printer-plot graphical output is also available with the microcomputer version, as well as tabulated results showing storage levels, releases, inflows, energy generation, power capacity, system losses and spills, demands, shortages, and flows in any reach of the system. MODSIM is also capable of modelling stream-aquifer systems for studying the conjunctive use of surface and groundwater, as reported in Labadie et al. (1986). A special output file of system flows is generated by MODSIM for input into a water quality routing model for river basins.

The underlying principle of MODSIM is that most physical water resource systems can be represented as capacitated flow networks which can be solved efficiently and rapidly with modern network flow computational algorithms. The term "capacitated" refers to the existence of strict bounds on each link. The components of the system are represented in the network as nodes: both storage (i.e. reservoirs) and nonstorage (i.e. river confluences, diversion points, points of inflow, and demand locations) and links or arcs (i.e. canals, pipelines, and natural river reaches). In order to consider demands, inflow, and desired reservoir operating rules, several artificial nodes and links must be constructed such that mass balance is satisfied throughout the network, as illustrated in Figure 3. These artificial nodes and links are created automatically by MODSIM, so the user need only be concerned with the actual system linkage.

MODSIM solves the following optimization problem:

$$\min \sum_{i=1}^{N} \sum_{j=1}^{N} c_{ij} \, q_{ij} \tag{5}$$

subject to:

$$\sum_{i=1}^{N} q_{ij} - \sum_{k=1}^{N} q_{jk} = 0 \text{ for } j=1,\ldots,N \tag{6}$$

$$\ell_{ij} \le q_{ij} \le u_{ij} \qquad \text{for } i,j=1,\ldots,N \tag{7}$$

where q_{ij} is the flow in link (i,j) defined by initial node i and terminal node j; c_{ij} is a unit cost or priority factor (negative cost represents a benefit) on flows in link (i,j); ℓ_{ij} is the lower limit on flow in link (i,j); u_{ij} is the upper limit; and N is the total number of nodes. Constraint (6) guarantees that mass balance is satisfied at every node.

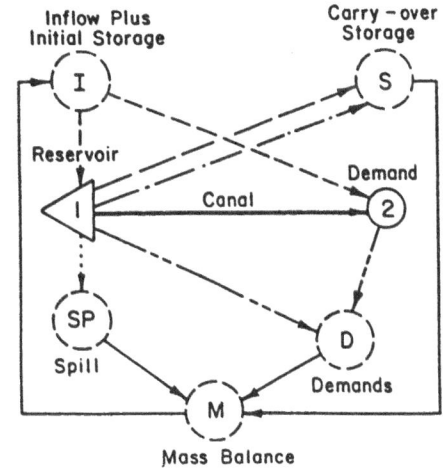

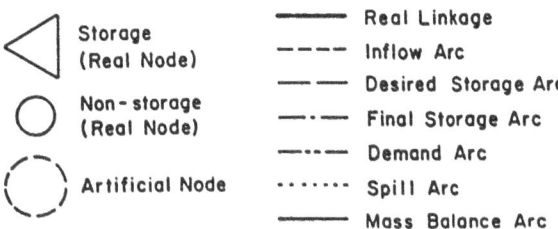

Figure 3. Illustration of a node-arc configuration of a network simulation model.

MODSIM allows the use of "costs" that can be real costs or benefits (i.e. negative costs), or simply operational priorities assigned to certain nodes and links that serve to rank operational alternatives. Again, most other available general-purpose river basin models do not have this capability. With these models, various demands and operating priorities cannot be ranked by the user.

Although technically speaking, MODSIM is an optimization model, it is actually using optimization to efficiently perform simulation in this case. The usual trial and error processes required to make sure that flows are allocated according to priority in a river basin are replaced with an efficient algorithm that converges rapidly to the desired solution. The network flow problem is solved iteratively in a sequential fashion over time, so it is not performing true, fully dynamic optimization.

MODSIM employs the out-of-kilter (OKM) algorithm (Clausen, 1968). Unlike other network programming-type algorithms, an initial feasible solution need not be provided when using the OKM. It is essentially a primal-dual linear programming algorithm specifically designed for efficient solution of minimum cost network flow problems. Studies performed by Glover et al. (1974) seemed to indicate that the OKM was much less efficient than specially designed primal methods. However, recent studies by Bertsekas and Tseng (1988) show that primal-dual methods similar to the OKM are still the most efficient approaches.

The current version of Program MODSIM is primarily intended for obtaining daily, weekly or monthly management guidelines over an entire river basin or selected subbasin. The model is not well suited to short term flood control operations requiring streamflow routing so its primary usage is in normal operations. The model is capable of generating operational plans that satisfy specified targets, priorities and constraints. It can also be used to evaluate tradeoffs between conflicting uses during periods of deficient water availability. This information can provide a rational, documentable basis for making difficult water allocation decisions.

4. Application to Nizao River Basin, Dominican Republic

The Nizao River Basin in the Dominican Republic includes important projects for hydropower production and water supply for irrigation (Figure 4). In addition, being relatively close to Santo Domingo, the capital of the country, there are plans for further development for domestic and industrial water supplies. The risk of hurricane activity in the Caribbean also makes flood control in the basin a high priority. The major storage project in the basin is Valdesia Reservoir. Its 60 MW power plant represents over 40% of the total rated hydropower capacity in the country and contributes 15% of total hydropower production. It also supplies water for irrigating over 10,000 Ha of important agricultural land. Unfortunately, since its completion in 1976, conflicts have arisen between energy and agricultural users of the system, resulting in unstable energy output and shortages to irrigation water supply.

A comprehensive reevaluation of normal operations in the Nizao Basin was undertaken to maximize capabilities in meeting energy and irrigation requirements during normal or nonemergency operating conditions and reduce conflicts among the various water uses. The need for such a reevaluation had become particularly evident since the destructive effects of Hurricane David in 1979. The study was a joint effort between the Instituto Nacional de Recursos Hidraulicos (INDHRI) of the Dominican Republic, the Instituto Interamericano de Ciencias Agricolas (IICA), and Colorado State University, with funding for the project provided by the World Bank. Detailed results of the study are described in Labadie et al. (1986).

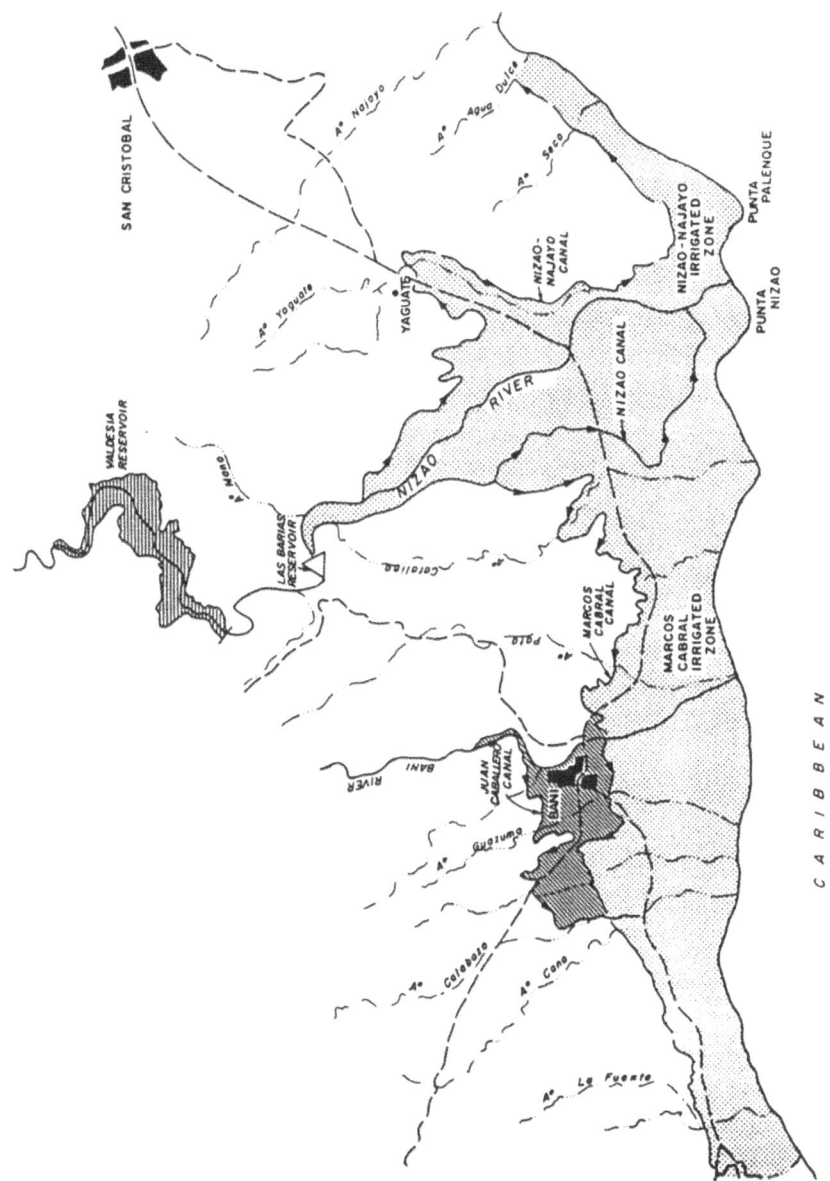

Figure 4. Nizao River Basin, Dominican Republic.

4.1. DESCRIPTION OF STUDY AREA

4.1.1. *Reservoir Subsystem.* The storage subsystem includes two projects: Valdesia Reservoir and Las Barias Reservoir immediately downstream. The construction of other major reservoirs upstream of Valdesia (Jiguey and Aguacate) has been planned, but these are not included in this present analysis. Valdesia Reservoir is located on the Nizao River Northwest of the City of Santo Domingo, capital of the Dominican Republic (Figure 4). The reservoir is impounded by a concrete dam designed for maximum storage of $153 \times 10^6 m^3$ at level 150 m.a.s.l. The spillway runs the entire length of the top of the dam and is controlled by five radial gates.

A tunnel discharges water at a maximum rate of 90 m³/s to a hydroelectric power plant with two Francis turbines rated at 30 MW each. Nonlinearities associated with hydropower calculations are indicated in Table 1, showing turbine efficiencies varying with head and discharge.

TABLE 1. Turbine efficiencies for Valdesia powerplant

Head (m)	Discharge (m³/s)								
	0	40	50	60	65	70	75	80	90
60	.0	.6442	.6893	.7085	.7190	.7131	.7133	.7046	.6797
64	.0	.6346	.6854	.7190	.7344	.7334	.7296	.7219	.6970
67	.0	.6346	.6893	.7258	.7373	.7430	.7440	.7411	.7430
71	.0	.6202	.6893	.7354	.7507	.7565	.7498	.7478	.7248
74	.0	.6144	.6874	.7402	.7526	.7632	.7613	.7574	.7315
77	.0	.6288	.6912	.7373	.7507	.7584	.7594	.7565	.7373
80	.0	.6422	.7037	.7421	.7526	.7622	.7670	.7613	.7334

Hourly energy generation data are compiled by the country's power company (CDE), as well as daily power discharge and water level measurements. This information was used in mass balance calculations to estimate reservoir inflows due to a lack of directly measured data. Valdesia Reservoir can provide flood control benefits through maintenance of a flood reserve during months where flood danger is high.

Las Barias Reservoir, about 15 km downstream of Valdesia, is of much smaller size ($3 \times 10^6 m^3$ maximum capacity at level 77 m.a.s.l.) with its primary purpose to reregulate daily peak period power releases from Valdesia Reservoir to provide stable discharges to the irrigation canals. Standardized operating criteria for these projects during both normal and flood emergency conditions were not available at the outset of this study, which has resulted in inconsistent energy production and untimely irrigation deficiencies.

Figure 5 shows the flood pool allocation for Valdesia Reservoir supplied from studies by Obeysekera et al. (1986b) which evaluated ideal flood pool levels based on a flood frequency analysis and subsequent routing through Valdesia Reservoir. Although there is no indication of any historical provision for flood control space in the reservoir, especially during periods of maximum flood danger, it was felt that some space should be made available to improve its flood protection capability.

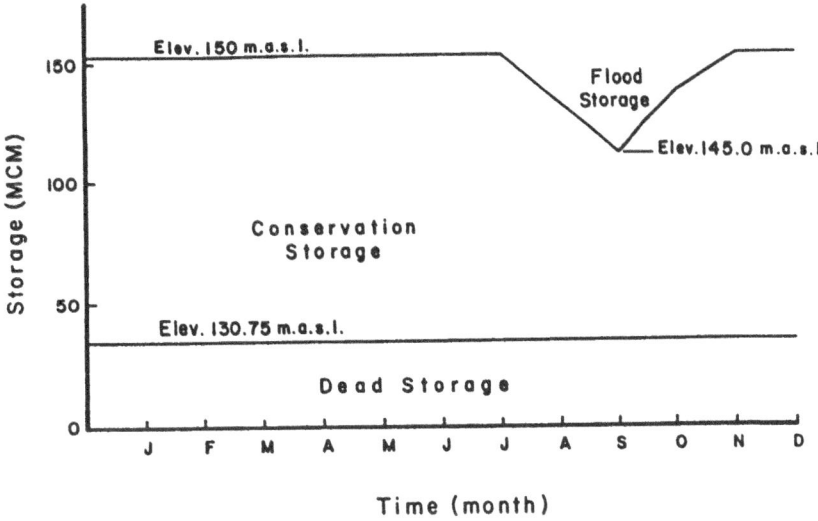

Figure 5. Designation of storage zones for Valdesia Reservoir

4.1.2. *Conveyance Subsystem.* Irrigation water supply from Valdesia and Las Barias Reservoirs is distributed through two major irrigation canals (Figure 4):

1. Marcos A. Cabral: which diverts water from the Nizao River at Las Barias Reservoir, conveying it 47 km to the West of the River, with a total irrigated command area of 8707 Ha. The canal has a maximum capacity of around 12 m³/s (cm) and includes two major laterals. Daily measured flows, with some interruptions, are available over a nine year history of the system.

2. Nizao-Najayo: which diverts flow 34 km to the East of the river and irrigates 1636 Ha. Maximum capacity is estimated at 2.8 m³/s. Daily flow data for this canal are also available over a four year period, but contain many gaps and inconsistencies. An additional canal, Juan Caballero, is currently separate from the Valdesia system, but may be connected in the future.

4.1.3. *Irrigation Subsystem.* Tropical and subtropical crops such as rice, sugarcane, small vegetables, and fruit (particularly large banana production) are grown in an average yearly temperature of 27°C with mean annual rainfall of 800 mm. Rainfall, however, is inadequate for meeting crop requirements and irrigation is needed year round, mainly for the variety of crops grown rather than multiple plantings of the same crop over the year. Overall irrigation efficiencies in the system are quite low. Based on a report by Frederiksen et al. (1985) and interviews with operational personnel for the Valdesia system, overall efficiencies of 35% are assumed. Future completion of lining of the main canals and possibly portions of the major laterals will of course improve efficiencies, even though on-farm application efficiencies may continue to be a problem. Other problems include the fact that farmers at the ends of the major canals tend to face consistent shortages, whereas those at the heads of the canals are not using all of the water available to them. Installation of control structures and siphons in the main canals should greatly improve equity in water distribution along the entire length of the canal.

The total irrigation cropping area has been divided into eight sectors as aggregated from 30 previous zone designations. The locations of the new sectors can be seen in Figure 6 (Fredericksen et al., 1985). The 35 or so crop types are combined into ten groups according to simularities in growing season.

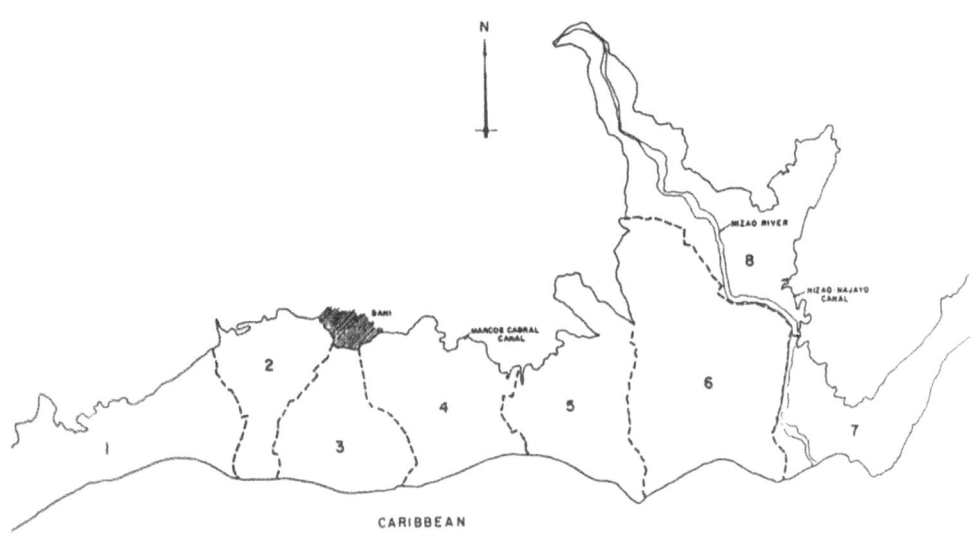

Figure 6. Distribution of new sectors for the Valdesia Irrigation Zone

Crop coefficient data were obtained from a table prepared by the Federal Water Resources Agency for the country, INDRHI (Instituto Nacional de Recursos Hidricos). Discussions with project engineers concerning certain crops not included in the table resulted in agreement that the FAO Manual (Doorenbos, 1975) should be consulted to complete this information.

The key to efficient water use is estimating as accurately as possible actual water needs and then minimizing both excesses and shortages. For this study, the modified Penman method (Kincaid and Heermann, 1974) was used for estimating crop water requirements, with precipitation contributions to meet crop water demands estimated using a procedure developed by Morel-Seytoux and Restrepo (1985). Use of these models requires extensive and accurate meteorological data, which were found to be severely deficient in this study. Meteorological data have been collected over limited time intervals at stations within or proximate to the irrigation zone. Though there are several years of daily climatological measurements available, processing is time consuming because the data are not in computer readable form. In spite of these deficiencies, it was decided that use of these models would provide impetus for improved data collection and processing. Calculated mean weekly consumptive use for the Valdesia irrigation subsystem were used as the basis for all subsequent modelling runs with CSUDP and MODSIM.

4.2. STOCHASTIC INFLOW CHARACTERISTICS

Due to the existence of strong monthly serial correlation among inflows to Valdesia Reservoir, the probabilities of inflow in a given month should be conditioned on measured inflows that occurred in the previous month. A set of discrete conditional probability distributions were developed for each month using a suite of stochastic data analysis and generation models developed under the direction of Dr. J. Salas at Colorado State University (Obeysekera et al., 1986a). Calculated distributions are in the form of discrete transition probability matrices $P(I_i|I_{i-1})$, where I_i is the inflow in month i and I_{i-1} is the inflow in the previous month.

Several transition matrices of varying order were calculated, with a 12x12 order ultimately selected for developing final operating rules. For the 12x12 transition matrices, inflows for both the current and previous month are divided into twelve classes. There are eleven limits for these twelve classes chosen as 5, 10, 20, 30, 40, 50, 60, 70, 80, 90 and 95 percentiles of the empirical accumulated probability distribution of generated inflow for each month, as illustrated in Table 2 for the month of March. The class mark of each class is taken as the mean value of all flow events in that class.

4.3. MODEL CALIBRATION FROM HISTORICAL DATA

Since MODSIM is to be used for weekly operational analysis, it must also be calibrated to weekly historical data. The approach taken here was to force MODSIM to maintain measured weekly storage levels over the nine year historical operating period, and then see if resulting canal releases and hydropower production calculations reasonably match the observed data.

The network configuration for the Valdesia system is shown in Figure 7. Two data files are required for running MODSIM. The ORGANIZ data file relates primarily to physical system features that will not change, though some editing will be required. This includes storage capacities, area-capacity-head tables, turbine efficiencies, discharge capacity, system link-node configuration, and link capacities. Also included are priority rankings for system demands. The actual demands are read into a different data file called ADATA, which also includes weekly historical inflows, net evaporation rates and hours of turbine generation for meeting energy demands.

The reservoir target storage levels and priorities are also input into MODSIM, with ranking factors between 1 and 99 also assigned to the reservoirs. For calibration purposes, maintenance of the target historical storage levels is given the highest ranking.

TABLE 2. Transition probability matrix for February-March

From February to March , 1100 data

Level	Mark	01	02	03	04	05	06	07	08	09	10	11	12
	Mark	3.9725	5.7109	7.1656	8.8120	10.275	12.016	13.763	15.501	17.678	21.527	26.341	34.457
01	4.0087	0.4364 (24)	0.2000 (11)	0.2545 (14)	0.0182 (1)	0.0182 (1)	0.0545 (3)	0.0000 (0)	0.0182 (1)	0.0000 (0)	0.0000 (0)	0.0000 (0)	0.0000 (0)
02	6.1114	0.2037 (11)	0.2037 (11)	0.2037 (11)	0.1667 (9)	0.1111 (6)	0.0556 (3)	0.0370 (2)	0.0000 (0)	0.0000 (0)	0.0185 (1)	0.0000 (0)	0.0000 (0)
03	7.8893	0.1171 (13)	0.1261 (14)	0.2252 (25)	0.1892 (21)	0.0991 (11)	0.0811 (9)	0.0541 (6)	0.0901 (10)	0.0180 (2)	0.0000 (0)	0.0000 (0)	0.0000 (0)
04	9.8973	0.0273 (3)	0.0455 (5)	0.1636 (18)	0.1909 (21)	0.2364 (26)	0.1545 (17)	0.0545 (6)	0.0909 (10)	0.0182 (2)	0.0182 (2)	0.0000 (0)	0.0000 (0)
05	11.740	0.0182 (2)	0.0727 (8)	0.1455 (16)	0.1182 (13)	0.2091 (23)	0.1091 (12)	0.1364 (15)	0.0909 (10)	0.0545 (6)	0.0455 (5)	0.0000 (0)	0.0000 (0)
06	13.528	0.0091 (1)	0.0545 (6)	0.1000 (11)	0.1455 (16)	0.1273 (14)	0.1455 (16)	0.1273 (14)	0.1364 (15)	0.0727 (8)	0.0636 (7)	0.0182 (2)	0.0000 (0)
07	15.542	0.0091 (1)	0.0000 (0)	0.0818 (9)	0.1182 (13)	0.1091 (12)	0.1545 (17)	0.1818 (20)	0.1182 (13)	0.1364 (15)	0.0909 (10)	0.0000 (0)	0.0000 (0)
08	17.757	0.0000 (0)	0.0000 (0)	0.0273 (3)	0.0818 (9)	0.1091 (12)	0.1273 (14)	0.1545 (17)	0.1545 (17)	0.1545 (17)	0.1364 (15)	0.0545 (6)	0.0000 (0)
09	20.398	0.0000 (0)	0.0000 (0)	0.0273 (3)	0.0455 (5)	0.0455 (5)	0.1091 (12)	0.1091 (12)	0.1545 (17)	0.2273 (25)	0.2091 (23)	0.0727 (8)	0.0000 (0)
10	24.194	0.0000 (0)	0.0000 (0)	0.0000 (0)	0.0182 (2)	0.0000 (0)	0.0545 (6)	0.1000 (11)	0.1273 (14)	0.1818 (20)	0.2273 (25)	0.1636 (18)	0.1273 (14)
11	29.196	0.0000 (0)	0.0000 (0)	0.0000 (0)	0.0000 (0)	0.0000 (0)	0.0182 (1)	0.0909 (5)	0.0545 (3)	0.1636 (9)	0.2364 (13)	0.1636 (9)	0.2727 (15)
12	38.625	0.0000 (0)	0.0000 (0)	0.0000 (0)	0.0000 (0)	0.0000 (0)	0.0000 (0)	0.0364 (2)	0.0000 (0)	0.1091 (6)	0.1636 (9)	0.2182 (12)	0.4727 (26)

Comparison of historical and computed canal flows showed good correspondence, except for a few weeks with suspect measured flows. Results comparing historical and computed flows in Marcos Cabral canal are given in Figure 8. A consistent overestimation of power output was corrected by a slight modification in turbine efficiencies (all values multiplied by 0.96). With these new values, more precise results were obtained. This is justified since the turbine efficiency table is based on newly installed units which can suffer loss in efficiency due to aging, etc.

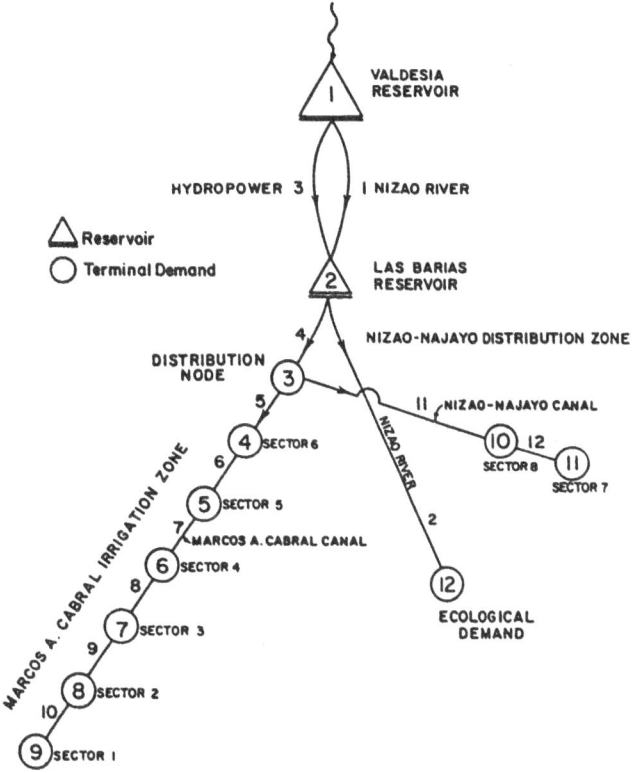

Figure 7. Link-node configuration for the Nizao River System

4.4. DEVELOPMENT OF OPTIMAL MONTHLY GUIDECURVES

Application of Program CSUDP for solving the stochastic dynamic programming problem of Equations 1 and 4 requires development of user-supplied FORTRAN subroutines STATE, OBJECT, READIN and the input data file. Subroutine READIN provides user interfacing for reading in data files for the particular application; i.e., in addition to the normal data provided by a dynamic programming analysis. Subroutine STATE defines the reservoir mass balance in either the inverted or noninverted forms (Equations 3 and 4), with the inverted form used for this particular study. The objective function is defined in Subroutine OBJECT. It was decided to initially select an objective function of maximizing total expected energy production, and then determine from the results if adequate releases would result in stable power capacity levels and reliable water supply for irrigation. If not, then further adjustments would be made, in consultation with the Valdesia Reservoir management staff.

364

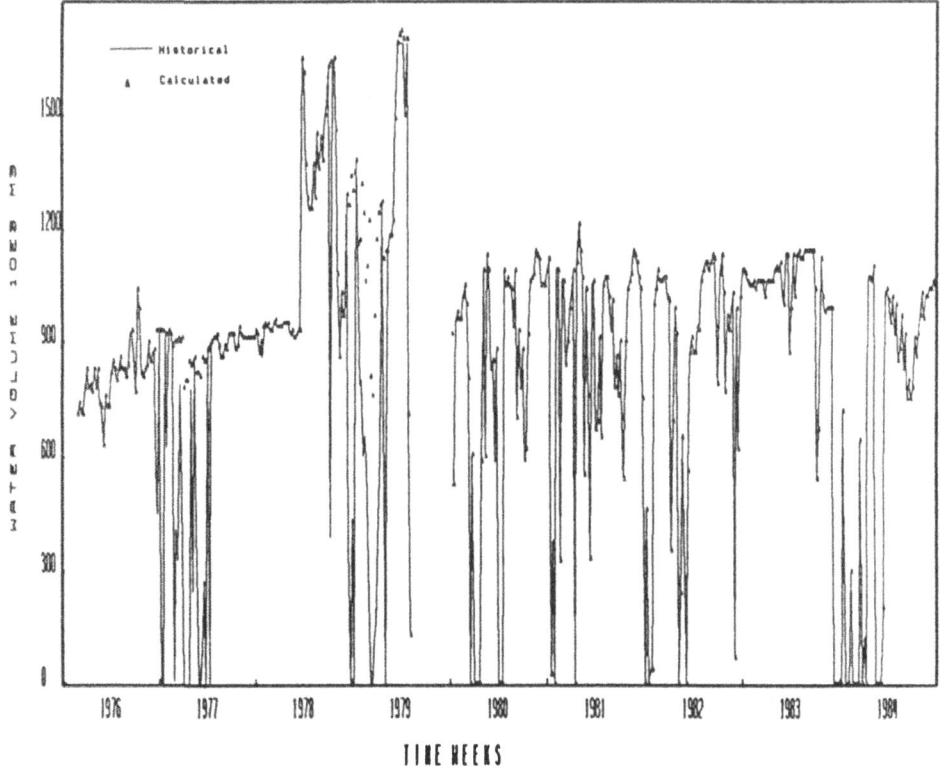

Figure 8. Comparison of historical and calculated flows in Marcos Cabral canal (1000 m³).

When it was attempted to include irrigation demands as explicit lower limits on reservoir releases, infeasibilities always occurred. It turns out that on rare occasions it was impossible to meet the irrigation requirements based on Valdesia release alone. Subsequent Monte Carlo analysis which included Las Barias Reservoir revealed that releases from Las Barias storage could meet these small shortages. Therefore, to encourage the model to attempt to meet irrigation demands if possible, a penalty term is subtracted from energy production if shortages occur.

Successful runs with CSUDP with seasonal inflow transition probability matrices resulted in a family of optimal target storage curves $V^*_{t+1}(V_t, I_{t-1})$ conditioned on current discrete storage levels V_t and discrete inflow classes I_{t-1} for the previous month. Four cyclic annual iterations were required for convergence to stationary target storage levels, for a total of 48 stages in the recursive calculations.

As an example, optimal storage rule curves for the month of October are shown in Figure 9. Colour coded plots of these rules were provided to the system operators, as well as more detailed tabular values to provide backup to graphical rules. Reservoir operators obtain optimal weekly target storage levels by first noting the initial storage level and then moving vertically to the appropriate plot corresponding to the previous month inflows. Some interpolation may be needed between two plots, or the operator can refer to the tabulated values. Subsequent horizontal movement to the ordinate gives the optimal end-of-month target storage level, which is then linearly interpolated into weekly targets based on the starting storage level. Storage levels are of course easily converted into reservoir level targets for ease of use by operational staff.

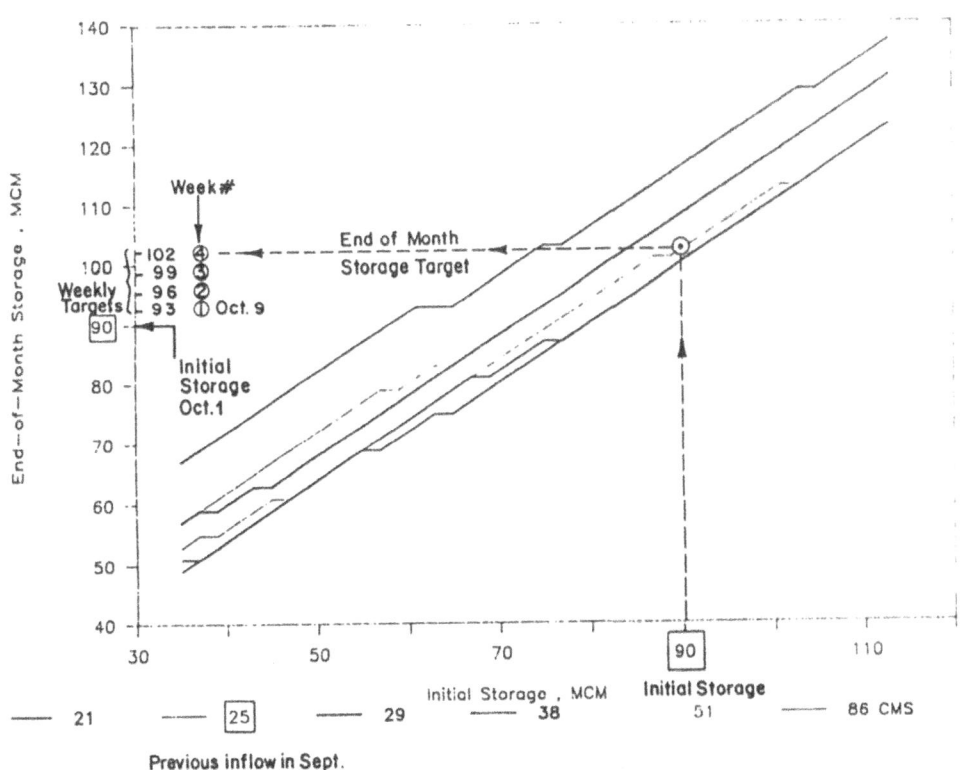

Figure 9. Optimal end-of-month storage guidecurves from "inverted" stochastic dynamic programming for the month of October with linear interpolation into weekly targets.

A Monte Carlo analysis was carried out whereby 400 years of monthly synthetically generated data for inflows and hours of turbine generation were input to Program MODSIM, along with the optimal end-of-month storage guidecurves from CSUDP. The results were extremely encouraging for irrigation supply, with virtually zero risk of incurring shortages under the optimum target storage policies.

4.5. COMPARISON OF HISTORICAL AND OPTIMAL OPERATIONS

The optimal stochastic DP operating rules were applied to historical data for the period from Hurricane David to 1984 in order to assess the value of using these operating rules as compared to what was actually done historically. The monthly DP storage guidecurves were broken into weekly targets by a simple linear interpolation method described previously. These storage guidecurves were then introduced into MODSIM and run over the historical period covering a 240-week period of August 12, 1980 to December 31, 1984 after the occurrence of Hurricane David.

It is important to note in this comparison that the optimal guidecurves do not benefit from any foreknowledge about the basin hydrology. Only probabilistic information is supplied to the stochastic dynamic programming optimization, as opposed to actually historic values that would be required for a deterministic analysis. Although one might argue that the transition probabilities represent some measure of foreknowledge, the stochastic analysis of inflows prior and subsequent to Hurricane David revealed little difference. Therefore, one could assume that the reservoir system operators could have had this information made available to them prior to the 240-week historical period and would have been able to employ the optimal guidecurves.

The results of the comparison are shown in the following figures. A comparison of storage levels is given in Figure 10, which indicates that the DP operating rule retains substantially higher storage levels. The lowest level occurs around week 216. Though maintenance of high storage levels is ideal for hydropower, these results suggest that there was sufficient water historically for additional deliveries for domestic supply, or perhaps increased irrigation demand, if the optimal rules were followed. The effects of these increases on energy production would have to be assessed. Notice in these results that care was taken to make sure that optimal DP storage targets began and ended with the same storage levels observed historically.

Figure 11 compares the average weekly power output. Table 3 compares the mean and standard deviations of the historical vs optimal power output and confirms a substantial increase as a result of the latter. Comparison of the mean and standard deviations for energy production are also given, and indicate an 11.3% increase in total energy production as a result of using the optimal DP rules, with very similar standard deviation. That is, the seasonal energy variations are quite comparable.

Table 3 also shows that there were no shortages produced as a result of the DP optimal operating rules, while showing that the historical shortages were substantial. Use of the DP optimal rules would obviously have greatly benefitted irrigation during this time.

4.6. ECONOMIC BENEFITS OF OPTIMAL OPERATION

Based on the economic analysis presented in Shaner (1985), and the foregoing results of the historical comparison, it is possible to provide some rough estimates as to the increased economic benefits that would have occurred over the historical period, assuming that the optimal DP operating guidecurves were followed.

For hydropower, analyses showed that if additional annual hydropower energy output of up to 8 GWH could be made available, then a replacement benefit of DR$0.53/kWh for diesel power plants could be accrued on the increased output. Notice from Table 3 that the increase in average annual energy as a result of following the optimal guidecurves is over 9 GWH. This translates into average annual increased benefits of DR$4.24 million.

TABLE 3. Comparison of historical and optimal operations (August 12, 1980 to December 31, 1984)

| | Historical | | Optimal | | % change |
	mean	standard deviation	mean	standard deviation	in mean
Power					
-weekly (MW)	30.6	2.7	37.8	4.4	+23.5%
Energy					
-weekly (MWH)	1698	1276	1832	1252	+ 7.9%
-annual (GWH)	88.3	-	95.3	-	+ 7.9%
Irrigation Shortages					
-frequency*	25.8%	-	0%	-	no
-ave. annual (m³/s)†	2.18	-	0	-	shortages

* based on the number of weeks shortages occurred divided by total number of weeks (221).

† only for weeks where shortages occurred.

For irrigation benefits, it was found that additional water supply would likely be used to expand the irrigated acreage. Under the optimal operating policy, it was found that all current irrigation demands could be met, with additional spills occurring at Las Barias. Analysis of these spills show that a 10% increase in irrigation supply is plausible, and from previous analysis, that there should be sufficient additional capacity in the canals to carry the increased flows. It was found, for the most part, that spills at Las Barias occur when there is excess capacity in Valdesia above the optimal guidecurve storage levels. This means that some of these spills could be temporarily stored in Valdesia for later release for meeting excess irrigation requirements. It is unlikely that this would greatly affect energy output.

Assuming that a 10% increase in water supply can correlate with a 10% increase in irrigated area, the economic studies attempted to translate this into associated values of crops at both the heads and ends of the canals. The result is an estimated increase in average annual benefits of DR$3.15 million/yr (for irrigation-headenders) and DR$0.54 million/yr (for irrigation-tailenders), for a total expected annual benefit of DR$7.93 million.

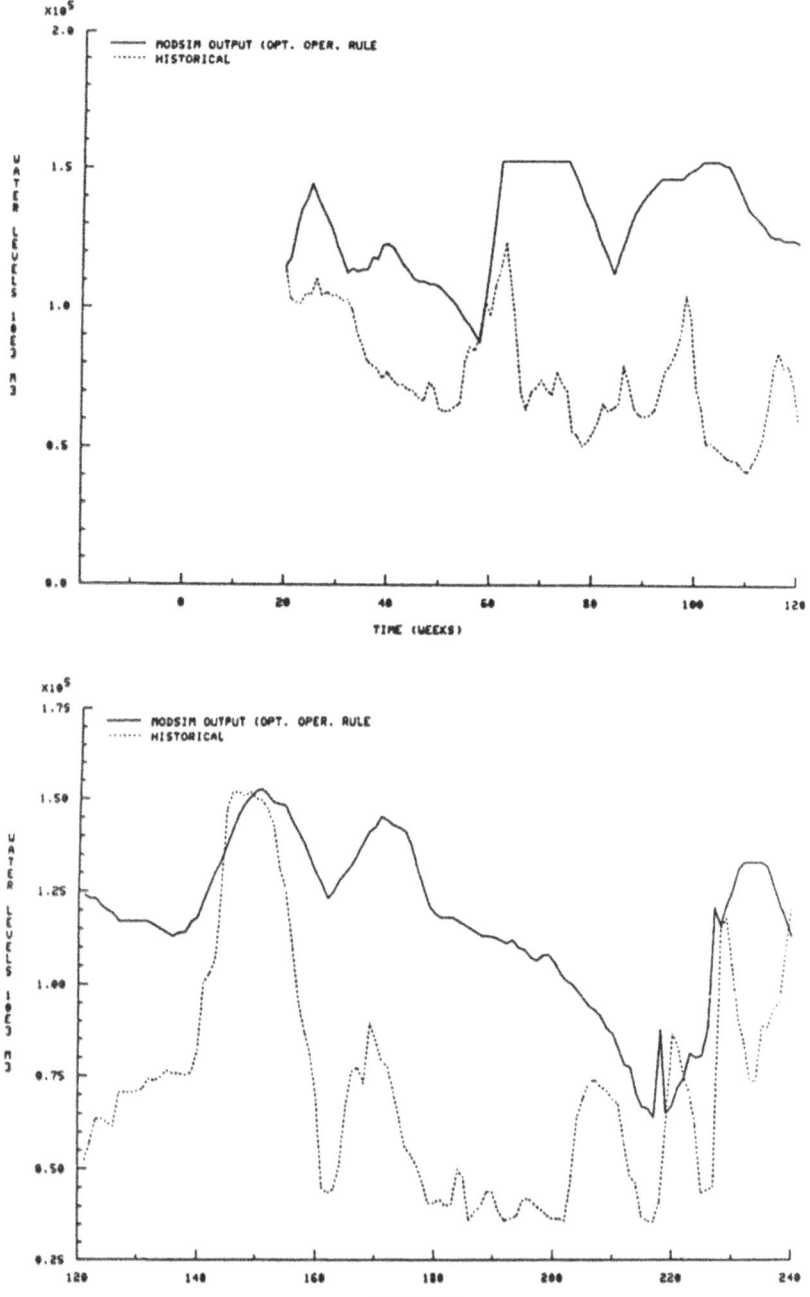

Figure 10. Historical vs optimal storage levels at Valdesia Reservoir.

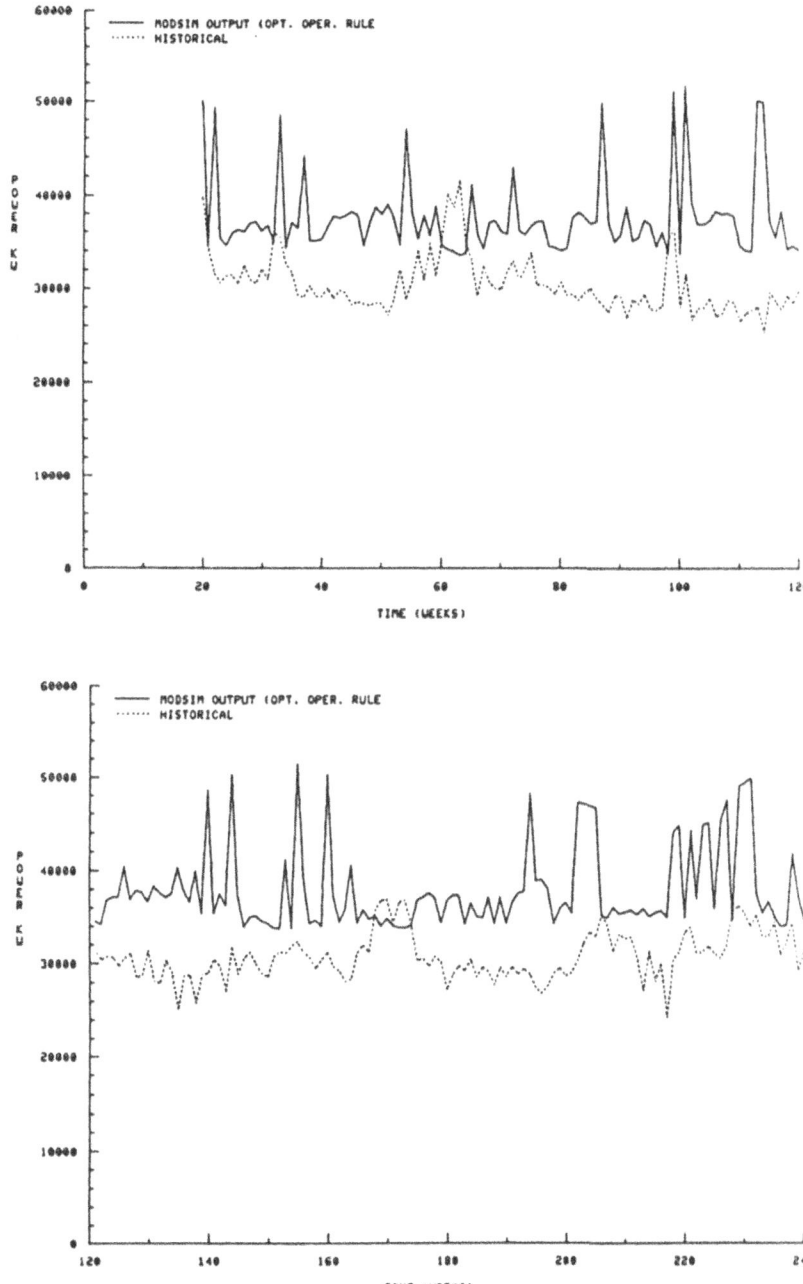

Figure 11. Historical vs optimal power capacity at Valdesia Powerplant.

5. References

Bertsekas, D. and Tseng, P. (1988) 'Relaxation methods for minimum cost ordinary and generalized network flow problems', Operations Research 36(1), pp 93–114.

Bureau of Reclamation (1985) 'CRSS system overview', Engineering and Research Center, Denver, Colorado.

Clausen, R. (1968) 'The numerical solution of network problems using the out-of-kilter algorithm', Rand Corporation, RM-5456-PR, Santa Monica, California.

Doorenbos, J. and Pruitt W. (1975) 'Crop water requirements', Irrigation and Drainage paper No. 24, FAO, Rome.

Dreyfus, S. and Law, A. (1977) 'The art and theory of dynamic programming', Academic Press, New York.

Eichert, B. (1989) 'Multi-purpose, multi-reservoir simulation on a PC', in J. Labadie et al. (eds), Computerized Decision Support Systems for Water Managers, ASCE, New York, pp 313–327.

Faux, J., Lazaro, C. and Labadie, J. (1986) 'Improving performance of irrigation/hydro projects', Journal of Water Resources Planning and Management 112(2), ASCE, pp 205–224.

Frederiksen, Kamine and Associates (1985) 'Informe preliminar de diagnostico: proyecto Nizao-Valdesia, Organización de Operación y Mantenimiento', Santo Domingo, the Dominican Republic.

Georgakakos, A. and Meeks, M. (1989) 'Stochastic control of hydropower systems', in J. Labadie et al. (eds), Computerized Decision Support Systems for Water Managers, ASCE, New York, pp 559–571.

Glover, F., Karney, D. and Klingman, D. (1974) 'Implementation and computational comparison of primal, dual and primal-dual computer codes for minimum cost network flow problems, Networks 4, pp 191–212.

Graham, L., Labadie, J., Hutchinson, I. and Ferguson, K. (1986) 'Allocation of augmented water supply under a priority water rights system', Water Resources Research 22(7), pp 1083–1094.

Harboe, R. (1977) 'A stochastic optimization model of the Lech River system', in L. Gottschalk et al. (eds), Stochastic Processes in Water Resources Engineering, Water Resources Publications, Littleton, Colorado.

Hydrologic Engineering Center (1988) 'Documentation and user manual for Program HEC 5: simulation of flood control and conservation systems', US Army Corps of Engineers, Davis, California.

Kincaid, D. and Heermann, D. (1974) 'Scheduling irrigations using a programmable calculator', Agricultural Research Service, ARS-NC-12, US Department of Agriculture, Fort Collins, Colorado.

Labadie, J., ed., (1980) 'Application of dynamic programming to water resources management', Notes for Short Course, Colorado State University, Fort Collins, Colorado.

Labadie, J. (1987) 'River basin network flow model: MODSIM', user's manual, Dept. of Civil Engineering, Colorado State University, Fort Collins, Colorado.

Labadie, J. (1983) 'Drought contingency model for water control at Corps reservoirs: Cumberland case study', Completion Report, Dept. of Civil Engineering, Colorado State University, Fort Collins, Colorado.

Labadie, J. (1990) 'Dynamic programming with the microcomputer', in A. Kent (ed.), Encyclopedia of Microcomputers, Marcel Dekker, New York.

Labadie, J., Phamwon, S. and Lazaro, R. (1986b) 'A river basin network model for conjunctive use of surface and groundwater', Proceedings of Water Forum '86, ASCE, New York.

Labadie, J., Fontane, D., Floris, V. and Chou, N.F. (1986a) 'Manuales de operacion de modelos computarizados para la operacion normal de sistemas de embalses', Dept. of Civil Engineering, Colorado State University, Fort Collins, Colorado.

Murtaugh, B. and Sanders, M. (1984) 'MINOS 5.0 user's guide', Systems Optimization Laboratory, Dept. of Operations Research, Stanford University, California.

Obeysekera, J., Tabios, G., Pons, F., Salas, J. and Shen, H.W. (1986a) 'Operational and safety studies of Valdesia Reservoir: hydrologic studies', Vol. I, Final Report, Contract IICA/INDRHI/CSU, Dept. of Civil Engineering, Colorado State University, Fort Collins, Colorado.

Obeysekera, J. Tabios, G., Pons, F., Salas, J. and Shen, H.W. (1986b) 'Operational and safety studies of Valdesia Reservoir: flood operation studies', Vol. III, Final Report, Contract IICA/INDRHI/CSU, Dept. of Civil Engineering, Colorado State University, Fort Collins, Colorado.

Roefs, T. (1968) 'Reservoir management: the state of the art', IBM Washington Scientific Center, Wheaton, Maryland.

Ross, S. (1983) 'Introduction to stochastic dynamic programming', Academic Press, New York.

Restrepo, J. and Morel-Seytoux, H.J. (1985) 'Subroutine RAIN', Dept. of Civil Engineering, Colorado State University, Fort Collins, Colorado.

Shaner, W. (1985) 'Esquema sobre los beneficios del estudio del embalse de Valdesia', Short Course Notes, Interamerican Institute of Cooperation for Agriculture, Santo Domingo, the Dominican Republic.

Sniedovich, M. (1979) 'On the reliability of reliability constraints', in E. McBean et al. (eds), Reliability in Water Resources Management, Water Resources Publications, Fort Collins, Colorado.

Stedinger, J., Sule, B. and Loucks, D. (1984) 'Stochastic dynamic programming for reservoir operation optimization', Water Resources Research 20(11), pp 1499-1505.

Trezos, T. and Yeh, W. (1989) 'Stochastic dynamic programming applied to multireservoir systems', in J. Labadie et al. (eds), Computerized Decision Support Systems for Water Managers, ASCE, New York, pp 559-571.

US Army Corps of Engineers (1987) 'Manual for SSARR model - streamflow synthesis and reservoir regulation', North Pacific Division, Portland, Oregon.

Yakowitz, S. (1982) 'Dynamic programming applications in water resources', Water Resources Research 18(4), pp 673-696.

DECISION SUPPORT SYSTEMS IN WATER RESOURCES

J.W. LABADIE
Department of Civil Engineering
Colorado State University
Fort Collins, Colorado 80523
USA

ABSTRACT. Computerized decision support systems can provide the foundation of maximizing the productivity, cost-effectiveness and reliability of complex water resource systems. The three fundamental subsystems of a decision support system (DSS) are: (i) dialogue subsystem, (ii) data subsystem, and (iii) models subsystem. The microcomputer revolution and the emergence of applications of artificial intelligence and expert systems have contributed greatly to development of DSS's which are powerful, low cost, and sensitive to the real issues facing water planners and managers. The explosion in development of DSS's in water resources planning and management was underscored by presentation of a wide range of implementations at a recent Workshop held at Colorado State University. As a case study, a successfully implemented DSS for investment planning in river basin salinity control is described.

1. Introduction

A controversial article by Rogers and Fiering (1986) entitled "Use of Systems Analysis in Water Management" presented a pessimistic view of the application of systems analysis in water resources planning and management. Based on interviews with the most important US federal agencies dealing with water resources issues, as well as consultation with certain state agencies and knowledgeable practitioners, they concluded that " ... US planning practice ... does not turn to systems analysis with impunity", and there is " ... ample reason to question the value of the continuing emphasis on training students, and promoting professors in water systems theory".
  These conclusions were based on two key assumptions enlisted by the authors: (i) " ... model implies optimization ..." and (ii) " ... techniques ... called 'operations research', 'systems analysis', 'management science', and 'cybernetics' ... are subsumed under the designation systems analysis". This is in direct contrast with definitions proposed by authors such as Hall and Dracup (1970), Loucks et al. (1981), and Votruba (1988) which are considerably broader than these terms. In fact, it is apparent that Rogers and Fiering (1986) actually have questioned the merits of applying the techniques of mathematical programming in water resources, rather than

373

J.B. Marco et al. (eds.),
Stochastic Hydrology and its Use in Water Resources Systems Simulation and Optimization, 373–396.
© 1993 *Kluwer Academic Publishers*.

systems analysis in general. It is a mistake to regard systems analysis studies as only those that utilize formalized mathematical programming and optimization techniques. The key is using tools that can treat water project features as a system, and thereby consider the complex interactions occurring between the various project components and objectives.

The criticisms of Rogers and Fiering (1986) appear to focus on two major points: (i) "...[systems] models are not usually concerned with what decision makers care about!", and (ii) "... optimization, as an element of a systematic solution, frequently is not so much misused as underused in terms of its ability to identify a wide range of alternative solutions ...". It is ironic that in the month prior to the publication of the article by Rogers and Fiering (1986), an entire issue of the Journal of Water Resources Planning and Management, American Society of Civil Engineers (ASCE), was devoted to the topic: "Computerized Decision Support Systems for Water Managers". The lead article in this issue by Labadie and Sullivan (1986) stressed that the tools and techniques of systems analysis should be embodied within broader scoped "decision support systems" if we are to see successful implementations. The remainder of the journal issue presented a variety of actual, real-world decision support systems, both for large and small-scale cases. An article by Allen and Bridgeman (1986) highlighted the successful application of mathematical programming (in this case, dynamic programming) to a number of hydropower system studies, one of which produced documented increased revenues of $5 million annually through improved system management.

As a follow-up to this dedicated issue, a Workshop was recently held at Colorado State University on this same topic (Labadie et al., 1989). This was the third in a series of Workshops sponsored by the Operations Management Technical Committee of ASCE. More than 200 public and private water management experts from the United States, Canada and Europe came to hear over 100 presentations organized into 20 technical sessions, including:

- Advanced Decision Support Systems for Water Control
- Application of Expert Systems in Water Management and Control
- Computer Control Applications in Water Supply and Distribution
- River System Modelling and Forecasting
- Optimization Models in Reservoir System Regulation
- Network Models for Large-scale Multireservoir Systems
- Computerized Flood Forecasting and Warning Systems
- Mainframe to Micro: Conversion of River Regulation Software
- Innovations in Information Processing and Data Base Management
- Computer Modelling and Automation in Irrigation
- Regulatory and Environmental Issues in Water Control

In addition to the technical sessions, special sessions were set aside for on-line demonstrations of decision support software and informal discussions among the participants on topics related to the sessions. Many found the latter to be the most important feature of the Workshop.

Although we have seen a dramatic acceptance of computer-aided engineering by water planners and managers in the past few years, the explosion in computer and information technology has left many bewildered about how best to incorporate this technology into the systems they are responsible for. The technology choices are enormous today, due to lack of standardization in the computer field. A primary purpose of the Workshop was to help water managers narrow down the choices and gain a better perspective on appropriate technology for their particular problems.

It is interesting to note the many successful applications of mathematical programming to a variety of complex water resource projects presented at the Workshop, which runs counter to the claims of Rogers and Fiering (1986). These implementations are summarized in Table 1 (Labadie et al., 1989). The key elements contributing to the success of these applications can be listed as follows:

1. Efforts in applying mathematical programming were not ends in themselves, but only a part of broader-scoped decision support systems involving several departments within the organization and emphasizing the practical usefulness of these tools for decision makers.

2. A strong commitment existed within upper level management to invest in and support the application of sophisticated system tools and computer technology within their organizations, recognizing that systems planning and management complexities can only intensify and that rational application of these tools is essential for future success.

3. There was an availability of a well trained in-house staff which, if not actually developing the decision support tools themselves, were sufficiently knowledgeable to provide intelligent guidance and direction to outside consultants.

4. All systems studies were provided with an effective foundation in data base management.

5. Powerful, user friendly computing resources were made available that allowed engineers to take major responsibility in building the decision support systems, rather than having to rely solely on computer specialists inexperienced in the technical and political challenges of water resources planning and management.

It is in the context of decision support systems that the tools of systems analysis can best be applied. Decision support systems provide the unstructured problem solving framework that allows generation of many alternative solutions. As stated by House (1983). "formal optimization turns out to be an important but not indispensable aspect of operations research/management science", and that "insight turns out to be more important than finding optimal problem solutions". Lacking this framework, the criticisms evoked by Fiering and Rogers (1986) can attain a certain validity.

The purpose of this paper is to provide a perspective on successful development and application of decision support systems in water resources planning and management. The general structure of a computerized decision support system is presented, followed by presentation of a successful DSS implementation for high level investment planning in river basin salinity control. A large federal agency has fully implemented the DSS on MS-DOS based microcomputers using generalized software developed at Colorado State University for project selection guidelines and Lotus 1-2-3 for data base management. Cost savings of over $200 million are attributed to the use of the DSS. An optimization model plays an important role in this DSS, but is sufficiently robust and flexible that the decision-makers are completely independent of outside consultants in its utilization.

TABLE 1. Working implementations of mathematical programming and optimization methods by public water agencies and authorities

Organization	Management Problem	Optimization Method
Severn Trent Water Authority, United Kingdom	Optimal pump scheduling in a complex water supply and distribution systems	Dynamic programming
California Department of Water Resources	Monthly operational planning of California State Water Project - complex system of storage, conveyance and energy production facilities	Incremental dynamic programming; out-of-kilter method
Ontario Hydro, Canada	Optimal operation of multireservoir hydropower system of Niagara River	Discrete differential dynamic programming
Western Area Power Admin., Loveland, Colo.	Hydropower scheduling for multireservoir system in North Platte River Basin	Nonlinear programming - generalized reduced gradient method
Western Area Power Admin., Sacramento, Ca.	Optimal capacity expansion of power facilities for Central Valley Project	Dynamic programming: generalized Benders decomposition
Tennessee Valley Authority	Weekly hydropower scheduling on the Tennessee and Cumberland River Systems	Linear goal programming; one dimensional search; dynamic programming
Hydro-Quebec, Canada	Medium-term hydropower scheduling for large, 18 reservoir systems	Nonlinear programming
Manitoba Hydro, Canada	Hydro-thermal scheduling for large-scale system	Linear programming
British Columbia Hydro, Canada	Revenue maximizing operations for Williston Lake, Peace River Basin	Stochastic dynamic programming
US Bureau of Reclamation, Denver, Colorado	Stochastic operational planning for the Colorado River System	Implicit stochastic dynamic programming
Northern Colorado Water Conservancy District, Loveland, Colo.	Operational planning for large-scale transbasin diversion project	Network flow optimization - out-of-kilter method

2. What is a Decision Support System?

The term "decision support" immediately suggests a practical focus; providing effective, modern, computer-based tools to help water resource system planners and managers in their complicated and challenging tasks.

Sprague (1983) claims that decision support systems (DSS's) should have the following attributes:

- "... attempt to combine the use of models or analytical techniques with traditional data access and retrieval functions"
- "... focus on features which make them easy to use by noncomputer people in an interactive mode"
- "... emphasize flexibility and adaptability to accommodate changes in the environment and the decision making approach of the user"

As illustrated in Figure 1, the three main components of an integrated DSS are:

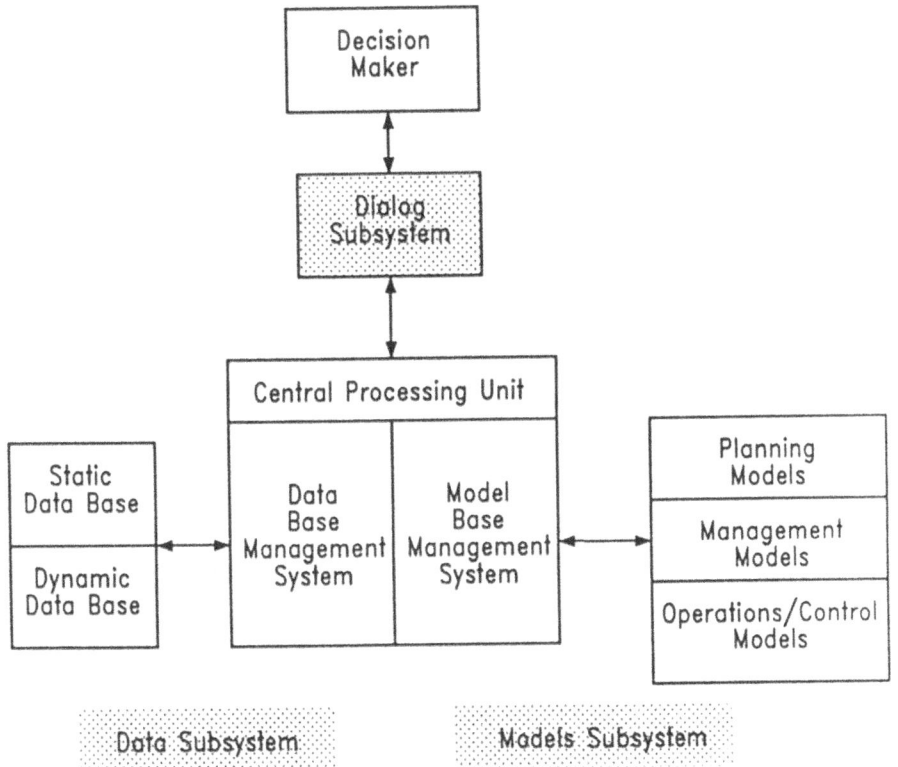

Figure 1. Illustration of the basic subsystems of a decision support system.

1. Data subsystem (acquisition, management, and processing).
2. Models subsystem (for analysis, prediction, and decision guidance).
3. Dialogue subsystem (for interactive man-machine coordination).

2.1. DATA SUBSYSTEM

The data subsystem can generally be categorized into static and dynamic elements. The static data base (or semi-static) includes physical features of the system, as well as economic, demographic, environmental, land use, and other pertinent information. Geographic information systems (GIS) combining powerful digital cartographic capabilities with relational data base management are finding increasing use in data base subsystems (Figure 2).

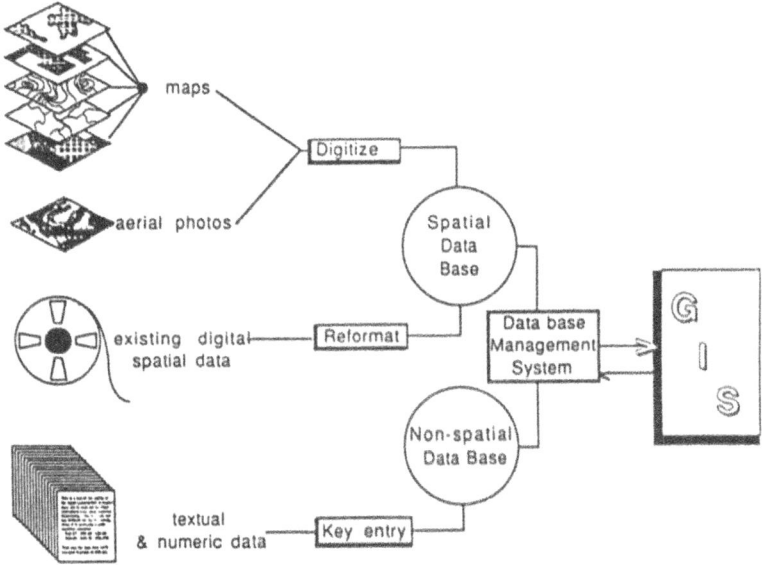

Figure 2. Data capture and formatting for GIS systems (Parker, 1988).

The dynamic data base includes time series information from hydrological, water quality and meteorological data sensing, and includes all associated telecommunication, data processing, and data base management. Identifying and modelling the stochastic features of a dynamic data base is particularly important, and includes both spatial and temporal aspects. Both data bases should allow efficient entry, retrieval, processing, sorting, error checking and storage of information.

In the past, the data subsystem required the use of expensive minicomputer systems specifically designed for these capabilities. Today, low cost microprocessors are able to perform these same functions with high speed, accuracy, and reliability. In addition, networking capabilities are now available for monitoring and controlling highly complex, spatially distributed water operations.

2.2 MODELS SUBSYSTEM

Although there are many implementations of automated data acquisition and processing systems in the water field, direct connection of this module to the models subsystem seems less common. Linkage of data base and analysis modules is critical for maximizing the effectiveness of the DSS. The models subsystem may include simulation models for predicting the impacts of proposed water plans and strategies. Models may also be employed in real-time control for providing operators with a rapid indication of how the system will perform under various control alternatives. Various levels of sophistication of optimization models can be employed for prescribing courses of action among a large number of alternatives. At the lowest level, certain elementary functions such as automated model calibration and low level decision-making can be carried out, with higher level decisions still performed by managers and operators.

Optimization can also be performed at higher levels, but still not automated; that is, managers and operators must decide whether or not to implement the decisions suggested by the optimizing models. The highest level is full closed loop, optimal control. It is likely that we must gain much more experience with DSS's in the water field before we see a proliferation of these kinds of implementations. At such levels, it is important that there is effective interfacing between optimization models and simulation models. Many organizations have developed their own "in-house" simulation models which they trust, but often rely on outside expertise for development of optimization models. Confirmation of the value and effectiveness of strategies and policies developed through formalized optimization should always be tested with appropriate simulation models. This includes simulation of the physical system, as well as stochastic simulation of random phenomena impacting the decision making process to assess risks of failure to meet project objectives.

The importance of generalization and standardization in modelling cannot be overstressed. Many water managers have been frustrated when a specialized model has been developed for a particular water system, and then the system objectives change or there are structural and operational modifications. Often, the original programmers of the model have left the organization and none of the current staff are sufficiently experienced to attempt to modify the model; or else the specialized model was not even developed in-house. A generalized, easy to use model which allows alterations via data input rather than reprogramming is essential. Models should ideally have a feedback capability, whereby decision adjustments can be made in appropriate response to perturbations in the system.

2.3. DIALOGUE SUBSYSTEM

The third module, dialogue management, provides the essential human-machine interface. Dramatic technological advances have been made to enhance interactive decision making, including high resolution colour graphics displays for rapid and effective presentation of information, touch screens, voice recognition, animation for dynamic systems, and various hardware peripherals for hardcopy output; also video image digitizing, cartographic software and digital mapping; CAD-CAM devices, sophisticated workstations, and menu driven software have greatly enhanced dialogue management.

There are two main factors that have heightened the interest of water planners and managers in computerized decision support systems: (1) the revolution which has occurred in microcomputer technology, (2) the increasing popularity of artificial intelligence techniques and expert systems among systems scientists and modellers.

The strong attraction of personal microcomputers has changed our society's attitude about computers from one of distrust to an increasing acceptance as a friendly tool for problem solving. Powerful software packages such as electronic spreadsheets and data base management have been designed with the computer neophyte in mind. Enhanced user interfacing through attractive peripheral hardware such as mouse, digitizing stylus and touch screen data entry, along with attractive colour display equipment with potent graphics software, have had a significant impact. This, combined with the dramatic cost reductions in advanced PC's and 32-bit workstations, has made powerful computing resources available to everyone. Although the concept of "personal computing" is giving way to more efficient microcomputing environments with multiuser, multitasking capabilities, there is no question that PC's provided a strong impetus for today's advances.

The other major contributing factor in the increasing popularity of computerized decision support systems is the growing use of artificial intelligence and expert systems among systems scientists and modellers. Water managers have always complained that their experience and understanding of water systems was given insufficient attention and not properly incorporated into the models developed by systems analysts and computer modelling specialists. With focus on development of expert systems, the opposite view is taken; the domain expert, such as an experienced water system manager, engineer, or operator, is in fact the key element in any modelling study. Knowledge engineers attempt to capture this experience base in an expert system and combine heuristic information with the traditional mathematical modelling structures (Figure 3). In effect, the advent of expert systems has produced a change in attitude among many system scientists and computer modellers which could set the stage for a dramatic increase in DSS implementations in the water field.

2.4. GRAPHICAL DISPLAY SYSTEM FOR INTEGRATED DECISION SUPPORT

One of the earliest pioneers in the development and application of interactive graphics-based systems to water resources management is D.P. Loucks at Cornell University (Loucks et al., 1985). The original "Interactive Water Resources Management Graphics Display System" developed at Cornell University under the direction of Dr. Loucks (French et al., 1979) has been adapted to the UNIX-based system at Colorado State University. All of the graphics programming in the original FORTRAN package have been rewritten in the "C" language. The package can run on virtually any machine connected to the token ring local area network called LANCE at CSU.

The GDS package has been modified to utilize a 3-D graphics library called GSR (Graphics Support Routines), which is similar to the emerging standard PHIGS (Programmer's Hierarchical Interactive Graphics System). Effort is currently being directed at implementing the GDS under X-Windows v.11, a public domain graphics environment developed under the DEC/IBM supported Athena Project at the Massachusetts Institute of Technology. Development of X-Windows is a major attempt to standardize graphics-based technology and maximize portability to any graphics display device connected to an X-Windows server.

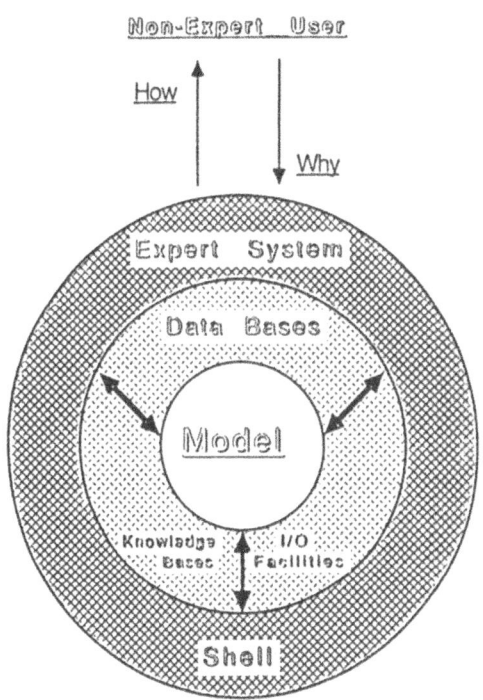

Figure 3. Expert system/model integration (Ludvigsen et al., 1987).

The GDS represents an integrated, graphics-based "shell" for incorporating all the necessary subsystems of a decision support system. Figure 4 shows the ideal GDS framework which is in the process of being constructed. The modular design of the package permits addition of various user-selectable analysis packages which can be connected to appropriate data bases. Knowledge-based preprocessing and postprocessing of information is possible under the GDS environment. Work to date has focused

382

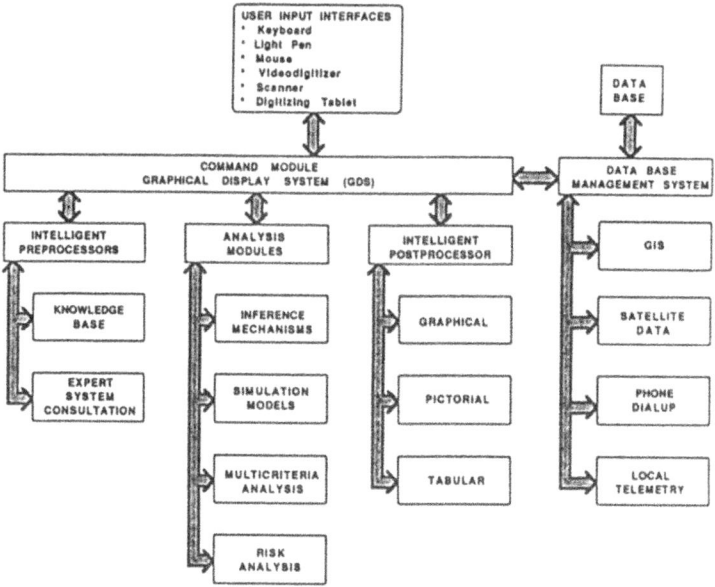

Figure 4. Schematic of interactive Graphical Display System (GDS) for a river basin decision support system.

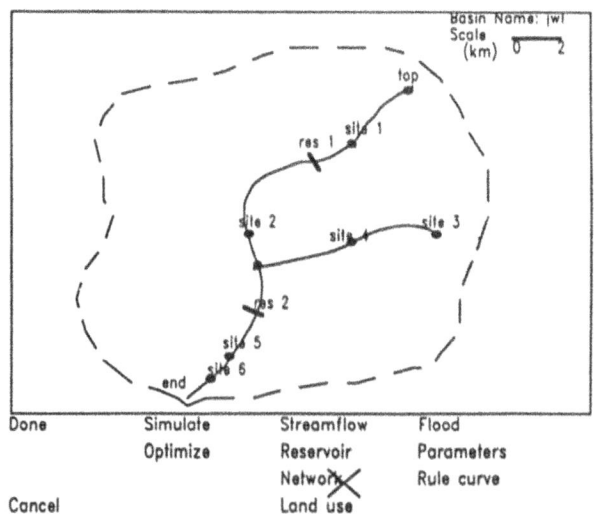

Figure 5. Data input menu and drawing features of the GDS.

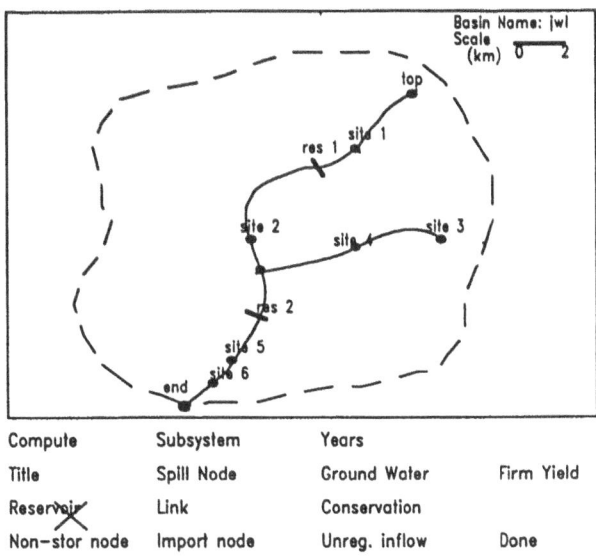

Figure 6. GDS menu features associated with data file development for MODSIM.

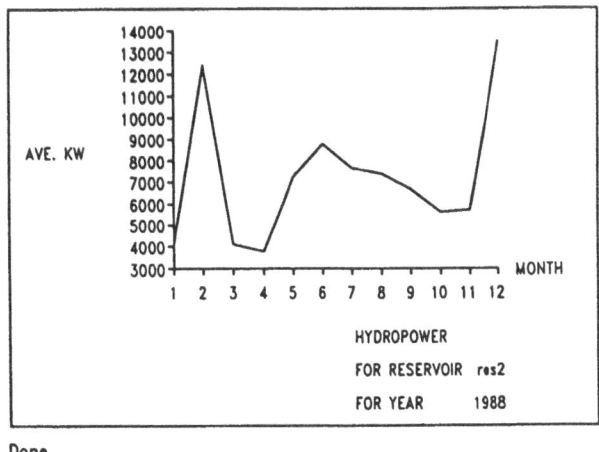

Figure 7. User selective graphical output display from MODSIM under GDS.

on attaching a variety of analysis modules to the package, including the generalized river basin network analysis model called MODSIM (Labadie et al., 1986; Brown and Law, 1989).

Figures 5, 6 and 7 are screen displays showing the use of the GDS shell for executing MODSIM. Users can digitize river basin spatial character- istics by digitizing the tablet or direct input of digital images. Well organized menus and help features provide maximum flexibility for data entry, error correction, processing, analysis, and colourgraphic display of model output. Zoom capabilities in the package provide the potential for analysis of large-scale systems through interconnection of river basin subsystems. MODSIM has also been successfully linked with a relational data base management system called BCS RIM (Boeing Computer Services, Inc., 1985) which is available on a wide range of computing equipment. BCS RIM is also the basis for MS-DOS microcomputer-based tools under the RBASE family developed by MicroRIM, Inc. Application of the integrated network analysis/relational data base management system to studies in the Rio Grande River Basin are described in Graham et al. (1986).

3. Decision Support System for Investment Planning in Salinity Control

3.1. PROBLEM STATEMENT

In response to concern about growing salinity levels in the lower portion of the Colorado River Basin, the Colorado River Water Quality Improvement Program was established through passage of Public Law 93-320 in 1974 by the US Congress. Figure 8 shows a total of 31 salinity control projects proposed under programmes conducted by the Bureau of Reclamation (BOR), US Department of the Interior, and the US Department of Agriculture (USDA), with cooperation from the Bureau of Land Management, the Geological Survey, the Fish and Wildlife Service and the Environmental Protection Agency. Major sources of salinity in the Colorado River Basin include:
- irrigation drainage from inefficient irrigation practice
- leakage from canals and laterals raising saline groundwater levels
- industrial and municipal point sources
- natural sources

The need for these projects is seen in Figure 9 which shows salinity projects without controls in relation to mandated standards. Salinity control measures include:
- improved efficiency in on-farm irrigation systems
- shallow well drainage collection
- brine outfalls
- deep well injection of brine
- canal and lateral lining
- point source collection and disposal

The success of the salinity control programmes is highly dependent on coordination efforts among the various agencies involved. This is accomplished through the Office of the Colorado River Salinity Control Basin Coordinator, located at the offices of the BOR, Denver, Colorado; and the Colorado River Basin Salinity Control Forum.

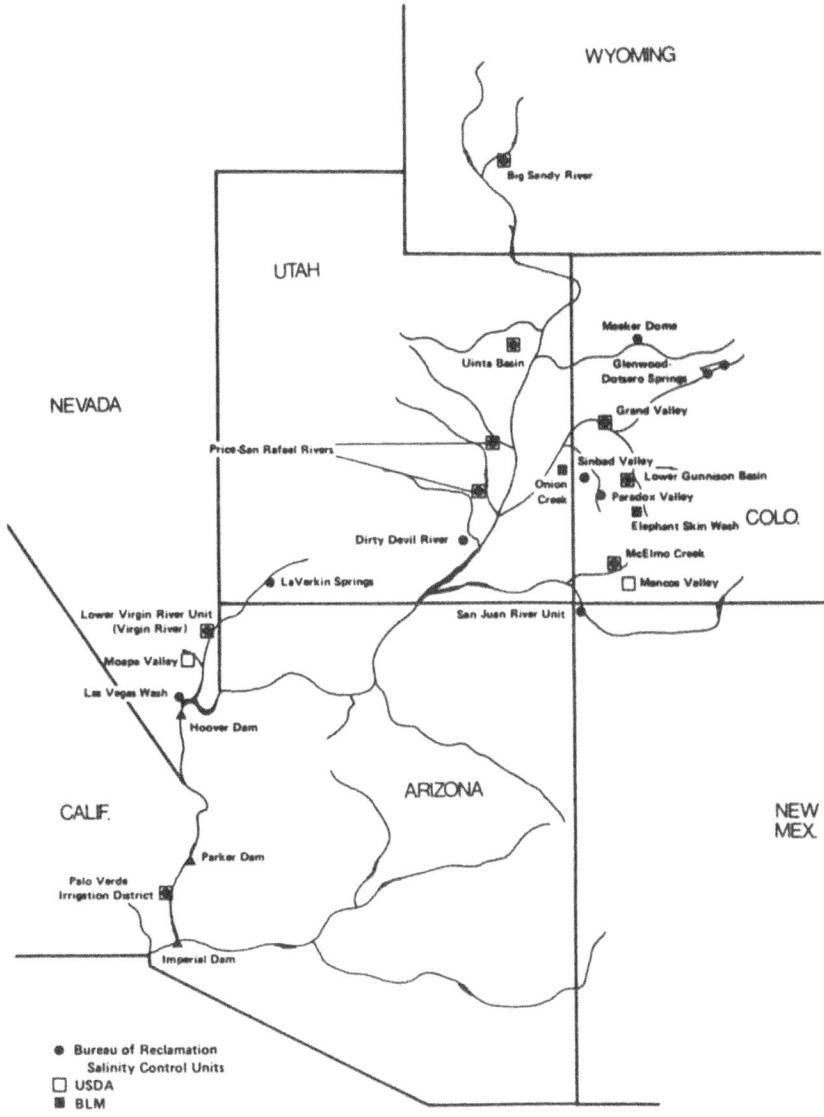

Figure 8. Proposed salinity control projects in Colorado River Basin.

Salinity Projections
without further controls
September 18, 1987

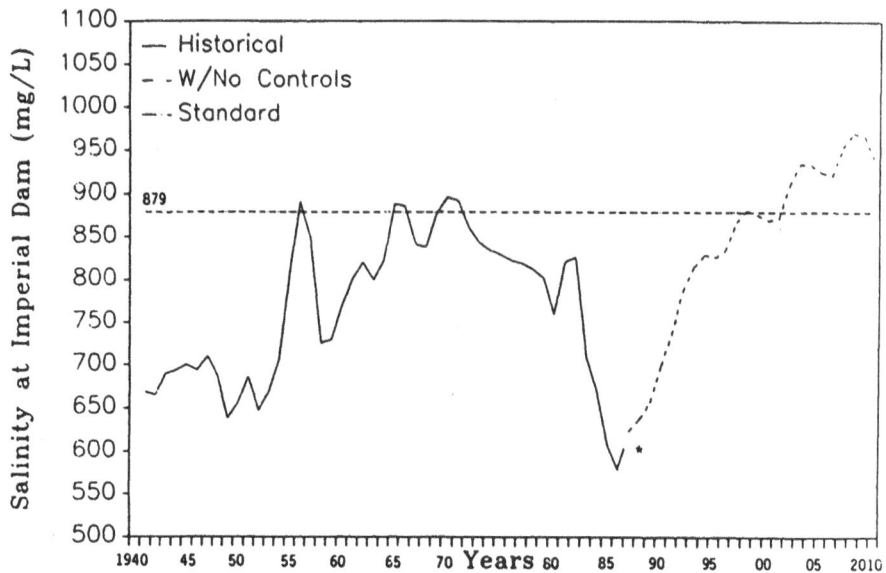

Figure 9. Salinity control projections at Imperial Dam without further controls.

The overall goal of these programmes is the selection and timing of those projects that will meet the basin salinity control standards at the least cost. Finding optimal investment strategies is complicated by the large number of proposed projects with varying load reduction capabilities and costs, and a long-time horizon covering up to 25 years for implementation. The selection process is considered to be highly dynamic, whereby initial choices may need to be updated and modified in the future as the performance of currently constructed projects is monitored and conditions change in the Basin. This requires more than just an optimization model to aid in selecting projects, but an entire decision support system which can be called upon continually by Project staff for updating recommended strategies. Total cost estimates to achieve the salinity control standards were originally $830 million. A study was initiated with Colorado State University to aid in the development of a decision support system for project selection and timing that could hopefully reduce these expected costs, while maintaining the required salinity control standards.

3.2. DATA SUBSYSTEM

A comprehensive data base has been compiled by the BOR and the USDA, in cooperation with other federal agencies, in support of the salinity control programmes. The static data base includes physical descriptions of the project areas, estimated salt load contribution, proposed implementation plan, both current (if already constructed) and future potential salt load reduction capabilities, and economic and financial analyses.

Water quantity/quality simulations were conducted using the Colorado River Simulation System (CRSS) Package developed by the BOR to estimate future salinity reduction targets needed for maintaining a mandated total dissolved solids (TDS) concentration of 879 mg/L at Imperial Dam. An extensive data base was utilized in support of these efforts, which involved back calculation of ungauged natural inflows to the basin using gaged records at Lees Ferry, Arizona, and adjusted upstream for estimated crop consumptive use, reservoir regulation, imports, exports, municipal and industrial depletions and other miscellaneous accretions/depletions.

HISTORICAL RECORD

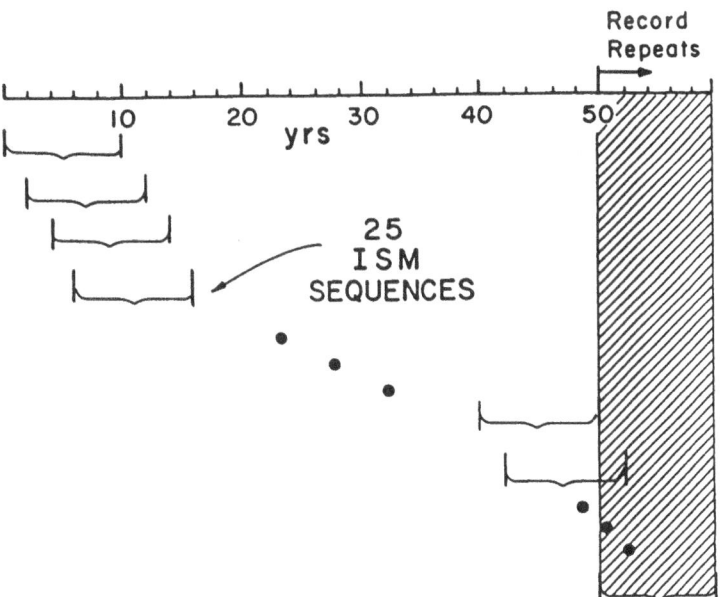

Figure 10. Illustration of indexed sequential method of data generation.

Once the "adjusted" natural inflow data base has been obtained, the BOR often employs a method called "indexed sequential hydrological modelling" (ISM) as an alternative to formal multivariate stochastic modelling techniques for synthetically generating flows in the Basin. This technique has been analyzed and compared in studies by Labadie et al. (1987) in the Central Valley of California, and by Frevert and Cheney (1989) in the Colorado River Basin. A series of hydrochemical traces (i.e. flows and associated TDS concentrations) are generated from the "adjusted" historical record by defining a series of overlapping sequences directly from the data, as illustrated in Figure 10. The aforementioned studies have shown reasonable correspondence with results produced from multivariate stochastic analysis. One advantage of this approach is the exact preservation of spatial correlation of inflows.

In this illustration the historical record is 50 years in length for all spatial locations over the same historical time frame. A series of overlapping ten-year sequences are sorted, with each sequence starting every other year. Monthly flows within each year are exactly preserved in the subsequences. The disadvantages relate to questions about the mutual dependence of the traces and the need to "wrap-around" the data, as shown in Figure 10.

Multiple runs with CRSS were made with the historically based inflow data, but utilizing future basin demand projections developed by the BOR. A best fit curve was then computed representing a mean estimate for target salinity reductions to maintain target concentrations at Imperial Dam. Project staff have developed microcomputer-based data file management structures using Lotus 1-2-3 which facilitate transfer and loading of the data base into the optimal investment timing model.

3.3. MODELS SUBSYSTEM

The least cost investment problem can be formulated as follows:

$$\min_{\substack{u_t, \ x_{t+1} \\ (t=1,\ldots,T)}} \sum_{t=1}^{T} \left[c_t(u_t) + \sum_{\tau=1}^{t} v_t(u_\tau) \right] \tag{1}$$

subject to:

$$c_t(u_t) + \sum_{\tau=1}^{t} v_t(u_\tau) \leq C_{max,t} \tag{2}$$

$$x_{t+1} = x_t + u_t \in E^N \ ; \ x_1 \text{ given} \tag{3}$$

$$0 \leq x_{t+1} \leq 1 \tag{4}$$

$$u_t \in S_t(x_t) \in E^N \tag{5}$$

$$\sum_{\tau=1}^{t} m_t(u_\tau) \geq M_t \tag{6}$$
$$(\text{for } t=1,\ldots,T)$$

where for N total proposed projects, $u_t = (u_{1t},\ldots,u_{Nt})$, with

$u_{it} = 1$ if project i is initiated in period t

$u_{it} = 0$ if project i is not initiated in period t
$$\tag{7}$$

and $c_t(u_t)$ is the total discounted capital cost of initiating projects u_t during period t; $v_t(u_\tau)$ is total discounted operations and maintenance costs during the current period associated with project selection u_τ in period $\tau \leq t$; $C_{t_{max}}$ is the maximum budget for projects selected in period t; x_t is the binary state variable indicating projects initiated up to period t; $S_t(x_t)$ is the set of feasible projects that can begin in period t, based on project contingencies, mutual exclusivity and other stipulated restrictions; $m_t(u_\tau)$ is salt mass load reduction in period t induced by construction of projects u_τ in period $\tau \leq t$; and M_t is the minimum salt mass load reduction for period t required to maintain acceptable salt concentrations at Imperial Dam.

For large values of N, this is a challenging combinatorial optimization problem. Dynamic programming is a popular approach to solving this class of problems, but Bellman's (1957) "course of dimensionality" continues to represent the greatest obstacle to full application of dynamic programming. A number of techniques have been proposed for ameliorating the dimensionality problems associated with dynamic programming, but are still sensitive to the dimensionality of the state-space. An approach called "objective-space dynamic programming", originally developed by Labadie et al. (1980), and later applied by Fontane et al. (1981) and Fontane et al. (1984), offer hope for this class of problems. A general mathematical formalism for the method is presented by Labadie and Fontane (1989).

Objective-space dynamic programming offers the advantage of being highly insensitive to the dimensionality of the state-space. This approach conditions solutions on the one-dimensional objective-space rather than the high dimensional state-space. Sufficient conditions for global optimality are defined based on certain uniqueness requirements in the optimization. Aside from specification of the countability of the finite subset of decision variables, no other assumptions on problem structure or functional characteristics are necessary, including differentiability, convexity, or even continuity.

Details on application of objective-space dynamic programming to this problem can be found in Fontane (1989). Application of the method begins with discretization of the objective-space into intervals Δf, rather than the normal approach of discretizing the state-space. A corridor bounded according to the budgetary constraints $C_{min,t}$ and $C_{max,t}$ (t=1,...,T) is defined, as shown in Figure 11. Although the lower bound $C_{min,t}$ is not included in the original formulation, it is reasonable to place an arbitrary lower bound since expenditures in the salinity control programme will in fact be made each year of the planning horizon.

Assuming $\overset{\Delta}{L_1} = 0$, we have:

$$\sum_{\tau=1}^{t} C_{min,\tau} \leq L_{t+1} \leq \sum_{\tau=1}^{t} C_{max,\tau} \qquad (8)$$

$$(t = 1,\ldots,T)$$

Starting with known initial state x_1, and $\overset{\Delta}{F_1(L_1)} = 0$, the following forward looking dynamic programming recursive relation is defined for stage t: for all discrete L_t in discrete intervals Δc, find

$$F_{t+1}(L_{t+1}) = \min_{L_t, u_t \in S_t(x_t)} [c_t(u_t) + V^*_t(L_t) + v_t(u_t) + F_t(L_t)] \qquad (9)$$

subject to:

$$c_t(u_t) + V^*_t(L_t) + v_t(u_t) + F_t(L_t) \geq L_{t+1} \qquad (10)$$

$$0 \leq x_{t+1} = x^*_t(L_t) + u_t \leq 1 \qquad (11)$$

$$\sum_{\tau=1}^{t} m_t(u_\tau) \geq M_t \qquad (12)$$

$$u_{it} = 0 \text{ or } 1 \ (i = 1,\ldots,N) \qquad (13)$$

$$(\text{for } t = 1,\ldots,T)$$

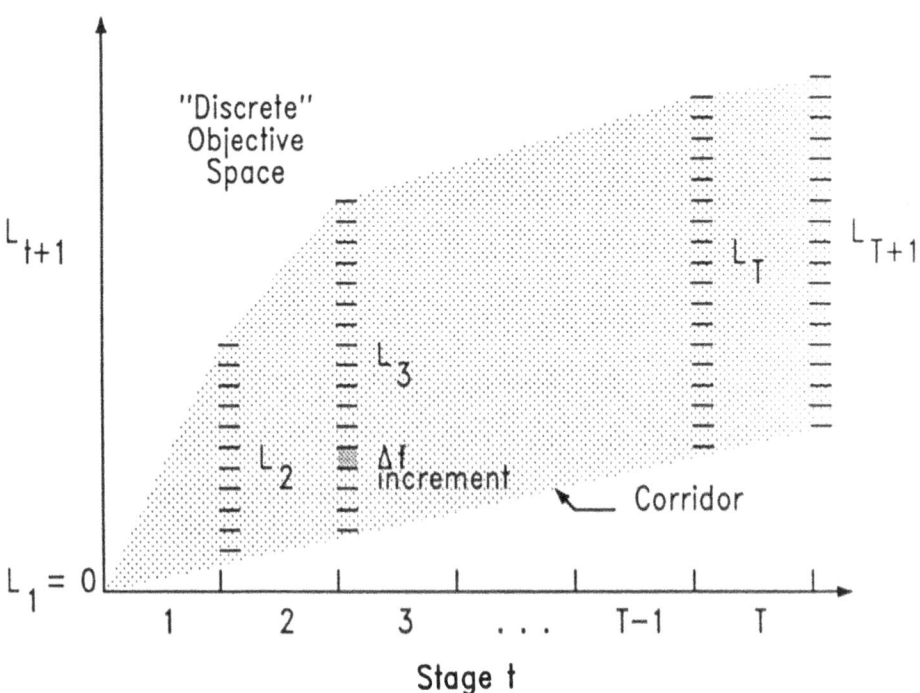

Figure 11. Discretization of objective-space within prescribed "corridor".

The term $V^*_t(L_t)$ is the accumulation of optimal operation and maintenance costs conditioned on bound L_t, and computed by tracing back through stored optimal policies $L^*_{\tau-1}(L_\tau)$ and $u^*_{\tau-1}(L_\tau)$ for $\tau = t, t-1, \ldots, 2$, and then computing

$$\sum_{\tau=1}^{t-1} v_t(u^*_\tau)$$

The uniqueness requirement for the successful use of objective-space dynamic programming requires that for given discrete bound L_{t+1}, a unique system state x_{t+1} can be found over all (possibly) nonunique optimal solutions L^*_2 and u^*_2. For each discrete bound L_{t+1}, optimal (and possibly nonunique) $L^*_t(L_{t+1})$, $u^*_t(L_{t+1})$ and unique optimal $x^*_{t+1}(L_{t+1})$ are obtained and stored in computer memory. The latter are particularly important since the "true" states of the system are needed for the next stage calculations for each discrete bound L_{t+1}.

Continuing in this manner through the remaining stages, the problem for stage T is solved for all discrete bounds L_{T+1}, resulting in function $F_{T+1}(L_{T+1})$ as illustrated in Figure 12.

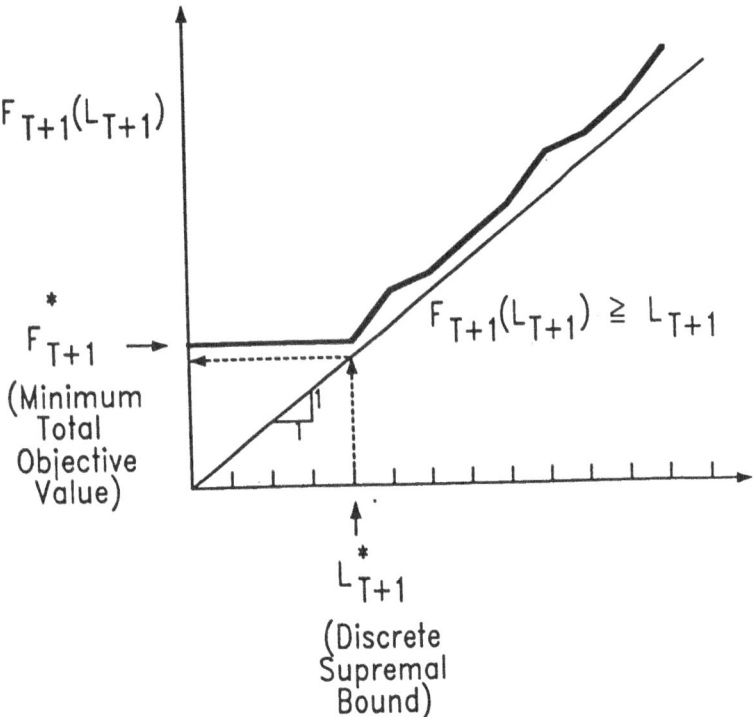

Figure 12. Illustration of final stage optimal cost functional conditioned on bound L_{T+1}.

The supremal bound L^*_{T+1} on the minimum total objective value can now be found by solving:

$$\sup_{L_{T+1}} \quad F_{T+1}(L_{T+1}) \tag{14}$$

where $F_{T+1}(L_{T+1}) \geq L^*_{T+1}$ is the minimum total objective value.

A traceback procedure is initiated to find the optimal decision policy. Having found $u^*_T(L^*_{T+1})$ and $L^*_T(L^*_{T+1})$, then find

$$u^*_{T-1}(L^*_T) \text{ and } L^*_{T-1}(L^*_T), \text{ then}$$

$$\cdot$$
$$\cdot$$
$$\cdot$$

$$u^*_2(L^*_3) \text{ and } L^*_2(L^*_3), \text{ then}$$

$$u^*_1(L^*_2)$$

A limited set of feedback decision policies can be computed for each stage from stored arrays $u^*_t(L_{t+1})$, $x^*_t+1(L_{t+1})$, and $L^*_t(L_{t+1})$. That is, we can define:

$$u^*_t(x_t) = u^*_t(x^*_t(L^*_t(L_{t+1}))) \text{ for all discrete } L_{t+1} \tag{15}$$

$$(t = 1,...T)$$

The objective-space formulation was coded using the generalized dynamic programming package CSUDP (Labadie and Shafer, 1979) as reported in Fontane et al. (1984). BOR staff converted the package to run on an IBM compatible microcomputer. All subsequent use of the model has been conducted by BOR staff, with no direct need for help by CSU staff.

3.4. DIALOGUE SUBSYSTEM

The MS-DOS microcomputer version CSUDP code includes a complete interactive, menu-driven environment for the user to create the subroutines and data files for any application. A built-in editor aids creation of the subroutines and provides automatic compilation and linkage using any of the popular FORTRAN compilers available for PC's under MS-DOS. Subroutine SELECT was designed as a generalized interface to the CSUDP package for the investment planning problem which is called as a user supplied subroutine by CSUDP. All required features of the investment planning problems are included in SELECT, such as:
- discounting of costs
- definition of which projects have been previously started and which are already completed
- target salinity reductions for each year of the planning horizon
- length of construction period for the projects
- costs and salinity reductions of each project, including delayed effects
- length of the pre-construction planning period required for each project

- consideration of certain projects which once initiated, are not interrupted until completion
- specification of certain projects with restricted startup dates
- discrimination between certain projects which provide no salinity reduction benefits until completed, while others can provide intermediate benefits

Any of these parameters can be easily altered and redefined by BOR staff, which provides the flexibility needed for this to be a true decision-support tool. The package also allows interesting studies on the benefits of relaxing some of constraints on annual budget and annual salinity targets, making sure that total budget and salinity controls are still maintained. In some cases, significant reductions in project costs are possible through exploration of some of these alternatives.

3.5. IMPLEMENTATION AND RESULTS

The most recent reporting of results from the use of the DSS for investment planning of salinity control can be found in the "1987 Joint Evaluation of

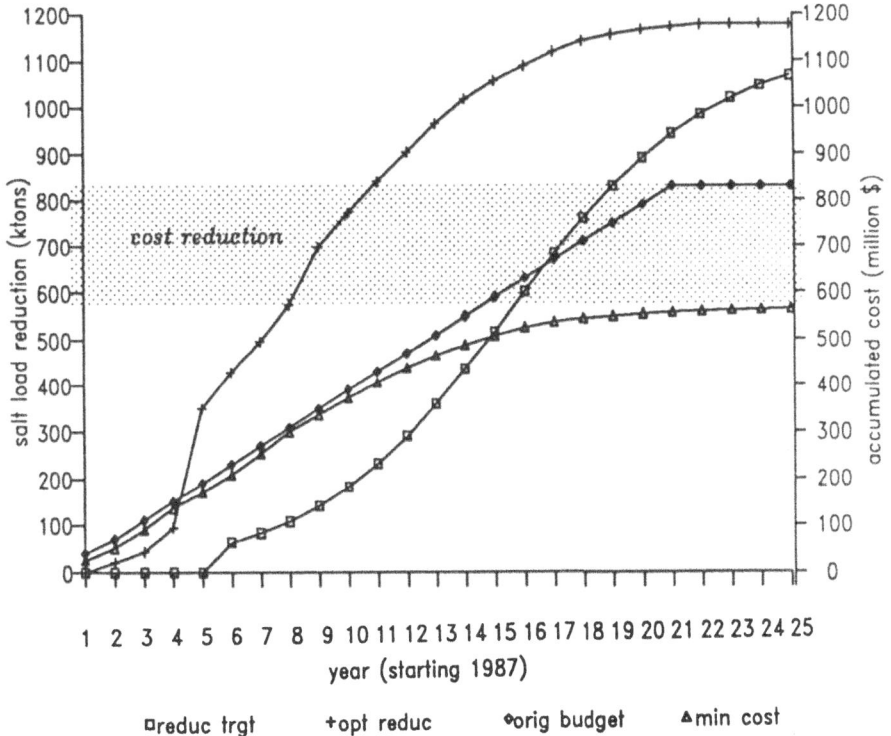

Figure 13. Comparison of original salt reduction target with original budget, with optimum salt reduction levels and minimum cost.

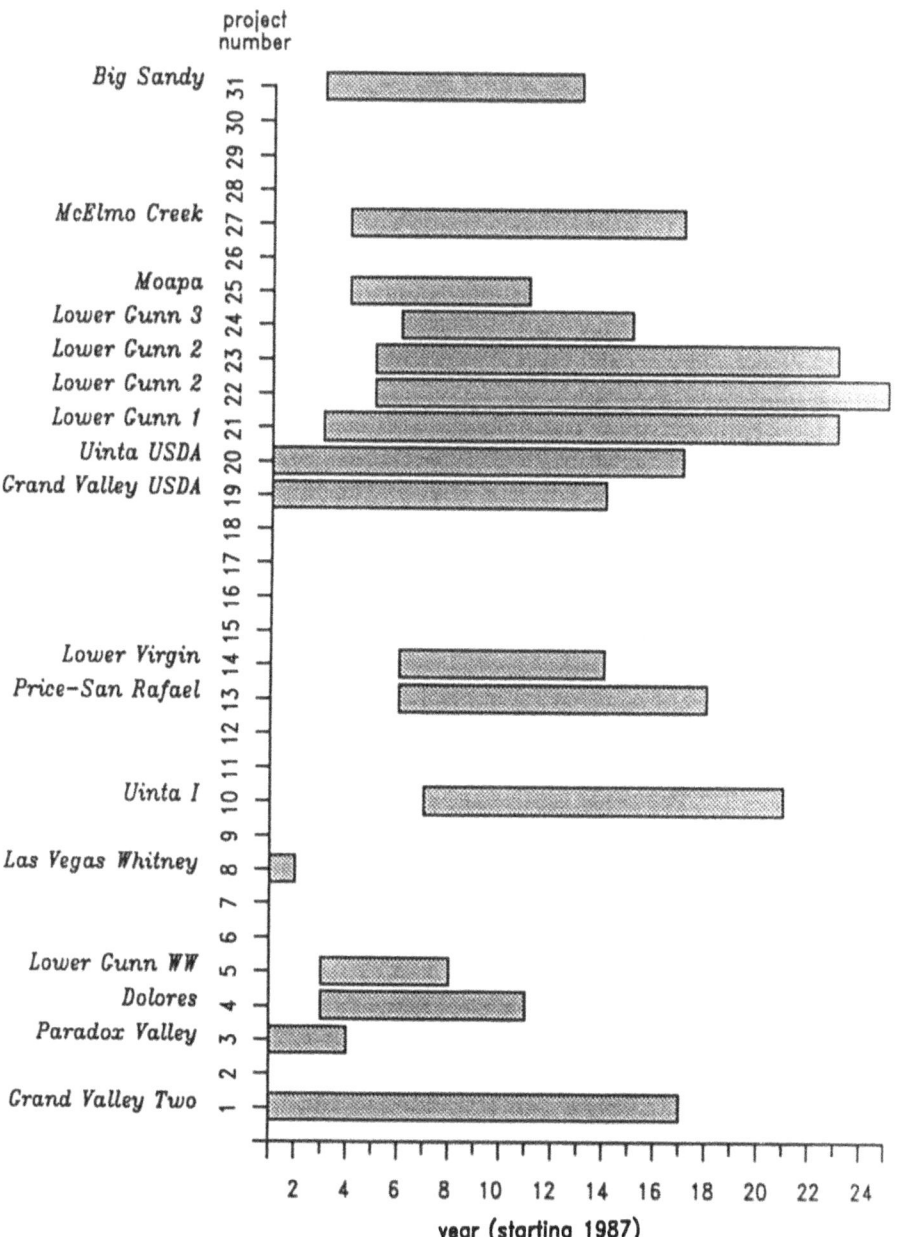

Figure 14. Optimal project selection and time diagram, including length of construction period.

Salinity Control Programs in the Colorado River Basin" (CRWQO and SCCC, 1987). The following results are taken from that report, although subsequent analyses and updates to the programme have been developed but not as yeat published.

Figure 13 includes the optimum salt load reductions computed by the dynamic programming model in relation to the minimum target reductions, along with associated costs have been reduced to $565 million as a result of using the optimal investment planning model, representing a 30% reduction. The optimal project selections are given in Figure 14, which display optimal timing for project initiation and time to completion.

4. References

Allen, R. and Bridgeman, S. (1986) 'Dynamic programming in hydropower scheduling', J. Water Resources Planning and Management 112(3), pp 339-352.

Bellman, R.E. (1957) 'Dynamic programming', Princeton University Press, Princeton, NJ.

Bertsekas, D.P. (1976) 'Dynamic programming and stochastic control', Academic Press, New York.

Boeing Computer Services, Inc. (1985) 'BCS RIM - Relational information management system', Verion 7.0 User Guide, Seattle, Washington.

Brown, M. and Law, J. (1989) 'Development of a large network model to evaluate yield of a proposed reservoir', in J. Labadie et al. (eds), Computerized Decision Support Systems for Water Managers, pp 621-631.

Colorado River Water Quality Office, Bureau of Reclamation, and Salinity Control Coordinating Committee, US Department of Agriculture (1987) '1987 joint evaluation of salinity control programs in the Colorado River Basin', Denver, Colorado.

Fontane, D., Labadie, J. and Loftis, B. (1981) 'Optimal control of reservoir discharge through selective withdrawal', Water Resources Research 17(6), pp 1594-1604.

Fontane, D., Loftis, B., Labadie, J. and Merritt, D. (1984) 'Implementation strategies for salinity control projects in the Colorado River Basin', Annual Convention, ASCE.

Fontane, D., Loftis, B., Labadie, J., and Merritt, D. (1989) 'Implementation strategies for salinity control projects in the Colorado River Basin', J. Water Resources Planning and Management 115(5), ASCE.

French, P. et al. (1979) 'Water resources planning using interactive computer graphics', Technical Report, Dept. of Environ. Engrg., Cornell University, Ithaca, New York.

Frevert, D. and Cheney, R.W. (1989) 'Alternative methods of generating hydrologic data for reservoir optimization', in J. Labadie et al. (eds), Computerized Decision Support Systems for Water Managers, ASCE, New York, pp 498-507.

Graham, L., Labadie, J., Hutchison, I. and Ferguson, K. (1986) 'Allocation of augmented water supply under a priority water rights system', Water Resources Research 22(7), pp 1083-1094.

Hall, W.A. and Dracup, J.A. (1970) 'Water resources systems engineering', McGraw-Hill, New York.

House, W. (1983) 'Introduction', in W. House (ed.), Decision Support

Systems, Petrocelli Books, New York, pp 3-14.

Labadie, J. (1989) 'Dynamic programming with the microcomputer', in Encyclopedia of Microcomputers, Marcel Dekker, New York.

Labadie, J. and Fontane, D. (1989) 'Objective-space dynamic programming approach to multidimensional problems in water resources', in A. Esogbue (ed.), Dynamic Programming for Optimal Water Resources Systems Analysis, Prentice-Hall, Englewood Cliffs, NJ.

Labadie, J. and Shafer, J. (1979) 'Documentation for generalized dynamic programming model CSUDP', Dept. of Civil Engineering, Colorado State University, Fort Collins, Colorado.

Labadie, J. and Sullivan, C. (1986) 'Computerized decision support systems for water managers', J. Water Resources Planning and Management 112(3), ASCE, pp 299-307.

Labadie, J., Bode, D. and Pineda, A. (1986) 'Network model for decision-support in municipal raw water supply', Water Resources Bulletin 22(6), pp 927-940.

Labadie, J., Fontane, D. and Loftis, B. (1980) 'Optimal control of reservoir discharge through selective withdrawal', in H. Stefen (ed.), Proceedings of the Symposium on Surface Water Impoundments, ASCE, pp 772-781.

Labadie, J., Fontane, D., Tabios, G. and Chou, N.F. (1987) 'Stochastic analysis of dependable hydropower capacity', J. Water Resources Planning and Management 113(3), ASCE, pp 422-437.

Labadie, J., Brazil, L., Corbu, I. and Johnson, L. (eds) (1989) 'Computerized decision support systems for water managers', ASCE, New York.

Loucks, D., Kindler, J. and Fedra, K. (1985) 'Interactive water resources modeling and model use: an overview', Water Resources Research 21(2), pp 95-104.

Loucks, D.P., Stedinger, J.R. and Haith, D.A. (1981) 'Water resources systems planning and analysis', Prentice-Hall, Englewood Cliffs, NJ.

Ludvigsen, P., Sims, R. and Grenney, W. (1987) 'A demonstration expert system to aid in assessing groundwater contamination potential by organic chemicals', Computing in Civil Engineering, ASCE.

Parker, D. (1988) 'GIS training handbook', Bureau of Land Management.

Rogers, P. and Fiering, M. (1986) 'Use of systems analysis in water management', Water Resources Research 22(9), 146S-159S, and Fiering 81986).

Sprague, R. (1983) 'A framework for the development of decision support systems', in W. House (ed.), Decision Support Systems, Petrocelli Books, New York, pp 85-123.

Votruba, L. (1988) 'Analysis of water resources systems', Elsevier, Amsterdam.

EXPERT SYSTEMS IN WATER RESOURCES SIMULATION AND OPTIMIZATION

K. FEDRA
Advanced Computer Applications
International Institute for Applied Systems Analysis (IIASA)
A-2361 Laxenburg
Austria

ABSTRACT.   Expert systems (ES) as a new and emerging technology of
information processing and decision support are becoming increasingly
useful tools in numerous application areas. Expert systems are man-machine
systems that perform problem-solving tasks in a specific domain. They use
rules, heuristics, and techniques such as first-order logic or semantic
networks, to represent knowledge, together with inference mechanisms, in
order to derive or deduce conclusions from stored and user-supplied
information.
   Application- and problem-oriented, rather than methodology-oriented,
systems are most often hybrid or embedded systems, where elements of AI
technology, and expert systems technology in particular, are combined with
more classical techniques of information processing and approaches of
operations research and systems analysis. Here traditional numerical data
processing is supplemented by symbolic elements, rules, and heuristics in
the various forms of knowledge representation.
   Applications containing only small knowledge bases, of at best a few
dozen to a hundred rules, can dramatically extend the scope of standard
computer applications in terms of application domains, as well as in terms
of an extended non-technical user community.
   This review covers a basic introduction to what expert systems and AI
methods are, what they can, and cannot do; the state of the art in
applications in the field of water resources systems analysis and modelling
and selected examples of expert and hybrid systems in the field that
integrates simulation modelling, optimization and AI technology.

1.   AI and Expert Systems

In discussing a domain as loosely defined as expert systems, it may be
useful to present a few definitions selected from the literature, to set
the stage and introduce the jargon.   Equally instructive are the
essentially graphic definitions that are available (Figures 1, 2).
"Artificial Intelligence is the part of computer science concerned with
designing intelligent computer systems, that is, systems that exhibit the
characteristics we associate with intelligence in human behaviour ..."

397

J.B. Marco et al. (eds.),
Stochastic Hydrology and its Use in Water Resources Systems Simulation and Optimization, 397–412.
© 1993 *Kluwer Academic Publishers.*

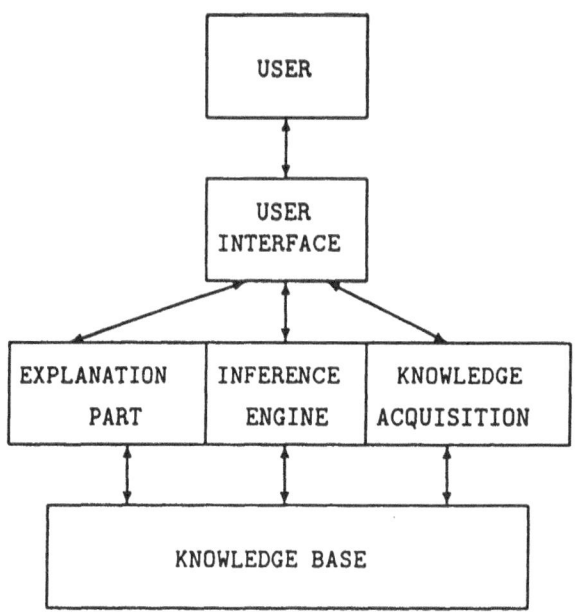

Figure 1. The five main components of an expert system (Trappl, 1985)

(Barr and Feigenbaum, 1981). Unfortunately, this does not really say much more than the name itself, other than indicating that we are talking about a branch of computer science. A somewhat more operational definition is: "Artificial Intelligence is that part of computer science that investigates symbolic, nonalgorithmic reasoning processes, and the representation of symbolic knowledge for use in machine inference" (Davis and Lenat, 1982).

Expert Systems, or Knowledge Based Systems, are a loosely defined class of computer software within the more general area of AI, that go beyond the traditional procedural, algorithmic, numerical and mathematical represent-ations or models, in that they contain large empirical knowledge e.g., in the form of rules or heuristics, and inference in mechanisms for utilizing this form of information to derive results by logical operations. They are fashioned along the lines of how an expert would go about solving a problem, and are designed to provide expert advice. Like any other model, they are sometimes extreme simplifications and caricatures of the real thing, i.e., the human expert.

However, definitions of functional descriptions of expert systems and claims to the expert system category of software cover a broad spectrum, ranging from fairly modest to rather optimistic parallels to human, or even super-human, performance:

"Most existing expert systems work in analytic domains, where problem solving consists of identifying the correct solution from a pre-specified finite list of potential answers ..." (Merry, 1985).

"Expert systems are computer programmes that apply artificial intel-ligence to narrow and clearly defined problems. They are named for their

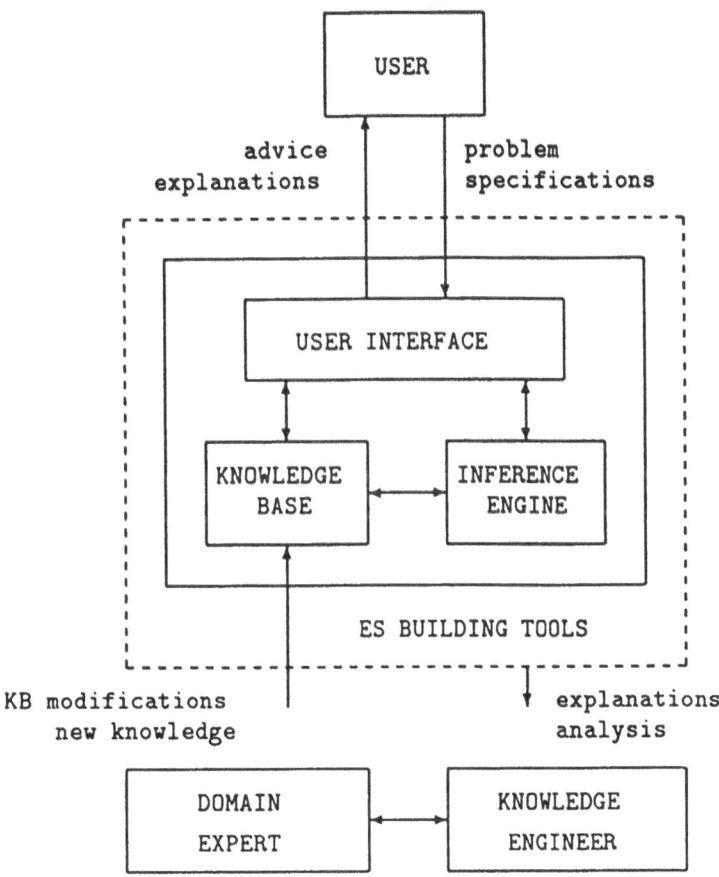

Figure 2. Interaction of knowledge engineer and domain expert with software tools that aid in building an expert system (Buchanan and Shortliffe, 1984).

essential characteristics: they provide advice in problem solving based on the knowledge of experts" (Ortolano and Perman, 1987).

"An expert system is a computer system that encapsulates specialist knowledge about a particular domain of expertise and is capable of making intelligent decisions within that domain" (Forsyth, 1984).

An expert system "handles real-world complex problems requiring an expert's interpretation [and] solves these problems using a computer model of expert human reasoning, reaching the same conclusions that the human expert would reach if faced with a comparable problem" (Weiss and Kulikowski, 1984).

There are, however, even more demanding definitions. In their description of MYCIN, one of the classic expert systems, Buchanan and Shortliffe argue that an expert system "... is an AI program designed (a) to provide

expert-level solutions to complex problems, (b) to be understandable, and (c) to be flexible enough to accommodate new knowledge easily" (Buchanan and Shortliffe, 1984). One of the more extensive definitions and more optimistic descriptions comes from Hayes-Roth: "An expert system is a knowledge-intensive programme that solves problems that normally require human expertise. It performs many secondary functions as an expert does, such as asking relevant questions and explaining its reasoning. Some characteristics common to expert systems include the following:
- They can solve very difficult problems as well as or better than human experts
- They reason heuristically, using what experts consider to be effective rules of thumb and they interact with humans in appropriate ways, including natural language
- They manipulate and reason about symbolic descriptions
- They can function with data which contains errors, using uncertain judgemental rules
- They can contemplate multiple, competing hypotheses simultaneously
- They can explain why they are asking a question
- They can justify their conclusions" (Hayes-Roth, 1984).
Obviously then, there seems to be no generally accepted definition of what exactly is or constitutes an expert system. Descriptions and definition in the literature range from rather narrow automata selecting predefined answers to better-than-human reasoning performance in complex problem domains. There is, however, general agreement that an expert system has to combine
- a knowledge base, that is a collection of domain specific information;
- an inference machine, which implements strategies to utilize the knowledge base and derive new conclusions from it (e.g., modus ponens, forward chaining, backward chaining);
- a knowledge acquisition mechanism that elicits information required from the user, but also from domain experts to initialize the knowledge base,
- and an explanation component, that can, on request, explain the systems inference procedures,
- and a conversational user interface that controls and guides the man-machine dialogue.
Obviously, an expert system must perform at a level comparable to that of a human expert in a non-trivial problem domain.
In summary, a very short description of AI would be "the art or science of making computers smart", and expert systems could be described as "smart problem-solving software".

2. Basic Concepts of Expert Systems

What makes expert systems different from ordinary models and computer programmes? Rather than trying to define differences in any formal way, it may help to introduce and discuss some of the basic concepts and approaches used in expert systems.
Expert systems are alternatively referred to as "knowledge based" systems. Knowledge representation, therefore, is one of the fundamental concepts and building blocks in expert systems.

Knowledge is represented in various forms and formats, following different paradigms. The more commonly used forms include rules, attribute-value lists, frames or schemata, and semantic networks. A brief comprehensive introduction to knowledge representation is given in Chapter 3 of Barr and Feigenbaum, 1981.

Probably the most widely used format, and also the most directly understandable form of knowledge representation are rules, also referred to as productions or production rules, or situation-action pairs. They are close to natural language in their structure, and they are familiar to programmers used to classical procedural languages such as FORTRAN or C: IF ... THEN ... ELSE is easy enough to understand. An example, borrowed from Beck, 1989, would be

    IF effluent ammonium concentration is high
       AND sludge wastage rate (control) is not high
       AND dissolved oxygen set-point (control) is low
    Then increase sludge wastage rate
       AND increase dissolved oxygen set-point

Obviously, the terms used in rules can be more-or-less cryptic (Doneker and Jirka, 1988) and require proper definition and interpretation in the system:

    IF vertical angle greater than 45 degrees
       AND length scale (1) is greater than 1.0
       AND length scale (2) is greater than 1.0
       AND length scale (3) is greater than 1.0
    THEN flow class is class V2.

Formal logic and propositional calculus offer another form of knowledge representation. Well defined syntax and semantics and expressive power make it an attractive option.

A "proposition", a statement about an object, is either TRUE or FALSE. Connectives permit the combination of simple propositions. The most commonly used connectives are:

AND	$\wedge$ or &
OR (inclusive)	$\vee$
NOT	$\neg$ or $\sim$
IMPLIES	$\rightarrow$ or $\supset$
EQUIVALENT	$\equiv$

Rules of inference, such as "modus ponens", allow the derivation of new statements from given ones: if X and X -> Y are TRUE, then Y is also TRUE:

$$(X \wedge (X \rightarrow Y)) \rightarrow Y$$

The rules of propositional calculus, extended by predicates, allowing more complex statements with more than one argument, quantifiers such as "for all" ($\forall$) and "there exists" ($\exists$) and inference rules for quantifiers, result in "predicate calculus" (Barr and Feigenbaum, 1981).

Adding the idea of operators or functions leads to first-order predicate logic, and this, restricted to so-called Horn clauses (which have at most one positive literal) corresponds to the syntax of Prolog (Clocksin and Mellish, 1984; Bratko, 1986), a powerful AI language popular at least in Europe and Japan.

Structured objects are another popular means of representation of information knowledge. They are known as "Schemas" (Bartlett, 1932); "Frames" (Minsky, 1975); "Prototypes" or "Units" (Bobrow and Winograd, 1977), or "Objects" in many languages or language extensions, e.g. SMALLTALK (Kay and Goldberg, 1977); LOOPS (Bobrow and Stefik, 1983); or FLAVORS (Moon and Weinreb, 1980).

Frames allow combinations of generic and specific information, where the former can be inherited within a hierarchy of frames, consisting of classes, super- and sub-classes, and instances.

As a data structure, frames for example can combine declarative and procedural components. Slots as units of descriptions can hold attribute-value pairs, but also function specifications and of course reference to other frames. The example below is part of the simplified description of a chemical required for water quality modelling:

```
(SUBSTANCE aniline
    ((SYNONYMS        (aminobenzene, benzeneamin, phenylamin, aminophen))
    (SUPERCLASSES     (amines))
    (DESCRIPTION      (oily, colourless, to brown liquid, irritative,
                       possible carcinogen))
    (ATTRIBUTES       (CAS#: 62.53-3)
                      (UN# : 1547)
                      (molecular weight:      93.1)
                      (melting point:         -6.)
                      (boiling point:         184)
                      (vapour pressure/35:     1.)
                      (solubility:         34,000)
                      (logPoct:         0.90/0.98)
                      (BOD5:                  1.5)
                      (biodegradation:       0.01)
                      ............
                      (toxicity-Daphnia:      0.4)
    (ACTIONS          (IF graphics-display
                       THEN draw-diagram (C6H5NH2)))
```

A more extensive definition and a description of a heterarchical hybrid information system for environmental chemicals is given in Fedra, Weigkricht and Winkelbauer, 1987.

Another form of representation is by means of "semantic networks", consisting of nodes, representing objects, concepts and events and links or arcs between the nodes, representing their interrelationships (Quillian, 1968). A well known example of an expert system using semantic networks is PROSPECTOR, dealing with mineral prospecting (Duda, Gashnig and Hart, 1979).

A specific and very important feature of expert systems is the inference engine, i.e., the part of the program that arrives at conclusions or new facts given the primary knowledge base and information supplied by the user. The basic principle was already hinted at above in the introduction to predicate calculus.

There are two basic strategies, namely forward and backward chaining. Forward chaining implies reasoning from data to hypothesis, while backward chaining attempts to find the data to prove, or disprove, a hypothesis, (Forsyth, 1984). Since both strategies have advantages as well as disadvantages, many systems use a mixture of both, e.g., the Rule Value approach (Naylor, 1983).

For many practical purposes, developers use expert systems shells and special development environments rather than basic languages such as LISP, PROLOG, or SMALLTALK. While they offer the advantage of easy use and ready-made structures and formats, they sometimes tend to restrict the user to specific forms of representations and, for the more complex and comprehensive ones, are expensive. For a more recent survey and discussion of selected software for expert systems development (see Ortolano and Perman, 1987).

3. Expert Systems in Water Resources Modelling

There is a rather extensive and very rapidly growing literature on AI and expert systems, starting from the by now almost classic three volume Handbook of Artificial Intelligence (Barr and Feigenbaum, 1981; Barr and Feigenbaum, 1982; Cohen and Feigenbaum, 1982). Recent review articles concentrating on environmental systems and engineering, and water resources in particular, include Ortolano and Steineman, 1987; Rossman and Siller, 1987; Hushon, 1987; Gray and Stokoe, 1988; Beck, 1989.

The number of expert systems described in the literature is large and growing rapidly. The number of operational systems, in practical use for practical purposes, however, seems to be rather small, in particular when looking at an area such as water resources systems analysis or engineering.

Of the 22 systems compiled in the following table, almost all at the R & D stage; little or no information exists on successful practical applications on a routine basis. This, however, does not make expert systems different from the vast majority of simulation and optimization models developed in the field.

Another obvious feature is that a large number of systems have been developed for operational applications, in particular in the wastewater treatment area, rather than planning. Groundwater systems, and in particular as related to hazardous waste management problems, are another obvious focus. Finally, there are several "Intelligent Front End" systems, i.e., model selection or parameter estimation tools.

4. Types of Applications

There are several types of expert systems applications in any particular domain: they range from purely knowledge driven systems or ES proper, to ES components in an intelligent front end, to fully embedded or hybrid systems. Each of these systems have their specific characteristics, use and problems. As with any attempt at classification, real things do not neatly fit into square boxes, but it helps to structure the discussion and seems to satisfy a basic need of the scientific mind.

APPLICATION DOMAIN	Contact of Reference
Trickling filter plants (sludge Cadet)	Catherine Perman, Department of Civil Engineering, Stanford University.
Anaerobic digester	Michael Barnett Dept. of Environmental Science and Engineering, Rice University
French water treatment plant	Pierre Lannuzel CERGRENE/ENPC
New York water treatment plant	Steve Nix Dept. of Civil Engineering, Syracuse University
Activated sludge plants	Deborah Helstrom Dept. of Civil and Environmental Engineering, Utah State University
Activated sludge diagnosis	Johnston, 1985
Water system loss	Steve Maloney, CERL
Sewer System Design	Lindberg and Nielsen, 1986
Multiple-use watershed management (MUMS)	Hushon, 1987
Groundwater flow analysis	Andrew Frank Dept. of Civil Engineering, University of Maine
Groundwater contamination (DEMOTOX)	Ludvigsen, Sims and Greeney, 1986
Groundwater vulnerability (AQUISYS)	Hushon, 1987
Well data analysis (ELAS)	Weiss, 1982
Water resources laboratory aide	Bob Carlson Dept. of Civil Engineering University of Alaska
Oil spill simulation	Antunes, Seixas, Camara et al., 1987
HSPF simulation advisor (HYDRO)	Gaschnig, Reboh and Reiter, 1981
Mixing zone analysis (CORMIX1)	Doneker and Jirka, 1988
Input parameter estimation for QUAL2E	Barnwell, Brown and Marek, 1986
Hydrological model calibration	J.W. Delleur School of Civil Engineering, Purdue University
Parameter estimation for runoff (EXSRM)	Engman, Rango and Martinec, 1986
Advisor for flood estimation (FLOOD ADVISOR)	Fayegh and Russell, 1986
Model selection for surface water acidification	Lam, Fraser and Bobba, 1987

Compiled from Ortolano and Steineman, 1987; Rossman and Siller, 1987; Hushon, 1987; Beck, 1989.

4.1. EXPERT SYSTEMS PROPER

An expert system proper would be a purely rule-based system, relying on a sizable knowledge base. As such, it is based on a largely empirical "model" or a qualitative, causal understanding of how things work. In the world of water resources modelling, that would put it in a class with the universal soil loss equation rather than a finite element model based on an albeit simplified version of the Navier-Stokes equations.

There are only a few purely knowledge based systems that do not contain a substantial conventional component. Some of the operation and control systems, in particular in the wastewater treatment area, seem to fit into this category. Also, a large number of systems are being developed for hazardous waste site assessment and related topics, such as permitting of waste site management, e.g., WA/WPM Generator (Paquette, Woodson and Bissex, 1986); RPI Site Assessment (Law, Zimmie and Chapman, 1986); GEOTOX (Mikroudis, Fang and Wilson, 1986; Wilson, Mikroudis and Fang, 1986); DEMOTOX (Ludvigsen, Sims and Greeney, 1986); or SEPIC (Hadden and Hadden, 1985). Reviews of these systems can be found in Ortolano and Steineman, 1987; Rossman and Siller, 1987; Hushon, 1987.

4.2. INTELLIGENT FRONT ENDS

"An intelligent front-end is a user-friendly interface to a software package, which enables the user to interact with the computer using his or her own terminology rather than that demanded by the package" (Bundy, 1984). What they can do, among other things, is to avoid or minimize misuse of complex models by less experienced users.

The QUAL2E Advisor, FLOOD ADVISOR, HYDRO, CORMIX1, or EXSRM are all examples of this type of application. Systems of this nature help a user to select the appropriate model to be used, assist in specifying input parameter values, and provide interpretation of the model's output (Rossman and Siller, 1987).

The QUAL2E Advisor (Barnwell, Brown and Marek, 1986) is a rule-based system, built with a commercial expert system shell, M.1. The system suggests appropriate parameter or input values for coefficients used in modelling stream temperature, the type of hydraulic model used and its associated coefficients, and biological oxygen demand removal, sediment oxygen demand and reaeration rate coefficients. Appropriate values are suggested in a question-and-answer session, where information about stream characteristics that can be easily obtained, e.g., by visual inspection, such as shape of channel cross-section, slope and depth, nature of stream bed, bank vegetation, are used to classify the river and estimate corresponding coefficients.

4.3. HYBRID SYSTEMS

The basic idea of an expert system is to incorporate into a software system expertise, i.e., data knowledge and heuristics, that are relevant to a given problem area. Application and problem-oriented rather than methodology-oriented systems are most often "hybrid" or "embedded" systems, where elements of AI technology are combined with more classical techniques of information processing and approaches of operations research and systems analysis. Here traditional numerical data processing is supplemented by symbolic elements, rules and heuristics in the various forms of knowledge representation.

There are numerous applications where the addition of quite a small amount of "knowledge" in the above sense, e.g., to an existing simulation model, may considerably extend its power and usefulness and at the same time make it much easier to use. Expert systems are not necessarily purely knowledge driven, relying on huge knowledge bases of thousand of rules. Applications containing only small knowledge bases, of at best a few dozen to a hundred rules, can dramatically extend the scope of standard computer applications in terms of application domains, as well as in terms of an enlarged non-technical user community.

Clearly, a model that "knows" about the limits of its applicability, what kind of input data it needs, how to estimate its parameters from easily available information, how to format its inputs, how to run it, and how to interpret its output will require not only less computer expertise from its user, it will also make less demands on its domain expertise. Water resources modelling usually deals with rather complex problems that touch upon many disciplines, and rarely will experts or a small group of individuals have all the necessary expertise at their disposal. The expert systems component of such systems can help to fill this gap and at the same time take over the role of a tutor.

Much of the above also holds true for the intelligent front end system, and any attempt at a clean cut classification will be found wanting; hybrid systems with embedded AI components would simply have several, in fact many, "micro expert systems" integrated into the overall software package. They rely on a number of disjunctive and specialized knowledge bases in different representation formats, depending on the domain and its most natural form of representation.

Several examples of integrated hybrid systems that also contain water resources models are described e.g., in Fedra, Weigkricht and Winkelbauer, 1987; Fedra, Li, Wang et al., 1987; Fedra, 1986; Fedra, 1988. The basic philosophy and early examples are described in Fedra and Loucks, 1985; Loucks, Kindler and Fedra, 1985; Loucks and Fedra, 1987.

These systems integrate several simulation and optimization models with a number of data bases and provide a menu-driven graphics interface (Figure 3). AI components are built into these systems in the form of small rule bases, e.g., to assist the user in editing his problem specifications, ensuring consistent and plausible input data. Expert system techniques such as a sideway chaining in a rule value approach are used in data base retrieval, assisting in efficient response even with incomplete user specifications.

Expert systems components also provide translations from a highly aggregated, symbolic problem description, e.g., in iconic form, into the necessary numerical specification required for the model. Similar techniques are used for the coupling of models, where the output from one, e.g. optimization model for industrial production, is used to provide water demand targets for the river basin simulator and water allocation model MITSIM (Fedra, Li, Wang, et al., 1987; Strzepek and Garcia, 1989).

A recent example is an interactive groundwater modelling system, based on a 2D finite-element simulator (Fedra and Diersch, 1989). By combining a finite-element model for flow and transport problems with an AI-based and symbolic, graphics user interface, the system is designed to allow the easy and efficient use of complex groundwater modelling technology in a problem- rather than model-oriented style.

Implemented on a colour-graphics engineering workstation, the system provides a problem manager that allows the selection of site-specific, as well as generic, groundwater problems from problem libraries, or the interactive design of a new problem. Using either satellite imagery (such as LANDSAT or SPOT) or Digital Line Graph (DLG) standard vector maps as a background, these problems can be edited and modified and then simulated under interactive user control.

The system also features an extensive Intelligent Computer Aided Design (ICAD) component that allows the user to design very efficiently, and parameterize, a new problem from the very beginning, using a map or an auxiliary grid as a background for the definition of problem geometry and hydrogeological parameters (Figure 3).

Building some of the knowledge of an experienced groundwater modeller into the software system and its interface through the rule bases driving, e.g., numerous transparent error-correction functions and an automatic mesh generator, allows for a very fast and efficient formulation of a new problem, even by an inexperienced user. Freed from the task of very detailed and demanding formalization of the computer representation of the problem, the user can thus concentrate on its management, regulatory, economic or technological aspects, for example. Designed for complex applications such as hazardous waste management, site evaluation, and groundwater contamination problems, ease of interactive use, responsiveness and efficiency in problem definition and comparative analysis are the most important characteristics of this hybrid simulation system.

5. Summary and Conclusions

AI and expert systems technology are certainly an intriguing new development in computer science that hold great promise for better applications. However, like any other method, they do not offer universal solutions and need a thorough understanding of their requirements and limitations for proper use.

By and large, expert systems are empirical systems, based on a more-or- less explicit, and usually qualitative, understanding of how things work. A perfect example of an ideal application area is law, or in the context of water resources, water rights and allocation problems. In water resources modelling, however, there is a substantial amount of physically based modelling, where an understanding of how things work can be expressed quantitatively. Much of our quantitative "understanding" is still empirical and not based on laws of nature (Darcy's law is an empirical formulation, but then physicists would argue that so is Schrödinger's equation).

However, it is important to realize that expert systems are cetainly no substitute for many time-tested methods and models, but should be seen as complementary techniques which can improve many of these models. Obvious applications related to numerical models are in data pre-processing, parameter estimation, the control of the user interface, and the interpretation of results. There are certainly enough arts and crafts

408

components in numerical modelling that open up attractive opportunities for AI techniques.

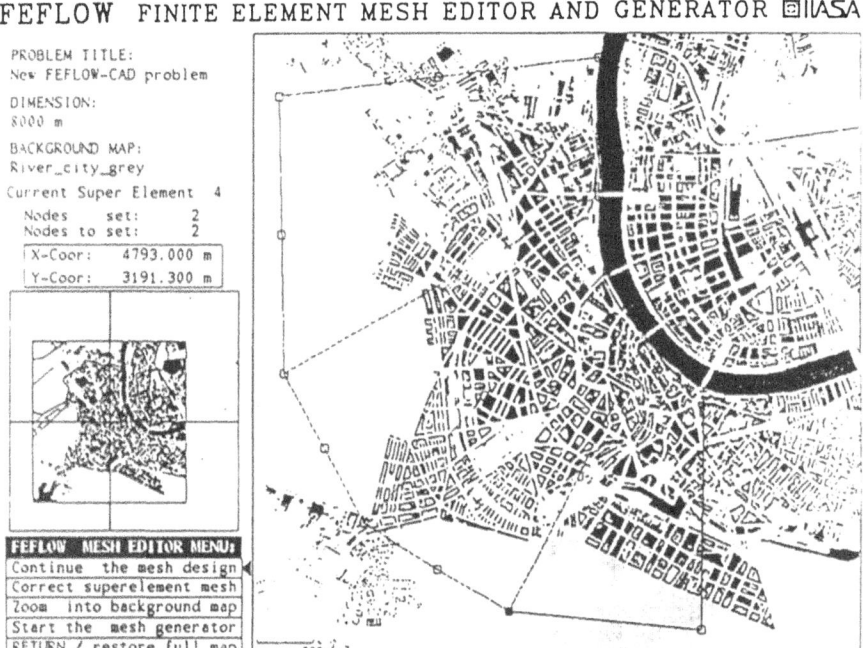

Figure 3. Defining a problem area in terms of super-elements (Fedra and Diersch, 1989).

While there is certainly some application potential for a purely knowledge-driven system in classifications and diagnosis tasks, the most promising area of application is in coupled, embedded, or hybrid systems, such as intelligent front ends, intelligent interfaces and modelling support, rather than new models themselves. When integrated with data base management and interactive colour graphics, AI concepts can help to shape a new generation of powerful but truly user-friendly "smart" software that actually gets used in planning and management.

AI applications are no longer restricted to expensive special-purpose hardware, but are increasingly supported on standard workstations and powerful PCs. With this wide accessibility, and an increasing number of affordable software tools, we may well be at the beginning of an exciting era of new developments and applications.

6. Rerefences

Antunes, M.P., Seixas, M.J., Camara, A.S. and Pinheiro, M. (1987), 'A new method of qualitative simulation of water resources systems', 2, Applications, pp 2019-2022.

Barnwell Jr., T.O., Brown, L.C. and Marek, W. (1986) 'Development of a prototype expert advisor for the enhanced stream water quality model QUAL2E', Internal Report, September, 1986, Environmental Research Laboratory, Office of Research and Development, US Environmental Protection Agency, Athens, Georgia 30613.

Barr, A. and Feigenbaun, E.A. (1981), 'The handbook of artificial intelligence', volume I, Pitman, London, 409 p.

Barr, A. and Feigenbaum, E.A. (1982), 'The handbook of artificial intelligence', volume II, Pitman, London, 428 p.

Bartlett, F. (1932), 'Remembering, a study in experimental and social psychology', Cambridge University Press, London.

Beck, M.B. (1989), 'Expert systems in environmental systems analysis and control', Dept. of Civil Engg. Imperial College, London, draft expert tutorial submitted to WHO, forthcoming.

Bobrow, D. and Stefik, M. (1983), 'The LOOPS manual', Xerox PARC, Palo Alto.

Bobrow, D.G. and Winograd, T. (1977), 'An overview of KRL, a knowledge representation language', Cognitive Science, 1:3-46.

Bratko, I. (1986), 'PROLOG Programming for Artificial Intelligence', Addison-Wesley, 423 p.

Buchanan, B.G. and Shortliffe, E.H. (1984), 'Rule-based expert systems. The MYCIN experiments of the Stanford Heuristic Programming Project', Addison Wesley, Reading, Massachusetts, 748 p.

Bundy, A. (1984), 'Intelligent front-ends', in J. Fox (ed.), expert systems, State of the Art Report, 12:7, Pergamon Infotech Ltd., Berkshire, England, pp 15-24.

Clocksin, W.F. and Mellish, C.S. (1984), 'Programming in PROLOG', Springer-Verlag, New York.

Cohen, P.R. and Feigenbaum, E.A. (1982), 'The handbook of artificial intelligence', vol. III, Pitman, London, 639 p.

Davis, R. and Lenat, D.B. (1982), 'Knowledge-based systems in artificial intelligence', Mc Graw-Hill, New York, 490 p.

Doneker, R.L. and Jirka, G.H. (1988), 'CORMIX1: an expert system for mixing zone analysis of conventional and toxic single port aquatic discharges', report EPA/600/3-88/013, Environmental Research Laboratory, USEPA, Athens, Georgia.

Duda, R. Gashnig, J. and Hart, P.E. (1979), 'Model design in the PROSPECTOR consultant system for mineral exploration', Rep. No. STAN-CS-76-552, Computer Science Department, Stanford University, Stanford, California.

Engman, E.T., Rango, A. and Martinec, J. (1986), 'An expert system for snowmelt runoff modeling and forecasting in proceedings', Water Forum '86, American Society of Chemical Engineers, New York, pp 174-80.

Fayegh, D. and Russell, S.O. (1986), 'An expert system for flood estimation', in C.N. Kostem and M.L. Maher (eds.), expert systems in Civil Engineering, ASCE, New York, pp 174-181.

Fedra, K. (1986), 'Integrated software for environmental assessment of industrial development in the Doon Valley', in V. Fedorov and R.E. Munn (eds), An Assessment of Environmental Impacts of Industrial Development with Special Reference to the Doon Valley, India, Phase I, Vol. III, Software and Data, SR-86-1.

Fedra, K. (1988). 'Information and decision support systems for risk analysis', in W.D. Ehrenberger ed. Safety of Computer Control Systems 1988 (SAFECOMP'88), Safety-related Computers in an Expanding Market, proceedings of the IFAC Symposium, Fulda, FRG, 9-11 November, pp 53-59.

Fedra, K. and Diersch, H.J. (1989), 'Interactive groundwater modelling: colour graphics', ICAD and AI, in proceedings of the International Symposium on Groundwater Management: Quantity and Quality, October 2-5, 1989, Benidorm, Spain, IAHS, forthcoming.

Fedra, K. and Loucks, D.P. (1985), 'Interactive computer technology for planning and policy modeling', Water Resources Research, 21/2, 114-122.

Fedra, K., Li, Z., Wang, Z. and Zhao, C. (1987), 'Expert systems for integrated development: a case study of Shanxi province, the People's Republic of China', SR-87-1, International Institute for Applied Systems Analysis, A-2361 Laxenburg, Austria, 76 p.

Fedra, K., Weigkricht, E., Winkelbauer, L (1987), 'A hybrid approach to information and decision-support systems: hazardous substances and industrial risk management, RR-87-12, International Institute for Applied Systems, A-2361 Laxenburg, Austria, reprinted from 'Economics and artificial intelligence', Pergamon Books Ltd.

Forsyth, R. (ed.) (1984), 'Expert systems. Principle and Case Studies', Chapman and Hall, London, 231 p.

Gaschnig, J., Reboh, R. and Reiter, J. (1981), 'Development of a knowledge-based expert system for water resource problems', Final Report, SRI Project 1619, SRI International, Menlo Park, California.

Gray, A. and Stokoe, P. (1988), 'Knowledge-based or expert systems and decision support tools for environmental assessment and management. Their potential and limitations'. School for Resource and Environmental Studies, Dalhousie University, Nova Scotia.

Hadden Jr., W.J. and Hadden, S.G. (1985), 'Expert systems for environmental regulation' (SEPIC), in K.N. Karna (ed.) expert systems in government symposium, MITRE Corp. Mc Lean, Va. and IEEE Computer Society, pp 558-566.

Hayes-Roth, F. (1984), 'Knowledge-based expert systems-the state of the art in US', in J. Fox (ed.) Expert Systems, State of the Art Report, 12:7, Pergamon Infotech Ltd. Berkshire, England, pp 49-62.

Hushon, J.M. (1987), 'Expert systems for environmental problems', Environ, Sci. Technol., vol 21, No. 9, pp 838-841.

Johnston, D.M. (1985), 'Activated sludge diagnosis', in proceedings, Computer Applications in Water Resources, American Society of Chemical Engineers, New York, pp 601-606.

Kay, A. and Goldberg, A. (1977), 'Personal Dynamic Media', pp 31-410.

Lam, D.C.L., Fraser, A.S. and Bobba, A.G. (1987), 'Simulation and analysis of watershed acidification', in M.B. Beck (ed.) Systems Analysis in Water Quality Management, Advances in Water Pollution Control, Pergamon, Oxfort, pp 85-96.

Law, K.H., Zimmie, T. and Chapman, D.R. (1986), 'An expert system for inactive hazardous waste site characterization', in C.N. Kostem and M.L. Maher (eds.) Expert Systems in Civil Engineering, American Society of Chemical Engineers, New York, pp 159-173.

Lindberg, S. and Nielsen, J.B. (1986), 'Modelling of urban sewer systems', in D. Sriram and R. Adey (eds.) Applications of Artificial Intelligence in Engineering Problems, Springer-Verlag.

Loucks, D.P. and Fedra, K. (1987), 'Impact of changing computer technology on hydrologic and water resource modeling', in Review of Geophysics, vol 25, No. 2, March.

Loucks, D.P., Kindler, J. and Fedra, K. (1985), 'Interactive water resources modeling and model use: an overview', Water Resources Research, 21:2, 95-102.

Ludvigsen, P.J., Sims, R.C. and Greeney, W.J. (1986), 'DEMOTOX: Development of a prototype expert system for assessing organic chemical mobility and degradation to determine soil treatment requirements', in Proceedings of the ASCE Fourth Conference of Computing in Civil Engineering, American Society of Chemical Engineers, New York, pp 687-98.

Merry, M. (1985), 'Expert systems - some problems and oportunities', in M. Merry (ed.) Expert Systems 85, Proceedings of the Fifth Technical Conference of the British Computer Society, Cambridge University Press, pp 1-7.

Mikroudis, G.K., Fang, H.Y. and Wilson, J.L. (1986), 'Development of GEOTOX expert system for assessment of hazardous waste sites', in Proceedings, 1st International Symposium on Environmental Geotechnology, Lehigh University, pp 223-232.

Minsky, M. (1975), 'A framework for representing knowledge', in P.H. Winston (ed.), The Psychology of Computer Vision, McGraw-Hill, pp 211-280.

Moon, D.A., and Weinreb, D. (1980), 'FLAVORS: Message-passing in the lisp machine', AI memo 602, MIT, Cambridge, Mass.

Naylor, C. (1983), Build Your Own Expert System, Sigma Technical Press, John Wiley, Chichester.

Ortolano, L. and Perman, C.D. (1987), 'Software for expert systems development', proc. American Society of Civil Engineers, pp 225-240.

Ortolano, L, and Steineman, A.C. (1987), 'New expert systems in environmental engineering', proceedings of the American Society of Civil Engineers, Jnl, Computing in Civil Engineering, 1(4), pp 298-302.

Paquette, J.S., Woodson, L. and Bissex, D.A. (1986), 'Improving the implementation of remedial investigation/feasibility studies using computerized expert systems', in Superfound '86, Hazardous Materials Control Research Institute.

Quillian, M.R. (1968), 'Semantic memory', in M. Minsky (ed.) Semantic Information Processing, MIT Press, Cambridge, Mass, pp 227-270

Rossman, L.A. and Siller, T.J. (1987), 'Expert systems in environmental engineering', in M.L. Maher (ed.) Expert Systems for Civil Engineers: Technology and Application, American Society of Civil Engineers, New York, pp 113-128.

Strzepek, K.M. and Garcia, L. (1989), 'User's manual for MITSIM 2.0-river basin simulation model', Technical Report No. 1, Center for Advanced Decision Support for Water and Environmental Systems, University of Colorado, Boulder, CO., Forthcoming.

Trappl, R. (ed.) (1985), 'Impacts of Artificial Intelligence', Elsevier Science Publishers BV, Amsterdam, The Netherlands, 279 p.

Weiss, S. et al. (1982), 'Building expert system for controlling complex programs, proceedings of the AAAI-82, American Association for Artificial Intelligence, pp 322-326.

Weiss, S.M. and Kulikowski, C.A. (1984), 'A practical guide to designing expert systems, Rowman & Allenheld, Totowa, NJ, 174p.

Wilson, J.L., Mikroudis, G.K. and Fang, H.Y. (1986), 'GEOTEX: a knowledge-based system for hazardous site evaluation, in D. Sriram and R. Adey (eds.) Applications of Artificial Intelligence in Engineering Problems, Springer-Verlag.

THE SEGURA RIVER BASIN MODEL.   DISAGGREGATION IN A SEMIARID ENVIRONMENT

JUAN B. MARCO
Catedratico de Ingenieria Hidraulica
Universidad Politecnica de Valencia
Valencia
Spain

1. The Segura River System

The Segura basin extends over 14,925 km^2 with a mean annual flow of 964 Hm^3. The river is located in southeastern Spain, the most arid zone of the country. The mean annual rainfall over the basin amounts to some 300 mm. From Roman times onwards an extremely good climate has produced intensive water use for irrigation and urban supply with the result that this is the most exploited river in the country. Only 60 Hm^3/year are delivered to the sea, most of this in the form of floods, so consumption amounts to 93% of the water resources.

To model this complex system and its management, statistical simulation was felt to be advisable (Andreu, 1989) since, owing to the significant memory system, the reservoir capacities are very high in comparison with the mean annual flow.

1.1. HYDROLOGY

A map of the basin is given in Figure 1 showing the subbasins and reservoirs. Fundamentally four hydrological regions can be distinguished.

The upper basin is located in the Segura and Cazorla mountain ranges where the River Segura and its main tributary, the Mundo, originate. The mean annual flows of these rivers leaving the mountains are 465 Hm^3 and 190 Hm^3 respectively. This means that 70% of the resources come from the mountains where rainfall reaches a mean 1000 mm. Some very important aquifers exist in this area since most of the mountain ranges are made of limestone. The Socovos aquifer is worthy of mention among these since its main outlet is the River Taibilla whose persistency over the year is very important even though its mean annual flow is only 60 Hm^3.

The middle basin runs through a valley in a semiarid region containing four small tributaries running from the South: the Moratalla, Argos, Quipar and Mula. These rivers are quasi-intermittent with a small memory and large changes in flow.

The River Guadalentin basin is the most arid region in Spain. Despite its large area, 3,225 km^2, its mean annual runoff only amounts to 31 Hm^3/year, most of which is produced by floods. Figure 2 shows its time

413

J.B. Marco et al. (eds.),
Stochastic Hydrology and its Use in Water Resources Systems Simulation and Optimization, 413–423.
© 1993 *Kluwer Academic Publishers.*

series and illustrates its extremely irregular behaviour.

The lower Segura basin is where most of the population and water consumption is concentrated. It is an alluvial plain with negligible water resources. Were it not for irrigation, it would be as arid as the Guadalentin Valley.

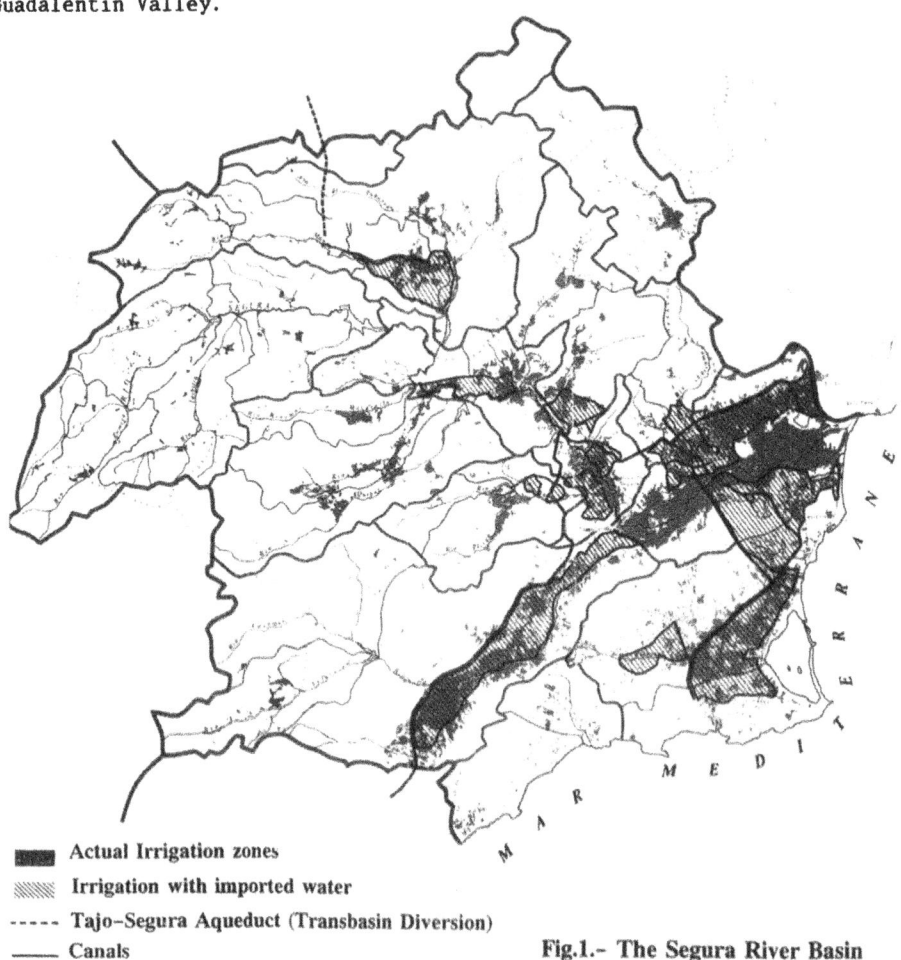

▰▰ Actual Irrigation zones

▨▨ Irrigation with imported water

----- Tajo-Segura Aqueduct (Transbasin Diversion)

——— Canals   Fig.1.- The Segura River Basin

1.2. THE DEMAND SYSTEM

The Segura river is the most exploited one in Spain. Its current deficit is 300 Hm³/year which is covered by mining the aquifers and a trans-basin diversion from the River Tagus basin.

The Tagus-Segura aqueduct can transport 36 m³/s over a distance of more than 250 km. It begins at the Entrepeñas Reservoir on the River Tagus and ends at the Talave Reservoir in the River Mundo. The mean annual flow of the Tagus is 1,461 Hm³/year and the existing storage capacity is 2,340 Hm³. The Tagus has a moderate memory, very similar to that of the River Segura. There are ten reservoirs in the Segura basin among which the four controlling the mountain ranges are worthy of mention. Their capacities and mean flows are given in Table 1. The remaining reservoirs control the tributaries for local use in the different subsystems.

TABLE 1. Mean annual flows and reservoir capacities

Series	Mean flow (Hm³)	Reservoir capacity (Hm³)
Tagus	1461	2340
Cenajo	465	472
Fuensanta	304	205
Talave	147	34
Camarillas	190	37
Moratalla	9	–
Argos	15	11
La Cierva	0	5
Alfonso XIII	20	22
Puentes	32	14
Valdeinfierno	8	14
Contraparada	875	–

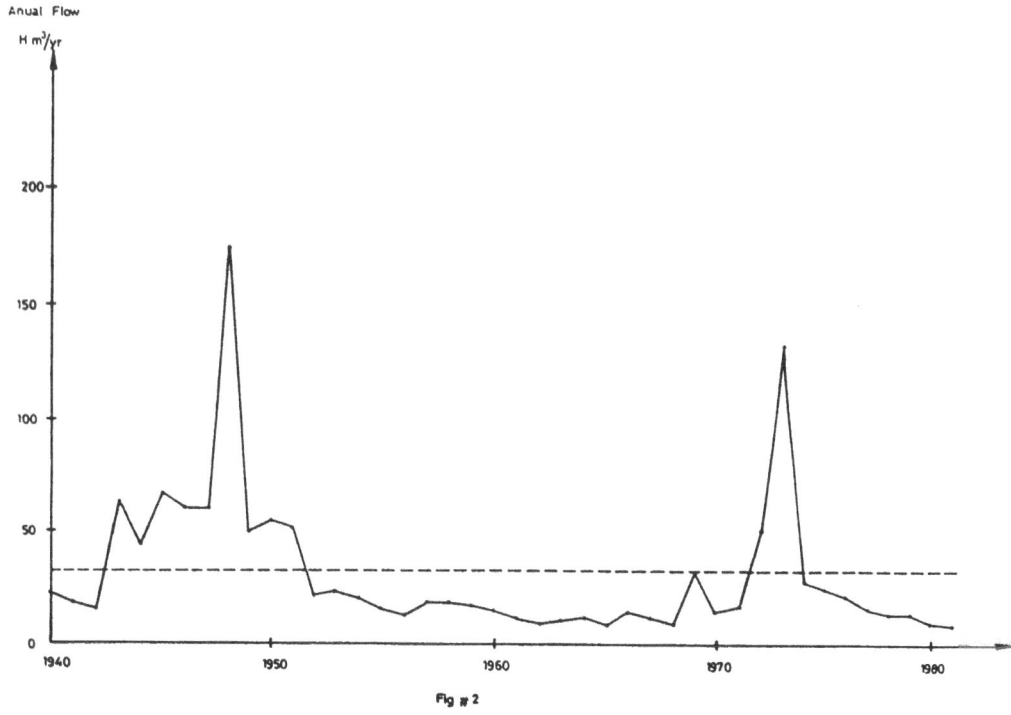

Fig # 2

Anual Series of the Guadalentin River

Figure 2. Anual Series of the Guadalentin River

The demand subsystem structure is extremely complex. There are three parallel distribution systems: the traditional irrigation channels, the new system built to convey water from the trans-basin diversion, and the urban supply coming from the River Taibilla and some other intakes, plus a series of pumping stations recovering the tail water of the system and drainage ditches for further use.

1.3. DATA STRUCTURE

The system model needed to be a very detailed one involving the largest number possible of generation points. Seven of the 20 flow measurement points had to be discarded as a result of problems over redundant information. In the end, eleven series from all the reservoirs in the system together with the Moratalla series and the River Segura at the Contraparada point were used. The Contraparada is where historically the river is allocated for distribution to the old channels. This is the longest and most accurate time series.

The data available for all thirteen generation points were 40 years of monthly flows.

The data quality was sufficient but the Taibilla series was suspected of being non-homogeneous.

2. Characteristics Required for the Stochastic Model

Operational model characteristics must fit not only well-known statistical requirements but also simulation requirements. This point is frequently overlooked.

It is clear that a multivariate 13-dimensional model is nonsensical. This number of parameters would be totally unrealistic. Furthermore it would reproduce relationships with negligible significance for the system management. Equally, the joint modelling of the time series of flows of different orders of magnitude entails placing them in the same hierarchy. The importance of this secondary series is purely local.

Spatial disaggregation was chosen as the most appropriate methodology for dealing with the problem on hand.

After making this decision, the type of time modelling then had to be decided. A monthly scale was the time resolution required since management decisions are taken on a monthly basis. Demand is highly seasonal. Flows are also highly seasonal with two strongly marked peaks occurring in spring and especially in autumn.

Stationarization by Fourier analysis (Yevjevich, 1972) was chosen as the appropriate method for several reasons.

First of all, the periodicity is highly significant, mostly in the series determining the overall performance of the system. There is a very small number of highly significant harmonics. There is a considerable saving in the number of parameters. Stationarization by marginal distribution decreases the sampling variability of some of the generated samples and perpetuates the extreme position within the year (Yevjevich and Harmancioglu, 1988). This is a very important matter for highly utilized systems such as this one.

Disaggregation in time (Valencia and Schaake, 1973) was not advisable since the correlations between the different months involved were very weak. Periodic parameter models, such as PARMA (Salas et al., 1986), were not necessary because the system memory stemmed from the aquifers with constant over-the-year discharge characteristics. No hydrological phenomena were involved, such as snowmelt, to produce periodic autocorrelation.

In addition, water use was very high so consequently the reservoirs ran out of water fairly quickly. Monthly management is the key problem.

Multivariate modelling was required for the main series as a result of the structure of the reservoir system. A crucial point needing investigation was the trans-basin diversion policy from the Tagus to the Segura. Another point to analyze was the coordination between both reservoir groups controlling the headwaters from the Mundo and Segura basins. What was therefore needed was to reproduce all the time and space relationships for a reduced set of flow series.

Stedinger et al. (1985) argue that univariate modelling plus spatial disaggregation can yield sufficient results. In this situation it was felt that all first-order cross-correlations should also be reproduced.

In short, the statistical model actually applied was characterized by the following procedures:

a) Fourier analysis deseasonalizing
b) multivariate modelling of the main time series
c) spatial disaggregation of the secondary series.

3. Stationarization

Fourier analysis was carried out on both the periodic mean and standard deviation for all the series. The results are shown in Table 2.

It can be observed that for the series of large, continuous flow, 95% of the periodic parameter variance can be explained by two harmonics at the most. This illustrates the high seasonality observed. The reduction of the parameters with respect to the marginal distribution stationarization is over 70%, with eight parameters actually in front of 24.

The method produced poorer results for the secondary series. The middle basin tributaries and the River Guadalentin are quasi-intermittent series. The mean flow did not have any clear physical significance. The explained variances were in all cases around 50%.

One problem still needing to be solved was the significant harmonic identification in a multivariate framework. Fisher's test on the periodic mean and the standard deviation were used, taking the same number of significant harmonics on both if the same harmonic proved to be significant in any one of them. This produced very good results and the series performed well, but sometimes the significant harmonics were not the same or did not reach significant levels, despite the obvious fact that the series were not stationary. Many of these problems were solved on the basis of physical consideration. However, this method, which generally produced very good results, needed more detailed analysis for regional and multivariate application.

TABLE 2. Fourier analysis

Series	№	Mean explained Variance	№	Standard Deviation Explained Variance
Cenajo	2	97,5	2	93,9
Camarillas	2	99,2	2	93,4
Tagus	1	94,6	1	92,8
Contraparada	2	99,4	2	98,5
Taibilla	1	83,1	0	0
Fuensanta	2	98,7	2	96,6
Talave	2	99,7	2	98,1
Moratalla	2	63,6	2	62,5
Argos	0	0	0	0
La Cierva	1	69,5	1	41,6
Alfonso XIII	1	53,0	1	53,0
Puentes	2	59,2	2	53,1
Valdeinfierno	2	72,5	2	65,9

4. Main Series Modelling

A prior problem consisted of defining the main series. This had to be done mainly on the basis of system awareness. The main series had to be characterized by possessing flows of a similar order of magnitude. They also had to represent the hydrology of a given region.

In the system the following series were chosen:

CENAJO
CAMARILLAS
TAGUS
CONTRAPARADA

In order to identify the multivariate model, the parsimony principle together with Anderson's test were employed. Three model structures were tried: ARMA(1,1), AR(1) and CAR(1). The results were perfectly clear. CAR(1) did not pass the independence tests for either auto- or cross-correlation.

AR(1) and ARMA(1,1) passed all the tests with statistically insignificant differences, but ARMA(1,1) had 42 parameters whereas AR (1) had only 26. The choice was therefore obvious. Table 3 gives some comparative results. The parameters were obtained by the moments method. The model employed therefore was:

$$X_t = \delta_1 X_{t-1} + \theta_0 \epsilon_t \qquad (1)$$

$$\underline{\delta}_1 = \begin{vmatrix} 0.339 & -0.055 & -0.137 & 0.077 \\ 0.030 & 0.412 & 0.124 & -0.024 \\ 0.041 & 0.085 & 0.630 & -0.059 \\ 0.164 & 0.156 & 0.097 & 0.133 \end{vmatrix}$$

$$\underline{\Theta}_0 = \begin{vmatrix} 0.892 & 0 & 0 & 0 \\ 0.472 & 0.735 & 0 & 0 \\ 0.308 & 0.136 & 0.664 & 0 \\ 0.713 & 0.234 & -0.042 & 0.469 \end{vmatrix}$$

TABLE 3. Multivariate model selection

Series	Explained Variance		
	CAR(1)	AR(1)	ARMA(1,1)
Cenajo	57,0	84,3	84,2
Camarillas	61,6	82,8	81,9
Tagus	74,3	65,8	65,6
Contraparada	57,2	84,1	83,7

A few of the parameters in these matrices could be set to zero without any serious loss. Some redundant information existed among the three series from the Segura basin (Nos. 1,2 and 4).

The final model was very close to the one by Young and Pisano (1968). Long-term persistence was not observed. It is very unlikely to find a long memory for medium-sized basins in semiarid counties like Spain.

5. Disaggregation

The Mejia and Rousselle (1976) model, as extended by Lane (1979), was chosen for the spatial disaggregation.

$$\underline{Y}_t = \underline{A}\,\underline{X}_t + \underline{C}\,\underline{Y}_{t-1} + \underline{B}\,\underline{\epsilon}_t \tag{2}$$

The objective was to preserve the relationship of the disaggregated series along with the main series vector with the previous value and among each other.

It is obvious that all the relationships among generating points are not relevant and should not be reproduced. Disaggregated series are secondary and, as such, their importance is purely local. Disaggregation was carried out in groups chosen in accordance with a common hydrology.

Consequently four disaggregation groups were modelled:

1) TAIBILLA
2) FUENSANTA
   TALAVE
3) MORATALLA
   ARGOS
   LA CIERVA
   ALFONSO XIII
4) VALDEINFIERNO
   PUENTES

The group composition was set up for practical reasons since the groups contained the series from homogeneous hydrological regions and the reservoir operation within a group was not independent.

The Taibilla series was disaggregated separately since it is the output from a large aquifer. Being a secondary series, it exhibited a long-term memory of up to five or six years. It clearly disrupted all the modelling schemes and capabilities and consequently produced poorer results.

In order to estimate the $\underline{A}$, $\underline{B}$ and $\underline{C}$ matrices, Mejia and Rousselle equations were used.

By multiplying $\underline{Y}_t^T$ and $\underline{Y}_{t-1}^T$ by $\underline{X}_t^T$, and taking the expected values, we obtain

$$S_{y,I} = A\, S_{I,I} + C\, S_{y-1,I} \tag{3}$$

$$S_{y,y-1} = A\, S_{I,y-1} + C\, S_{y,y}$$

and solving

$$A = [S_{y,I} - S_{y,y-1}\, S_{y,y}^{-1}\, S_{y-1,I}] \cdot [S_{I,I} - S_{I,y-1}\, S_{y,y}^{-1}\, S_{y-1,I}]^{-1} \tag{4}$$

$$C = [S_{y,y-1} - A\, S_{I,y-1}]\, S_{y,y}^{-1}$$

and

$$BB^T = S_{y,y} - AS_{I,I}A^T - S_{I,y-1}C^T - CS_{y-1,I}A^T - CS_{y,y}C^t$$

Note that $\underline{\epsilon}_t$ was assumed to be serially independent but provision was not made to ensure that this was so. The model as written can be considered an ARMAX(1,0,0) but no identification was carried out.

Consequently, residual independence could or could not be achieved. For the River Segura basin three out of four disaggregation groups achieved independence. Good results were obtained for the middle-basin tributaries and the River Guadalentin, that is, the series with weak structure and low correlation with the main ones. Problems arose, however, with the Fuensanta and Talave series where residual autocorrelation, albeit small but significant, had to be neglected.

If the model is interpreted as an ARMAX model, canonical forms should be introduced.

The Mejia and Rousselle model only partially reproduces the serial dependence from the disaggregated series. This was shown by Lane (1980,

1982) and Stedinger and Vogel (1985). As this is the case if an AR(1) is used for $\underline{X}_t$ modelling, then

$$E [X_t Y_{t-1}^T] = \underline{\delta} E [X_t Y_t^T] \tag{5}$$

But since the following equation is used for parameter estimation

$$E [x_t y_{t-1}^T] \neq \underline{\delta} E [x_t y_t^T] \tag{6}$$

The structure of the main series is then reflected on the secondary ones. Stedinger and Vogel propose

$$\underline{Y}_t = \underline{A} \ \underline{X}_t + \underline{W}_t \tag{7}$$

where $\underline{W}_t$ allows for a structure to avoid that problem. This could be termed a transfer function model. In any case, regardless of whether a TFM or ARMAX approach is used, proper order identification and canonical setting could provide a more general approach.

In the River Segura basin model only the Mejia-Rousselle model was used. The overall results were reasonably good but further development along the lines suggested could improve them.

6. Final Results

The model as built explains a high percentage of time series variance. Table 4 gives the explained variances for all the series. It is to be noted that the poorest results correspond to Group 2 where residual independence for disaggregation was not achieved.

TABLE 4. Model adequacy

Series	Explained Variance	Anderson's Test	Residual Normality
Cenajo	84,3	YES	YES
Camarillas	82,8	YES	YES
Tagus	65,8	YES	YES
Contraparada	84,1	YES	NO
Taibilla	54,1	NO	NO
Fuensanta	45,7	NO	YES
Talave	45,9	NO	YES
Moratalla	79,5	YES	NO
Argos	76,3	YES	NO
La Cierva	79,1	NO	NO
Alfonso XIII	63,9	NO	NO
Valdeinfierno	89,7	YES	NO
Puentes	86,8	YES	NO

Residual normality was also an important factor. Several normalizing transformations were tried since most of the middle-basin tributaries and the River Guadalentin series were highly skewed.

The results were negative. The secondary series improved slightly with logarithmic transformation but dramatically with square root transformation. Nevertheless, the main series that had normal residuals started to show a negative skew. Furthermore, if normalization was carried out, significant departures from the sample mean appeared in the samples generated. The additive property for disaggregation was also lost.

The secondary series of the middle basin and the River Guadalentin were quasi-intermittent and clearly followed a Gamma distribution, even taking up an inverted J shape. For this reason their behaviour improved with square-root transformation. But the problem stems from the intermittency. To date, intermittency modelling for multivariate series remains an unsolved problem which cannot be resolved within the framework of traditional models.

It was consequently decided not to transform the series for skewness reproduction. Naturally this has an important effect on the management conclusions for some subsystems. Methods like those proposed by Todini (1980) or Hoshi and Burges (1979) are too complex to apply here, given the nature of the problem. The Wilson-Hilferthy transformation involving periodic parameters could prove of use, but the origin of the problem lies in the intermittency and there have been very few advances in this field.

This is a problem affecting semiarid regions and hydrologists pay very little attention to it.

7. References

Andreu, J. (1989) 'Simulation models for short and long term management. The Segura River basin example', NATO-ASI on Stochastic Hydrology, Peñiscola.

Hoshi, K. and Burges, S.J. (1979) 'Disaggregation of streamflow volumes', J. Hydraul. Div., ASCE 105 (HY1), pp 27-41.

Lane, W.L. (1979) 'Applied stochastic techniques: user manual', Div. of Plann. Tech. Serv., Water and Power Resour. Serv., Denver, Colorado.

Lane, W.L. (1980) 'Applied stochastic techniques: user manual (first revision)', Div. of Plann. Tech. Serv., Water and Power Resour. Serv., Denver, Colorado.

Lane, W.L. (1982) 'Corrected parameter estimates for disaggregation schemes', in Statistical Analysis of Rainfall and Runoff. V. Singh Ed., Water Resources Publications, Littleton, Colorado.

Mejia, J.M. and Rousselle, J. (1976) 'Disaggregation models in hydrology revisited', Water Resources Research 12(2), pp 185-186.

Salas, J.D., Tabios, G.Q. and Bartolini, P. (1985) 'Approaches to multivariate modeling of water resources time series', Water Resources Bulletin V21(4), pp 683-708.

Stedinger, J.R., Lettenmaier, D.P. and Vogel, R.M. (1985) 'Multisite ARMA(1,1) and disaggregation models for annual streamflow generation', Water Resources Research 21(4), pp 497-509.

Stedinger, J.R. and Vogel, R.M. (1984) 'Disaggregation procedures for generating serially correlated flow vectors', Water Resources Research

20(1) pp 47-56.

Todini, E. (1980) 'The preservation of skewness in linear disaggregation schemes', Journal of Hydroloy 47, pp 199-214.

Valencia, R.D. and Schaake, J.C. (1973) 'Disaggregation processes in stochastic hydrology', Water Resources Research 9(3), pp 580-585.

Yevjevich, V. (1972) 'Stochastic processes in hydrology', Water Resources Publications, Fort Collins, Colorado.

Yevjevich, V. and Harmacioglu, N.B. (1988) 'Description of periodic variation in parameters of hydrological time series', Water Resources Research 25(3), pp 421-428.

Young, G.K. and Pisano, W.C. (1968) 'Operational hydrology using residuals', ASCE Jour. of the Hydraulic Div. 94 (HY4), pp 909-923.

OPTIMIZATION AND SIMULATION MODELS APPLIED TO THE SEGURA WATER RESOURCES SYSTEM IN SPAIN

J. ANDREU* and J. CAPILLA**
* Departamento de Ingeniería Hidráulica y Medio Ambiente
** Departamento de Física Aplicada
Universidad Politécnica de Valencia
46071 Valencia
Spain

ABSTRACT. The River Segura basin, in southeastern Spain, is a semiarid Mediterranean region with a highly developed agriculture. The demand of water far exceeds the renewable surface and ground water resources. Mining of aquifers has been a practice for many years and still continues. There is also an external source of water: the Tagus-Segura interbasin transfer. Proper management is crucial in order to avoid or minimize shortages and conflicts. Systems analysis techniques are being used and a hierarchy of models has been developed to be used as tools for the planning and management of the basin. These include models for operational hydrology, as well as optimization and simulation models. This paper introduces the OPTIRED optimization model and the SGC simulation model. The models have been designed as general models so they can be used for other water resources systems.

1. Introduction

The Segura River basin, which will be described later, is probably the most complex water resources system in Spain. The complexity is owing to the following reasons:
  - The basin is a semiarid region, as detailed elsewhere (Marco, 1989), with mean annual precipitation of 376 mm.
  - Irrigated land has been increasing owing to high value crop production. At this point the water demanded, and used, per year is about 160% of the renewable resources of the basin.
  - As a consequence overexploitation of the aquifers is taking place and imports of water from other basins is necessary.
  - Different water conveyance and distribution systems are superimposed in the basin.
  - Conflicts arise and will keep arising among different groups of water users with different priorities and different origins for their water.

In order to plan the facilities and to find the long term operation rules for the system, two models developed at the Universidad Politecnica de Valencia (UPV) are being used in a study carried out by the consulting firm

425

J.B. Marco et al. (eds.),
Stochastic Hydrology and its Use in Water Resources Systems Simulation and Optimization, 425–437.
© 1993 Kluwer Academic Publishers.

EPTISA and the Department of Hydraulic and Environmental Engineering of the UPV for the Confederacion Hidrografica del Jucar River Basin Authority.

The first model is a network optimization model which has been designed for the general purpose of optimizing simplified schemes of water resources systems over a variable number of years using the month as the smallest division of time. The model was developed in 1986 and has been used for different water resources systems throughout Spain (Ebro-Navarre, Ebro-Rioja, Segura, Mijares).

The second model is a simulation model, also on a monthly basis, which uses a network optimization algorithm as a tool for the decision taking process in order to meet demands in accordance with operating rules.

Even though the model is a general use model it was developed initially for use in this study and, therefore, has some features which probably would not have been needed in other water resources systems, but since they can be used or not according to the user's wishes, the model is still one for general application.

The approach utilizes the optimization model as a screening model for simplified schemes of the Segura System and to devise the operating rules that best meet the objectives from the results obtained. Then, with a synthetically generated hydrological series, and with the historical series, simulations are carried out in order to check whether the operating rules perform adequately, or whether they need to be refined, or the optimization procedures reconsidered.

2. The Segura Water Resources System

The Segura WRS is located in the South-East of Spain and has an extension of 18,630 km^2, most of it being the Segura River Basin. The region has a mostly subtropical and semiarid Mediterranean climate with a mean annual precipitation of 376 mm per year, ranging from 1000 mm/year in the northwestern part to 200 mm/year in the South-East.

Water is used in the region for three purposes: irrigation, water supplies for municipal and industrial uses and hydroelectric production. The most important use, in terms of the amount of water demanded, is irrigation. At present 197,000 Ha are being irrigated. The distribution of this surface by the origin of the water being used is given in Table 1. The main crops are vegetables, citrus and other fruit.

In ten years' time the irrigated surface will be 250,000 Ha, and in 20 years 295,000 Ha. The population is about 134,000 inhabitants, most of them concentrated along the coast and along the lower basin of the River Segura. In the system there are 23 working hydroelectric plant and seven more are under study or at the project stage, involving a total energy production of 195 MWh/year at the present plus 451 MWh/year that would be produced by the seven new plant.

Also there is great concern over flood control. In some places the rainfall in one day has been greater than the mean annual precipitation.

Even though the demand of water is up to 1368 Hm^3/year for agricultural uses and 195 Hm^3/year for non-agricultural uses, the water actually used is 1052 Hm^3/year for agricultural purposes and 169 Hm^3/year for non-agricultural uses.

Figure 1. Location of the River Segura Water Resources System.

TABLE 1. Distribution of the Irrigated Surface by the Origin of the Water

Origin of Water	Ha	%
Surface water only	58,013	29
Groundwater only	71,577	36
Mixed	57,676	30
Re-use	9,612	5
Total	196,878	100

The sources of water are:
- surface water from the River Segura and its tributaries
- groundwater
- imported water

The hydrology of the River Segura (see Fig. 2) has been described and studied elsewhere (Marco, 1989), the main feature being that the mean annual flow of the river is 957 Hm³/year.

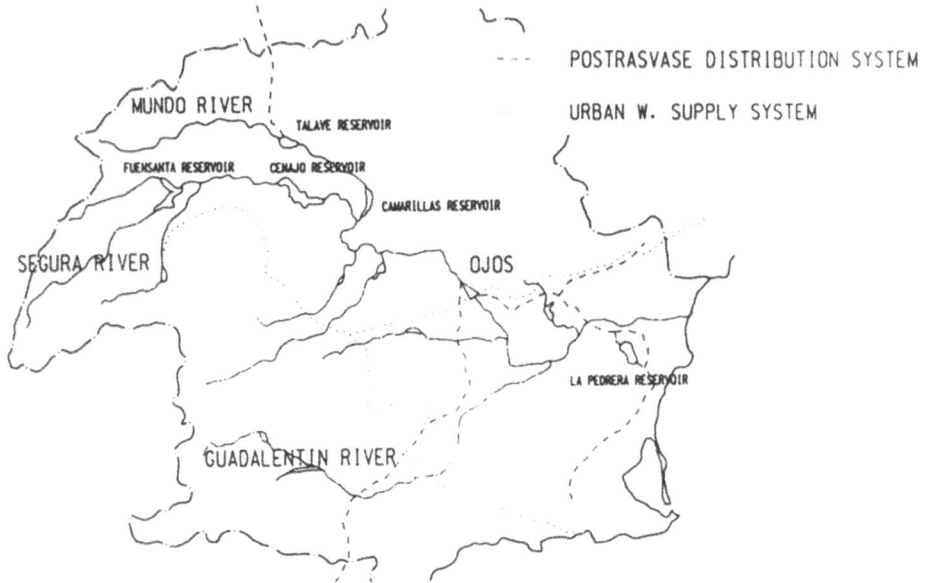

Figure 2.   The River Segura and main water conveyance systems.

The geology of the region is very complex and so, therefore, is the hydrogeology.   There are many aquifer units, the most important being the ones shown in Figure 3.   The renewable resources of the aquifers have been estimated as 317 Hm³/year, while the withdrawals at present are about 589 Hm³/year.   As a consequence, there is an overdraft of about 272 Hm³/year. There are five important aquifer units that have been or will therefore be legally declared as overexploited aquifers.

In order to import water from the Tagus Basin an aqueduct has been constructed which is over 250 km long, from a complex of three reservoirs in the River Tagus (namely Entrepeñas, Buendia and Bolarque) with a lift of 262 m to the canal, to the Talave Reservoir in the River Mundo, the main tributary of the Segura.   The capacity of the canal is 33 m³/sec.   The amount of water theoretically allocated to the diversion is 600 Hm³/year at the present and 1000 Hm³/year in the year 2010.   But up to now the highest diversion value has been 372 Hm³ in one year.   This is due to a severe drought period in both basins, and consequently to sociological opposition from the regions in the Tagus River basin to export water.

There are 11 main reservoirs in the basin.   The most important are the following (see Fig. 2):

  - Talave Reservoir, in the River Mundo, with 34 Hm³ capacity.
  - Camarillas Reservoir, in the River Mundo, with 36.5 Hm³ capacity.
  - Fuensanta Reservoir, in the River Segura, with 204.8 Hm³ capacity.
  - Cenajo Reservoir, in the River Segura, with 472 Hm³ capacity
  - La Pedrera Reservoir, which is in the lower basin, away from the actual river, with a capacity of 220 Hm³.

Figure 3. Main aquifers in the Segura Water Resources System.

There are three main distribution systems in the basin. One is the traditional irrigation network of the River Segura and its tributaries. The second is the urban supply system (Mancomunidad de Taibilla), deriving from the River Taibilla (see Figure 1) and serving most of the population centres in the region. The third is the new system (Postrasvase) of canals with a total length of 315 km in order to distribute the imported water. The La Pedrera Reservoir is in the line of the last system. There are some connections among the systems, for instance, the urban supply network has intakes from other tributaries of the Segura and from the Postrasvase. Also, the Postrasvase shares the Mundo River reservoirs and the bed of the Mundo and Segura Rivers from them to the Ojos diversion point.

We can therefore say that there are four different demand subsystems involved, with their own conveyance and distribution network, but interconnected. The fourth demand subsystem would be the portion of lands irrigated with ground water, in some instances pumping being done several kilometres away from the irrigated zones and conveying the water through pipes.

The main water-related problems of the region are the following (setting aside water quality problems):

- The shortages that are experienced when a situation of drought occurs.
- The overdraft of aquifers, which allowed the irrigated land to expand and, therefore, an important development of the region to take place. This is producing undesirable effects such as important drawdowns and future extinction of the reserves.

- It is difficult to obtain the water transferred from the Tagus basin when there is a drought situation in the Segura basin. It usually coincides with low flows in the River Tagus given the relatively high correlation between the two flows.

- The flood control issue, which will not be covered in this work.

In order to deal with the first three issues, a study of the management of the resources in the region has been undertaken. There follows a description of the application of the models.

3. OPTIRED Optimization Model and Its Application

OPTIRED is an optimization model capable of optimizing the management of a water resources network for a given time period of N years. For that period the model gives the optimal flows in the network as well as the optimal monthly amount of water stored in the reservoirs at the beginning of each month. This is a general model developed in order to deal with simplified schemes of water resources systems and used as a tool to infer management policies in line with the objectives. The model accepts a definition of the system which might include the following elements:

- Reservoirs, with their monthly storage level.

- Conduits, which can be stream reaches, canals, or pipes. A maximum flow capacity can be specified, as well as a minimum (e.g. ecological) flow. The minimum flow is taken as an objective and can have a specified level of priority.

- Inflows at a given point of the scheme.

- Demands, with their monthly distribution and an associated level of priority.

The objective function is

$$\min \sum_{t=1}^{N.12} \left[\sum_{i=1}^{n_c} \alpha_i d_{i,t} + \sum_{i=1}^{n_c} \tau_i , tq_{i,t} + \sum_{i=1}^{n_z} \beta_i d'_{i,t} \right] \tag{1}$$

subject to physical constraints. This is a linear objective function where
N   is the number of years in the time horizon to be optimized,
n_c is the number of conduits,
n_z is the number of demand zones,
$\alpha_i = K_\alpha - (100 . p_i)$
$\beta_i = K_\beta - (100 . p'_i)$
where p_i and p'_i are the priority numbers assigned by the user to the minimum flows and demand zones respectively, and $\tau_{i,t}$ is a dummy cost defined by the user for the flow through a conduit during the month corresponding to the index t.

The optimization problem is solved using the out-of-kilter algorithm (Bazaraa and Jarvis, 1977). Since the algorithm is for conservative network flows, the model itself automatically builds an internal representation of the problem in the form of a conservative flow network by adding balance nodes and arcs and expanding the network in order to cover the time dimension of the problem.

The same technique is used in similar models that have been reported in the literature, either specific designed for a case (Hamdan and Meredith, 1975; Andreu et al, 1986) or for general use (Martin, 1981). So, the

resulting model is a general one which is very simple and easy to understand and to use, and it is completely documented in Spanish (Andreu,
1989). In fact it has been used at the Universidad Politecnica de Valencia
to analyze water resources systems in the Ebro, Mijares and Segura basins.

In order to apply the model to the present case study, simplified schemes
of the system were used. The scheme corresponding to one of the
alternatives can be seen in Figure 4. Optimizations were performed using
historical data.

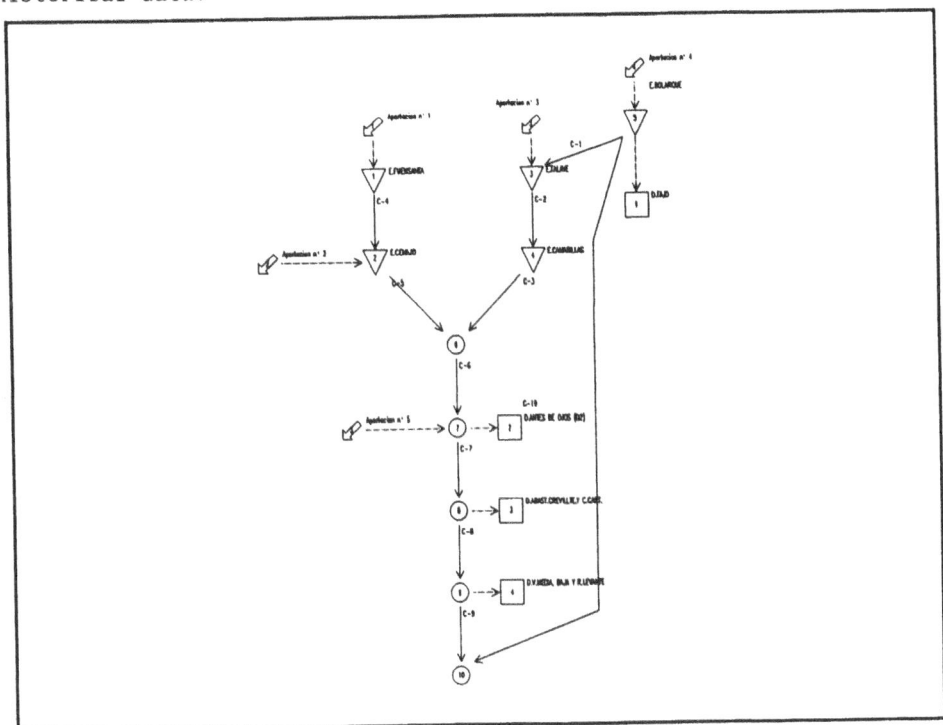

Figure 4. User's scheme for the OPTIRED model

The analysis of the results of the model is facilitated by the option of
obtaining a summary output which includes mean values of the variables as
well as some other indicators such as reliability in terms of percentage
of failures, volumetric reliability, frequencies of full and empty
reservoirs, and the maximum amount of the deficits. Also, in order to
analyze the results, there is an option for obtaining files with the
monthly values of the variables. These files can be directly imported by
Lotus-123. Then, figures showing the optimal behaviour of the elements can
be obtained. In fact, a program using macrocommands of the Lotus-123 has
been set up and is utilized to import the files and plot the results
automatically, either in the form of monthly evolution for the optimized
horizon or in the form of mean monthly values for that horizon. So the
analysis and also its report are made much easier.

As a result of the application of the model, the main useful management
indicators obtained were:

- Operating rules for the Fuensanta and Cenajo Reservoirs in the form of target releases (standard operating rule).

- Operating rules for the Talave and Camarillas Reservoirs in the form of target storages (rule curves).

- Operating rules for the transfer of water from the River Tagus to the Segura basin in order to minimize conflicts.   This was obtained by assigning costs (i.e. $\tau_{i,t}$ in Equation 1) for the flow in the transbasin diversion inversely proportional to the mean monthly inflows to the Tagus reservoirs.

- The degree of utility of several facilities planned in the basin such as new hydropower plant, canals and reservoirs.

- The tradeoffs in hydroelectricity production and irrigation, as well as in ecological flows and irrigation.

The rules and conclusions obtained proved to be quite robust when their sensitivity to changes in some key data was tested.

4. The SGC Simulation Model and Its Application

In order to simulate the management of the system with a high degree of detail, a simulation model was designed.   The model goes along the same lines as other models existing in the field of water resources systems simulation, for instance SIM-V (Martin, 1982) and MODSIM (Labadie, 1987). Like the above mentioned models, SGC utilizes network flow optimization as the tool to allocate the water to users and the storage of water to reservoirs, taking into account the priorities, the reservoir operating rules and the physical constraints involved.

The model accepts the following elements in order to represent a water resources system:

- Reservoirs.   The user supplies the physical parameters such as evaporation coefficients for the infiltration equation and capacity of the release devices.   The management parameters are also needed in order to define reservoir zoning and priorities.

- Inflows to the system.

- Conduits.   The river reaches, canals and pipes are included under this term.   Three types of conduits are provided:

. Type 1, simple conduits.

. Type 2, with infiltration losses, l, given by the equation $l=a+b.q^c$, where a is a constant factor, b and c are coefficients, and q is the flow through the reach of river or channel.

. Type 3, with hydraulic connection to aquifers.   In this case the river reach can have either losses or gains, depending on the state of the aquifer.   The amount of interflow is obtained from the simulation of the aquifers that will be described later.

For any of the preceding types the maximum monthly capacity of flow can be defined, as well as a minimum monthly flow.

- Consumptive demands, as, for instance, irrigated zones or municipal and industrial supply.   The zone has global monthly demands which can be supplied from up to three intakes from the surface system, each intake with a maximum monthly and yearly supply involving different irrigation efficiencies and surface irrigation returns to different points of the system.   Also, the zone may have the possibility of pumping from an aquifer up to a given pumping capacity. The user also assigns a priority number to the zone. Different zones with the same priority number will belong to the same group of users. The model will try to allocate the water in the same group proportionally to the demand of each user.

- Return elements. These are simply definitions of points where water from consumptive demands can be incorporated back into the system.

- Non-consumptive demands, this is the case of the hydropower stations. They make use of the water, but no significant amount is consumed. They are defined by their maximum flow capacity and the parameters needed to calculate the production of electricity, as well as by their target monthly flows.

- Artificial recharge facilities. These are flows that enter the aquifers. They are defined by their maximum capacity of recharge.

- Additional pumping facilities. The wells incorporating the pumped water into the surface conveyance facilities are included under this term.

- Aquifers. The model allows for the inclusion of any of the following types:

. Reservoir type. There is no other discharge from the aquifer but the pumped water.

. Aquifer with discharge through a spring.

. Aquifer hydraulically connected to a surface stream, modelled as a single cell aquifer.

. Aquifer hydraulically connected to a surface stream, modelled as a multiple cell aquifer.

. Aquifer modelled as distributed, using the eigenvalues approach (Andreu and Sahuquillo, 1987).

In order to use the model, the user must compound a scheme of the water resources system using the above elements and connecting nodes. A node must be located in any of the following points:

- reservoirs
- junctions of stream reaches or conduits
- diversion points
- intake points for demand zones (in reality they are also diversion points)
- entry points for the inflow to the system
- any point at which a stream or a conveyance facility changes its characteristics.

Even though the model is a general one, and can be used to analyze any system that fits into the configurations allowed, it has some features included specific to the Segura system. Such characteristics are:

- It is possible to define two separate networks that coexist and share some elements. In this way the traditional distribution and demand systems and the new distribution and demand systems are simulated together. But, losses of water owing to evaporation and filtrations in reservoirs, as well as river reaches, are distributed over the two systems proportionally to the water stored for each one in the common reservoirs or to the part of the flow through the reach of river belonging to each system.

- Another type of aquifer model is included. It is a nonlinear model with three outputs. The first output is in the form of evapotranspiration. The second is through a drainage system. Finally, the third is a stream-aquifer relationship. The discharge through each output is proportional to the volume of the aquifer between the piezometric level and the corresponding outlet.

Using the data supplied by the user concerning the scheme of the system, the model elaborates an internal conservative flow network. Each element of the user's scheme produces a set of arcs and nodes designed to simulate the physical characteristics of the element, as well as the management characteristics. Therefore, the internal network, which is not seen at all by the user unless he/she wants to, has many more arcs and nodes than the user's scheme.

The internal network is optimized using the out-of-kilter algorithm. The equivalent objective function proves to be:

$$\text{Min } [T_R + T_{R1} + T_{R2} + T_{R3} + T_{CD} + T_{ND} + T_{AR} + T_{AP}] \tag{2}$$

where T_R, T_{R1}, T_{R2}, T_{R3}, T_{CD}, T_{ND}, T_{AR}, and T_{AP} are terms corresponding to reservoirs, reaches of conduits type 1, 2 and 3, consumptive demands, non-consumptive demands, artificial recharges, and additional pumps respectively. The equation for each term is a linear expression and they will not be reproduced here for the sake of clarity.

The results of the optimization are used to allocate water to users, to decide how much water is pumped from the aquifer, as well as to evaluate the recharge to the aquifers owing to deep infiltration from irrigation, reservoirs and river reaches. The simulation of the aquifers is carried out at this stage. Since the simulation of the aquifers gives the values for the surface-groundwater relationship, these are updated in the network and this is optimized again, continuing this iterative process until adequate convergence is reached. The iterative process allows other nonlinear processes to be included such as evaporation and infiltration from reservoirs and river reaches.

As described in the optimization model, the SGC model produces a complete output of results for every month of the entire simulated time horizon. It also produces a summary of the simulation including monthly values and reliability indicators. Optionally, only the summary is obtained. Also, in order to facilitate the analysis of the results, files are produced which contain the values of the variables and which can be directly imported by Lotus-123 in order to produce graphs of the evolution of the variables over time, and their mean monthly values. This model is completely documented in Spanish (Andreu et al, 1989).

In order to apply the model to the Segura system, a user's scheme was designed involving 14 reservoirs, 18 inflows to the system, 92 conduits (i.e. including river reaches), 50 consumptive demands, three hydropower facilities, five additional pumping facilities and 19 aquifers. Figure 5 shows the aspect of the user's scheme. The figure is presented only to demonstrate the complexity of the scheme.

5. Conclusions

At the present stage the study is at the point of using the simulation model to validate and refine the operating rules derived from the optimization model. Synthetic hydrological series are used (Marco, 1989).

Even though no final conclusions have been drawn as yet, some observations can be made:

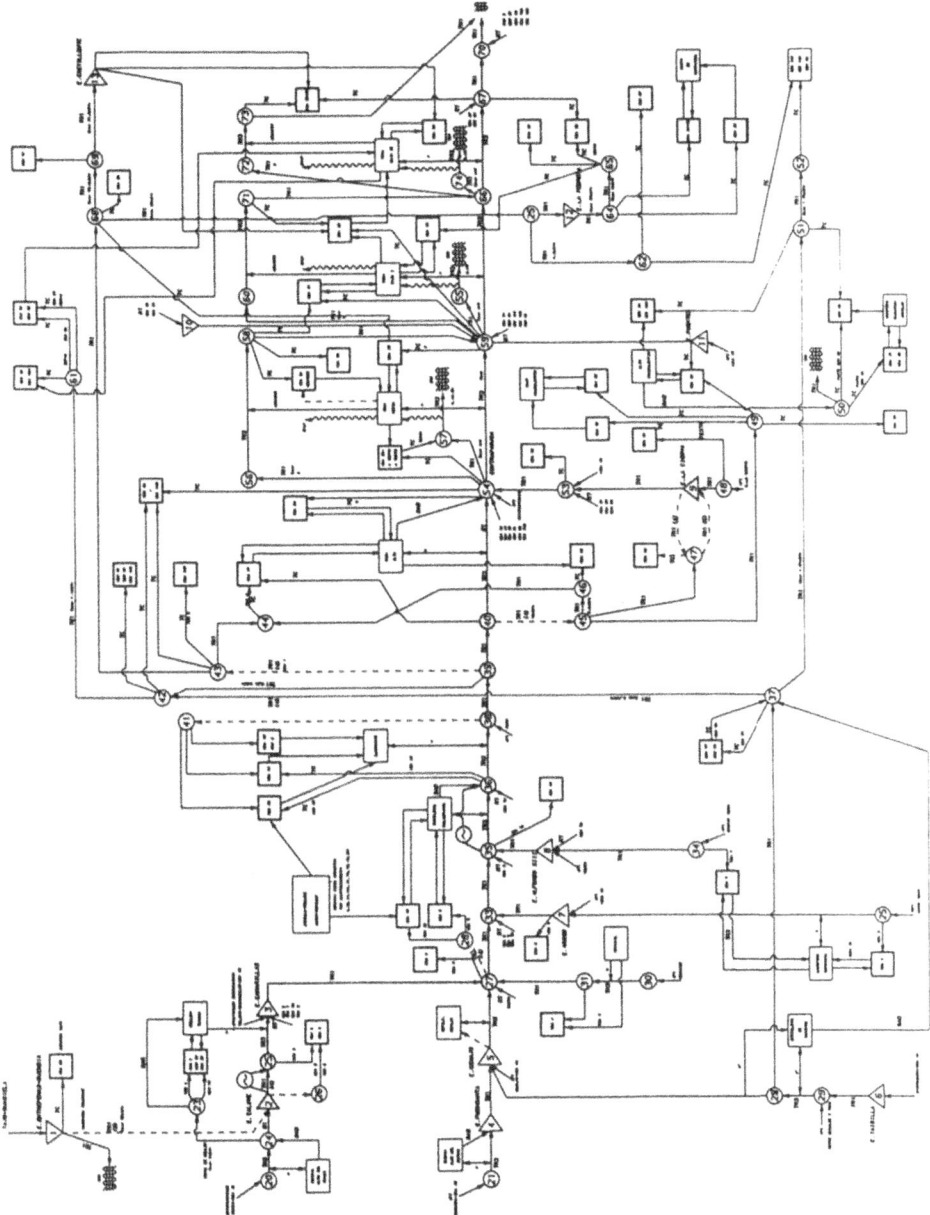

Figure 5. User's scheme for the SGC model

- The main problem with a detailed scheme for simulation, such as the ones that are being utilized in this study, is the amount of data needed. Therefore, as is usual everywhere, it is sometimes difficult to obtain so much information of the necessary quality for the parameter estimation. The model has proved to be one of easy calibration with respect to the physical behaviour of the system. The difficulty of the calibration does not lie in the model, but in the system itself. For instance, more data are needed concerting the degree of re-use of the water. These could be obtained with the installation of gauges in the drainage network and distribution networks.

- The main guidelines for management obtained from the optimization model perform very well when a more detailed simulation is carried out. In fact, only minor modifications were needed.

- The model is very well suited to defining the differences among alternative plans.

- The length of the synthetic series has some influence on the results. This is due to the fact that aquifers are being mined. The simulation model includes a limit on aquifer exploitation. When the limit is reached, no further pumpages are authorized until the aquifer goes back to a better condition. In some instances the limits are not reached if short series (i.e. 50 years) are used, while they are reached with long series. Even if the limit is reached in both cases, the reliability indicators are lower for long series than for short ones since the amount of months with availability of resources is the same in both cases, but the amount of months with no availability is larger, the longer the series.

6. References

Andreu, J. (1989) 'Manual del usuario del modelo OPTIRED de optimización de sistemas de recursos hidraulicos', Depto. de Ing. Hidraulica y Medio Ambiente, Universidad Politecnica de Valencia, Valencia, Spain

Andreu, J. and Sahuquillo A. (1987) 'Efficient aquifer simulation in complex systems', Journ. of Water Res., Planning and Manag., Vol. 113, No. 1, pp 110-129.

Andreu, J., Gomez, J. and Marco, J. (1986) 'Un modelo de optimizacion para el sistema de recursos hidraulicos superficiales Jucar-Turia', Revista de Obras Publicas, February, pp 123-132.

Andreu, J., Capilla, J. and Ferrer J. (1989) 'Manual del usuario del modelo SGC de simulacion de sistemas de recursos hidraulicos', Depto. de Ing. Hidraulica y Medio Ambiente, Universidad Politecnica de Valencia, Valencia, Spain.

Bazaraa, M.S. and Jarvis, J. (1977) 'Linear programming and network flows', John Wiley & Sons.

Hamdan, A.S. and Meredith, D.D. (1975) 'Screening model for conjunctive use water systems', Journ. of Hydraulics Division, ASCE, Vol. 101, No. HY10, October.

Labadie, J. (1987) 'River basin network flow model: MODSIM', User's Manual, Dept. of Civil Engineering, Colorado State University, Fort Collins, Colorado, USA.

Labadie, J. (1989) 'Combining simulation and optimization in river basin management', in Stochastic Hydrology in Water Resources Systems:

Simulation and Optimization, NATO ASI, Peñiscola, Spain.

Loucks, D.P. (1989) 'Implicit stochastic optimization and simulation', in Stochastic Hydrology in Water Resources Systems: Simulation and Optimization, NATO ASI, Peñiscola, Spain.

Marco, J. (1989) 'Models for forecasting: transfer function and ARMAX techniques', in Stochastic Hydrology in Water Resources Systems: Simulation and Optimization, NATO ASI, Peñiscola, Spain.

Martin, Q.W. (1981) 'Surface water resources allocation model AL-V: program documentation and user's manual', Rep. UM-35, Texas Depart. of Water Resources, Austin, Texas, USA.

Martin, Q.W. (1982) 'Multireservoir simulation and optimization model SIM-V: program documentation and user's manual', Rep. UM-38, Texas Depart. of Water Resources, Austin, Texas, USA.

ITERATIVE DYNAMIC PROGRAMMING TECHNIQUES: A CASE STUDY

M. BAYAZIT
Istanbul Technical University
Istanbul,
Turkey

1. Introduction

The optimization of the operation of a complex water resources system comprising a large number of reservoirs is a difficult task. Several optimization models have been developed for this purpose, based on various mathematical programming techniques (Yeh, 1985). Dynamic programming (DP) excels others as the technique that best suits the problem of finding the optimal operating rules of a reservoir system (Yakowitz, 1982).

DP decomposes a highly complex problem with a large number of variables into a sequence of subproblems with fewer variables which are solved recursively. In reservoir operation, the original problem of a multi-stage decision process is simplified into a series of single-stage problems, the stage corresponding to the time period. Thus, memory requirement and computer time are significantly reduced.

Consider the optimization of the operation of a single reservoir. The objective function can be written as:

$$\max_{y_1,\ldots,y_T} V(y_1,y_2,\ldots,y_T) \tag{1}$$

where y is the reservoir release in the time period j (j = 1,2,...,T), and V is a function of the form of either

$$V = \sum_{j=1}^{T} v_j(y_j) \tag{2a}$$

or

$$V = \min_j [v_j(y_j)] \tag{2b}$$

or

$$V = \prod_{j=1}^{T} v_j(y_j) \tag{2c}$$

439

J.B. Marco et al. (eds.),
Stochastic Hydrology and its Use in Water Resources Systems Simulation and Optimization, 439–451.
© 1993 *Kluwer Academic Publishers.*

Equation 2a applies when it is desired to maximize the total water (energy) supply along the operation horizon T, whereas Equation 2b corresponds to the maximization of the firm water (energy) supply. Equation 2c is applicable when the objective function is multiplicative in form (particularly for irrigation water).

This problem will be solved subject to the constraints below for j = 1,2,...,T:

$$S_{j+1} = S_j + x_j - y_j - e_j \tag{3}$$

$$S_{min} \le S_j \le S_{max} \tag{4}$$

where S_j is the reservoir water volume at the beginning of the j-th period, x_j is the volume of flow into the reservoir in that period, e_j is the evaporation loss from the reservoir, S_{min} and S_{max} are the minimum and maximum storages, respectively, allowed in the reservoir. The number of state variables (S_j) and the number of decision variables (y_j) is each equal to T (the total number of time periods in the operation horizon).

DP reduces this T-dimensional decision problem into T problems, each with only one decision variable, y_j, through the use of the recursive equation, which is of the form of either

$$f_j(S_j) = \max_{y_j} [v_j(y_j) + f_{j+1}(S_{j+1})] \tag{5a}$$

or
$$f_j(S_j) = \max_{y_j} \{\min [v_j(y_j), f_{j+1}(S_{j+1})]\} \tag{5b}$$

or
$$f_j(S_j) = \max_{y_j} \{v_j(y_j) \cdot f_{j+1}(S_{j+1})\} \tag{5c}$$

where S_{j+1} is to be expressed in terms of S_j and y_j on the basis of the water balance equation (Eq. 3).

In solving the problem of reservoir operation optimization by DP, both the state and decision variables are discretized (DDP, discrete dynamic programming). For the single reservoir problem, computer time requirement is proportional to m^2, m is the number of discrete states, or of discrete decisions. Therefore, there are m^2 combinations of state and decision variable values at which f_j is to be evaluated.

When it is required to solve a similar problem for a system of n reservoirs, it can easily be seen that the computer time will grow exponentially with the number of reservoirs, since the number of combinations to be compared is now m^{2n} (assuming that m is the same for all reservoirs). For example, for a system of two reservoirs in series, the recursive equation can be written as either

$$f_j(S_{1j},S_{2j}) = \max_{y_{1j},y_{2j}} \{[v_{1j}(y_{1j})+v_{2j}(y_{2j})] + f_{j+1}(S_{1,j+1}+S_{2,j+1})\} \tag{6a}$$

or
$$f_j(S_{1j},S_{2j}) = \max_{y_{1j},y_{2j}} \{\min[v_{1j}(y_{1j})+v_{2j}(y_{2j})], f_{j+1}(S_{1,j+1}+S_{2,j+1})]\} \tag{6b}$$

or $\quad f_j(S_{1j},S_{2j}) = \max_{y_{1j},y_{2j}}\{[v_{1j}(y_{1j})+v_{2j}(y_{2j})] \cdot f_{j+1}(S_{1,j+1}+S_{2,j+1})\}$ (6c)

For this problem the number of combinations (the number of state transitions) equals m^4 since the recursive equation must be evaluated at m^2 points for each combination of state variables (of which there exist m^2 couples) in each stage.

This so called "curse of dimensionality" prohibits the use of DDP in practice when the number of reservoirs exceeds two (or three). In order to reduce the computational burden, a coarse discretization (small m) could be considered. But the practical applications have shown that incorrect answers are obtained when m is too small. In other words, it is necessary to consider a rather large number of states in each reservoir in order to obtain a solution of reasonable accuracy.

2.   Iterative Techniques

In order to overcome this difficulty, various approaches have been proposed. These successive approximation techniques can be broadly classified into two groups.

1. An initial estimate of the operating policy is chosen for each reservoir. Then, in each iteration, the optimal policy is sought only in a corridor of a certain width centred about the initial policy (Fig. 1). If a better policy can be found, it replaces the initial policy, and the next iteration starts. The procedure is continued until a certain convergence criterion is satisfied.

This approach reduces the number of combinations to be tested in each stage. If r discrete states are considered for each state and decision variable, then the computer time per iteration will be proportional to r^{2n} (usually r equals 3, i.e., only the states that are just above and below the initial policy are considered). Although the computational burden is drastically reduced (3^{2n} combinations instead of m^{2n}), the problem of dimensionality still exists, i.e., the computer time grows exponentially with the number of reservoirs.

This algorithm was first proposed by Hall et al. (1969), who called it "incremental DP", to maximize the firm energy of two parallel reservoirs. Later Heidari et al. (1971) referred to this scheme as "discrete differential DP (DDDP)".

2. In another approach the operation of only one of the reservoirs is optimized in each iteration while the contents of the other reservoirs are not changed. Thus the original n-dimensional problem is broken down into a series of one-dimensional subproblems. In the next iterations the reservoir to be optimized is changed successively until all the reservoirs in the system are optimized once, after this, the optimization is continued in the same sequence as before if necessary.

The advantage of this scheme is that the problem of dimensionality is overcome. In each iteration the computer time is proportional to m^2, just like in the case of a single reservoir.

This method was first mentioned by Larson (1968) as "DP with successive approximations", and applied to a system of four reservoirs for hydropower and irrigation purposes. Trott and Yeh (1973) applied it to the optimization of multiple reservoirs. Yeh (1985) refers to this scheme as "incremental DP with successive approximations (IDPSA)". Boehle et al. (1982) used a similar scheme called "sequential optimization" in a study of a system of multiple reservoirs with the main purpose of low flow augmentation. Reservoirs are optimized by DP in a certain sequence to obtain a low flow target for each reservoir. Releases of this target are then considered in the optimization of the downstream reservoir.

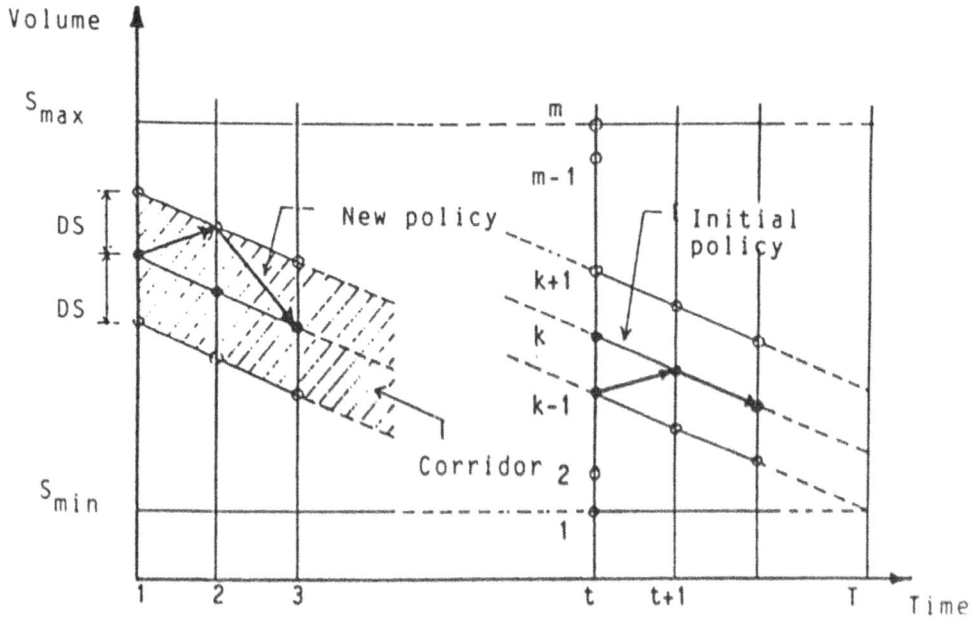

Figure 1. Basic concepts of the DDDP technique.

The problem with both of the above methods of successive approximation is that the convergence to the global optimum cannot be guaranteed (a local optimum can be reached or there may be no convergence). Yakowitz (1982) states that under various smoothness assumptions both methods converge to stationary policies with decreasing discretization grid size (increasing m). The convergence rate, however, is linear, which implies that the convergence will be rather slow. On the other hand, the convergence of the solution to the true (global) optimum cannot be proved. Different initial policies should be tried and the final policies should be compared. Turgeon (1982) argued that the state increment (corridor width) should be varied along the stages in order to avoid a local optimum. Hall et al. (1969) showed that it is more efficient to reduce the state increments as the iterations proceed.

3. Case Study

The system shown in Figure 2 consisting of five reservoirs, is planned to supply water to Ankara, capital city of Turkey, in the future. In the year 2030 it is estimated that Ankara will have a population of 4,300,000, and the annual water demand will rise to 690 million m^3.

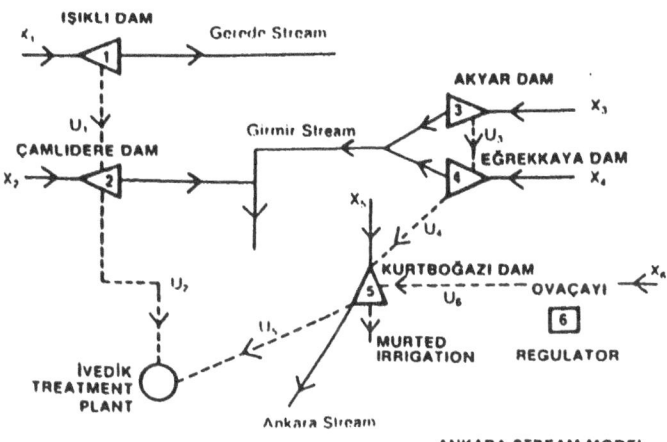

Figure 2. System of reservoirs in the case study.

The main characteristics of the reservoirs are shown in Table I. One of the reservoirs (Res. 1) is in a different basin and has a rather small capacity. It works as a diversion dam to transmit the inflows to the large reservoir (Res. 2) in the neighbouring basin. All the remaining reservoirs are in the same basin but they are linked to some of the others by pipes and tunnels in a rather complicated configuration.

TABLE I

Reservoir Number	Reservoir Name	Active Storage (10^6 m^3)	Average Annual Inflow (10^6 m^3)
1	Işikli	100	354
2	Çamlidere	1076	171
3	Akyar	47	57
4	Eğrekkaya	85.5	91
5	Kurtboğazi	83.5	52

It is required to determine the optimal operating policy of the reservoirs in the system so that the firm water supply is maximized. Reservoir 1 is not included in the optimization procedure since its regulating capacity is very low; it is assumed that this reservoir simply diverts the inflows into Reservoir 2. Thus we are faced with the problem of optimizing the operation of a rather complex system of four reservoirs.

3.1. DP FORMULATION OF THE PROBLEM

The objective function of the problem is

$$\max V = \min_{j} \left[(u_{2j} + u_{5j})/\beta_j \right] \tag{7}$$

where V is the annual firm water that can be supplied by the system, u_{ij} is the volume of water supplied by Reservoir i in the month j, β_j is the monthly distribution coefficient (percentage of annual water demand required in the month j). u_{ij} can be written as

$$u_{ij} = \min \left[y_{ij}, K_j(S_{ij}, S_{i,j+1}) \right] \tag{8}$$

where y_{ij} is the release from the reservoir i in the month j, K_j is the capacity of the supply line from the i reservoir as a function of the storage in that reservoir.

The constraints of the problem have the following forms:

$$S_{i,j+1} = S_{ij} + x_{ij} + u_{i-1,j} - y_{ij} - e_{ij} \tag{9}$$

$$S_{min,i} \leq S_{ij} \leq S_{max,i} \tag{10}$$

for Reservoirs 2 to 5 (i = 2-5). For Reservoir 3, $u_{i-1,j} = 0$, since there is no reservoir upstream. For Reservoir 5 water diverted from the No. 6 regulator and the irrigation water demand should also be considered in the water balance equation.

In the backward dynamic programming formulation of the problem the recursive equation is

$$f_j(S_{ij}) = \max_{y_{ij}} \left\{ \min \left[V_j, f_{j+1}(S_{i,j+1}) \right] \right\} \tag{11}$$

where the subscript i takes the values 2, 3, 4 and 5. V_j represents the water supplied by the system in month j (expressed in terms of the annual supply):

$$V_j = (u_{2j} + u_{5j}) / \beta_j \tag{12}$$

f_j is the firm water supply of the system for the last T-j+1 months of the operation period. $S_{i,j+1}$ (i = 2-5) are related to S_{ij} by means of the water balance equations (Eq. 9).

Monthly streamflow records are available for a period of 21 years. The operation of the system is optimized first with the historical inflows, then stochastic optimization is attempted.

3.2. DETERMINISTIC OPTIMIZATION

Although the historical inflows will certainly not be repeated in the future, it is advantageous to start the optimization by using the observed historical record in the deterministic sense in order to gain insight into the behaviour of the system. Therefore the operation of the water supply system for the city of Ankara was first optimized by DP using the historical monthly streamflows.

Since the number of reservoirs to be optimized is four, classical DP cannot be used, and it is necessary to apply an iterative method. First an attempt was made to use DDDP.

The magnitude of the state increment (grid size) DS, corresponding to the half-width of the corridor in DDDP, should be chosen such that transitions are possible between all the states in successive stages (Turgeon, 1982). This requires that DS is not greater than the volume of inflow during a time period. In order to accelerate the convergence, a large DS value was chosen in the initial iterations (1/20 - 1/40th of the reservoir capacity), making it possible to screen a larger part of the state space in the optimization. The DS value was reduced (by about 25%) as soon as no significant increase was observed in the objective function at two consecutive iterations.

The choice of the initial policies has a strong effect on the rate of convergence. Three different types of initial policies were tried:

a. a uniform policy where storage is constant (reservoirs half-full) throughout the period of operation,

b. a linear policy which starts with a full reservoir emptying at the end of the period of operation,

c. a combined policy starting with a full reservoir at the beginning of the estimated critical period, emptying at the end of it, and then starting to fill again.

The initial policies described above did not provide rapid convergence. The annual firm water supply could not be raised above 550 million m³ with these policies at the end of ten iterations. A somewhat better result was obtained when the optimal operating policy of each reservoir obtained by DP for that single reservoir was assumed to be the initial policy. This led to a firm water supply of 600 million m³ after 15 iterations.

A closer examination of the system showed that Reservoir 2 has a very large capacity in comparison with the other reservoirs. Therefore, this reservoir provides over-year storage, whereas others mainly regulate within-year fluctuations. In order to take this into account, the initial policy for Reservoir 2 was taken such that it was full at the beginning, emptied at the end of the critical period, and then started to fill again. The smaller remaining reservoirs in the system, however, were assumed to be full at the end of the wet period each year, and 40% full at the end of the dry season of the year. This set of initial policies (which corresponds to 490 million m³ of firm water supply per year) provided 670 million m³/yr by DDDP at the end of 19 iterations. In spite of the improvements obtained by a very careful selection of the initial policies, the convergence rate was slow and the computer time required was prohibitive.

In order to accelerate the convergence, a new technique was introduced, which is a combination of DDDP and IDPSA. This method, to be called "incremental sequential DP (ISDP)", has the following properties:

1. The reservoirs in the system are numbered in the direction of stream-flow (upstream reservoirs having smaller numbers). In the case of parallel reservoirs, those with a larger capacity have priority in numbering.

2. Reservoirs are optimized by DDDP one by one in the sequence determined in the previous step. The entire system is taken into consideration in the optimization, operating all the reservoirs except the one to be optimized

with their assumed initial policies, and including their releases in the objective function. Having obtained the new (better) policy of the reservoir being optimized, that reservoir is then operated with that policy. Once all the reservoirs in the system have been optimized in this manner, the next iteration starts from the first reservoir.

3. In order to achieve a rapid convergence, the number of states considered in DDDP is increased. Instead of three states, r = 5-11 states are considered in the corridor. It is observed that a suitable choice of r drastically reduces the required number of iterations. Although the computer time for each iteration is increased, the total time to reach the optimum is reduced. The value of r should be decreased in the latter stages of iteration when the operating policy approaches the optimum.

4. In case more than one policy is found to correspond to the same value of the objective function at the end of the optimization of a reservoir, the new policy should be chosen as the one that is closest to the previous policy in order to accelerate the convergence.

5. In choosing the initial policy for a complex system, the dynamic behaviour of the reservoirs in the system should be considered carefully. This has a significant effect on the rate of convergence as explained before.

6. State increment DS can be chosen at a rather large value in the first iterations, and then gradually reduced.

The water supply system of Ankara was optimized by ISDP. Details of the iterations are shown in Table II. It can be seen that an annual firm water volume of 690 million m^3 is reached at the end of eight iterations. The supply is 675 and 685 million m^3 at the end of one and three iterations, respectively, showing that the convergence is very rapid (Fig. 3). DS is chosen as 1/20 - 1/50 th of the reservoir capacity in the earlier iterations and reduced to 1/200 - 1/400 of the capacity as the iterations progress. The number of states in the corridor is r = 7 in the beginning, which is gradually decreased gradually to r = 3.

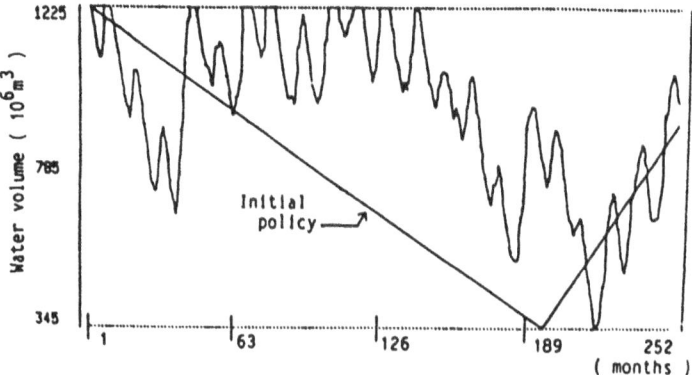

Figure 3. Optimal storages of Reservoir 2 obtained by ISDP compared with the initial policy.

TABLE II

Iteration No.	DS	r	Firm Water $(10^6 m^3)$
1	1/50	7	675
2	1/50	7	679
3	1/50	7	685
4	1/200	5	686
5	1/200	5	687
6	1/200	5	688
7	1/200	5	689
8	1/400	3	690

It can be seen that ISDP drastically reduces the number of iterations required to reach the optimum as compared with DDDP, on condition that the initial operating policies of the individual reservoirs are chosen in accordance with the dynamic behaviour of the whole system. Computer time and memory requirements of various methods are compared in Table III. ISDP requires a much smaller computer time per iteration compared with DDDP. As an example, the ratio of computer time requirements for DDDP (r = 3) and ISDP (r = 7) is $3^{2 \times 4}$: 4×7^2 = 33.5 per iteration for a system of four reservoirs.

TABLE III

Method of Optimization	Computer Time per Iteration	Computer Memory
Dynamic programming (DP)	CTm^{2n}	$2m^{2n}+Tnm^n$
Discrete differential DP (DDDP)	$CT3^{2n}$	$2 \times 3^{2n}+Tn3^n$
Incremental DP with succ.app.(IDPSA)	$CTnm^2$	$2m^2+Tm$
Incremental sequential DP (ISDP)	$CTnr^2$	$2r^2+Tr$

C : a constant for a certain type of computer
T : No. of time periods in the operation horizon
m : No. of states in the reservoir
r : No. of states considered in the corridor
n : No. of reservoirs in the system

3.3. STOCHASTIC OPTIMIZATION

Implicit stochastic optimization was used in which the optimum policy was determined by ISDP for each of several generated (synthetic) streamflow sequences, and then the set of optimum policies obtained in this manner was examined by multivariate analysis to arrive at an operating strategy.

Streamflow sequences were simulated through a two-station model as described by Kottegoda and Yevjevich (1977). The monthly inflows to Reservoir 2 (the main reservoir in the system) were generated by the Thomas-Fiering model. Then the inflows to each of the remaining reservoirs were generated by regression on the inflows to Reservoir 2 using Model II of the above study. This model preserves the mean, standard deviation, lag-one autocorrelation and lag-zero cross-correlation coefficients of the flows (cross-correlations between the pair of stations one of which is not the main station, are not intended to be preserved).

In the stochastic optimization by ISDP, the initial policies of the reservoirs were decided using the regression equations based on the results of deterministic optimization. These equations provide the volume of stored water in a reservoir as a linear function of the sum of the storage in the previous month and the monthly inflow. A regression equation was obtained for each reservoir and for each month of the year ($k = 1,2,\ldots12$):

$$S_{ij} = a_{ik} + b_{ik} (S_{i,j-1} + x_{ij}) \tag{13}$$

Regression coefficients a_{ik} and b_{ik} were determined using 21-year long results of deterministic optimization studies. (Correlation coefficients ranged between 0.64 and 0.99).

On a DEC MicroVax-II computer, the total time required for the optimization with one 21-year long sequence of monthly flows was an average of 80 minutes, during which about 12 iterations were performed.

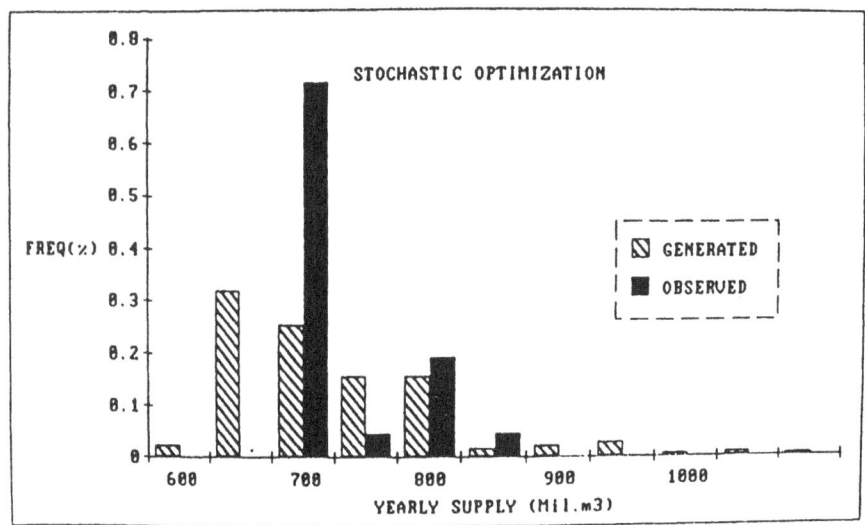

Figure 4. Frequency histogram of water supply based on normal synthetic flows compared with that of observed flows.

The synthetic flow generation model described above was used first with the flows, and then their log-transforms. The distribution of the observed monthly flows followed the lognormal law more closely.

The frequency histogram of the annual water supply determined by the optimization of the operation of the reservoir system using the normally distributed synthetic flow series is shown in Figure 4, together with the histogram based on historical flows. Corresponding histograms for the results of stochastic optimization based on the lognormally distributed synthetic flow sequences are given in Figure 5.

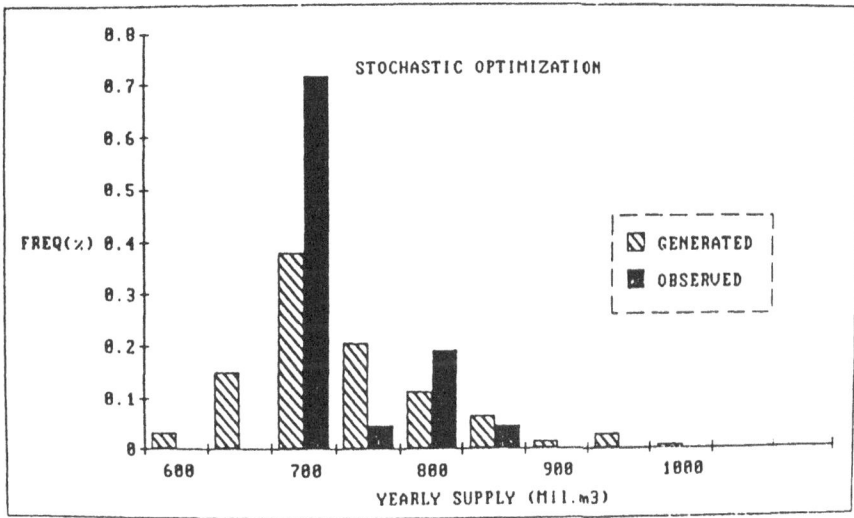

Figure 5. Frequency histogram of water supply based on lognormal synthetic flows compared with that of observed flows.

The statistics of annual water supply based on observed, normal and lognormal synthetic flows are given in Table IV. It can be seen that the assumption of normal distribution results in a larger scatter (higher standard deviation and wider range) and larger skewness, compared with the case of lognormally distributed inflows. Although both the minimum and maximum water supply volumes are higher in the normal case, the mean supply is slightly lower. The percentages of supply volumes lower than 690 million m^3 (firm water supply of the observed flow sequence) are 34 and 19, respectively, for normal and lognormal flows. Thus, in this case study the assumption of normal distribution leads to a greater value of risk compared with the assumption of lognormal distribution.

TABLE IV

Flows	Mean ($10^6 m^3$)	Std.Dev. ($10^6 m^3$)	Min. ($10^6 m^3$)	Max. ($10^6 m^3$)	Skewness
Observed	727.6	41.8	690.5	813.2	1.08
Normal synthetic	728.0	74.4	634.7	1031.5	1.44
Lognormal synthetic	729.9	63.1	614.7	936.8	1.11

Finally, the operation rules of the reservoirs were determined according to Equation 13. Coefficients a_{ik}, b_{ik} in this equation were now determined using the results of stochastic optimization study. As an example, coefficients a_{2k}, b_{2k} for Reservoir 2 are shown in Table V together with their values based on deterministic optimization, which were used in selecting the initial policies.

TABLE V

| k | Deterministic Optimization | | | Stochastic Optimization | | |
	a_{2k}	b_{2k}	St.error	a_{2k}	b_{2k}	St.error
1	-55.7	1.012	6.6	-45.2	0.999	4.9
2	-50.8	1.009	4.6	-40.8	0.996	3.4
3	-50.7	1.018	10.0	-39.4	1.003	4.7
4	-49.2	1.018	11.6	-41.0	1.008	5.5
5	-24.3	0.992	13.3	-27.0	0.993	7.8
6	43.8	0.917	31.0	22.6	0.940	10.2
7	61.7	0.900	31.3	70.3	0.894	17.1
8	38.9	0.915	21.2	51.8	0.907	16.5
9	-15.5	0.964	10.4	-29.4	0.979	5.2
10	-38.3	0.986	7.4	-40.3	0.988	2.5
11	-47.5	0.997	4.7	-45.1	0.993	3.8
12	-49.1	1.001	1.1	-47.7	0.997	7.7

4. Conclusions

The operation of a complex reservoir system can be optimized by incremental sequential dynamic programming, which has significant advantages over discrete differential dynamic programming with respect to the computer time and memory requirements. It is important to choose the initial operating policies of the reservoirs in accordance with the dynamic behaviour of the system. The same technique can be employed in implicit stochastic optimization where the optimum policies obtained with several synthetic streamflow sequences are analyzed to obtain the operating rules of the reservoirs.

5. References

Bayazit, M., and Duranyildiz, I. (1987) 'An iterative method to optimize the operation of reservoir systems', Water Resour. Management 1, pp 255-266.

Boehle, W., Harboe, R., and Schultz, A. (1982) 'Sequential optimization of a multipurpose reservoir system', in T.E. Unny and E.A. McBean (Eds.), Decision Making for Hydrosystems: Forecasting and Operation, Water Resources Publications, pp 203-219.

Hall, W.A., Harboe, R.C., Yeh, W.W.G., and Askew, A.S. (1969) 'Optimum firm power output from a two reservoir system by incremental dynamic programming', Contribution 130, Water Resources Center, University of California, Los Angeles.

Heidari, M., Chow, V.T., Kokotivic, P.V., and Meredith, D.D. (1971) 'Discrete differential dynamic programming approach to water resources system optimization', Water Res., 7, pp 273-283.

Larson, R. (1968) State Increment Dynamic Programming, Elsevier, NY.

Trott, W.J., and Yeh, W.W.G. (1971) 'Optimization of multiple reservoir systems', J. Hydraulics Div. Amer. Soc. Civil Eng. 99, pp 1865-1884.

Turgeon, A. (1982) 'Incremental dynamic programming may yield nonoptimal solutions', Water Resour. Res. 18, pp 1599-1604.

Yakowitz, S. (1982) 'Dynamic programming applications in water resources', Water Resour. Res. 18, pp 673-696.

Yeh, W.W.G. (1985) 'Reservoir management and operation models. A state-of-the-art review', Water Resour. Res. 21, pp 1791-1818.

USE AND NON-USE OF WATER MANAGEMENT MODELS IN SWEDEN: EXPERIENCES FROM MODEL DEVELOPMENT TESTS

L. de MARE
Jordbruks Verket
Vallgatan 8
551 82 Jönköping
Sweden

ABSTRACT. Three modelling experiences performed at IIASA in the Regional Water Management Case Study on Western Skåne are described: Irrigational Demand, Nutrient Leaching, and Cost Allocation. The model tests proved to be valuable at the time, but since then there have been very few Swedish examples of water management based on simulation or optimization models. This is partly due to difficulties in data collection, but more to the Swedish decision-making system involving many independent local decision-makers. The currently existing and foreseen water problems, quantity and quality, do however call for an improved water management based on scientific models.

1. General Background

Swedish water managers are generally facing a situation where water is relatively abundant. Quantitative resource management has thus not been a major issue except in the water power sector where benefits from optimal reservoir operation are obvious. In other water management sectors the main problems are connected with water quality. Due to the Swedish decision-making administration with far reaching local self-dependency, most problems are solved locally based on administrative regulations with little incentives for regional solutions (de Maré, 1977).

However, in the early 70's a series of dry years occurred which called for improved water management practices, especially in the agricultural regions. The most pronounced agricultural region is the province of Skåne. Table 1 shows some indicators on the relative importance of the Skåne region compared to national totals (Andersson et al., 1979).

The severity of the dry period was enhanced by the 1965 forecast for municipal and industrial water use which pointed to considerable increase for the Skåne region. This increase alone motivated a decision taken in 1973 to construct a large water transfer scheme for the future water supply in Skåne.

This situation with foreseen increasing water use in the municipal and industrial sectors as well as in the agricultural sector, and with an already initiated interregional water transfer project, formed the back-

453

J.B. Marco et al. (eds.),
Stochastic Hydrology and its Use in Water Resources Systems Simulation and Optimization, 453–459.
© 1993 *Kluwer Academic Publishers.*

TABLE 1. Percentage shares for Skåne in relation
to the Swedish totals (1976)

Land area	2.7%
Population	12.3%
Agricultural area	16.7%
Irrigated area	24.6%
Irrigation water amount	33.0%
Potential water resources	1.1%

ground to the Regional Water Management Case Study of Western Skåne performed at the International Institute for Applied Systems Analysis (IIASA) in the late 70's/early 80's. In the project a broad spectrum of models were developed, all focusing on one or more of the four core topics which were derived from the different kinds of problems encountered: (1) criteria of choice among alternative courses of action, (2) conflict over resource allocation and use, (3) uncertainty and risk in water management planning, and (4) institutional framework of regional water management.

In the next three sections of the paper three modelling experiences will be described. For an excellent summary of the full project exercise, see Kindler (1982).

2. Irrigational Water Demand

During and immediately after the dry period, irrigation developed rapidly in Sweden, especially in Skåne. In view of the possible importance of irrigation demands to the water resource situation in the region, Arthur (1980) analyzed the effects of supplementary irrigation on the total amount of water available to crops. The analysis was based on a simulation model of the actual irrigation practices recommended to the farmers at the time, assuming an unlimited supply of water for irrigation.

The irrigation practices are based on daily measurements of rainfall recorded by the farmers and on the fact that evapotranspiration amounts to roughly 3.5 mm per day in May–July and 2.5 mm per day in August–September. The amount of water needed is thus set to 3.5 mm per day as an average for the early summer. For sandy soils the longest period without water application is set to seven days.

Starting from the beginning of the irrigation season, and with potatoes on sandy soil as an example, rainfall is recorded during seven days. If, on the seventh day, no rainfall has occurred, 25 mm (7 times 3.5 mm) of irrigation water are applied. If rainfall has been recorded, say 15 mm, the irrigation day is postponed by four days ($\approx$ 15/3.5) and consecutively further if more rain falls during the postponement period.

With a data series of daily rainfall, the simulation of the irrigation practice is simple and straightforward. The amount of irrigation water needed during each season is then just the number of irrigation days times 25 mm (18 mm for irrigation days in August–September).

Estimates of irrigation water needs are summarized in Table 2. The estimates stem from simulation based on a 75-year series of daily precipitation data recorded at Lund.

TABLE 2. Estimates of irrigated water needs based on 75 simulated seasons

Crop	Soil Type	Irrigated Season	Irrigation mm Average	Range
Potatoes	Sandy	June 16–Aug 31	98	18–179
Grain	Sandy	June 1–July 15	88	25–150
Grain	Clay	June 1–July 15	90	35–140
Vegetable oil	Clay	May 16–July 15	127	35–175
Sugar beet	Clay	July 1–Sep 15	86	25–155
Ley, harvested	Clay	May 16–Aug 15	161	70–235
Ley, grazing	Clay	May 16–Sep 15	184	70–285

For potatoes the average amount of irrigation water corresponds to four irrigation days. In the "wettest" year only one irrigation day is needed, falling in August, and the "driest" year needs eight irrigation days.

Table 3 shows, for three selected crops, the irrigation effects on the total amounts of water applied: average for the 75 seasons, standard deviation (SD), and range, all with and without irrigation.

TABLE 3. Water available to crops, mm, based on 75 simulated seasons

Crop	Average	SD	Range
Potatoes			
without irrigation	176	56	50–350
with irrigation	274	33	200–400
Grain			
without irrigation	83	34	20–200
with irrigation	172	24	120–260
Harvested lay			
without irrigation	179	49	50–400
with irrigation	340	32	280–460

The main result from Table 3 is that the standard deviation and the spread of the distributions of seasonal water totals have been considerably reduced with irrigation compared to only rainfall application. In fact, for all crops simulated, the minimum water supply with irrigation in any season is greater than the average supply without irrigation. Apart from the obvious benefits from a consistent supply of water to the crops thoughout the season, there is thus also an often overlooked benefit in terms of reducing risk from reduced supply variability.

In a companion study, Anderson (1980) showed that reduced supply variability as such is economically beneficial to the farmer even if the mean water input stays unchanged. Based on Arthur's simulation results, the reduction in variance of the water input distribution accounts for 10-20% of the net benefits to the farmers obtained by the supplementary irrigation.

3. Nutrient Leaching

The problem of nutrient leaching has for several years been a primary concern to the Swedish environmental authorities. Numerous studies of the river transport of chemical substances indicate that leaching from agricultural land is a main source for nutrients, especially nitrate. High nitrate concentrations are found in all water bodies surrounded by agricultural land, even in relatively deep municipal ground-water wells. The only countermeasures of a long-term nature seem to be changes in agricultural management practices.

In relation to the irrigation studies described in Section 2, the effects of irrigation on the leaching of nitrate was investigated using the CREAMS model (de Maré, 1982). The CREAMS model (Chemicals, Runoff, and Erosion from Agricultural Management Systems) is a field scale model consisting of several components representing different aspects of the pollution load from agricultural fields. The sub-models, governing equations and interrelationships, are thoroughly described in Knisel (1980).

Growing potatoes on sandy soil, fairly common in Skåne, is regarded as the most severe practice with respect to nitrate leaching. However, according to agricultural statistics for the region, wheat is, by far, the most common crop. It was decided, therefore, to compare the situation of wheat on clay soil with that of potatoes on sandy soil. Since the climatic, topographic, and soil conditions are such that only negligable amounts of surface runoff occurs, it was decided to base the simulation on only the percolation part of the hydrology sub-model and the nutrient sub-model.

The two years of 1976 and 1977 were simulated. The climatic input for the model consisted of daily precipitation, monthly mean temperature, monthly mean radiation, and a characteristic value for the nitrogen content of rainfall, all recorded in the region. Soil data were based on characteristic textural compositions for sandy and clay soils in Skåne. From these, input parameters such as soil porosity, saturated hydraulic conductivity, field capacity and wilting point could be calculated. Crop data and farming operation information included planting and harvesting dates, leaf area indices, fertilizer application rates and dates, tillage dates, and optimum yields. Irrigation amounts and application dates, finally, were derived from the irrigation practice mentioned in Section 2.

The simulation results are summarized in Table 4. The table shows, for the two years and the two crops studied, yearly totals of water application, evapotranspiration, percolation, and nitrate leached.

The results in Table 4 clearly show that the nitrate leaching from potato fields can be substantial and increased by irrigation. The increase is, however, occasional and then usually the effect of heavy rainfall, or worse, of irrigation and heavy rainfall immediately following fertilizer application. In fact, preliminary simulation runs on a 20-year data series indicate that appropriate irrigation applications may reduce nitrate leaching on a long-term basis due to higher amounts of nutrient uptake by the crop.

4. Cost Allocation

The third modelling exercise addresses a completely different question,

TABLE 4. Simulation results using the CREAMS model for 1976 and 1977, annual totals

Crop		Water Input mm	Evapo- transpiration mm	Perco- lation mm	Nitrate leached kg/ha
Potatoes					
without irrigation	1976	543	417	44	29
	1977	667	525	142	21
with irrigation	1976	768	660	72	36
	1977	892	635	248	36
Wheat					
without irrigation	1976	543	455	83	10
	1977	667	500	172	11
with irrigation	1976	768	640	111	11
	1977	892	645	236	11

namely how the total cost of a joint project, in this case the intermunicipal water transfer scheme, should be allocated among the municipalities. For this purpose, Young et al. (1982) compared the practical behaviour of different models from the game theory literature to the commonly used proportional methods, for example proportional to population or to water use.

The practical behaviour was defined against certain common sense principles. These were two obvious rationality principles: individual rationality (no municipality should pay more in the joint project than by going alone) and group rationality (no municipality should pay less than the marginal cost of including it). Another important principle is that of monotonicity (no municipality should pay less if the total cost goes up or more if the cost goes down). The model should also be insensitive to the definition of direct costs, since in practice this definition is always somewhat arbitrary. Also, there should be no payments to dummies, i.e. municipalities that contribute nothing to savings should go alone. Finally, the method should be computationally simple.

The model comparison is summarized in Table 5, which shows the principles that are or can be violated by the different allocation methods. The methods compared are:

> P = Proportional
> SCRB = Separable Cost/Remaining Benefit
> SV = Shapley Value
> N = Nucleolus
> WN = Weak Nucleolus
> PN = Proportional Nucleolus

Based on the principles of Table 5 the authors advocate the proportional nucleolus. This method is also unique in the sense that it divides the un- foreseen costs in proportion to the saving each municipality makes by join- ing the coalition. However, because of the information needs of cost func- tions, alternative costs, optimal scales, etc., it was not surprising

TABLE 5. Principles violated by six cost allocation methods

Principle	Method					
	P	SCRB	SV	N	WN	PN
Individual rationality	x	x			x	
Group rationality	x	x	x			
Direct costs	x	x				
Monotonicity		x		x		
Dummy	x				x	
Simplicity		x	x	x	x	x

to the authors to find that the proportional to population method was the one actually chosen.

An interesting follow up to test this theoretical study was a gaming experiment performed by Ståhl (1980). The game was set up in exactly the same manner as the theoretical case and played by Swedish water managers. The outcome of the game did not coincide with any of the methods, but was closer to the game theory models than to the proportional methods. There seemed to be a tendency to form successive coalitions and for each new member to share the total savings equally. This is closest to the Shapley Value model.

5. Conclusion

Water managers and physical planners in Skåne showed great interest in the IIASA modelling activities. As indicated by the examples there have, however, not been many incentives to use the models or further test the methods in real world planning. The examples to be seen stem from the universities and certain central agencies such as the Swedish Meteorological and Hydrological Institute.

One obvious reason is that scientifically based simulation or optimization models usually require input data that are difficult to collect on a local or regional level. Another reason is that most decisions concerning water management and physical planning are taken at municipal level. Only when the problems become really severe for a whole region, like for example the 1989 drought in the beginning of the growing season, is there a demand for regional solutions and proper management models.

With such situations occurring, with the now intensive debate about our increasing and constantly arising new environmental problems, and with a generation of decision-makers more acquainted with mathematical modelling, more use of simulation and optimization models can be foreseen for the Swedish water management.

6. References

De Maré, L. (1977), 'A water quality management model for the Rönne River', Dept. of Water Resources Engineering, Lund Institute of Technology, Report № 3012.

Andersson, Å.E., Hjorth, P., de Maré, L., and Thelander, A.L. (1979), 'Water resources planning and management in advanced economies: the case study of Western Skåne, Sweden - A background report', IIASA, WP 1979 (23).

Kindler, J. (1982), 'Introduction', in J. Kindler (ed.), 'Issues in regional water management - A case study of Southwestern Skåne', IIASA, pp 1-16.

Arthur, S. (1980) 'Irrigation in Skåne - estimated water needs and effects on water available to crops', IIASA, WP 1980 (12).

Anderson, R.J. (1980) 'The probability distribution of water inputs and the economic benefits of supplementary irrigation', IIASA, WP 1980 (165).

de Maré, L. (1982) 'Applications of the CREAMS model: Western Skåne, Sweden', in V. Svetlosanov and W.G. Knisel (eds), European and United States Case Studies in Application of the CREAMS Model, IIASA, CP 1982 (S11), pp 71-82.

Knisel, W.G. (1980) 'CREAMS: a field-scale model for chemicals, runoff, and erosion from agricultural management systems', US Dept. of Agriculture, Science and Education Administration, Conservation Research Report No. 26.

Young, H.P., Okada, N. and Hashimoto, T. (1982) 'Cost allocation in water resources development', Water Resources Research, 3.

Ståhl, I. (1980) 'A gaming experiment on cost allocation in water resources development', IIASA, WP 1980 (38).

PART IV

CONTRIBUTED PAPERS ON WATER RESOURCES SYSTEMS
OPTIMIZATION AND SIMULATION

PRACTICAL IMPLEMENTATION OF WATER RESOURCE SYSTEMS ANALYSIS IN A WATER UTILITY IN THE UK

SUSAN WALKER
Northwest Water Ltd
Warrington WA5   3LW
United Kingdom

1.   Introduction

There has been considerable emphasis since the 1960's on the development of sophisticated methods of stochastic hydrology and water resources analysis. At technical and scientific meetings and in publications it is understandable that the emphasis will be on the development of these techniques and models. However, concern is being expressed that, except for a few notable exceptions, the practical application of such methods in the day to day business of a water utility is not always achieved.

As a hydrologist working within a water utility in the United Kingdom, my definition of "practical application" is the actual use of these methods to solve practical problems facing water supply and planning managers in the performance of their duties rather than the application of such techniques to theoretical analysis of albeit real water resource systems.

In this paper I would like to outline my perspective of some of the reasons why sophisticated methods of stochastic hydrology and water resources systems analysis are not applied in practice. Then I would like to explain how such methods have been introduced and accepted within a water utility in the UK.

2.   The Role of Research

A major reason for the lack of success in the application of water resources systems techniques stems from the differing perspectives of research workers and practitioners. Practitioners would see the role of research as the development of tools which would allow them to do their job better, rather than the development of the science of hydrology for its own sake. Obviously, speculative research has a role to play in the advancement of science. However, certainly in the UK, it is difficult to obtain research monies for projects which are neither operational nor perceived to have a real world application. To have a real chance of applicability, research should begin with the problem definition and then identify the appropriate method to solve the problem. Too often, research seems to start with the development of the method and then a search is made

463

J.B. Marco et al. (eds.),
Stochastic Hydrology and its Use in Water Resources Systems Simulation and Optimization, 463–467.
© 1993 *Kluwer Academic Publishers.*

for potential applications.

In assessing why water resources methods are not applied, it is worth considering that a water resources research project which may be extremely competent science may not have wide applicability in the real world. There is often a conflict between a project which will produce a good masters or doctorate thesis and the product a water utility may require from a research project it funds. For the most part, utilities fund research in order to address a particular problem they face. They are interested in the practical solution of the problem, rather than the sophistication of the model developed. If a simple model provides the answer, then they are not interested in a more aesoteric 'n' parameter model. Utility staff are interested in models which will help them to do their job more effectively, more efficiently and where possible with a cost saving. For them, if a simple model gives an answer which provides 95% of the explanation, this is acceptable. Perfect solutions are not a requirement. This may upset the sensibilities of researchers but, if a researcher wants to contribute to real world applications, this must be accepted.

3. **Role of the Hydrological Services Section in North West Water Limited**

North West Water Ltd provides potable water and sewerage services to a population of around 7 million people in the North West of England. A Hydrological Services Section provides "in house" expertise in the application of hydrological methods in water resources, water supply, sewerage and sewage treatment. This covers a range of applications from feasibility studies and planning to operational issues (Walsh and Walker, 1989) and budgetary control.

A variety of hydrological analyses have been developed and used over the last twenty years. In the water resources and supply fields these include:
- yield assessments (NWW, 1981)
- calculation of supply availability from different types of surface and groundwater sources (Pearson and Walsh, 1982)
- derivation of strategic operating policies using a range of techniques including the analysis of critical periods in the historic flow record (Walsh, Walker and Pearson, 1988), simulation (Walker, 1985), linear programming and stochastic dynamic programming (Walker, Walsh and Wyatt, 1989)
- budgetary control
- drought management (Walker, 1990).

The hydrological methods used have been predominantly developed "in house" but also include methods developed on behalf of the utility by consultants and researchers. In the latter case the Hydrological Services Section provided the technical audit of the methods and formed the bridge between the outside organisations and the other departments within the utility. Close liaison between all parties is essential, especially when some of the work is carried out by external organisations.

4. Implementation of Water Resource Systems Analysis Techniques

A key to the acceptance of new approaches in water resources management by water supply managers and other non hydrological staff in the organisation has been the foundation of successful introduction and application of methods over many years. Since the 1960's new methods have been introduced, evaluated and shown to work in practice by hydrological staff in conjunction with water supply and other function managers. Each new method builds on the sound basis of the last. New ideas have always been introduced in a controlled and unhurried way. This requires patience but ensures that non hydrologists fully appreciate the philosophies of the techniques rather than being overwhelmed by too many new ideas at one time. Care is taken to ensure that they have sufficient opportunity to question and query the results of the methods and compare them with a base case, say, how they would have previously operated.

It is important to accept that not all good ideas work in practice so sometimes one has to be prepared to accept that a model, however well founded in theory, has to be discarded.

An important element in the successful introduction of new methods has been stable staffing over a number of years which has allowed an excellent relationship to develop between hydrological services staff and other technical staff in the organisation. This, together with the joint experience of operation with the hydrological methods, has created sufficient trust for them to be receptive to new ideas and to be willing to experiment with approaches.

The need to establish close communication links with other staff in the organisation is essential for the successful implementation of the new methods of analysis. This has led recently to a member of staff from the water supply department being transferred into the hydrology section. This allows him to gain more of an appreciation of the techniques involved and allows the hydrological team to learn from him more about the practical issues facing water supply managers.

In developing models, it is important to appreciate that water supply managers often have many years' experience in operating their systems. This is an asset which should be used. Well-founded operational experience should always, where possible, be included in the model. For instance, in North West Water Ltd, care has been taken to introduce into the modelling the degree of complexity that the water supply staff feel is appropriate, above and beyond that which is necessary from a hydrological viewpoint. For example, simulation models used either independently or in conjunction with optimisation routines include the water supply managers' interpretation of the operational realities of the water resources system. After initial results of the model have been produced, these are presented to water supply managers and discussed at length to ensure that the results are sensible and workable. The model would be revised if they were not.

Usually the results of the model would be compared with some acceptable comparison standard eg. the output from previous operating policies. Similar discussions would take place throughout the development phases. This gives water supply managers confidence in the results and hence, say, the operating policies which come out of the analysis. In this way they become committed to the policies. These discussions also act as a catalyst for them to consider how in practice they might be able to operate the system differently to achieve greater efficiency. Their suggestions can then be modelled to show their effect. Initially, it is important to use the model to answer the questions which the water supply and planning managers ask, but by close dialogue it may then be possible for hydrological staff to suggest what might have been more appropriate questions.

It is essential that results of the modelling and analysis are presented in a concise and easy to understand way in terms which water supply managers understand. Efforts are continually being made to update the methods of presentation in the light of technical developments such as enhanced computer graphics and expert systems, in order to ensure that the information which is obtained from the modelling can be used quickly and efficiently by hydrologists and water supply managers alike, in the management of water resource systems.

5. Conclusion

Whilst the need for innovation in water resource systems analysis is recognised, the full potential of such methods will not be fully realised, unless the practical problems facing water resources managers are addressed.

The need for good, sound methodologies is essential, but sound science will not alone ensure acceptability and use in practice. The need for constructive dialogue between modellers, operating personnel and planners should not be underestimated. Without regular discourse during the development of the new techniques, it is unlikely that the models will address the issues which are important to the managers and the model will probably not truly describe the practicalities of the real-world water resources system. As a result, it will be extremely difficult to get such methods accepted and used in the real world. The potential for major advance in water resource systems analysis has been made in recent years. It is important that these methods move out of the realms of academic theory and into the real world.

6. References

Walsh, P.D. and Walker, S. (1989) 'Decision support systems as and in the operational management of multiple water source systems', in: Closing the Gap Between Theory and Practice, ed. Loucks D.P., IAHS Pub., No. 180.

North West Water Authority (1981) Survey of Existing Water Sources, North West Water, Warrington.

Pearson, D. and Walsh, P.D. (1982) 'The derivation and use of control curves for the regional allocation of water resources', in: Optimal Allocation of Water Resources, ed. Lowing M.J., IAHS Pub. No. 135.

Walsh, P.D., Walker, S. and Pearson, D. (1988) 'A decade of developments in the derivation of operating policies for surface water sources in North West Water', J. Inst. Wat. & Environ. Management, 2(1) 51.

Walker, S. (1985) 'Economic operation of Lake District sources', North West

Water, Warrington.
Walker, S., Walsh, P.D. and Wyatt, T. (1989) 'Derivation and application
    of medium term operating policies for the Northern Command Zone of North
    West Water', in: Closing the Gap Between Theory and Practice, ed. Loucks
    D.P., IAHS Pub. No. 180.
Walker, S. (accepted paper, 1990) 'Practical application of hydrological
    analysis in drought management in North West Water Ltd (UK)',
    International Symposium on Water Resource Systems Application, Winnipeg,
    Canada, June, 1990.

EXPERT SYSTEMS FOR FLOOD HYDROLOGICAL FORECASTING AND DECISION SUPPORT

ENRIQUE CIFRES
Confederacion Hidrografica del Jucar
(Proyecto SAIH)
Blasco Ibañez, 48
Valencia
Spain

JOSE CUENA, LUIS GARROTE
Lab. Sistemas Inteligentes
(Proyecto SAIH)
Univ. Politecnica de Madrid
28040 Madrid
Spain

ABSTRACT. The paper presents two knowledge representation environments, CYRAH and SIRAH, to be used for decision-making support in flood forecasting in Mediterranean basins.

First, a general approach for decision-support systems during flood is presented. Then an overview of the environments is produced as two different concepts of implementation of the same line of reasoning.

The environments embody not only classical problem-solving paradigms (rules and frames), but also the professional-reasoning paradigms used by hydrologists, such as traditional simulation models in numerical or qualitative version.

The particular requirements involved in hydrological analysis have made it necessary to develop of a specific software environment especially adapted to the purposes of the project.

1. Introduction

The SAIH (Automatic Hydrological Information System) is a project promoted by the Spanish Ministry of Public Works with the global goal of supporting the decision-making process related to flood forecasting and water resource management. The main tasks achieved by the system are:
- automatic data collection of hydrological variables such as rainfall intensity, water levels, etc.
- control, processing and storage of data using microprocessors in remote stations
- transmission of elaborated data through a telecommunications network to a data processing centre
- data interpretation and aid in decision-making in the data processing centre.

This paper presents some aspects related to the decision-aid systems developed for the data processing centre in the Jucar basin. Flood forecasting in the Mediterranean area is greatly influenced by the meteorological, topographical and geomorphological features of its watersheds. Short rivers, steep slopes, deforestation and torrential rains are the main causes for a fast and intense response to rainfall in the zone

469

J.B. Marco et al. (eds.),
Stochastic Hydrology and its Use in Water Resources Systems Simulation and Optimization, 469–480.
© 1993 *Kluwer Academic Publishers.*

and the consequent flood. A warning system must meet strong requirements: prompt response of relative accuracy under exacting conditions, such as unpredictable meteorological situations or ignorance of the basin's physical behaviour. It should be noted that, while most real-time hydrological forecasting systems measure their time scale in days, in this particular case the time scale must be hours since the gap between the starting of rains and the flood is just a few hours.

An expert system approach was considered for aid in decision making (Cuena, 1983), due to the need for handling qualitative knowledge and criteria about physical phenomena. The proposed architecture for these expert systems integrates traditional simulation models in numerical or qualitative version together with symbolic reasoning elements like rules and frames. Physical processes should be properly represented, and therefore the knowledge base and inference engine of this type of expert system must be specially adapted to meet this requirement. Due to this fact, the classical methods and software environments for expert systems management may be insufficient or may produce very naive models, and thus an adapted software for knowledge representation has been developed.

2. The Problem of Flood Decision Support Systems

The need for anticipating as much as possible the discharges expected in problematical areas has encouraged the development of a great number of mathematical models to simulate rainfall-runoff processes and flood propagation along rivers. Flood managers keep using empirical methods based on their personal knowledge and intuitions, though the dramatic development of computer equipment during the last few years has enabled them to enhance their forecasting systems with such models (O'Connell, 1988).

It can be inferred that management of simulation models in short-term forecasting has a number of inconveniences that set practical limitations to their real-time application, such as the following (Schultz, 1986):

- data are usually inadequate and insufficient, and must be checked for accuracy before they are accepted as inputs to the models
- there are uncertainties with respect to the physical properties of the basin and the different possibilities must be considered
- physical processes are only partially represented by the models
- for a given state of the situation, a number of future evolutions can be expected and must be taken into account
- outputs of the models must be interpreted in order to outline the adequate conclusions.

At the heart of the problem lies the fact that there is a certain gap between the subsequent links of the water resource management chain: data collection, modelling and decision making (Duckstein, 1985). Complex mathematical models are usually developed in an academic environment under optimal conditions, and their jump from the laboratory to the field is often traumatic. Data requirements for the models usually exceed the possibilities of real data acquisition systems, parameter estimation methods or calibration procedures, mainly due to budget limitations in real-life applications. Models that are conceptually perfect cannot expand easily from laboratory test to big scale operation because they require excessive investments in information. Therefore, real-time management of hydrological models is almost impossible without the direct intervention of experts to fill the gap between the data available and the information required and to interpret the output in order to draw significant conclusions.

Experts are not easily available for day-to-day operation, and thus decision aid systems for the SAIH in the Jucar basin are defined under the expert system approach in order to be able to represent the reasoning process that experts follow during a flood event. On-line intelligent systems can integrate mathematical numerical models and heuristic criteria using symbolic reasoning in a way that makes possible the automatic interpretation of the great amount of data collected by a network such as the SAIH.

The decision-support system is conceived as an intelligent environment capable of automatically identifying the problems that might occur consistent with the set of data received during a flood event. The ability to direct the search for problems with a rough engineering model and then refine and adjust the evaluation using numerical analysis is a classic approach to problem solving in engineering. This approach is the basis for the overall structure of the environments within the scope of real-time flood forecasting. Much emphasis is placed on the fact that the inputs and the physical behaviour of the system are only partially known and therefore the uncertainty on the data should be assumed.

Two knowledge representation environments, CYRAH (Calculo y RAzonamiento Hidrologico) and SIRAH (Sistema Inteligente de RAzonamiento Hidrologico) are currently under development with the main goal of supplying the practitioners with adequate tools to represent their way of understanding the decision-making process. In order to represent the behaviour of the physical system, a reasoning line has to be assumed, but most of the knowledge handled by the environment is intended to be declarative, that is, defined by the user to adapt to his particular interest.

CYRAH and SIRAH are different concepts of implementation of the same reasoning line: in the first phase, the current situation is analyzed. The expert is able to focus his attention only in the cases that are or might be problematic in the future. His knowledge of past events enables him to make conjectures on plausible evolutions of the meteorological situation and evaluate roughly the response in the different basins to future rains. As a result of this analysis, a limited number of cases are selected for further attention. In the second phase, quantitative or conjectural models are used to estimate discharges and levels in those cases. The uncertainty in the data and the descriptors of the physical structure is overcome by repeating the simulations using representative combinations of inputs and interpreting the different outputs obtained. The chances of problems arising in the near future are deduced in the third phase from the discharges and levels estimated.

3. The CYRAH Environment

The CYRAH expert system is an intermediate application that stands between the classical numerical simulation systems, based on mathematical models, and the artificial intelligence systems, based on symbolic knowledge representation and inference procedures.

3.1. COGNITIVE ELEMENTS

The reasoning about a flooding process can be structured into four levels:
- The analysis of the meteorological situation. Meteorological zones should be defined in order to reason on past and future rainfall distribution.
- The watersheds and the fast transport network. The area of study should be divided into independent basins and rainfall-runoff models used to compute discharges in points of interest.
- The floodable river reach, where backwater effects are significant and, thus, the Saint Venant equations could be applied. A model of the reach should be built, specifying the points where water stages are relevant to deduce the flooded areas.
- The problem detection in urban areas and communication networks due to water stages or discharges in floodable areas.

The CYRAH environment offers two types of knowledge representation: classic quantitative simulation models and knowledge bases to formalize the reasoning process about these topics.
* Physical processes of runoff generation and river flow are represented by two mathematical models:
   (1) The SAVEL model computes the response of a basin with a number of homogeneous watersheds connected to a network of steep-sloped channels.
   (2) The REBOLSA model computes the evolution of water stages in a river reach of low slope.
* Knowledge related to conjectures, reasonings, estimations or qualitative syntheses is defined by rules using a classical knowledge representation environment.

3.2. INFERENCE PROCEDURE

The management of mathematical models that an experienced hydrologist would carry out is represented by the reasoning line of the CYRAH system, as shown in Figure 1.

3.2.1. *Envisagement of the Cases Requiring Numerical Analysis.* The first phase is a global analysis of the situation, following two parallel lines:
- estimation of the possible evolution of rainfall distribution
- identification of the levels of rainfall intensity that might become problematical.

From a synthesis between both lines, the cases requiring accurate numerical analysis are selected.

a) *Generation of hypotheses of plausible patterns in the evolution of the meteorological situation*: the first step is a prior analysis of the meteorological situation. The goal of this analysis is to identify some patterns of recent evolution in the rainfall and, using the knowledge deduced from past storms, to generate some future alternatives of spatial and temporal distribution of rainfall.

A rule base is used to formalize the process. The premises for the knowledge base are obtained from numerical data provided by the pluviometre

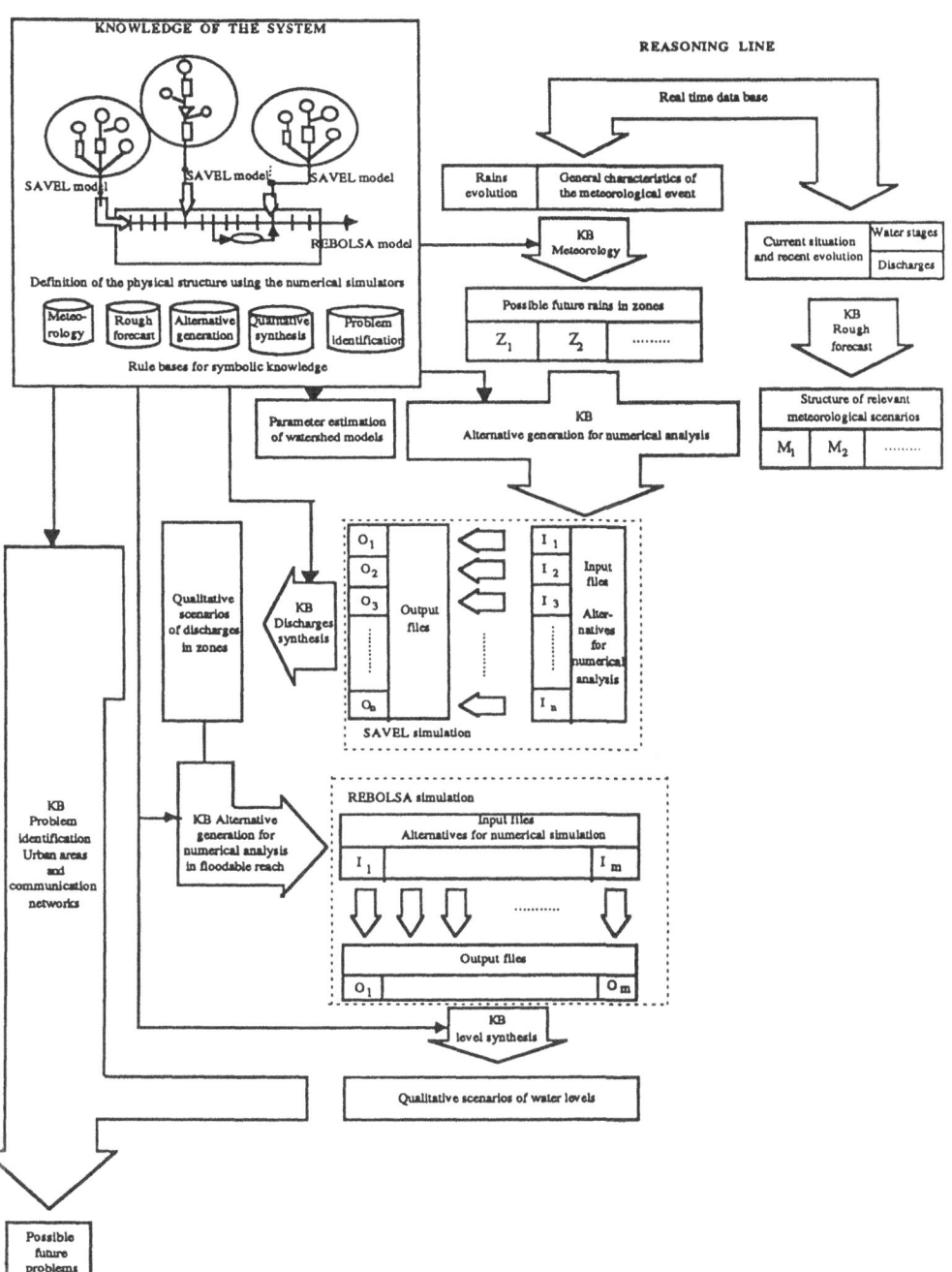

Figure 1. Knowledge and reasoning line of CYRAH.

network and from qualitative data deduced from meteorological maps and satellite photographs. The inference engine links several steps of reasoning through three different levels (the global situation, the zone and the subzone) and concludes on a number (between one and three) of possible patterns of future evolution in rainfall intensity for every subzone. Each pattern is affected by a certainty factor which represents the feasibility of its occurrence.

b) *Identification of the level of rainfall intensity that might become dangerous*: all the points where discharges are relevant in order to deduce problems are checked in this step using a simple reasoning model. Given the rain gauged in each basin over the last hours, the question analyzed is: what is the level of rainfall intensity that might become dangerous during the next hours?

Future discharges in every outlet are computed with the rainfall-runoff model using the rain gauged in the past and no rain in the future. The hydrograph thus obtained represents the future discharges that can be expected due to the antecedent rain. The difference between the discharge considered dangerous and this hydrograph is used as an input to a simple basin model based on shallow reasoning in order to deduce the threshold of future rain in each meteorological zone that might raise the discharges or water stages to a dangerous level.

c) *Definition of pertinent scenarios for numerical analysis*: a process of comparison between the possible patterns of rainfall evolution in each meteorological zone and the dangerous rain for each basin gives as an output the list of cases that need additional analysis. At this stage, a final synthesis is needed to select among all the possible combinations a set of future scenarios for numerical analysis that representatively include the conceivable optimistic and pessimistic situations. This requires a reasoning step to model the criteria used by an expert when defining the relevant scenarios so that the great number of combinations can be reduced logically.

3.2.2. *Inference Step Based on Mathematical Models.* The cases selected in the previous phase are computed using the numerical models to include the different foreseeable situations. The program first consults the information system and infers the initial state in each basin or river reach. Then it controls several executions of the model with the inputs corresponding to the different scenarios selected and makes a synthesis of the outputs which concludes with the assignment of intervals of probable discharges or water stages to the points of interest.

The rainfall-runoff model is applied first because the discharges that it computes are the inputs to the floodable reach. The qualitative syntheses of hydrographs in the tributaries are the base for a process of generation of scenarios for the computation of water stages and lateral outflows.

The tasks completed by the system in order to provide the models with the appropriate input files are implemented in different ways. Whenever it is possible, the data are taken directly from the information system through a parametrized consultation to a data base. When the data are not directly controlled by the network, some related measurements are used to deduce their value through declarative or procedural methods. The most important ones, such as the deduction of the antecedent moisture condition, use knowledge bases to permit their definition by the user. There are some programmed procedures to compute some quantities or to select the best option among a set supplied by the user. Default values must be defined for the least relevant magnitudes that can be neither measured nor deduced.

3.2.3. *Problem Detection*. The definition of the probable values of discharges and water stages leads to the identification of future problems using a rule base. A number of floodable areas are defined, each one related to discharges or levels computed in the previous phase. If knowledge is available, the state of flooding of each area can be deduced in a first step. The problem detection rules represent the knowledge deduced from past experience on the relation between such states of flooding and:
- flooding of selected urban areas
- highways interrupted
- degree of isolation of uncommunicated towns
- alternative routes to closed highway trunks.

3.3. PRACTICAL LIMITATIONS

The use of procedural numerical models in CYRAH to simulate the behaviour of parts of the system places limitations on the operation of the environment:
- Although the expert makes his reasoning as a whole, accounting for the different possibilities globally, numerical models simulate only one event under a unique set of conditions. To account for the different possibilities, a number of options must be selected for simulation.
- Numerical computation is slower than symbolic reasoning and, thus, only a limited number of cases can be simulated in the lapse of time available for operation.
- There is a need for qualitative synthesis after numerical simulation to interpret the outputs and to integrate them in the subsequent symbolic reasoning steps.

As a result of the above limitations, the actual reasoning of the user was not represented easily in this environment and more elaborated tools had to be developed.

4. The SIRAH Environment

The CYRAH environment was conceived as a first implementation for illustrative purposes and as a development tool of the definitive environment, SIRAH. While the former is based on numerical computation to represent the physical behaviour of the basins and rivers, the latter includes knowledge-representation techniques that allow the user to specify his professional knowledge about the behaviour of different physical elements involved in the response of a hydrological basin.

4.1. REASONING OF PHYSICAL BEHAVIOUR IN SIRAH

The general reasoning process for decision making during a flood event has as main steps (Cuena, 1989): (1) prediction of the evolution of external

476

actions in the near time horizon, (2) identification of possible future state changes predictions by the predictable external actions, assuming that the present control strategy is maintained, (3) identification of present and predictable problems based on the possible states deduced from (2), and, finally, (4) inference of possible new decisions to change the present control strategy to solve or decrease the problems detected.

It is difficult to formulate general knowledge modules for the (3) and (4) aspects, but it is possible to define modules for the (1) and (2) steps in cases where the general behaviour may be explained by a flow of actions among components (in the flood case the flow is water in different states from precipitation to flows and water levels). In these flow cases, the predicted behaviour can be described by a hierarchy of task applications.

Two basic task types can be considered:

- Basic Simulation Tasks (BST), representing the response of the different physical elements integrating the flow system. The internal knowledge of a BST is a qualitative model relating the input flow scenario to the possible output flow.

- Scenario Generation Tasks (SGT), representing the knowledge to generate the input flow scenarios for a given type of BST from the output scenarios generated by other BST's upstream.

The knowledge in an SGT that must be taken into account are (1) the feasibility of combinations among the flows generated by other BST's, and (2) the relevance of the combinations to be selected for the answers about patterns of behaviour whose feasibility is sought. These behaviour patterns represent features relevant for the evaluation of the current state and the control strategy (i.e., in the case of the flood problems, the relevant behaviours are those that can produce flood when there is a normal situation or can reduce floods in a flooded situation).

Figure 2 shows an example of a flow structure with five components. Components C1, C2, C3 are of the same type and their general behaviour is represented by the task module BST1, so that the instances BST1.1, BST1.2 and BST1.3 represent the knowledge for the prediction of behaviour of C1, C2, C3. BST2.1 is the instance representing C4 behaviour and BST3.1 is the instance for C5 behaviour. Three corresponding SGT's must be defined.

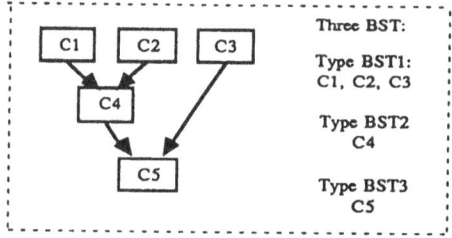

Figure 2. A flow system.

Figure 3 partially shows the reasoning tree of behaviour generation through this flow system.

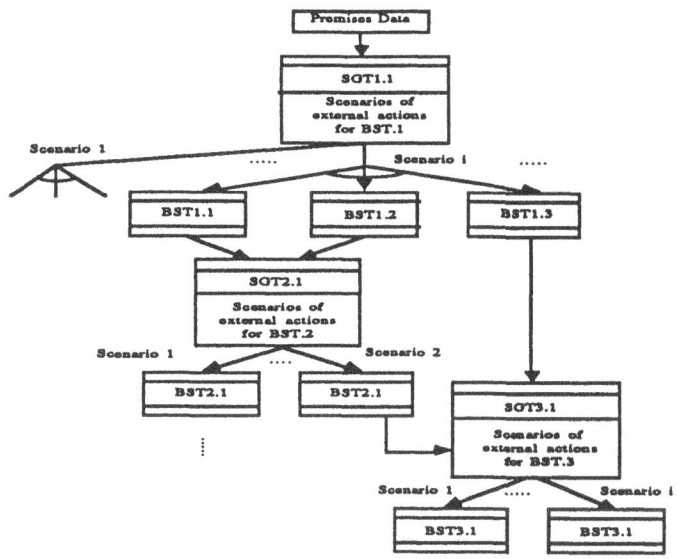

Figure 3. Reasoning of behaviour prediction.

The exploration tree shows the role played by the SGT's in limiting their combinatory expansion by producing only feasible goal-relevant behaviours. However, this may still be cumbersome because many relevant behaviours may be only slightly different. This is the reason for introducing an additional filter to ensure that the proposed behaviours are sufficiently diverse to pay for the effort of their generation in the first depth strategy. To meet this objective, an intelligent backtracking must be done once an alternative behaviour is generated to an intermediate SGT in order to select a new scenario that can produce feasible and significant goal behaviours that are also diverse enough with respect to the already generated behaviours.

In the next sections a version of these ideas is presented, applied to the SIRAH case.

4.2. COGNITIVE ELEMENTS

The previous ideas are the base of the modelling used for the SIRAH environment where three basic simulation tasks have been defined:

- *Reception Area (RA)* to model the behaviour of a basin area submitted to rainfall and producing as a result flows to be drained by the transport network.

- *Upper Basin Network (UBN)* to model the fast flow river network behaviour in the upper basins where the flows generated in the tributary reception areas are concentrated, producing surface flows. These flows produce impacts in the low basin floodable river links.

- *Floodable River Link (FRL)* to model the water level behaviour in the low basin floodable river links submitted to different flows produced by the upper basin networks.

Every task module is defined by a frame with:

- The attribute premises for the internal reasoning process (permanent attributes and time variables attributes).

- The attribute conclusions, set of possible values of state variables to be used by subsequent SGT's.

- A knowledge base defined by rules and constraint confluences for the qualitative modelling of behaviour.

- An inference engine to reason, using the knowledge base, from premises to conclusions. There may be a knowledge base to help the creation phase also to infer advanced characteristics in attributes from more primary attributes.

Three Scenario Generation Task types have been defined:

- The meteorological scenario generation task that infers possible significant goal combinations of local rainfall predictions acting upon the set of reception areas.

- The flow transport scenario generation task that infers the combinations of input flows to the upper basin networks.

- The floodable river link scenario generation task that infers the combination of output flows from the upper basin river networks with a floodable river link.

The SGT's have a general structure based on:

- a frame with: (1) attribute premises: possible values of every scenario component and significant values of every pattern of final behaviour to be regressed; (2) attribute results: lists of every feasible combination that makes a scenario

- two knowledge bases: (1) modelling conditions to constrain the scenarios to be relevant for the final behaviour prefixed patterns: (2) modelling conditions for the spatio-temporal consistency of the appropriate predictions

- an inference engine: for reasoning in two steps: (1) to generate, using both knowledge bases, consistent sets of possible values of every scenario variable; (2) to generate scenarios defined by combinations of the inferred possible values of the scenario variables that are consistent with both knowledge bases.

4.3. INFERENCE PROCEDURE

The general reasoning paradigm using these basic concepts is an adaptation of the general behaviour reasoning scheme presented in Figure 3, where the main steps are:

a) Use of an ad hoc knowledge base for real-time data interpretation to identify the present problems and the potential future problems. To explore the feasibility of these future problems, this knowledge base defines patterns of the final behaviour to be explored by the predicture reasoning (i.e., potential problems to be explored may be defined if in some places levels are high but still not problematic or they are low but the rate of level change is important, etc.).

b) Use of the meteorological scenario generation task to propose possible future rainfall scenarios on the surface of the basin. These scenarios are generated in order of proximity to the existing registered trend and only the first is retained.

c) A first depth process of qualitative simulations along the different

physical elements is produced:
* The first meteorological scenario is introduced as input to the set of reception area task modules.
* A first scenario of input flows is generated from the predicted flows drained from the reception areas for the UBN's:
    - The scenario of upper basin network inflows is introduced as input to the set of river networks in upper basins.
    - A set of upper basins outflow scenarios is generated:
        . to be used for problem identification in the cases where the use of flows alone is possible
        . to be introduced as input to floodable river link tasks in the cases where water levels data are available.

Once a set of behaviours (flows or levels) is obtained in this process, the exploration of the next scenarios for rain or upper basin flows or outflows is decided, based on a metacontrol reasoning step that uses as premises the resulting behaviours and the current patterns of the final behaviour to be explored. This metacontrol step is produced by a general control task and generates additional constraints to the SGT's in such a way that, when these SGT's are fired with the new constraints, they produce a first scenario which, once simulated, is expected to produce sufficiently different patterns of behaviour from the ones already inferred. With the scenarios generated in this way, a new first depth traversal is carried out.

This process is repeated until some SGT's generate no feasible scenarios or a previously defined time period or number of iterations is met.

5. Conclusions

Two expert systems for decision support have been presented. Both represent a first approach to the problem of decision making during floods using knowledge-based systems.

Decision making during a flood event involves a number of steps that clearly exceed the possibilities currently offered by classic mathematical models. The decision usually takes place in a professional environment where uncertainty places practical limitations on the use of simulation models. Knowledge-based systems offer the possibility of representing the actual reasonings and judgements made by the expert to build a model that embodies the professional knowledge usually applied in these cases. The implementation of the data processing centre in Valencia will demonstrate the actual capabilities of the environment during real-time operation.

This expert system approach for the reasoning of physical systems requires special knowledge-representation techniques. Classic paradigms based on rules or frames may be insufficient to represent the professional knowledge involved in real-time flood analysis and therefore an adapted environment must be specially developed for this purpose.

6. Acknowledgements

The project is currently under development with the sponsorship of the Spanish Ministry of Public Works. The authors are grateful to all the experts who participated in the knowledge engineering sessions, and whose ideas have been the basis for the systems, to Mr Manuel Alonso and Mr Martin Molina, responsible for the software analysis and implementation of the environments, and to Miss Reyes Riera whose patience in typing made this paper possible.

7. References

Cuena, J. (1983) 'The use of simulation models and human advice to build an expert system for the defense and control of river floods', IJCAI-83, Karlsruhe, Kaufmann.

Cuena, J. (1989) 'Knowledge-based systems for aid in decision making: methodology and examples', part of Perspectives in Artificial Intelligence, Vol. 1, pp 73-92, Campbell, J.A., Cuena, J. (Eds), Ellis Horwood.

Duckstein, L., Ambrus, S. and Davis, D. (1985) 'Management forecasting requirements', part of Hydrological Forecasting, pp 559-585, Anderson, M.G. and Burt, T.P. (Eds), Willey.

O'Connell, P.E. et al. (1988) 'IKBS for real-time management of risk processes', Laboratoire d'Hydraulique de France.

Shultz, G.A. (1986) 'Relationship between theory and practice of real-time river flow forecasting', part of River Flow Modelling and Forecasting, pp 181-193, Kraijenhoff, D.A. and Moll, J.R. (Eds), Reidel.

SUBJECT INDEX

The manufacturer's authorised representative in the EU is Springer
Nature Customer Service Centre GmbH, Europaplatz 3, 69115 Heidelberg,
Germany. If you have any concerns regarding our products, please
contact ProductSafety@springernature.com

Printed and bound by CPI Group (UK) Ltd, Croydon, CR0 4YY

24/04/2026

02096348-0011